Cultured Microalgae for the Food Industry

Current and Potential Applications

Cultured Microalgae for the Food Industry

Current and Potential Applications

Edited by

Tomás Lafarga
Department of Chemical Engineering, University of Almería,
Almería, Spain

Gabriel Acién
Department of Chemical Engineering, University of Almería,
Almería, Spain

ELSEVIER

ACADEMIC PRESS
An imprint of Elsevier

Academic Press is an imprint of Elsevier
125 London Wall, London EC2Y 5AS, United Kingdom
525 B Street, Suite 1650, San Diego, CA 92101, United States
50 Hampshire Street, 5th Floor, Cambridge, MA 02139, United States
The Boulevard, Langford Lane, Kidlington, Oxford OX5 1GB, United Kingdom

British Library Cataloguing-in-Publication Data
A catalogue record for this book is available from the British Library

Library of Congress Cataloging-in-Publication Data
A catalog record for this book is available from the Library of Congress

ISBN: 978-0-12-821080-2

For Information on all Academic Press publications
visit our website at https://www.elsevier.com/books-and-journals

Publisher: Charlotte Cockle
Acquisitions Editor: Nina Rosa de Araujo Bandeira
Editorial Project Manager: Billie Jean Fernandez
Production Project Manager: Paul Prasad Chandramohan
Cover Designer: Miles Hitchen

Typeset by MPS Limited, Chennai, India

Contents

List of contributors

Caleb Acquah School of Nutrition Sciences, Faculty of Health Sciences, University of Ottawa, Ottawa, ON, Canada

Ingrid Aguiló-Aguayo Institut de Recerca i Tecnologia Agroalimentàries (IRTA), XaRTA-Postharvest, Parc Científic i Tecnològic Agroalimentari de Lleida, Parc de Gardeny, Edifici Fruitcentre, Lleida, Spain

Luísa Barreira CCMAR-Centre of Marine Sciences, University of the Algarve, Gambelas Campus, Faro, Portugal

Juan Roberto Benavente-Valdés Department of Chemical Engineering, Faculty of Chemical Sciences, Autonomous University of Coahuila, Saltillo, México

Tom M.M. Bernaerts Laboratory of Food Technology (Member of Leuven Food Science and Nutrition Research Centre, LFoRCe), Department of Microbial and Molecular Systems (M^2S), KU Leuven, Leuven, Belgium

Nigel Brunton University College Dublin, School of Agriculture and Food Science, Dublin, Ireland

Berinka Araceli Carreón-González Faculty of Biological Sciences, Autonomous University of Coahuila, Torreón, México

Imene Chentir Higher School of Food Science and Agri-Food Industry, Algiers, Algeria; Laboratory of Enzyme Engineering and Microbiology, National School of Engineering of Sfax, University of Sfax, Sfax, Tunisia

Jorge A.V. Costa College of Chemistry and Food Engineering, Federal University of Rio Grande, Laboratory of Biochemical Engineering, Rio Grande, RS, Brazil

Margarida Costa R&D Department, Allmicroalgae Natural Products S.A., Pataias, Portugal

Christine Edwards School of Pharmacy and Life Sciences, Robert Gordon University, Aberdeen, United Kingdom

Flora-Glad Ekezie School of Nutrition Sciences, Faculty of Health Sciences, University of Ottawa, Ottawa, ON, Canada; School of Food Science and Engineering, South China University of Technology, Guangzhou, P.R. China

Ana F. Esteves LEPABE—Laboratory for Process Engineering, Environment, Biotechnology and Energy, Faculty of Engineering, University of Porto, Porto, Portugal

Alice Ferreira Bioenergy Unit, LNEG, National Laboratory of Energy and Geology I.P., Lisbon, Portugal

Katkam N. Gangadhar CCMAR-Centre of Marine Sciences, University of the Algarve, Gambelas Campus, Faro, Portugal

Marco García-Vaquero University College Dublin, School of Agriculture and Food Science, Dublin, Ireland

Ana L. Gonçalves LEPABE—Laboratory for Process Engineering, Environment, Biotechnology and Energy, Faculty of Engineering, University of Porto, Porto, Portugal

Luisa Gouveia Bioenergy Unit, LNEG, National Laboratory of Energy and Geology I.P., Lisbon, Portugal

Inês Guerra R&D Department, Allmicroalgae Natural Products S.A., Pataias, Portugal

Benoit Guieysse School of Food & Advanced Technology, Massey University, Palmerston North, New Zealand

Israel Hernández-López Faculty of Biological Sciences, Autonomous University of Coahuila, Torreón, México

Tomás Lafarga Department of Chemical Engineering, University of Almería, Almería, Spain

Inês Maia CCMAR-Centre of Marine Sciences, University of the Algarve, Gambelas Campus, Faro, Portugal

Giorgos Markou Institute of Technology of Agricultural Products, Hellenic Agricultural Organization-Demeter (HAO-Demeter), Athens, Greece

Gregory Martin Algal Processing Group, Department of Chemical and Biomolecular Engineering, The University of Melbourne, Parkville, VIC, Australia

J. Masojídek Institute of Microbiology of the CAS, Centre Algatech, Laboratory of Algal Biotechnology, Opatovický mlýn, Třeboň, Czech Republic; University of South Bohemia, Faculty of Science, České Budějovice, Czech Republic

Alan Eduardo Medina-Antillón Faculty of Agrotechnological Science, Autonomous University of Chihuahua, Chihuahua, Mexico

Alejandro Méndez-Zavala Department of Chemical Engineering, Faculty of Chemical Sciences, Autonomous University of Coahuila, Saltillo, México

Julio César Montañez-Saénz Department of Chemical Engineering, Faculty of Chemical Sciences, Autonomous University of Coahuila, Saltillo, México

Etiele G. Morais CCMAR-Centre of Marine Sciences, University of the Algarve, Gambelas Campus, Faro, Portugal

Michele G. Morais College of Chemistry and Food Engineering, Federal University of Rio Grande, Laboratory of Microbiology and Biochemistry, Rio Grande, RS, Brazil

Lourdes Morales-Oyervides Department of Chemical Engineering, Faculty of Chemical Sciences, Autonomous University of Coahuila, Saltillo, México

Dolores Camacho Muñoz School of Pharmacy and Life Sciences, Robert Gordon University, Aberdeen, United Kingdom

Joseph Samuel Palmer School of Pharmacy and Life Sciences, Robert Gordon University, Aberdeen, United Kingdom; ScotBio, BioCity Scotland, Newhouse, United Kingdom

Hugo Pereira CCMAR-Centre of Marine Sciences, University of the Algarve, Gambelas Campus, Faro, Portugal

José C.M. Pires LEPABE—Laboratory for Process Engineering, Environment, Biotechnology and Energy, Faculty of Engineering, University of Porto, Porto, Portugal

Maxence Plouviez School of Food & Advanced Technology, Massey University, Palmerston North, New Zealand

Gaurav Rajauria School of Agriculture and Food Science, University College Dublin, Dublin, Ireland

K. Ranglova Institute of Microbiology of the CAS, Centre Algatech, Laboratory of Algal Biotechnology, Opatovický mlýn, Třeboň, Czech Republic; University of South Bohemia, Faculty of Agriculture, České Budějovice, Czech Republic

Rajeev Ravindran School of Agriculture and Food Science, University College Dublin, Dublin, Ireland

María Janeth Rodríguez-Roque Faculty of Agrotechnological Science, Autonomous University of Chihuahua, Chihuahua, Mexico

Nora Aideé Salas-Salazar Faculty of Agrotechnological Science, Autonomous University of Chihuahua, Chihuahua, Mexico

Rogelio Sánchez-Vega Faculty of Zootechnics and Ecology, Autonomous University of Chihuahua, Chihuahua, Mexico

Lisa Schüler CCMAR-Centre of Marine Sciences, University of the Algarve, Gambelas Campus, Faro, Portugal

Joana Silva R&D Department, Allmicroalgae Natural Products S.A., Pataias, Portugal

A.M. Silva Benavides Escuela de Biologia, CIMAR—Centro de Investigación en Ciencias del Mar y Limnología, Universidad de Costa Rica, San Pedro, San José, Costa Rica

Mayra Cristina Soto-Caballero Faculty of Agrotechnological Science, Autonomous University of Chihuahua, Chihuahua, Mexico

Giuseppe Torzillo CNR—Institute of Bioeconomy, via Madonna del Piano 10, Sesto Fiorentino, Italy; Escuela de Biologia, CIMAR—Centro de Investigación en Ciencias del Mar y Limnología, Universidad de Costa Rica, San Pedro, San José, Costa Rica

Ioannis Tzovenis MicroPhykos, Athens, Greece

Chibuike C. Udenigwe School of Nutrition Sciences, Faculty of Health Sciences, University of Ottawa, Ottawa, ON, Canada; Department of Chemistry and Biomolecular Sciences, University of Ottawa, Ottawa, ON, Canada

Carlos Guillermo Valdivia-Nájar Department of Food Technology, Centro de Investigación y Asistencia en Tecnología y Diseño del Estado de Jalisco (CIATEJ), Zapopan, Mexico

Ann M. Van Loey Laboratory of Food Technology (Member of Leuven Food Science and Nutrition Research Centre, LFoRCe), Department of Microbial and Molecular Systems (M^2S), KU Leuven, Leuven, Belgium

João Varela CCMAR-Centre of Marine Sciences, University of the Algarve, Gambelas Campus, Faro, Portugal

María Elena Velázquez-Arellano Department of Animal Nutrition, The Antonio Narro Agrarian Autonomous University, Saltillo, México

Silvia Villaró Food Technology Department, University of Lleida, Lleida, Spain; Department of Chemical Engineering, University of Almería, Almería, Spain

Inmaculada Viñas Food Technology Department, University of Lleida, Lleida, Spain

Bhagya Yatipanthalawa Algal Processing Group, Department of Chemical and Biomolecular Engineering, The University of Melbourne, Parkville, VIC, Australia

G. Chini Zittelli CNR—Institute of Bioeconomy, via Madonna del Piano 10, Sesto Fiorentino, Italy

Chapter 1

Culturing of microalgae for food applications

Giuseppe Torzillo[1,2,*], G. Chini Zittelli[1], A.M. Silva Benavides[2], K. Ranglova[3,4] and J. Masojidek[3,5]

[1]CNR—Institute of Bioeconomy, via Madonna del Piano 10, Sesto Fiorentino, Italy, [2]Escuela de Biologia, CIMAR—Centro de Investigación en Ciencias del Mar y Limnología, Universidad de Costa Rica, San Pedro, San José, Costa Rica, [3]Institute of Microbiology of the CAS, Centre Algatech, Laboratory of Algal Biotechnology, Opatovický mlýn, Třeboň, Czech Republic, [4]University of South Bohemia, Faculty of Agriculture, České Budějovice, Czech Republic, [5]University of South Bohemia, Faculty of Science, České Budějovice, Czech Republic

**Corresponding author. e-mail: giuseppe.torzillo@cnr.it*

1.1 Introduction

Microalgae can play an important role in the modern bioeconomy as they can be used as sources of important building blocks such as amino acids, polyunsaturated fatty acids, pigments, vitamins, and antioxidants. To date, the number of species being cultured on a large scale for commercial purposes remains limited, with the most used being *Arthrospira platensis* (Spirulina) and *Chlorella* spp., produced mainly for use as nutritional supplements, and *Dunaliella* and *Haematococcus* as sources of carotenoids. On a much smaller scale, although still commercial, is the production of *Porphyridium* for the production of polysaccharides and phycobilin pigments (Arad Malis & van Moppers, 2013). Most of the commercial-scale cultures are carried out in open ponds, especially paddle wheel-driven raceway systems, while only a few (e.g., *Haematococcus* and *Porphyridium*) are produced in closed systems. Fig. 1.1 shows the amount of biomass required to produce 1 ton of end product. However, a commercial plant needs to be able to produce at least 10 tons of high-value products per annum, or more than 1000 tons of commodity products such as proteins (Borowitska & Vonshak, 2017).

Current commercial culture system units range from about 1000 L in aquaculture feed, to about 3×10^4 L in raceway ponds for algal biomass production, to about 1×10^4 L in tubular reactors, and up to 1×10^9 L in the extensive ponds used for culturing *Dunaliella* in Australia (Borowitska, 2016).

Cultured Microalgae for the Food Industry. DOI: https://doi.org/10.1016/B978-0-12-821080-2.00002-2

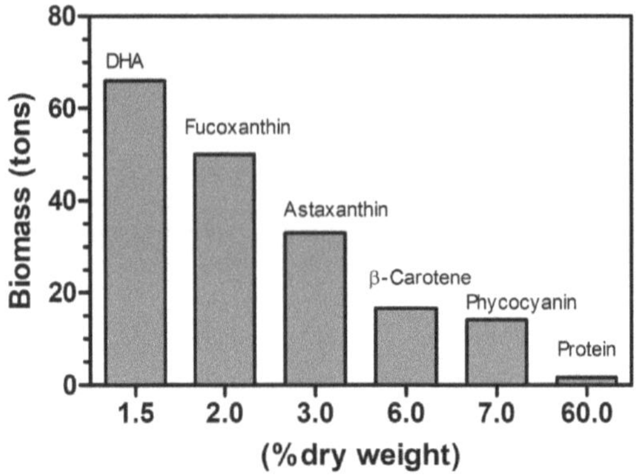

FIGURE 1.1 Amount of microalgal biomass required to produce 1 ton of a given product according to the percentage present in the dry matter (based on data from Borowitska & Vonshak, 2017).

The current microalgal biomass market is dominated by a limited number of microalgal species: (1) *Arthrospira platensis* with more than 18,000 tons of biomass being produced per year—of which about two-thirds is produced in China (Cheng et al., 2016); (2) *Chlorella*, with about 9500 tons per year; (3) *Dunaliella salina*, mainly marketed for its high β-carotene content, with an annual production of 1700 tons; (4) *Aphanizomenon* for the extraction of phycocyanin with 1500 tons of biomass per year; and (5) *Haematococcus* for the production of astaxanthin with about 300 tons of biomass being produced per year (da Silva Vaz, Botelho Moreira, Greque de Morais, & Costa, 2016; Hamed, 2016; Spolaore, Joannis-cassan, Duran, & Isambert, 2006). Reported figures must be regarded as purely indicative of the microalgal market size, as in several cases farmers do not wish to disclose their production commercial size which can be subject to variations over the years. It should be underlined that these species benefit from the Generally Recognized As Safe (GRAS) status, that is, generally regarded as safe, given by Food and Drug Administration (FDA) for food purposes and are thus considered safe and fit for human consumption in the United States (Hamed, 2016; Spolaore et al., 2006; Wells et al., 2017). In addition to their use in human food, microalgae have shown a growing interest in animal nutrition as feed additives. Microalgal biomass can be incorporated into the diet of a wide variety of animals ranging from fish to pets and livestock (Borowitska, 2013; Pulz & Gross, 2004; Spolaore et al., 2006). They are mainly used to improve not only the physiology of animals but also the organoleptic properties of the final consumer product (Hamed, 2016). Currently, about 30% of current global microalgal production is sold to the feed industries (Milledge, 2012; Raja, Hemaiswarya, Kumar, Sridhar, & Rengasamy, 2008; Spolaore et al., 2006). The opportunities offered by microalgae as food and feed are related to their diversity of biomass composition, which can be achieved either by strain selection or by manipulating their growth conditions. Indeed, biochemical composition can vary greatly in response to a change in the environment and growth conditions, which creates an unbalance between light absorption and reductant utilization. These are usually referred to as stressing procedures, and examples are nutrient depletion, high light irradiance, extreme pH, temperature, and high salinity.

The problems related to the scaling up of algal cultures have been recently revised by Borowitska and Vonshak (2017). In order to fully exploit the potential of microalgae, a good understanding of the physiology, biochemistry, and ecology of microalgae in industrial-scale cultures is key to achieving reliable cultures and to optimizing productivity and enhancing the production of the desired compounds. In this chapter we summarize the results of different studies on the effects of environmental factors such as light, temperature, and oxygen concentration on productivity and biomass composition, with special emphasis on large-scale microalgal cultures.

1.2 Light

The production of sufficient amounts of biomass is a prerequisite for meeting large-scale demand and reaching economic viability. To date, in industrial-scale microalgal production systems, light conversion efficiency ranges between 1% and 2% on a solar basis, far short of the theoretical maximum of about 10%, owing to energy loss at all stages of the process. Table 1.1 shows a summary of the expected energy losses of total incident solar radiation during the biomass production process.

It is estimated that about 10% of photosynthetically active radiation (PAR) is reflected by the cultures. Microalgae, when grown under high light intensities, also synthesize some protective pigments (xanthophyll cycle pigments, in green algae) that dissipate excess light as heat (Masojídek et al., 1999). This photosynthetically inactive light absorption, to the extent that it occurs, further lowers the photosynthetic conversion efficiency. Furthermore, photons above 700 nm and below 400 nm are not utilized by microalgae pigments (as well as by higher plants), and thus about 55% of incident solar light is unavailable to drive photosynthesis. The sum of losses by reflection and, more importantly, the amount of light out of the PAR range, reduces the available light for photosynthesis to about 41%. The energy required to drive a charge separation event in PSII is approximately 176 kJ mol^{-1} (i.e., equal to the energy of a 680 nm wavelength photon), and 171 kJ mol^{-1} (i.e., the energy content of a photon at 700 nm) for PSI. If we assume that the mean energy content of photons in the 400–700 nm range is about 218 kJ mol^{-1}, then the loss of energy between absorption and charge separation in the two photosystems will be approximately $\{[218-(171+176)/2]/218\} \times 100$. This means that approximately 20.4% of the incident solar energy is irretrievably lost as heat in the process, because of the relaxation of higher excited states of chlorophyll (Chl) to the first excited singlet

TABLE 1.1 Summary of expected energy losses of total incident solar radiation energy in outdoor microalgal cultures.

Process	Energy radiation losses (%)	Remaining energy (%)
Total solar radiation	–	100
Reflection/scattering	10	90
Radiation outside PAR	55	41
Loss of useful absorbed PAR energy at 680 nm (PSII) and 700 nm (PSI)[a] due to nonphotochemical processes	20[a]	32.8
Conversion energy to biomass	71.3[b]	9.4
Photosynthesis saturation and photoinhibition	80[c]	1.88
Cell maintenance in cultures subjected to diurnal cycle	10	1.7
Actual light conversion efficiency under solar light		1.7

[a]*The energy required to drive a charge separation event in PSII is approximately 176 kJ mol^{-1} (i.e., equal to the energy of a 680 nm photon) and 171 kJ mol^{-1} (i.e., at 700 nm) for PSI. Therefore the average for the two photosystems is 171 + 176/2 = 173.5 kJ mol^{-1}. If we assume that the average energy content of photons in the 400–700 nm range is 218 kJ mol^{-1}, then the loss of absorbed energy due to nonphotochemical processes is 218–173.5/218 × 100 = 20.4%. For blue photons the degradation loss increases to 40%.*
[b]*Conversion energy to biomass: Assuming that for light conversion to biomass (i.e., 1 mol of hexose yields 2808 kJ mol^{-1}, and that a minimum of 9.4 mol of quanta are required to release 1 mol of O_2, (2808 kJ mol^{-1}/(173.5 kJ mol^{-1} × 6 (9.4 mol quanta) × 100 = 28.7% (i.e., an energy loss of 71.3%).*
[c]*Mean loss of energy evaluated through measurements of the effective quantum yield of PSII ($\Delta F/F_m'$) at different times of day in summer.*

state (Zhu, Long, & Ort, 2008). For light conversion to biomass, considering that 1 mol of hexose yields 2808 kJ mol^{-1}, and that a minimum of 9.4 mol of quanta are required to release 1 mol of O_2 [2808 kJ mol^{-1}/173.5 kJ mol^{-1} × 6 (9.4 mol quanta) × 100 = 28.7%] (i.e., an energy loss of 71.3%), then consequently the theoretical maximum light conversion efficiency for biomass production is about 9.4% of solar light. Measurements of the effective quantum yield of PSII using Chl fluorescence, carried out at midday in summer, usually shows a strong reduction indicating that up to 90% of the light energy is lost via nonphotochemical processes, mainly as heat (see Section 1.9). This drops the light conversion efficiency usually attainable in large scale to less than 2%. Finally, although at a much lower extent, the biomass loss occurring during the night must be considered, involving dissipation of energy through the respiration usually at the expense of carbohydrates (or other stored products) synthesized during the light period (Torzillo, Sacchi, Materassi, & Richmond, 1991). Bearing this information in mind, it is possible to estimate the amount of biomass attainable in a precise location with known light energy available, provided that neither technological aspects (such as mixing, cell concentration), nor environmental constraints (such as temperature, pH, O_2) are limiting. For example, with an average mean solar energy of 18 MJ m^{-2} day^{-1}, and a cultivation period of 210 days, with a light conversion efficiency of 1.7%, the energy content of biomass being 20 kJ g^{-1}, the corresponding biomass yield will be about 32 tons ha^{-1} $year^{-1}$. The corresponding mean yield per square meter of the reactor will be about 15 g m^{-2} day^{-1} (mean over the cultivation period of 210 days). However, if the theoretical efficiency could be reached, then the mean biomass per square meter would raise to about 83 g m^{-2} day^{-1}. Unfortunately, neither with laboratory nor outdoor cultures, has an efficiency of 9.4% ever been attained. Continuous cultures of *Synechocystis* reached an efficiency of 12.5% on PAR, corresponding to 5.6% on a solar basis (Touloupakis, Cicchi, & Torzillo, 2015). It must be pointed out that the first calculations of photosynthetic efficiency coming from laboratory experiments where microalgae were usually exposed to light irradiance within the PAR gave results that must not be confused with those obtained in sunlight (i.e., full light spectra) which are 55% lower. As can be argued from Table 1.1, the greatest energy losses are due to the conversion of solar energy into biomass (about 71%) and to light saturation of photosynthesis (80%). Obviously, while little can be done to improve the light conversion into biomass, a strong enhancement of yield could be attained if the problem of light saturation could be reduced.

To attenuate the effects of this physical constraint that acts as a strong drawback in algal biotechnology, some strategies have been proposed: (1) fast mixing in order to induce rapid light/dark cycles within the culture depth of dense cultures (Giannelli, Scoma, & Torzillo, 2009); (2) use of photobioreactor (PBR) designs in which it is possible to dilute light over a larger surface of the culture (Wijffels & Barbosa, 2010); and (3) the use of organisms with small antenna size. All these strategies aim at improving light conversion efficiency by enhancing the number of cells having access to light and reducing as much as possible the time that cells spend in the dark. Therefore the optimization of the reactor design cannot exclude an appropriate fluid dynamics study.

1.3 Mixing

One of the key problems related to the illumination of microalgal cultures is that light is not a miscible nutrient as other substrates are, but rather a spatial, external, physics-dependent, self-distributing element. Light penetration in dense cultures of microalgae is subjected to rapid attenuation due to scattering and absorption from cell pigments, thus partitioning the culture in several compartments with two extremes: a highly illuminated external algal layer in which cultures are very often subjected to light irradiance superior to that required for saturation, and a bottom one where cells are in the dark.

Between these two extreme light field conditions, there is a layer in which cell photosynthesis spans from light limitation to light saturation (Fig. 1.2).

A debate has developed concerning the potential enhancement of productivity due to the increasing mixing rate of culture. Richmond and Becker (1986) stated that, when nutritional requirements of cultures are satisfied and environmental conditions, such as temperature, are not limiting, then mixing designed to create an optimal light–dark regime for the cells represents the most important requisite to consistently increase the culture yield. Mixing affects productivity in different ways: (1) prevents cell sedimentation; (2) increases mass transfer due to higher mixing rates, preventing the formation of nutritional and gaseous gradients, and lowering the level of dissolved oxygen in the culture that is usually much higher than the air saturation level as a result of photosynthesis; and (3) moves the cells through an optically dense gradient with variation in the quantity and quality of the light. Therefore the assessment of the effect of mixing on culture productivity is complex, making it difficult to separate the contribution of the different factors influenced by mixing, particularly in mass cultures where it is more difficult to separate the effect of single environmental factors.

Following inoculation of a culture operated in a batch, the cells will be subjected to a transition from the high light, acclimating to low light conditions as the culture density increases (Grobbelaar, Nedbal, Tichy, & Setlik, 1995). This condition entails an increase in cell pigments with a consequent

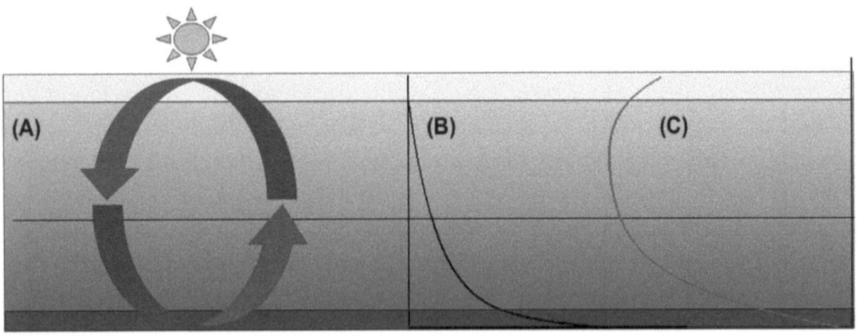

FIGURE 1.2 Simplified scheme of light distribution across the culture layer. In dense algal cultures, the optical path can be schematically divided into four zones (A), created by the attenuation of light with depth (B). The upper layer can be subjected to light irradiance above the saturation, which may induce photoinhibition (C). Below this layer, light irradiance is still saturating the photosynthesis process (C), while a further decrease in irradiance due to its extinction by cell pigments, creating a zone where the photosynthetic activity is light limited [linear part of *P/I* curve in (C)]. At the bottom part of the culture there is a layer where light does not penetrate or its intensity is below the compensation point of photosynthesis (i.e., no net oxygen release). Mixing of the culture, indicated with arrows, moves the cells across the light gradient, and thus creates a certain homogeneity of exposure of the culture to light. *Adapted from Grobbelaar, J. U. (2010). Microalgal biomass production: Challenges and realities. Photosynthesis Research 106, 201–206.*

reduction of light penetration. The goal of increasing turbulence should be reducing the level of dark acclimation by preventing/retarding the increase in cell pigmentation.

An interesting laboratory experiment on the effect of different light patterns on the acclimation of cells of *Dunaliella tertiolecta* was carried out by Havelková-Doušová et al. (2004). They compared four different patterns of illumination: (1) constant light, typical of laboratory cultures; (2) sinusoidal pattern typical of solar illumination in perfectly stratified water columns; and (3) two fluctuating light patterns simulating vertical mixing having the same average light intensities. They concluded that cultures subjected to simulated vertical mixing acclimate to a light intensity which is about three times lower than the average light experienced by the cells. Therefore it might be argued that outdoor dense cultures subjected to intermittent illumination caused by mixing will acclimate to a light intensity much lower than the average calculated light and indicating that probably more than two-thirds of the light received by the cells is dissipated by nonphotochemical quenching (NPQ).

1.3.1 Intermittent light

There is a general consensus that because of photosynthesis saturation, a large part of the light energy (up to 90% in summer) captured by the cell pigments is lost because it is above the level of photosynthesis saturation. This fact becomes increasingly important as the cultures acclimate to low light conditions as a result of the increasing concentration, and their maximum photosynthesis rate (P_{max}) further decreases. This problem has been well known since the start of algal biotechnology (Burlew, 1953). One of the strategies to reduce the effect of saturation in cultures exposed to high light is to supply light in an intermittent manner. Kok (1953) used the term "*flashing light*" to indicate the lapse of time during which the cells are exposed to light. In his experiment cells were exposed to a flash frequency ranging within more than one order of magnitude, from 60 Hz and 0.5 Hz. Flash times ranged between 3 and 300 ms. The duty cycle, that is, the ratio between total time and the flash time (Tf + Td/Tf) was maintained at a constant 5.5. The incident artificial light intensity was close to solar light attained at Palo Alto, California (1400 μmol photons m^{-2} s^{-1}), while the integrated light intensity remained constant at about 240 μmol photons m^{-2} s^{-1}. He found that when cells were exposed to flashes of 3 ms followed by dark periods of 13.5 ms (i.e., 60 Hz), cells maintained 100% of their initial photosynthesis rate. In other words, the photosynthetic activity of the cells remained constant as they were exposed to constant light of 1400 μmol photons m^{-2} s^{-1}. This is not surprising if we consider that the turnover time of the plastoquinone, from a reduced to an oxidized state and then reoxidized again, is close to 3 ms. However, as soon as the flash increased to 150 ms and the frequency was decreased to 1.2 Hz the photosynthetic activity reduced to about 30% of the initial value. While flash frequencies of 60 Hz are beyond any engineering possibility, flash frequencies close to 1 Hz seem to be compatible with those attainable in thin-layer cultures and small tube diameters (Grobbelaar, 2009, 2010).

Grobbelaar, Ladislav, and Tichy (1996), carried out a detailed study on the effect of intermittent light on cultures of *Chlorella vulgaris*. Cultures were either irradiated by continuous white light or under light/dark frequencies between 0.05 and 5000 Hz. The cycles were either of equal duration (1L/1D) or with a dark period double the light period (1L/2D). The incident light delivered by the light source (LEDs) was 800 μmol photons $m^{-2} s^{-1}$; the integrated light intensity (averaged light) was 400 μmol photons $m^{-2} s^{-1}$ in cells exposed to 1L/1D, while it was about 260 μmol photons $m^{-2} s^{-1}$ received by the 1L/2D. It was found that none of the photosynthesis rates in fluctuating light treatments came near to continuous illumination. Photosynthetic rates for the equal light/dark cycle (1L/1D) were equal to those in continuous integrated light at about 50–100 ms L/D fluctuations. At longer time durations the photosynthesis rates decreased, and vice versa. This was also true for the L/D fluctuations where the dark time was double that of the light period (1L/2D). In summary, Grobbelaar et al. (1996) came to the following conclusions: (1) the photosynthetic rates increased exponentially with increasing light/dark frequency; (2) a longer dark period relative to the light period does not necessarily lead to higher photosynthetic rates; and (3) acclimation to a specific frequency, in the range and with the resolution tested, did not take place. Another interesting conclusion was that in continuous light, high light acclimated cells had the highest maximal photosynthesis rates, while the converse was observed in modulated light, where low light acclimated cells had the higher photosynthesis rates. The authors concluded that the most important factor influencing photosynthesis rates, either in continuous or intermittent light, is whether the cells show characteristics of light or dark acclimation. Microalgal cells exposed to high-intensity intermittent light increase their photosynthesis rates and acquire high light acclimation characteristics (Grobbelaar et al., 1996). The effect of intermittent light on the productivity of *Nannochloropsis* cultures was also investigated by Sforza, Simionato, Giacometti, Bertucco, and Morosinotto (2012). Cultures were subjected to either 1200 or 350 μmol photons $m^{-2} s^{-1}$ and frequencies ranging between 1 and 30 Hz. Cultures exposed to 350 μmol photons $m^{-2} s^{-1}$ and 30 Hz, or 1200 μmol photons $m^{-2} s^{-1}$ at 10 Hz performed similarly to those exposed to continuous light of the same integrated intensity (120 μmol photons $m^{-2} s^{-1}$).

1.3.2 Intermittent light application in the real world

Grobbelaar (1989) pointed out that microalgae grown in dense mass cultures are subjected to three ranges of intermittent light, namely:

1. Low-frequency cycles of hours, days, or years, determined by diurnal and seasonal variations mainly influencing the periodicity of cell division;

2. Medium-frequency fluctuations of seconds to minutes which are dominant in open pond raceway mass algal production systems;
3. High-frequency fluctuations of 100 ms (10 Hz) or less which may take place in cultures with a very low optical path (about 1 cm).

For convenience a distinction of the culture system can be made according to their optical path, namely:

1. Short light path (SLP): flat PBRs, immobilized systems such as thin-layer cascades (TLCs), flat plates, immobilized cell systems, and narrow-diameter tubular reactors with less than 4.0 cm and mostly 1.0–2.5 cm;
2. Medium light path (MLP): systems such as open pond raceways and large tubular PBRs typically 10–15 cm deep;
3. Long light path: extensive production ponds (maturation ponds and some aquaculture systems and wastewater treatment systems usually 30 cm deep).

Raceway pond cultures (MLP) are usually mixed with paddle wheels impressing a superficial velocity ranging between 10 and 30 cm s^{-1}, which may promote vertical movement of cells within a range of seconds to minutes. Their frequency cycle is much lower than 1 Hz and therefore unable to benefit from intermittent light. On the other hand, the possibility of reaching a high culture speed is technically impossible, and it is energetically not achievable in large commercial-scale ponds. The attempt by Law, Terry, Wickman, and Chalup (1983) to promote vertical mixing by placing aerofoil types of devices in flowing cultures, which apparently strongly increased the light conversion efficiency up to 10% (solar basis) of the culture, was then reinterpreted as due to resuspension of settled cells and breaking down of nutrient gradients rather than a "flashing effect" (Law & Berning, 1991). Cheng, Yang, Ye, Zou, and Cen (2015) placed up–down chute baffles in a laboratory-operated raceway pond to generate a clockwise and anticlockwise vortex to promote vertical mixing. These devices allowed a reduction of the ponds' mixing time by 41% and an increase in the mass transfer coefficient of 25%. The dark–light cycle was strongly reduced to one-twelfth as compared to a control culture (without chute baffles), which promoted an increase in productivity of about 33%, and this increase was ascribed to the flashing light effect. These experiments demonstrate that it is possible to enhance the culture productivity by promoting vertical mixing, but it remains highly questionable whether these technologies can be profitably transferred to large-scale ponds.

TLS (Thin-Layer System) systems (Setlik, Sust, & Malek, 1970), where the culture velocity is about 0.3 m s^{-1}, are probably amenable to benefit from light intermittence and thus attain higher biomass yields as compared to open ponds. An example of a pilot TLC is shown in Fig. 1.3A. The unit is made up of two identical platforms divided longitudinally in three tracks

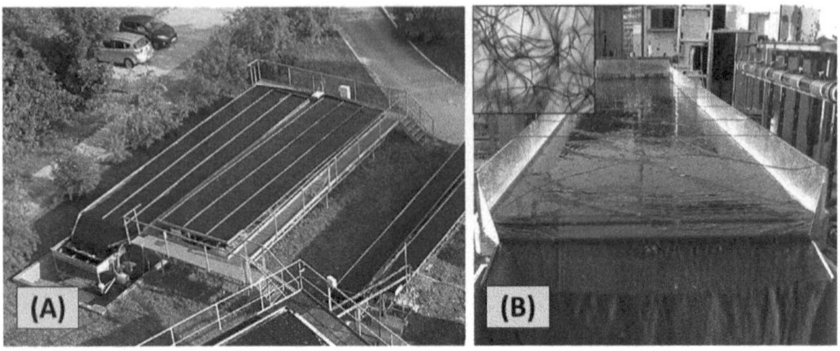

FIGURE 1.3 (A) Areal view of a thin-layer cascade constructed as a system of connected sloping planes (90 m^2) that can be operated with a volume between 500 and 1000 L (Masojídek, Sergejevová, Malapascua & Kopecký, 2015; Grivalský et al., 2019; patent pending) at the Institute of Microbiology, Třeboň, Czech Republic. (B) Thin-layer culture system (10 m^2) operated with a culture of *Arthrospira platensis* strain M2 (shown on the top left corner) of the culture collection CNR-IBE-Florence, during summer 2016 at the CNR-IBE (Florence, latitude 43°47′14″64N). The culture was 1 cm deep, and operated in a semicontinuous way with a mean morning biomass concentration of 5 g L^{-1} (Silva et al., 2017). Cultures were circulated night and day using a special type of horizontal centrifugal pump that does not break *Arthrospira* trichomes even when operated at the maximum speed (up to 1000 rpm). It was made of PVC equipped with a paddle wheel enclosed in the housing. The design of the pump has been described elsewhere (Bocci, Torzillo, Vincenzini, & Materassi, 1988). *Photographs by Dr. D. Kubáč.*

where the microalgae culture is exposed to sunlight in a south–north orientation. It was designed to minimize the dark volume of the circulated microalgae culture as the exposed surface to total volume ratio can be operated in the range between 60 and 180 m^{-1}. The slope of each platform is (independently) vertically adjustable between 0.5 and 3 degrees, which in combination with a variable-flow (20–60 cm s^{-1}) makes it possible to change layer thickness (between 5 and 15 mm) at varying biomass density. This is advantageous to regulate the light supply to the microalgae culture according to its physiological demands to maintain an optimum irradiance regime.

According to Doucha and Livansky (1995), the turbulence is reached with a Reynolds number of 3000, which may result in flashing light of low frequency around 1 Hz. Another observation supporting the influence of intermittent yield in TLC is the fact that cultures maintain linear productivity within a vast range of concentrations (up to 30 g/L of dry weight) indicating that culture concentration remained optimal during the whole run, possibly as an effect of intermittent light (Doucha & Livansky, 2014). These findings were recently confirmed by Silva, Ranglova, Malapasqua, Masojidek & Torzillo (2017) who compared the productivity of Spirulina cultures grown on TLC (1 cm depth) and ponds (10 cm depth; Fig. 1.3B). Biomass productivity over 20 g m^{-2} day^{-1} was attained in the TLC at the optimal temperature, while at the suboptimal temperature the productivity decreased by 20%. In the open circular pond (OCP), the biomass productivity at both temperatures was about one-third lower when compared to those in the TLC. The better culture performance in the TLC was mainly ascribed to the shorter light path that promoted much faster light/dark cycles favorable for photosynthesis, as well as the faster warming of the cultures in the morning as compared to the OCP cultures.

A clear improvement in culture productivity was also observed with cultures of Spirulina grown in tubular reactors of different diameters (Table 1.2). A significant increase in the yield was observed when the diameter was consistently reduced, from 14 to 2.5 cm.

These observations support the fact that a TLC can benefit from intermittent light. However, being subjected to relatively low frequency, the benefit, although significant, is much lower than that observed by Kok (1953), realizing a benefit ranging between 20% and 30% higher than cultures in open ponds, that is, much lower than what could be expected by Kok experiments in which the yield of culture could increase of a factor of 5.5.

1.4 Optimal biomass concentration

The volumetric productivity of the culture is expressed as follows:

$$P = \mu \times V\ (\text{g L}^{-1}\ \text{day}^{-1}),$$

While for mass culture it can be written as follows:

$$P = \mu \times d\ (\text{g m}^{-2}\ \text{day}^{-1})$$

where d is the culture depth.

The growth rate is a dependent variable of the concentration of the culture: the higher the concentration, the lower the growth rate. For a given culture system, there is an optimal concentration at which culture productivity reaches the maximum, below or higher a certain value where productivity decreases. Therefore in order to maximize daily productivity, the concentration should be maintained as close as possible to the optimal value by harvesting the excess of biomass. However, harvesting very likely cannot be carried out daily because in many cases the change of concentration within 24 h is negligible, particularly in open pond cultures usually operated at 15–30 cm depth. Alternatively, cultures can be operated according to a fixed dilution rate, with a constant volume of culture being removed daily. In this case, a fixed growth rate is imposed on the culture. Both strategies require an experienced culture manager. Maintaining the concentration within a given range requires taking into account the changes in light irradiance occurring seasonally. In summer the concentration is usually higher than in autumn or spring. However, in summer the concentration can change little if the culture layer is increased, usually to reduce the risk of overheating of the culture. The adoption of fixed dilution rates must also take into account the

TABLE 1.2 Productivity of *Arthrospira plantensis* cultures grown outdoors in photobioreactors made with tubes of different diameters.

Tube internal diameter (cm)	$L\ m^{-1}$ [a]	$L\ m^{-2}$ [b]	Mean areal density ($g\ m^{-2}$)[b]	*S/V* (m^{-1})[b]	Optimal biomass concentration ($g\ L^{-1}$)[b]	Mean net volumetric productivity ($g\ L^{-1}\ d^{-1}$)	Mean areal yield ($g\ m^{-2}\ day^{-1}$)[b]
14.0	15.4	110	50–70	9	0.4–0.6	0.20	22
13.1	13.3	102	60–80	10	0.6–0.8	0.23	23
7.4	4.3	58	70–80-	17	1.2–1.4	0.40	23
5.0	1.96	39	70–90	25	1.4–2.0	0.65	25
2.5	0.49	19.6	70–120	50	3.5–6.0	1.4	27

[a] *Volume per linear meter.*
[b] *Surface area refers to the ground occupied by the tubes calculated considering the sum of diameters.*

changes in light irradiance occurring within the season. In general, if the chosen dilution rate is too high this will entail a reduction in the concentration of the culture with the risk of exposing cells to photoinhibition, while if it is too low the excessive rise in concentration reduces productivity.

The importance of an optimal biomass concentration as a practical tool for maximizing culture productivity outdoors has been extensively debated by Richmond (2013). The best harvesting strategy is to carry out harvesting during the late afternoon when light intensity declines significantly, or at night, as this can reduce the risk of photoinhibition and reduce the amount of biomass lost during the night (Torzillo et al., 1991). However, the composition of the biomass can change even from the evening to the following morning, and this represents another variable to be considered. For example, Spirulina biomass at the end of day is richer in carbohydrates than that harvested in the morning, while protein shows the opposite behavior (Torzillo et al., 1991). Similarly, in organisms accumulating lipids, such as *Nannochloropsis*, some loss of lipid can occur because of night respiration, and internal changes in the ratio between saturated and unsaturated fatty acid have been reported. Usually, biomass harvested in the morning contains a higher percentage of eicosapentaenoic acid (EPA) compared to that harvested in the evening (Carneiro et al., 2020).

1.5 Culture depth

Culture depth is another important parameter that can affect the productivity of the culture. It is a parameter that is related in an inverse manner to the biomass concentration, that is, the higher the depth of the culture the lower is the optimal biomass concentration. Because of this interdependence between depth and biomass concentration it is useful to compare the different culture depths and culture systems by adopting a common parameter, that is, the biomass footprint, expressed as g of biomass m^{-2} (g dry weight m^{-2}), also indicated as areal density (Soeder, 1980; Hartig et al., 1980). A detailed study on the effect of culture depth on productivity was carried out by Moheimani and Borowitska (2006). It was found that in ponds in autumn, when light irradiance and temperature are lower, productivity of the coccolithophorid alga *Pleurochrysis carterae* was higher when the depths were maintained at 13 and 16 cm, while the opposite was found in summer during which optimal depth increased up to 21 cm. The culture depth influences the surface area to volume ratio (*S*/*V*) in a direct manner, as it varies the amount of energy received per unit of culture volume. In a light-limited system, like mass cultures are, the higher the *S*/*V* ratio the higher the volumetric productivity ($g\,L^{-1}\,day^{-1}$). Venancio, Cella, Garcia Lopes, and Bianchini Derner (2020) recently studied the effect of *S*/*V* ratios on TLCs. The authors of that study compared two depths, 0.5 cm and 1.0 cm, corresponding to *S*/*V* ratios of

$80\ m^{-1}$ and $60\ m^{-1}$, respectively, using *Scenedesmus obliquus* cultures. The higher *S/V* system resulted in a higher culture concentration of $20.14\ g\ L^{-1}$ ($S/V = 80\ m^{-1}$) compared to *S/V* $60\ m^{-1}$ ($14.6\ g\ L^{-1}$), and differences in daily productivities of $2.42\ g\ L^{-1}\ day^{-1}$ (*S/V* $80\ m^{-1}$) versus $1.85\ g\ L^{-1}\ day^{-1}$ ($S/V = 60\ m^{-1}$). When comparing different culture depths in a closed system it is important not to alter the percentage of culture volume in the dark (mostly in the tubes for recycling the culture, and retention tank) which can influence the dark biomass loss. Decreasing the culture depth is limited by two factors, friction and temperature instability (Béchet et al., 2013; Oswald, 1988). Shallow raceway ponds require more paddle wheels per square meter of culture surface to compensate for frictional loss. Oswald (1988) calculated the maximum distance between paddle wheels and concluded that by increasing the culture depth from 10 to 30 cm, the maximum distance between them increased almost 12-fold. Chiaromonti et al. (2013) determined that the total energy required to operate a raceway pond at 5 cm was 50% less than operating at 20 cm, taking into account friction and reduced operating volume, which would reduce harvesting costs by processing less water. Finally, the correct management of the culture depth is particularly useful to compensate for seasonal changes in irradiance and temperature, thus maximizing the yield of the system.

1.6 Temperature

Because of seasonal and diurnal fluctuations, temperature represents one of the major biological limitations for biomass production of microalgae. As pointed out by Borowitska (2016), because of the important role played by temperature on productivity, it should probably be the first parameter for strain selection. According to Vonshak and Richmond (1988), high net biomass output may not be achieved in open ponds, mainly due to the difficulties in maintaining the optimal temperature throughout the day and around the year. Indeed, even during summer days when temperatures reach maximum values, the temperature of the culture in the morning, almost 10°C below the optimum, prevents full exploitation of the photosynthetic capacity of the culture for a few hours. In a comparative study on the yield of Spirulina achieved in open ponds and in PBRs, the better yield in the latter was attributed to the better temperature profile maintained inside the PBRs (Torzillo et al., 1986). Laboratory experiments have shown that the maximum biomass yield is obtained when Spirulina is grown at the optimal temperature of 35°C (Payer et al., 1980; Tomaselli et al., 1987). In general, growth versus temperature curves are asymmetrical, with an asymptotic increase on the left side of the curve and an abrupt decrease on the right hand side (Ras, Steyer, & Bernard, 2013). A narrow range between optimal and lethal can indicate that the specie is sensitive to heat stress, while a wide range may indicate that the specie is able to acclimate to high temperature. Temperature tolerance of microalgae is large below the optimal value but very small above the optimum. Long-term acclimation strategies based on generational sequences during progressive temperature increase have shown that some species are able to adapt to supraoptimal temperature (Ras et al., 2013).

Algal cultures grown outdoors are usually exposed to a combination of environmental stresses. The most common combination is high light and low temperature. It happens while fluctuations in light intensity occur in a range of 1–2 h, the increase of temperature is a much slower process and takes about 4–5 h. This kind of desynchronization between the two most important environmental factors, which affect photosynthesis and growth of outdoor algal cultures, results in a unique stress condition under which photoinhibition may indeed be induced at relatively low light intensity due to the suboptimal temperature conditions (Vonshak, Torzillo, Masojídek, & Boussiba, 2001). Acclimation to low temperature of two *A. platensis* strains, M2 and Kenya, was studied by Vonshak and Novoplansky (2008). Both strains showed similar growth rates when grown at 30°C, but once acclimated to low temperature, the M2 strain showed a greater decline in growth, suggesting that the Kenya strain acclimated better to low temperature by downregulating its photosynthetic activity through (1) decreasing antenna size, (2) decreasing the reaction center density, and (3) increasing the energy dissipation. In the same study, the importance of screening and selecting strains for mass culture on the basis of their ability to withstand and acclimate to low temperature was pointed out. Indeed, this parameter can fluctuate in open ponds from 15°C in the early morning to 35°C in the middle of day, causing a significant reduction in productivity and changes to biomass composition (Torzillo & Vonshak, 1994; Vonshak et al., 2001).

The importance of using different strains adapted to different temperatures at different times of the year can be useful to extend the cultivation period and attain higher productivities round year. This strategy has been successfully tested with Spirulina at Earthrise Farm in California, United States (Belay, 1997).

Culture temperature is also strongly modified by the culture system. For example, in cultures grown in closed systems, due to the greenhouse effect, the temperature rise in the morning is much faster that in open ponds (Torzillo et al., 1986). However, in a closed system the temperature can exceed the optimal in the middle of the day and employing a cooling system may become mandatory. A similar situation can be experienced by cultures in ponds located inside greenhouses. Indeed, in contrast to cultures in open ponds outdoors, where cooling is promoted by heat exchange and evaporative cooling, in the case of cultivation of microalgae in greenhouses, the evaporation and heat exchange with the atmosphere are limited and the temperature of the culture can reach as much as 42°C, which usually has lethal results for the overwhelming majority of microalgae species including the mesophilic cyanobacterium Spirulina. Therefore cultivation of microalgae in ponds in greenhouses is not a trivial task. For optimal design of the greenhouse, it is of seminal importance that a careful preliminary study of the installation site climatic conditions be carried out, together with a proper design of the raceways ponds, considering various factors such as: (1) the proper choice of the style and size of the greenhouse; (2) doors and designing structure; (3) covering material; and (4) cooling and ventilation. The latter is probably one of the most important factors for microalgae cultivation in greenhouses. There is a variety of cooling systems, including options for mechanical ventilation, natural ventilation, heat exchange, and shading. It is beyond the scope of this chapter to analyze in detail the design of a greenhouse for microalgal cultivation.

Fig. 1.4 shows two examples of temperature profiles of Spirulina cultures recorded in raceway ponds in differently designed and managed greenhouses.

In Fig. 1.4A, ponds were placed in transparent plastic tunnels such as those used for crops, with little possibility of controlling the air temperature. In the center of Italy, during summer, the air temperature in a tunnel reaches about 45°C, and the corresponding temperature of the culture ranges between 37°C and 40°C. Between 11 a.m. and 6 p.m. the culture temperature is above the optimal (35°C). One option to reduce the culture temperature is spraying water on the surface of the tunnel, and in this way it is possible to attain a temperature reduction of about 2°C–3°C. Obviously, this cooling system entails a consistent increase of the water footprint of the process. The case of cultivation of microalgae in a high-tech greenhouse with optimal design and in which it is possible to achieve an automatic control of the air temperature and thus of the pond cultures is different (Fig. 1.4B), with the possibility also to take advantage of an intelligent decision support system to assist the microalgal

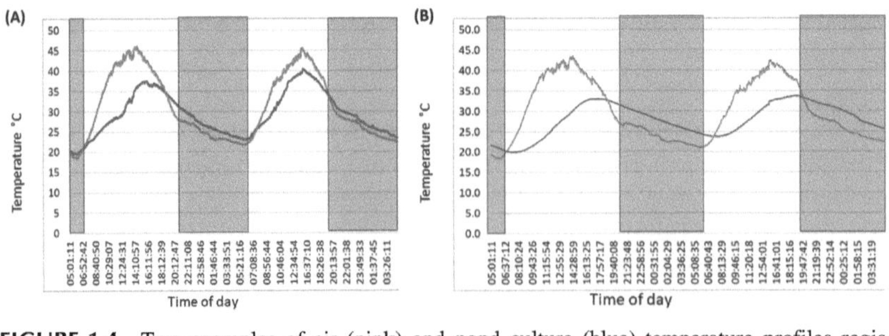

FIGURE 1.4 Two examples of air (pink) and pond culture (blue) temperature profiles registered in two differently designed and managed greenhouses located in central Italy (41°6′54″36N). In Fig. 1.4A, *Arthrospira platensis* ponds were placed in a transparent plastic tunnel without any possibility of controlling the air temperature, during the hottest period, except by spraying water on the surface of the tunnel, while in Fig. 1.4B, there is fully controlled temperature in greenhouse properly designed for the cultivation of microalgae. *Courtesy Ing. M. Diano, M2M Engineering.*

farmer. In this high-tech greenhouse, it was possible to keep the culture temperature within the optimum value even in the hottest season, when the ambient temperature reached as high as 38°C. It is expected that in this case the cost of biomass will be significantly higher compared to open-air production, and therefore the product would be destined for the health food market, pharmaceuticals, food ingredients, organic microalgal biomass, and other more stringent uses. On the other hand, cultivation in greenhouses allows to extend the cultivation period, reduce the risk of contamination of the culture, and produce biomass with an almost constant biochemical composition.

Finally, the culture layer and tube diameter of a PBR and its orientation can strongly influence the temperature profile of the culture. For Spirulina a sharp decrease in the photosynthetic activity was found above 35°C (Torzillo & Vonshak, 1994). Heat stress causes changes in the reduction–oxidation properties of PSII acceptors and reduces the efficiency of photosynthetic electron transport in both photosystems (Mathur, Agrawal, & Jajoo, 2014). High-temperature stress also influences the shape of the O–J–I–P curve (see Section 1.9 for explanations), decreasing F_m and increasing F_0. The increase in F_0 may be due to the release of light harversting complex II (LHC II) from the PSII complex, inactivation of the PSII photochemical reaction, or inhibition of electron flow due to the reduced transfer of Q_A to Q_B electron transporter.

The influence of culture temperature on the productivity and biochemical composition of *Arthrospira platensis* was studied in outdoor thermostated tubular PBRs (Torzillo et al., 1991). During a 5-month experiment performed in summer, the average net biomass synthesis during the daylight period in the culture grown at the optimal temperature of 35°C was superior by 14% to that in the culture grown at 25°C (Table 1.3).

TABLE 1.3 Effect of temperature on the productivity of *Arthrospira platensis* strain M2 during the daylight period and over 24 h.

Months	Daylight productivity (mg L^{-1} day^{-1})		Net productivity (mg L^{-1} day^{-1})	
	35°C	25°C	35°C	25°C
May	390 ± 34	350 ± 38	360 ± 38	280 ± 26
June	450 ± 45	400 ± 44	400 ± 39	275 ± 20
July	525 ± 46	425 ± 40	450 ± 43	330 ± 31
August	454 ± 43	400 ± 35	425 ± 40	325 ± 28
September	400 ± 40	350 ± 32	320 ± 41	225 ± 23

Ammonia was used as a source of nitrogen. Data are mean values ± SD of 30 dry weight measurements performed each month.

It was also observed that the average biomass loss during the night was significantly higher in the culture grown at 25°C (7.6% of total dry weight reached in the evening) than in the culture grown at 35°C (5% of dry weight). This difference in biomass loss accounted for the more pronounced differences in the net productivity of the two cultures (23% average); that is, productivity over 24 h. The night biomass loss appeared to be influenced by both temperature and the quantity of light irradiance received by the cells in the preceding day; in fact, higher rates of night biomass loss were observed, especially in the culture grown at 25°C, during the months of June and July when the total light irradiance was higher. The influence of the daily light irradiance received by the cells on the night biomass loss became even more evident in another experiment in cultures grown at different concentrations and thermostated at 35°C during the day and 20°C during the night. Night biomass loss ranged from 14% of the total dry weight reached in the evening in the culture grown at the lowest biomass concentration, to 5.1% in the one grown at the highest biomass concentration (Table 1.4).

Biomass composition of Spirulina can change in function of the concentration at which cultures are grown. Cultures grown at the suboptimal concentration are normally characterized by a lower protein content which is counterbalanced by a higher carbohydrate content (Table 1.5). Similarly, biomass harvested at sunset is usually lower in protein, and higher in carbohydrate, while lipid content shows no significant change (Table 1.6).

Changes in the chemical biomass composition can occur not only in small-scale reactors as previously described but even in Spirulina industrial ponds. Hidasi and Belay (2018) monitored the biomass composition during the day in a Spirulina culture grown in a 4000 m^2 pond. A decline of the protein content was found at the end of day. However, because of the nocturnal metabolism of carbohydrates, the biomass composition returned to a similar

TABLE 1.4 Effect of biomass concentration on the productivity and night biomass loss in *Arthrospira platensis* strain M2 cultures grown outdoors in a tubular photobioreactor.

Biomass concentration (mg L^{-1})		Daylight productivity (mg L^{-1} day^{-1})		Net productivity (mg L^{-1} day^{-1})		Night biomass loss			
						A		B	
710	± 28	492	± 24	323	± 48	14.0	± 2.6	34.4	± 7.2
1290	± 80	496	± 25	396	± 56	6.6	± 1.7	22.0	± 7.9
2169	± 398	434	± 26	302	± 62	5.1	± 1.3	31.0	± 8.5

A = as % of the dry weight reached at the end of daylight period; B = as % of daylight productivity. Data are the mean values ± SD of a week experiment. Light energy throughout the experiments was 21.35 ± 1.3 MJ m^{-2} day^{-1}.

TABLE 1.5 Influence of biomass concentration on protein and carbohydrate contents of *Arthrospira platensis* harvested at sunset.

Biomass concentration (mg L^{-1})		Protein (%)		Carbohydrate (%)	
710	±28	55.4	±0.67	32.8	±0.64
1290	±80	61.3	±0.49	27.8	±0.87
2169	±398	64.3	±0.63	25.4	±1.70

Data are mean value ± SD (n = 4).

TABLE 1.6 Biochemical composition of *Arthrospira platensis* strain M2 biomass harvested at sunset and sunrise.

Component	Percentage of ash-free dry weight	
	Sunset	Sunrise
Protein ($N \times 6.25$)	57.15 ± 0.69	70.90 ± 2.37
Carbohydrate	33.81 ± 0.66	19.48 ± 1.48
Lipid	9.05 ± 0.61	9.61 ± 0.90
C	49.53 ± 0.32	50.91 ± 0.29
H	7.52 ± 0.18	7.67 ± 0.13
N	9.14 ± 0.11	11.34 ± 0.38
O	33.85 ± 0.31	30.07 ± 0.29
Reduction degree	39.85 ± 0.31	41.72 ± 0.03
Heat of combustion	5.342 ± 0.03	5.632 ± 0.008

level the following day. A similar pattern was observed in phycocyanin content, which was about 20% lower at the end of the day, particularly in summer (Hidasi & Belay, 2018).

Temperature can also affect the lipid composition in *Spirulina platensis* and particularly the amount of γ-linolenic acid (GLA). Higher contents of GLA (17% of total fatty acids) have been recorded in cultures grown at sub-optimal temperatures of 20°C. The GLA content was found to decrease linearly with the increase in temperature with a minimum (11.2%) at 40°C (Oliveira, Monteiro, Robbs, & Leite, 1999; Tomaselli, Giovannetti, & Torzillo, 1993; Tomaselli, Giovannetti, Sacchi, & Bocci, 1988). An opposite behavior was found with *Spirulina maxima*, which had 12.8% at 20°C and rose to 15% at 35°C and 40°C. The two species also showed different optimal temperatures for growth, 30°C with *Arthrospira maxima* and 35°C for *Arthrospira platensis*. Moreover, the decline in growth rate above the optimum was less abrupt with *S. platensis*, indicating a higher capacity of this strain to acclimate to high temperature as compared to *A. maxima* (Oliveira et al., 1999). Changes in biomass composition are also well documented with oleaginous species such as *Nannochloropsis*. A higher eicosapentaenoic acid (EPA) content was observed in the culture maintained at the suboptimal temperature of 18°C by day and by night; EPA reached 20% of the total fatty acids at the end of the light period and rose up to 27% the following morning (Carneiro et al., 2020; Chini Zittelli et al., 1999). This difference can significantly contribute to increasing the value of the biomass, particularly when harvested during the morning. Nonetheless, attention must be paid to the balance between biomass productivity and EPA content of the biomass, in order to obtain the highest EPA productivity possible.

1.7 Oxygen

The photosynthetic electron transport system represents the major source of reactive oxygen species (ROS) having the potential to generate singlet oxygen, hydrogen peroxide, and the superoxide radical. The generation of ROS is enhanced when the photosynthetic apparatus absorbs excess light or under conditions where CO_2 or NADP is limited (Asada, 1994; Nishiyama & Murata, 2014; Takahashi & Murata, 2008). According to Nishiyama, Allakhverdiev, Yamamoto, Hayashio, and Murata (2004), ROS primarily affect the synthesis of the D1 protein de novo at the translation elongation step and thus inhibit the repair of PSII. The negative effect of oxygen on culture productivity of mass cultures has been overlooked for many years, and it appeared evident when cultivation of microalgae in closed systems was proposed (Torzillo et al., 1986). Only a few studies have addressed the potential role of singlet oxygen as a marker for resilience to excess light environments, and to find out whether singlet oxygen resilience correlates with a strain's robustness in excess light environments. Recently, it was demonstrated that Rose Bengal resilient strains showed increased photosynthetic activity in excess light environments (Sheleg et al., 2019).

The yields and biomass compositions of *S. maxima* cultures grown in two PBRs of 250 and 500 m length were compared. The yields in the 500 m long PBRs were 30% lower and the biomass produced in the long PBR had a lower protein content, counterbalanced by a higher carbohydrate content (Torzillo et al., 1986). Taking into account that no significant differences in the temperature profile or in biomass concentrations occurred in the two PBRs, the influence of the length of the tubular circuit on the yield of biomass and on its biochemical composition was attributed mostly to the adverse effect of high oxygen concentration on protein synthesis in Spirulina as previously described by the authors

(Torzillo, Giovannetti, Bocci, & Materassi, 1984). Oxygen accumulation represents a major obstacle in the scaling up of tubular reactors by increasing the tube length. The oxygen production of an algal culture can be calculated according to the following equation (Pirt et al., 1983):

$$O_2(\mathrm{g\,L^{-1}}) = (\mu X\,T_c)/Y_O \tag{1.1}$$

where μ is the growth rate of the culture (h^{-1}), X is the dry weight ($g\,L^{-1}$), T_c is the time cycle of the reactor (h), that is, the time interval between two degassing stations, Y_O (−) is the ratio between the biomass synthesized and the oxygen released. For Spirulina grown with nitrate as the nitrogen source, $Y_O = 0.507$ (i.e., 1.97 g of O_2 are released per g of biomass synthesized). From Eq. 1.1 it is possible to calculate the maximum permissible tube length for a given culture velocity, which allows to maintain the level of oxygen below the inhibition threshold for the culture. The experience gained by this author and others (Molina Grima et al., 1999) indicated that the optimal length for tube diameters within 5 and 10 cm ranges between 80 and 120 m. This length also seems adequate to minimize the CO_2 loss. Torzillo et al. (1984) reported evidence that, in Spirulina cultures, a concentration of oxygen above 30 mg L^{-1} had a negative effect on both growth and biomass protein content. For example, in PBRs made with tubes having an internal diameter of about 5 cm, in well-growing Spirulina cultures, the oxygen concentration can increase at a rate of 2–3 mg L^{-1} min^{-1}. This can result in an oxygen concentration of up to 70–80 mg L^{-1} in the middle of day, even with a gas exchange with air every 50 s and with a high turbulence rate (Torzillo, Bernardini, & Masojidek, 1998). The effect of oxygen on productivity of *Arthrospira platensis* strain M2 cultures is influenced by the acclimation state of the cells. Fig. 1.6 shows the productivities of *Arthrospira* cultures grown under high (about 60 mg L^{-1}) oxygen content, both at the optimal and suboptimal temperatures. In high light acclimated cultures (HLAC), the high oxygen cultures showed a 30% reduction in productivity compared to low oxygen cultures (about 20 mg L^{-1} dissolved oxygen). The difference in productivity increases to 70% if high oxygen is superimposed to low temperature stress. In low light acclimated cultures (LLAC), the productivities were lower than attained in HLAC, however, the difference in productivity was similar (30% lower) to that observed in HLAC (Fig. 1.5). The productivity significantly decreases if oxygen stress is superimposed at low temperature stress. In this case the loss of productivity rises up to 80% compared to low oxygen and optimal temperature cultures.

Analogous results were achieved by Moheimani and Borowitska (2006) with cultures of *Pleurochrysis carterae*. Laboratory cultures, acclimated to lower light as compared to outdoors ones, showed a strong inhibiting effect of high oxygen concentration on photosynthesis rates. The higher effects in LLAC could be due to higher absorption capacity of the photosynthetic apparatus, a lower turnover rate of the D1 protein, and a reduced dissipation

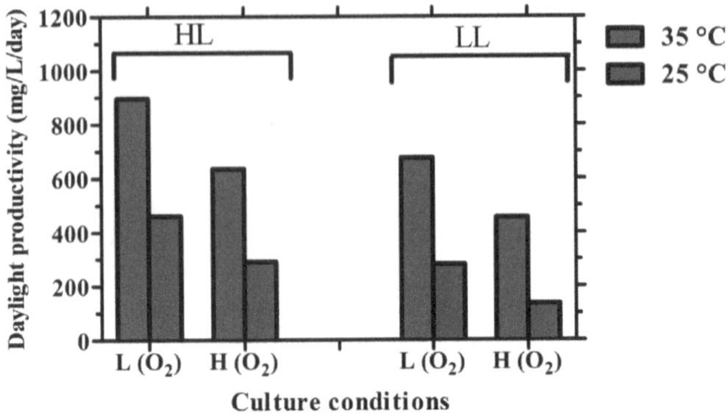

FIGURE 1.5 Effect of oxygen concentration and low temperature on productivities of *Arthrospira platensis* cultures previously acclimated to high light (HL) and low light (LL). These conditions were achieved by growing the cultures for about 3 days at a low biomass concentration (HL) about 300 mg L^{-1}, and at a high biomass concentration, about 3000 mg L^{-1} (LL). In high oxygen cultures H (O_2), the dissolved oxygen concentration was allowed to increase as a result of photosynthetic activity and reached up to 80 mg L^{-1} in the middle of the day. In low oxygen cultures L (O_2) the concentration of oxygen was maintained at about 20 mg L^{-1} by bubbling nitrogen in the reactor degasser. Both the experiments with HL and LL cultures were performed with an initial (morning) biomass concentration of 300 mg L^{-1}.

capacity of excitation energy via NPQ. A 30% reduction in productivity was also found with cultures of *Chlorella vulgaris* when the dissolved oxygen concentration was above 30 mg L^{-1} (Kazbar et al., 2019). An air-lift system with efficient gas–liquid mass transfer ($K_La > 7\ h^{-1}$) was found to be not subjected to accumulation of oxygen in the cultures. Reductions in growth and Chl fluorescence caused by high oxygen concentration (above 200%) have been reported in laboratory cultures of *Chlorella sorokiniana* (Ugwu, Aoyagi, & Uchiyama, 2007).

The combination of high oxygen and low temperature in outdoor cultures can occur at the beginning of the cold season, when the culture temperature drops greatly below the optimum but the irradiance is high enough to drive photosynthesis. Indeed, it is common for light fluctuations to occur within a range of 1–2 h, while the increase in temperature is a much slower process and takes about 4–5 h. Low temperature and high irradiance may alter the balance between the absorbed light energy by pigment antenna and the ability to use it. The situation deteriorates further when these environmental factors are coupled with high dissolved oxygen concentration in the medium (Torzillo et al., 1998). Such conditions may result in an excessive excitation pressure on PSII that results in damage to the photosynthetic apparatus (Wilson et al., 2006). In cyanobacteria, in particular, since they lack the xanthophyll cycle, an increased turnover of D1 and/or an increase in the dissipation of energy via photochemistry may represent the most important ways to cope with outdoor stress conditions. In this respect Chl fluorescence can be used as a good tool to get timely information on the physiological status of the culture and to obtain quantification of the damage caused by the stress as well as the reduction of photosynthesis and productivity (see Section 1.9).

1.7.1 Oxygen stress at low light intensity

Photoinhibition is thought to occur mainly in cultures exposed to supraoptimal light conditions, particularly in combination with other stress such as suboptimal temperature and/or high oxygen concentrations. The incidence of photoinhibition in mass culture has been investigated since the 1990s (Vonshak, Torzillo, & Tomaselli, 1994), while the incidence of photorespiration on microalgal mass culture has been overlooked. The incidence of photorespiration on cultures of *Euglena gracilis* and *Chlamydomonas reinhardtii* was evidenced by Merrett and Armitage (1982). Cultures of *Euglena*, exposed to oxygen concentrations from 21% to 2% (vol./vol.) gave a two-fold increase in dry weight yield for *Euglena gracilis*, while no effect was found in *Chlamydomonas reinhardtii*. This difference in behavior was ascribed to the fact that a CO_2-concentrating mechanism has been demonstrated in low-CO_2-grown *Chlamydomonas* cells (Beardall & Raven, 1981), whereas with *Euglena* a similar CO_2-concentrating mechanism may not be operative. The effect of high O_2:CO_2 ratios was investigated in turbidostatic *Chlamydomonas* cultures exposed to subsaturating light irradiance (Kliphuis, Martens, Janssen, & Wijffels, 2011). The authors concluded that a high O_2:CO_2 ratio caused a reduction in the biomass productivity within 20%–30%. They attributed this reduction mainly to photorespiration since conditions stimulating Melher reactions (i.e., reduced pool of the NADP/NADPH) and photoinhibition (high light conditions) could be ruled out. During photorespiration both CO_2 and NH_4^+ are lost and their refixation needs additional ATP and reductant NADPH which represent a waste of energy.

The effect of oxygen partial pressure on growth of *Neochloris oleabundans* was studied at subsaturating conditions in a fully controlled PBR (Sousa, de Winter, Janssen, Vermuë, & Wijffels, 2012). It was found that by significantly increasing the CO_2:O_2 ratio it was possible to mitigate the negative effect of photorespiration on growth. Raso, van Genugten, Vermuë, and Wijffels (2012) using a culture of *Nannochloropsis* grown under subsaturating light intensity, and exposed to O_2 concentrations above 75% air saturation, proved that increased oxygen concentration in the culture led to a linear reduction of the specific growth rate. At an O_2 concentration 250% above air saturation the specific growth rate dropped by 65%.

Large raceway ponds are characterized by insufficient turbulence attained by paddle wheel mixing, especially in Spirulina cultures where too vigorous mixing would cause breakage of filaments, the oxygen removal is limited, and it can reach very high concentrations (300%–400% of air saturation). Mass transfer in each section of the reactor is quite different, K_La (h^{-1}) is higher in close proximity to the paddle wheel (about 164) while in the channel and bends it drops below 1. This is a clear indication that cultures are poorly mixed to allow adequate stripping of oxygen (Mendoza et al., 2013). Oxygen stripping can be improved by placing a sump where air is injected to reduce the oxygen content. In the sump a K_La of about 64 can be achieved. According to Acien-Fernandez, the sump capacity should be about 20% of the total culture volume (personal communication). However, the significant volume of the culture standing in the sump, that is, with no or scarce illumination, may enhance the incidence of dark volume load and thus enhance the loss of biomass due to respiration. Moreover, injecting air needs further absorption of energy which can enhance the energetic cost of mixing. Therefore a balance between the advantages due to increased yield and higher energy requirement must be carefully considered.

1.8 Contamination

Because biomass algal production is usually carried out with a few selected microalgal strains there is increased potential for parasites to have a devastating effect on commercial-scale monocultures. Only three algal species are cultivated at an industrial level: *Arthrospira*, *Dunaliella*, and *Chlorella*. The first two species are able to grow in a very selective medium which mitigates the risk of contamination with other species of microalgae, protozoa, and aquatic fungi. *Chlorella* sp. (*C. sorokiniana* and *C. vulgaris*) is cultured in a nonselective medium but it is characterized by a high growth rate, and is able to grow in a very wide range of environmental conditions. It is not a coincidence that this organism is the most used for the treatment of wastewater. However, most algal species of economic interest are not extremophiles.

As commercial algal production continues to expand with a widening variety of applications, including biofuel, food, and pharmaceuticals, the parasites associated with microalgae will become of greater importance and of potential economic impact. A number of important microalgal parasites have been identified in algal mass culture systems in the last few years and this number is likely to grow as the number of commercial algal ventures increases (Carney & Lane, 2014). For example, some species of fungi belonging to class of *Chytrids* can attack cultures of microalgae such as *Haematococcus* (Hoffman et al., 2008) and cause the loss of the culture (Fig. 1.6A).

Some parasite-forming spores are difficult to eradicate even with aggressive pesticide use. The use of pulsed electric fields has been proposed for the control of predators in industrial-scale microalgae (Rego et al., 2015). Routine monitoring and early detection of pest species is the most important requirement for large-scale cultivation. Knowledge gained from long-term experience operations may allow for the rapid identification of common pest species and the environmental conditions promoting their presence. When possible, molecular-based techniques are the most reliable and sensitive for

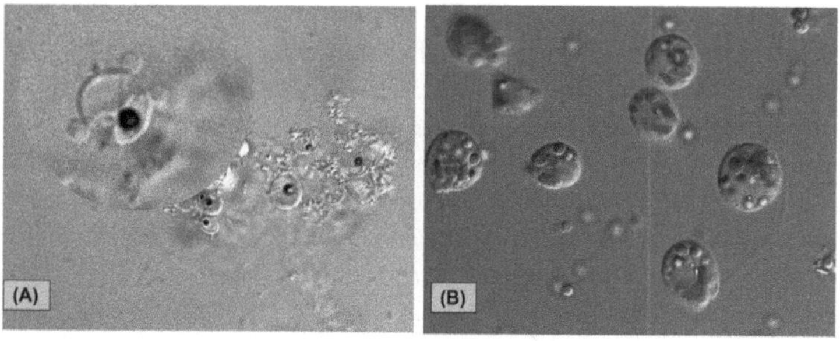

FIGURE 1.6 (A) Light microscopic photograph of Chytrid fungi infecting a *Chlorella vulgaris* culture in a closed system. The magnification clearly shows the sporangia (red) inside the microalgal cells and the pin-shaped rhizoids (colorless) perforating the cell wall. (B) Light microscopic photograph of *Poterioochromonas* cells with ingested *Synhechocystis* cells. On the left side corner is visible a *Poteriochromonas* cell during division phase.

the purpose of identifying which parasite may be present using Sanger, shotgun, or next-generation sequencing and then monitoring for these specifically using quantitative polymerase chain reaction (qPCR) or phylochip technology (Carney & Lane, 2014). FlowCAM, a flow cytometer that employ continuous digital imaging to measure the number, size, and shape of microscopic particles in a fluid medium, has been proposed to provide a warning signal of ciliate contamination in *Nannochloropsis* cultures (Day, Thomas, Achilles-Day, & Leakey, 2012). To accomplish mitigation of contamination for any culture system it is required to have: (1) information on the means by which a contaminant enters the bioreactor (vector); (2) information on the amount (rate) of contaminant entry through the vector (i.e., inoculum size); and (3) information on the time (lag phase) it takes for a contaminant to be detected under a light microscope. Despite a paucity of information on the economic impact of parasitism on the nascent algae biomass industry, there is a general consensus that parasites constitute a serious threat to the industrial development of microalgal cultures. At this stage of knowledge, in many cases the easiest solution to parasite contamination and the least technologically demanding and least satisfying response to parasite contamination is the salvage harvest, which must be done upon detection of a parasite species and prior to serious biomass losses. The success of salvage harvest is dependent on both the early detection and quantification of the contaminating parasite and the operator experience necessary to find the optimal compromise between maximizing biomass yield against the potential catastrophic loss.

Contrary to what is generally believed, even cultures growing in a closed system are subjected to contamination by other microalgae (Carney & Lane, 2014; Forehead & O'Kelly, 2013; Hoffman et al., 2008). In fact, cultures grown in closed systems are usually affected by contaminants in spite of their protection from the outside atmosphere. In many cases the vehicle of contamination is the water used for medium preparation. Surprisingly, in some cases the contamination is higher in a tubular system than in open ponds (Torzillo, 1980). This problem can become particularly serious with aged plastic tubes which have reduced smoothness. For PBRs designed as a loop, removal of the biofilm can be achieved by introducing in the tubes a brush or a sponge, of the same diameter as the tubes, which is pushed by the flow of the culture. This technique has been demonstrated to be very effective in removing biofilm of *Leptolyngbya* which is then collected at the exit of the circuit (Torzillo, 1980).

One of the greatest dangers experienced by our research group in the outdoor mass cultivation of *Synechocystis* was represented by flagellates belonging to the species of *Poterioochromonas* (Synurophyceae; Fig. 1.6B). These are common members of freshwater planktonic communities and possess digestive vacuoles in order to use phagotrophy to supplement phototrophic growth. They are very efficient phagotrophs, showing high growth rates, and have been identified as potential contaminants in *Synechocystis* mass cultures (Holen & Boraas, 1995). Their growth is supported by a heterotrophic metabolism, with cyanobacterial cells representing the feeding prey. In addition to the feeding mechanism, this kind of contamination has what can be defined as a killing effect on the culture, because when these flagellates are phagotrophically active, they produce toxins with a strong antibiotic effect (Blom & Pernthaler, 2010; Leeper & Porter, 1995).

We have reported evidence that, in the case of contamination by *Poterioochromonas* sp. in an outdoor culture of *Synechocystis*, the complete arrest of such contaminants is achieved by increasing the pH value to 11 (Touloupakis, Cicchi, Silva Benavides, & Torzillo, 2016). In these conditions, *Poterioochromonas* does not survive and *Synechocystis* is still able to grow, although with a 10% reduction in productivity, and no alteration in the biochemical composition of the biomass was observed. This strategy was validated by growing cultures in a large outdoor PBR (1300 L). By simply switching off the pH control unit, the culture's pH naturally rises to values close to 11 during the day, as a result of CO_2 uptake during photosynthesis. Night respiration entails a reduction of pH to 9.0–9.5, which must be prevented by keeping the pH at 11 by automatic addition of NaOH solution. Dissolved CO_2 concentration has also resulted in effective control of *Poterioochromonas* in a pilot-scale culture of *Chlorella sorokiniana*. It was argued that increasing the CO_2 partial pressure in the culture medium may enhance diffusion of CO_2 into the cytoplasm of *Poterioochromonas*, thus lowering the intracellular pH and resulting in cell death (Ma, Yuan, Yue, Park, & Gong, 2017). Recently, Hue, Vandamme, and Muylaert (2020), reported that low doses of garlic oil were capable of eradicating the ciliate *Oxytricha* from contaminated *Chlamydomonas* cultures without affecting the growth of the culture. Garlic oil has the advantage of being a low-cost natural product with well-known biological activity against several protozoa.

In general, microalgal cultures become more prone to contamination and predation if they are not growing under optimal conditions. This can happen either after several days of adverse environmental conditions or because of inaccurate management of the cultures, such as faulty pH control, poor mixing, nutrient deficiency, high organic load in the pond, or culture concentration far from the optimum. Therefore the key factor for minimizing the risk of contamination and the consequent loss of culture is to keep the culture conditions as close to the optimum as possible. Last but not least, it is strongly advisable to start with appropriate strain selection, choosing one which is able to manage well within the minimum and maximum temperatures of the site where production will take place.

1.9 Monitoring the physiological state of cultures through chlorophyll fluorescence

Successful cultivation requires not only continuous monitoring of a culture's physicochemical variables, namely its pH, temperature, dissolved oxygen concentration, and nutrient status, but also monitoring photochemical activities. One, direct method is to measure photosynthesis online/in situ during the diel cycle using Chl fluorescence to estimate the actual physiological state of a culture. The other possibility is to work offline using microalgae samples to measure photosynthetic variables in detail. This approach is qualitatively different from previously used physiological methods upgrading to molecular level.

Until the late 1980s mass microalgae cultures were usually optimized semi-empirically, leading to disputes between biotechnologists, physiologists, and photosynthesists discussing classical approaches versus photosynthesis monitoring. Based on previous experience on Chl fluorescence measurements on higher plants, we have extended this technique to microalgae cultures (Knoppová, Masojídek, & Pokorný, 1993; Vonshak et al., 1994). Soon after, in the 1990s–2000s, the Chl fluorescence technique become one of the most popular and useful approaches used for monitoring of microalgae mass cultures due to its noninvasiveness, ease of use, sensitivity, and the availability of commercial fluorimeters that are easily utilized in situ as well as its prompt provision of results (Masojídek et al., 1999, 2000; Torzillo, Accolla, Pinzani, & Masojidek, 1996; Torzillo et al., 1998). At present the theory is well described, making the interpretation of fluorescence data, even in various microalgae groups, easier (Campbell, Hurry, Clarke, Gustafsson, & Öquist, 1998; Masojídek, Vonshak, & Torzillo, 2011b; Schreiber, Endo, Mi, & Asada, 1995; Strasser, Srivastava, & Govindjee, 1995).

Chl fluorescence measurements make it possible to follow the energy distribution between the photochemical and nonphotochemical processes in photosynthesis. Presently, two basic Chl fluorescence techniques are used for monitoring photosynthetic efficiency in microalgae mass cultures: (1) rapid

Chl fluorescence induction kinetics and (2) saturating pulse analysis of fluorescence quenching. While the rapid fluorescence induction kinetics provide information on the redox status of the photosynthetic electron transport chain, the pulse-amplitude-modulation technique (PAM) gives information on the balance between photosynthetic electron transport and nonphotochemical energy quenching (dissipation). It is beyond the scope of this chapter to discuss in detail the basic principles of the two fluorescence approaches which can be found in several excellent reviews (Baker, 2008; Maxwell & Johnson, 2000; Strasser, Tsimili-Michael, & Srivastava, 2004).

In a model experiment in a TLC, the outdoor culture of *Chlorella* sp. strain R-117 was exposed to nitrogen limitation in a 1-week trial. The fluorescence induction curves (Kautsky curve) at midday samples on day 1 were not stressed as they showed values of the *J* and *I* steps of 0.37 and 0.59, respectively (Fig. 1.7). In the course of the trial the cultures became nitrogen limited and this stress was reflected by increased values of V_j and V_i up to 0.55 and 0.81, respectively on day 7. This increase corresponds to an overreduction of the PSII electron acceptors due to the decrease in the photosynthetic activity.

In the other approach—saturation pulse analysis of fluorescence quenching—a weak modulated excitation beam is employed, and the fluorescence

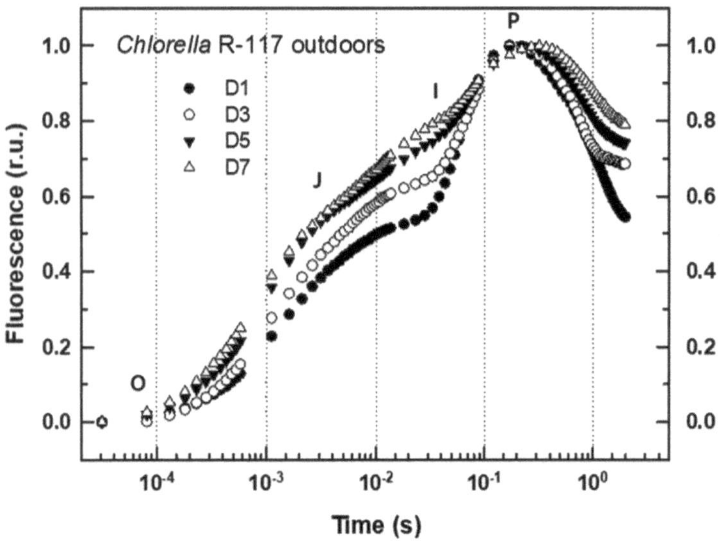

FIGURE 1.7 Rapid chlorophyll fluorescence induction kinetics (OJIP test) of *Chlorella* sp. (strain R-117) grown under nitrogen limitation in an outdoor thin-layer cascade during a 1-week trial. The OJIP curves were measured at midday on days 1, 3, 5, and 7 and double-normalized to F_0 and F_m to distinguish the differences in the inflections *J* and *I* representing various reduction states of the PS II electron carriers. Values are mean ± SE ($n = 3$). Temperature was kept between 20°C (morning) and 35°C (midday) with maxima of about 2 mmol photons m^{-2} s^{-1}.

signal is processed by a selective amplifier which rejects all nonmodulated signals (Schreiber, Schliwa, & Bilger, 1986). The principle of in situ modulated Chl fluorescence measurement in turbulent microalgal cultures is based on the assumption that signal modulation is two orders of magnitude faster than the suspension movement in a cultivation unit. As compared to induction fluorimeters, PAM devices have been a significant step forward due to the following advantages: low intensities of modulated measuring light do not excite the PSII reaction centers, that is, the basic F_0-level can be continuously monitored; photochemical and NPQ can be clearly separated and quantified using the saturation-pulse method in situ; and the signal is not disturbed by ambient light, such that experiments can be carried out continuously, even in full sunlight in the field. A measurement in situ can be made by simply pointing a fluorimeter fiber-optic at a PBR, or by submerging it into the suspension using an ambient irradiance as actinic light (Fig. 1.8).

Several physiologically useful variables have been derived from induction curves and saturation-pulse analysis (Table 1.7), which comprehensively describe the fate of the excitation energy in PSII—photochemical and nonphotochemical energy conversion (reviewed by Schreiber, 2004; Baker, 2008).

The maximal quantum yield F_v/F_m (variable to maximum fluorescence yield after dark adaptation) is considered a measure of the maximal

FIGURE 1.8 Chlorophyll fluorescence measurements in situ using various fluorimeters with fiber-optics in a microalgae culture. The trials were carried out in: (A) tubular indoor PBR, Institute of Physical Biology, Nové Hrady, Czech Republic; (B) tubular outdoor PBR, Institute of Bioeconomy, CNR, Florence, Italy; (C) outdoor TLC, Algatech Center, Institute of Microbiology, Třeboň; and (D) outdoor TLC when fluorescence signal and irradiance intensity were measured in parallel, Algatech Center, Institute of Microbiology, Třeboň.

TABLE 1.7 Important variables calculated from Chl fluorescence measurements.

Variable	Symbol	Formula
Maximum photochemical yield of PSII	F_v/F_m	$F_v/F_m = (F_m - F_0)/F_m$
Effective photochemical yield of PSII (i.e., photochemical quenching)	$Y_{II} = \Delta F'/F_m'$	$Y_{II} = (F_m' - F')/F_m'$
Relative electron transport rate through PSII (rate of photochemistry)	rETR	$\text{rETR} = Y_{II} \times E_{PAR}$
Actual electron transport rate through PSII (correlated with primary productivity)	ETR	$\text{ETR} = Y_{II} \times E_{PAR} \times a^* \times f_{PSII}$
Stern–Volmer coefficient of nonphotochemical quenching	NPQ	$\text{NPQ} = (F_m - F_m')/F_m'$

F_0, F_v, F_m, minimum, variable, and maximum fluorescence in dark-adapted state; F', F_v', F_m', steady-state, variable, and maximum fluorescence in light-adapted state; E_{PAR}, light intensity; a^* is the Chl-specific absorption cross-sections; and f_{PSII} is the fraction of absorbed quanta by PSII (Johnsen & Sakshaug, 2007).

photochemical PSII efficiency, an estimate of the physiological state of the culture. In healthy microalgal cultures, F_v/F_m ranges from 0.6 for cyanobacteria, to 0.8 for green algae, and varies during the diurnal cycle, depending on the irradiance regime and conditions. The fraction of absorbed light utilized in electron transport is given by the effective PSII photochemical yield, designated as $\Delta F'/F_m'$ that is an indicator of the actual acclimation of PSII as it depends on the redox state of the reaction centers and the electron transport rate (ETR) through PSII, used as a proxy of the photosynthetic capacity and productivity (Figueroa, Jerez, & Korbee, 2013; Kromkamp et al., 2009).

The open reaction center acts as a fluorescence quencher and the fluorescence yield rises proportionally with the level of PSII closure. This phenomenon is called photochemical quenching (actual PSII photochemical yield $\Delta F'/F_m'$). On the other hand, the so-called NPQ indicates an increased heat dissipation of absorbed energy. In principle, NPQ is inversely related to photochemistry, and is considered a safety valve protecting PSII reaction centers from damage by excess irradiance, especially under unfavorable conditions. Stress-induced damage of the photosynthetic apparatus is often reflected by an increase of NPQ which can compensate for a decrease of photochemistry ($\Delta F'/F_m'$).

A simple and reliable variable to estimate the photosynthetic performance of outdoor microalgae cultures is the measure of the ETR through PSII that is obtained as the product of $\Delta F'/F_m'$ multiplied by the actual intensity of PAR E_{PAR} in the culture expressed in μmol photons m^{-2} s^{-1} (Maxwell & Johnson, 2000; Schreiber, 2004). A more accurate measure of the ETR requires (Table 1.1) to include the specific absorption cross-sections a^* of Chl a and f_{PSII} as the fraction of absorbed quanta by PSII, which depends on the PSI/PSII ratio, which for higher plants and microalgae is assumed to be 1, while with cyanobacteria it can be considerably higher. Moreover, measurement of a^* is based on the extinction coefficient of Chl a and therefore cannot be considered specific for PSII (Szabó et al., 2014), however, it allows for more accurate measurement of ETR in microalgal suspensions compared to that directly extracted from the PAM software which was developed to perform measurement in leaves, and the results are easier to compare with those obtained with the *P/I* curve measurements which are also normalized on a Chl a basis (Vonshak et al., 2020). Experiments with Spirulina cultures cultivated under unfavorable conditions of suboptimum temperature and high dissolved oxygen concentration showed a good correlation between daily sum of rETR and biomass productivity since the trends of these two variables were similar showing the same percentage of decrease (Torzillo et al., 1996, 1998).

The so-called rapid light-response curves (RLC; the dependence of photosynthetic ETR on the irradiance intensity) measured by Chl fluorescence are used as the measuring method that provides information on the actual photosynthetic performance of microalgae cultures (Ralph & Gademann, 2005). These curves can be used for calculation of maximum ETR through the PSII complex ($rETR_{max}$; e.g., maximum ETR in $\mu mol\ e^{-1}\ m^{-2}\ s^{-1}$), quantum yield of photosynthesis (α; initial slope of RLC which reflects the quantum efficiency of photosynthetic electron transport in electrons/photon), and irradiance intensity saturating photosynthesis E_k (for a review see Enriquez & Borowitzka, 2011). The E_k value is determined from the interception point of the α slope with the maximum level of $rETR_{max}$; it defines the onset of light saturation. The NPQ coefficient indicates a part of absorbed light which is dissipated as heat. In principle, NPQ is inversely related to photochemistry, and is considered a safety valve protecting PSII reaction centers from damage by excess irradiance (overreduction of the photosynthetic apparatus).

Compared to the steady-state or "slow" light response curve (LRC), the RLC is measured at increasing irradiance intensities within short time periods. Each actinic light intensity can last for 10–30 s to obtain the balanced fluorescence level F' and then a saturating pulse is triggered to reach the maximum F_m'. RLC of the ETR versus irradiance curve (ETR/*E* curve) can be completed within a few minutes compared to the several hours required for LRC (Ralph & Gademann, 2005; Ritchie & Larkum, 2012). As an example, we show RLC records of outdoor *Chlorella* cultures grown in the TLC during a 1-week nitrogen limitation experiment. Such a stress is often used to induce production of certain compounds, for example, starch or carotenoids. In this experiment, cultures of *Chlorella* R-117 showed decreases by 60% and 80% of F_v/F_m and rETR, respectively, on day 7 compared to the start (Fig. 1.9). In contrast, the NPQ values increased 3.4-fold after 7 days in the nitrogen-limited cultures.

Simple and fast procedures (manuals) for microalgae biotechnologists have been worked out based on Chl fluorescence measurements that can be used to monitor changes in photosynthetic activity and physiological status of microalgae cultures in order to optimize growth (Malapascua, Jerez, Sergejevová, Figueroa, & Masojídek, 2014; Masojídek et al., 2011b; Torzillo et al., 1996). Recently, we have worked out a rapid, 1-day test based on photosynthesis measurements to estimate a suitable growth regime for microalgae strains. It employs three measuring techniques—oxygen production/respiration, saturating pulse analysis of fluorescence quenching, and fast fluorescence induction kinetics to estimate the suitable growth regime to reach a high growth rate (Ranglová, Lakatos, Câmara Manoel, Grivalský, & Masojídek, 2019).

Chl fluorescence measurements in our experiments showed that changes to some fluorescence variables can be well correlated with changes in cultivation conditions, physiological status, and the growth of a given microalgae strain in selected cultivation system. Chl fluorescence measurements in situ have indicated that changes to the daily integrated values of the

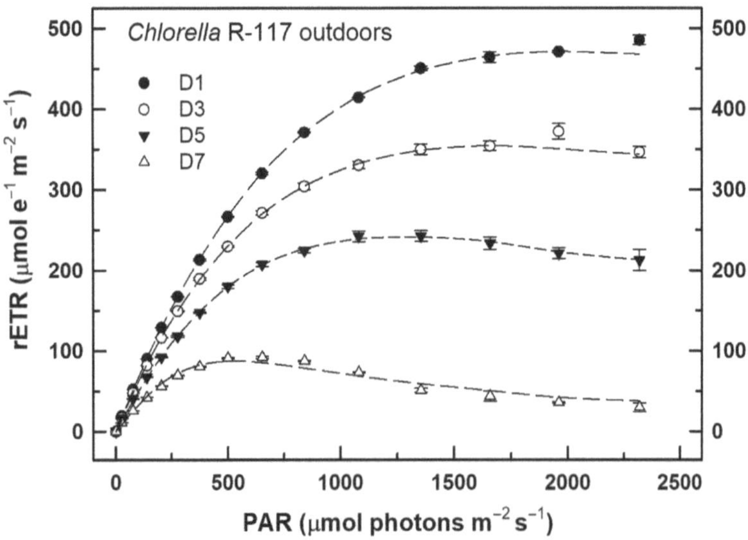

FIGURE 1.9 Rapid light-response curves (RLCs) of *Chlorella* sp. R-117 grown under nitrogen limitation in an outdoor thin-layer cascade during a 1-week experiment. RLCs (rETR vs. irradiance) were measured in culture samples taken at midday on days 1, 3, 5, and 7. The measurements were carried out offline in a light-protected measuring chamber with mixing using a pulse-amplitude-modulation fluorimeter (PAM-2500, H. Walz, Germany). A series of stepwise increasing irradiance intensities (red LEDs; 0–2300 μmol photons m^{-2} s^{-1}) were applied in 20-s intervals to obtain the steady-state fluorescence level F' in the light-adapted state at respective irradiance level and then a saturating pulse (> 10,000 μmol photons m^{-2} s^{-1}, 0.6 s duration) was flashed.

relative PSII electron transport can be well correlated with analogous changes in the daily productivities of the cultures grown under different conditions (Masojídek et al., 2000; Masojídek, Kopecký, Giannelli, & Torzillo, 2011a; Torzillo et al., 1998).

The increased interest in microalgae biotechnology, aimed at the production of biomass, high-value products, or even biofuels, has prompted the application of online measurements for monitoring growth and obtaining rapid evidence of unfavorable conditions affecting the performance of outdoor cultures. For these applications the use of fluorimeters which monitor the actual photochemical quantum yield in situ at a given light intensity during the day is mandatory. We have found that a midday depression of PSII photochemical yields between 20% and30%—as compared with maximal morning values—is physiological for well—performing microalgae cultures (Malapascua et al., 2014; Masojídek et al., 2011a). A lower or higher decrease in photochemical yields indicated low-light acclimated or photoinhibited cultures, respectively. These results are important from a biotechnological point of view when the growth of outdoor microalgae mass cultures is optimized under varying climatic conditions.

Finally, some recommendations are necessary. (1) Though Chl fluorescence represents a rapid technique for stress detection in plants and microalgal cultures, it must always be accompanied by other physiological measurements (e.g., oxygen evolution, dry weight). (2) It is relatively easy to generate fluorescence data, although care must always be taken to select and calculate sensible variables. As long as this is kept in mind, Chl fluorescence represents a powerful technique allowing rapid monitoring of physiological status.

1.10 Good manufacturing practices

Microalgae are a promising and sustainable feedstock for alternative foods. The production scheme for microalgae production comprises several stages and processing steps. It starts with the selection of suitable strains and their cultivation, followed by biomass harvesting and consequent thickening, dewatering, drying, and finally the extraction of the target product. Independently of the strain to be produced, the overall production system must be approved for the "food industry"; this compliance assessment involves the materials, systems, and protocols used during the production process. The safety of the products must be assessed before their launch on the markets, and regulatory requirements are necessary prior to commercialization.

1.10.1 Best practices for the cultivation of microalgae of interest as food sources

Microalgae production is a process that must be adequately planned and performed. Major steps involved in any microalgal production process include

(1) preparation of culture medium, (2) production of biomass, (3) harvesting of biomass, (4) treatment of used water for recirculation or disposal, and (5) stabilization of produced biomass or transformation into end products. The core of the process is the cultivation system and open ponds are currently still the preferred choice for microalgae production on a large scale. However, due to the requirement of "Good Manufacturing Practices" (GMPs) guidelines, production of high-value products from microalgae seems feasible in well-controlled PBRs which ensure strict control of culture parameters and reduce contamination (Torzillo & Chini Zittelli, 2015). For example, tubular PBRs can satisfy the GMP requirements (Fernández et al., 2012), and large-scale microalgae cultivation in tubular systems is a well-established and reliable technology. In recent years, large-scale tubular PBRs have been constructed and operated in Germany and Israel for the production of *Chlorella* and *Haematococcus*, respectively, for the production of high-value products for human nutrition, cosmetics, and pharmaceutical applications (Chini Zittelli, Biondi, Rodolfi, & Tredici, 2013).

For many markets and certifying authorities, GMPs certification ISO 9001–2000 and other ISO quality management system standards are essential. There are also other marketing considerations. Specific markets may also require certification by a recognized authority, for example, that the product is kosher or halal; and organic certification is often desirable. These regulations must thus be considered carefully when developing a production process and, depending on the application of the final product, a careful selection of the algal strain is also required (Borowitzka, 2016; Champenois, Marfaing, & Pierre, 2015). The selected species must be food-grade, not produce toxins or allergens, and be robust and easy to adapt to environmental conditions.

The term "Good Manufacturing Practices" describes the key requirements for all aspects of commercial food production, storage, and distribution necessary to ensure a consistent product that conforms to all food safety, quality, and consumer attribute requirements. GMPs are the basic operational and environmental conditions required to produce safe foods. They ensure that ingredients, products, and packaging materials are handled safely and that food products are processed in a suitable environment. GMPs address the hazards associated with personnel and environment during food production. They provide a foundation for any food safety system. Once GMPs are in place, processors can implement a Hazard Analysis Critical Control Point system to control hazards that may affect the ingredients and packaging material during food processing (https://www.gov.mb.ca/agriculture/food-safety/at-the-food-processor/food-safety-program/pubs/fs_15.pdf). GMPs are made up of a set of rules that describe methods, equipment, means, and production management to ensure appropriate quality standards. The principles and guidelines of the GMP are set out in two documents: Commission Directive 2003/94/EC and Commission Directive 91/412/EEC.

Several companies have already obtained the appropriate regulatory approvals for their products. Among the most relevant: Algatechnologies (AstaPure, www.algatech.com), Earthrise (Spirulina, https://www.earthrise.com/certifications), Cyanotech (Spirulina, https://www.cyanotech.com), and Roquette Klötze GmbH & Co. (Chlorella, www.algomed.de).

1.10.2 Safety and regulatory issues

Although the use of microalgae as food has been carried out for centuries, only a few species are currently approved and applied in food production, with the majority being considered Novel Foods (NF) in the EU and thus subject to regulation procedures and safety assessments in order to be marketed. There are two EU regulations that most affect the production and commercialization of microalgae-based products for food: The Regulation on Food Safety and the Regulation on Novel Food and Novel Food Ingredients (Vigani et al., 2015). These will be discussed in detail in later chapters of this book.

A number of conditions are still to be met to exploit the production and commercial potential of microalgae products in the EU. From a regulatory point of view, the European regulations on novel foods and novel food ingredients, on food safety, and on nutrition and health benefit claims affect the marketing of microalgae products. The safety of products must be assessed before their launch on the market, but restrictive regulatory requirements can slow down the steps to commercialization. In 2002 the European Parliament and the Council adopted the General Food Law Regulation (Regulation EC No 178/2002) laying down the general principles and requirements of food law. It sets out a framework for the development of food legislation both at the EU and national levels. It also set up an independent agency responsible for scientific advice and support, the European Food Safety Authority (EFSA).

A Novel Food is defined as a food that has not been consumed to a significant degree by humans in the EU prior to May 15, 1997, when the first Regulation n 258/97 (Regulation EC No 258/97) on novel foods came into force. Up to date, 11 macroalgal and 5 microalgal species have been authorized as foods or food ingredients and listed as "not novel" (or old) in the EU Novel Food Catalog. The microalgae used prior to May 1997 in Europe and thus authorized as food in the EU are *Aphanizomenon flos-aquae* from Klamath Lake, *Arthrospira platensis*, *Chlorella luteoviridis*, *Chlorella pyrenoidosa*, and *Chlorella vulgaris* (European Union, Novel Food catalog). The diatom *Odontella aurita* was authorized in 2005 (European Union, 2005), and, in 2009, Docosahexaenoic acid- rich (DHA) oil from *Ulkenia* sp. was approved as a novel food ingredient (European Union, 2009). In 2014 *Tetraselmis chuii* and astaxanthin from *Haematococcus pluvialis* were approved as novel food ingredients (AECOSAN, 2014; EFSA European Food Safety Authority, 2014).

In 2015 oil from *Schizochytrium* sp. was also authorized (European Union, 2015a,b) (Table 1.8).

A new Novel Food regulation for the EU (European Union, 2015c; Regulation EU 2015/2283), repealing Regulation EC No 258/97, was adopted on November 2015, coming into effect in January 2018, with the purpose of guaranteeing shorter and less expensive authorization processes, while still maintaining a high level of consumer protection. The new EU Regulation introduces the following: (1) authorization procedure centralized at the EU level with fixed time limits; (2) all applications for NF authorization are submitted to the EC which may require a risk assessment to the EFSA. The EU bases the authorization decisions on the outcome of the EFSA's evaluations; (3) a simplified procedure for traditional foods consumed to a significant degree in third countries but not in the EU prior to 1997; (4) a "Union List," set up and updated by the EU, that is NF authorized or notified to be placed on the EU market; (5) improvement of efficiency and transparency by establishing deadlines for the safety evaluation and authorization procedure, thus reducing the overall time spent in the authorization process; (6) generic Novel Food authorizations which remove the need for a separate application seeking to demonstrate substantial equivalence to an authorized Novel Food; and (7) a maximum 5-year period (from the date of authorization) of intellectual property protection for new scientific evidence and data produced in support of applications. Fig. 1.10 shows a step-by-step diagram of the new procedure.

TABLE 1.8 Microalgae approved for food use in the EU. Microalgae that dominate the market are in bold

Microalga scientific name	Year of EU approval as food
Arthrospira platensis (Spirulina)	Used prior to May 1997 in EU NF catalog
Chlorella luteoviridis	Used prior to May 1997 in EU NF catalog
Chlorella pyrenoidosa	Used prior to May 1997 in EU NF catalog
Chlorella vulgaris	Used prior to May 1997 in EU NF catalog
Aphanizomenon flos-aquae (Alga Klamat)	Used prior to May 1997 in EU NF catalog
Odontella aurita	2005
Ulkenia sp.	2009
Tetraselmis chuii	2014
Haematococcus pluvialis	2014
Schizochytrium sp.	2015

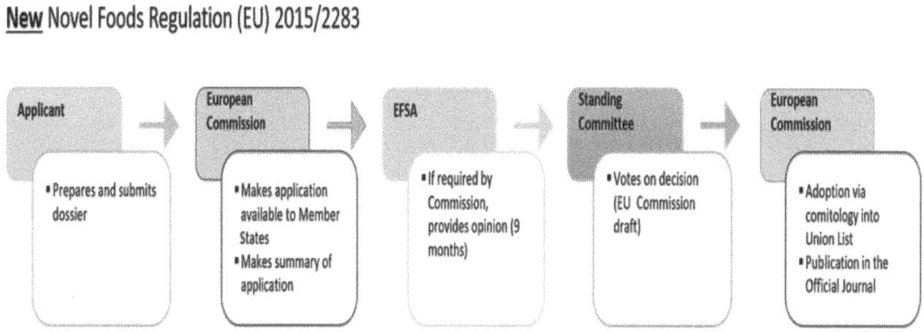

FIGURE 1.10 A step-by-step simplified scheme of the new Novel Food regulation procedure, based on Regulation EU2015/2283.

When dealing with novel matrices for food production, one of the most important requirements is to demonstrate the absence of toxicity in the range of concentrations at which that food or food ingredient is intended for use. Possible risks associated with algae include excess intake of toxic metals, allergenicity, cyanotoxins, and certain secondary metabolites as well as contamination with pathogens (Van der Spiegel, Noordam, & van der Fels-Klerx, 2013; Wells et al., 2017). Hence, toxicological studies and the correct and undoubtful identification of the algae strains are a fundamental step to ensure the safety of large-scale industrial production of algae (Enzing, Ploeg, Barbosa, & Sijtsm, 2014; EFSA European Food Safety Authority, 2016).

In addition to the EU, other developed countries have specific regulations (Niccolai et al., 2016). In the United States the FDA regulates the safety of food including algal products. According to the classification of the Center for Food Safety and Applied Nutrition, algal biomass is classified as "other dietary supplement." *Spirulina*, *Chlorella*, *Dunaliella*, *Haematococcus*, *Schizochytrium*, *P. cruentum*, and *C. cohnii* are classified as food sources falling into the GRAS category. Other products derived from microalgae classified by the FDA as GRAS are oils obtained from *Schyzochitrium* and *Ulkenia*, as well as a whole microalgal protein powder and a lipid ingredient derived from *Chlorella* (FDA, GRAS Notices).

The Food Standards Australia New Zealand (FSANZ) (2016) considers as novel food all nontraditional foods or their derivatives and requires safety assessment. In Australia, all *Chlorella* (including *Chlorella sorokiniana*) and *Arthrospira* species and derivatives for which a novel food application was submitted have been so far considered as traditional food, whereas *A. flos-aquae* was considered a novel food and safety assessment was required due to the potential presence of cyanobacterial toxins such as microcystins and nodularin (FSANZ, 2016).

All the approved microalgae (and their products) represent a very small number of species compared to the high number present in nature (Guiry, 2012). Other species could be suitable ingredients in innovative functional foods and as nutraceuticals (Batista et al., 2019; Niccolai, Chini Zittelli, Rodolfi, Biondi, & Tredici, 2019).

1.10.3 Cultivation of organic microalgae

One of the current challenges that microalgae producers and biotechnologists are facing concerns the organic certification of food-grade microalgae to respond to the growing demand for organic and fully traceable plant-based ingredients. Although the regulation currently in force (Regulation EC No 889/2008) clarifies that the production rules for "seaweed" must also be applied to the production of microalgae for further use as food, there are several practical issues to overcome. For instance, no detailed production rules have been defined for the production system, although it is clear that closed systems are a better solution for controlled cultivation. Moreover, the addition of minerals hinders organic certification, so there are many essential nutrients and micronutrients specific to microalgal growth that cannot be used, while the use of other organic fertilizers, such as manure, is not an option due to the risk of water and biomass contamination.

The new "BIO" regulation 848/2018 that will come into force in January 2021 incorporates the regulatory framework outlined in the previous Reg. 889/2008 and confirms that all algae, including macroalgae, microalgae, and cyanobacteria, are subject to the same discipline. With regard to algal cultivation, this would represent a very heavy hindrance, in particular the impossibility of using soluble salts as a source of nutrients. If the implementing regulations do not deviate from those currently in force (Regulation EC No 889/2008), where the use of any form of nitrogen and phosphorus usable by microalgae is not allowed, it will be practically impossible to produce "organic" algae. To overcome this major limitation, the use of Nitro from Chile as a source of nitrogen has been proposed to the competent authorities, as already provided in the "private rules for the production of organic Spirulina" approved by MIPAAF (Nota Mipaaf, 2012).

1.11 Concluding remarks

Microalgal cultures outdoors are subjected to continuous changes in light intensity, which cause strong variations in the efficiency of light utilization and biomass composition. A better understanding of the mechanisms that regulate light utilization/dissipation may shed light on the potential methods for optimizing productivity and product quality. However, these studies are complicated by the fact that outdoors very often environmental factors act in synergy. Therefore a pure laboratory study, on a commercial microalgal species, although very important, must be considered carefully and always confirmed outdoors. On the other hand, in order to better understand the physiological basis of stress phenomena occurring outdoors, it is necessary to return to the laboratory where the effect of environmental factors can be studied separately, under well defined conditions. Table 1.9 summarizes the main challenges to implementation and the workable solutions for a successful algal biotechnology. The critical factor which limits the productivity of cultures is the low efficiency of light conversion. Although photosynthesis has been optimized over three billion years of evolution, it remains inefficient at converting solar energy into chemical energy and biomass. In fact, the light conversion efficiency is only 1%–2%, while the theoretical maximum is about 10%. Microalgae, unlike higher plants in which the amplitude of NPQ is proportional to the light intensity at which they are exposed, strongly upregulate their ability to NPQ even with moderate light irradiance, causing a remarkable reduction in the light conversion efficiency (Bonente, Pippa, Castellano, Bassi, & Ballottari, 2012). Microalgae with truncated antenna size may substantially increase the biomass productivity. This approach represents a synthetic variation of the natural photo-adaptive mechanism of microalgae in which pigments reorganize in response to high light exposure (Vonshak et al., 2020). Unfortunately, mutants generated by random UV or chemical mutagenesis are likely to induce multiple mutations, and among them is the increased respiration rate of cells, which reduces the benefit of the reduced light absorption of the truncated antenna size cells. Therefore in these studies, an important parameter that needs to be addressed, other than the lower Chl level per cell and reduced PSII antenna size, is the P_{max}/respiration ratio which should not be affected significantly.

TABLE 1.9 **Main aspects to be considered in future research for improving the productivity of algae cultures outdoors**

Challenges for implementation	Possible solutions
High oxygen concentrations inhibit photosynthesis	1. Create/select oxygen-tolerant strains (resistant to ROS) 2. Create/select strains able to cope with the combined effect of high oxygen and low temperature
Low light conversion efficiency due to light saturation of photosynthesis	1. Create/select strains with higher saturation level of photosynthesis 2. Reduce antenna size, without affecting the respiration rate
Low productivity due to scarce resistance to sub- and supraoptimal temperatures	Select phenotypes able to acclimate to high or low temperatures
Susceptibility to pollution	Select strains growing under selective environmental (medium) conditions

Several high-value microalgal products are well established in the market, and there are good opportunities for additional new products. However, the overall production system must be approved for the "food industry"; this regulatory compliance involves the materials, systems, and protocols used during the production process. It is also important to consider that regulation and permits take time.

Good science and scientists are essential conditions for the progress of algal biotechnology, and to make algal biomass a commodity traded in large quantities. We believe that the progress in algal biotechnology could be greatly accelerated if basic research into microalgae was carried out taking into account the new necessities pointed out by applied research, and vice versa: in other words, biologists should understand more about bioengineering and engineers should understand more about biology.

References

AECOSAN (Agencia Española de Consumo, Seguridad Alimentaria y Nutrición). (2014). Authorisation to market *Tetraselmis chuii*. Ministry of Health, Social Services and Equality.

Arad Malis, S., & van Moppers, D. (2013). Novel sulphated polysaccharides of red microalgae; basics and applications. In A. Richmond, & Q. Hu (Eds.), *Handbook of microalgal culture: Applied phycology and biotechnology* (pp. 406–416). Chichester: Wiley-Blackwell.

Asada, K. (1994). Mechanisms for scavenging reactive molecules generated in chloroplasts under light stress. In N. B. Baker, & J. R. Bowyer (Eds.), *Photoinhibition of photosynthesis: From molecular mechanisms to the field* (pp. 129–142). Oxford: Bios Scientific Publishers Ltd.

Baker, N. R. (2008). Chlorophyll fluorescence: A probe of photosynthesis in vivo. *Annual Review of Plant Biology*, *59*, 89–113.

Batista, A. P., Niccolai, A., Bursic, I., Sousa, I., Raymundo, A., Rodolfi, L., ... Tredici, M. R. (2019). Microalgae as functional ingredients in savory food products: application to wheat crackers. *Foods*, *8*, 611.

Beardall, J., & Raven, J. A. (1981). Transport of inorganic carbon and the "CO2 concentrating mechanism" in *Chlorella emersonii* (Chlorophyceae). *Journal of Phycology*, *17*, 134–141.

Béchet, Q., Shilton, A., & Guieysse, B. (2013). Modeling the effects of light and temperature on algae growth: State of the art and critical assessment for productivity prediction during outdoor cultivation. *Biotechnology Advances*, *31*, 1648–1663.

Belay, A. (1997). Mass culture of Spirulina outdoors—the Earthrise farm experience. In A. Voshak (Ed.), Spirulina platensis (Arthrospira): *Physiology, cell-biology and biotechnology* (pp. 131–158). London: Taylor and Francis.

Blom, J. F., & Pernthaler, J. (2010). Antibiotic effects of three strains of chrysophytes (*Ochromonas, Poterioochromonas*) on freshwater bacterial isolates. *FEMS Microbiology Ecology*, *71*(2), 281–290.

Bocci, F., Torzillo, G., Vincenzini, M., & Materassi, R. (1988). Growth physiology of *Spirulina platensis* in tubular photobioreactor under natural light. In T. Stadler, J. Mellion, M. C. Verdus, Y. Karamanos, H. Morvan, & D. Christiaen (Eds.), *Algal biotechnology* (pp. 303–314). London: Elsevier Applied Science.

Bonente, G., Pippa, S., Castellano, S., Bassi, R., & Ballottari, M. (2012). Acclimation of *Chlamydomonas reinhardtii* to different growth irradiances. *The Journal of Biological Chemistry, 287*, 5833–5847.

Borowitska, M. A. (2013). High-value products from microalgae—Their development and commercialization. *Journal of Applied Phycology, 25*, 743–756.

Borowitska, M. A. (2016). Algal physiology and largescale outdoor cultures of microalgae. In M. A. Borowistka, J. Beardall, & J. A. Raven (Eds.), *The physiology of microalgae, developments in applied phycology 6* (pp. 601–652). Switzerland: Springer International Publishing.

Borowitska, M. A., & Vonshak, A. (2017). Scaling up microalgal cultures to commercial scale. *European Journal of Phycology, 52*(4), 407–418.

Borowitzka, M. A. (2016). Systematics, taxonomy and species names: do they matter? In M. A. Borowitzka, J. Beardall, & J. A. Raven (Eds.), *The physiology of microalgae* (pp. 655–681). Dordrecht: Springer.

Burlew, J. S. (1953). *Algae culture from laboratory to pilot plant.* Washington, DC: Carnegie Institution of Washington.

Campbell, D., Hurry, V., Clarke, A., Gustafsson, P., & Öquist, G. (1998). Chlorophyll fluorescence analysis of cyanobacterial photosynthesis and acclimation. *Microbiology and Molecular Biology Reviews: MMBR, 62*, 667–683.

Carneiro, M., Cicchi, B., Maia, I. B., Pereira, H., Chini Zittelli, G., Varela, J., ... Torzillo, G. (2020). Effect of temperature on growth, photosynthesis and biochemical composition of *Nannochloropsis oceanica*, grown outdoors in tubular photobioreactors. *Algal Research, 49*, 101923.

Carney, L. T., & Lane, T. W. (2014). Parasites in algal mass culture. *Frontiers in Microbiology, 5*(278), 1–8.

Champenois, J., Marfaing, H., & Pierre, R. (2015). Review of the taxonomic revision of *Chlorella* and consequences for its food uses in Europe. *Journal of Applied Phycology, 27*, 1845–1851.

Cheng, J., Yan Wang, Y., Benemann, J. R., Zhang, X., Hu, H., & Qin, S. (2016). Microalgal industry in China: Challenges and prospects. *Journal of Applied Phycology, 28*, 715–725.

Cheng, J., Yang, Z., Ye, Q., Zou, J., & Cen, K. (2015). Enhanced flashing light effect with up-down chute baffles to improve microalgal growth in a raceway pond. *Bioresource Technology, 190*, 29–35.

Chiaromonti, D., Prussi, M., Casini, D., Tredici, M. R., Rodolfi, L., Bassi, N., ... Bondioli, P. (2013). Review of energy balance in raceway ponds for microalgae cultivation: Re-thinking a traditional system is possible. *Applied Energy, 102*, 101–111.

Chini Zittelli, G., Biondi, N., Rodolfi, L., & Tredici, M. R. (2013). Photobioreactors for mass production of microalgae. In A. Richmond, & Q. Hu (Eds.), *Handbook of microalgal culture: Applied phycology and biotechnology* (2nd edition, pp. 225–266). Oxford: Wiley.

Chini Zittelli, G., Lavista, F., Bastianini, A., Rodolfi, L., Vincenzini, M., & Tredici, M. R. (1999). Production of eicosapentaenoic acid by *Nannochloropsis* sp. cultures in outdoor tubular photobioreactors. *Journal of Biotechnology, 70*, 299–312.

Commission Directive 2003/94/EC—good manufacturing practice for medicines and investigational medicines for human use.

Commission Directive 91/412/EEC of 23 July 1991 laying down the principles and guidelines of good manufacturing practice for veterinary medicinal product.

da Silva Vaz, B., Botelho Moreira, J., Greque de Morais, M., & Costa, V. (2016). Microalgae as a new source of bioactive compounds in food supplements. *Current Opinion in Food Science, 7*, 73–77.

Day, J. G., Thomas, N. J., Achilles-Day, U. E. M., & Leakey, R. J. G. (2012). Early detection of protozoan grazers in algal biofuel cultures. *Bioresource Technology, 114*, 715–719.

Doucha, J., & Livansky, K. (1995). Novel outdoor thin-layer high density microalgal culture systems: Productivity and operational parameters. *Archiv für Hydrobiologie, 76*(Suppl. 106), 129–147, Algological Studies.

Doucha, J., & Livansky, K. (2014). High density outdoor microalgal culture. In R. Balpai, A. Prokop, & M. Zappi (Eds.), *Algal biorefineries* (vol. 1, pp. 147–173). Dordrecht: Springer, Cultivation of cells and products.

EFSA (European Food Safety Authority). (2014). Scientific opinion on the safety of astaxanthin-rich ingredients (AstaREAL A1010 and AstaREAL L10) as novel food ingredients. *EFSA Journal, 12*, 3757.

EFSA (European Food Safety Authority). (2016). Guidance on the preparation and presentation of an application for authorization of a novel food in the context of regulation (EU) 2015/2283. *EFSA Journal, 14*(11), 4594, 24 pp.

Enriquez, S., & Borowitzka, M. A. (2011). The use of the fluorescence signal in studies of seagrasses and macroalgae. In D. J. Suggett, O. Prášil, & M. A. Borowitzka (Eds.), *Chlorophyll a fluorescence in aquatic sciences: Methods and applications* (pp. 187–208). Dordrecht: Springer.

Enzing, C., Ploeg, M., Barbosa, M., Sijtsm, L. (2014). Microalgae based products for the food and feed sector: an outlook for Europe. In: Vigani, M., Parisi, C., Rodriguez Cerezo, E. (Eds.), JRC scientific and policy reports. <https://publications.jrc.ec.europa.eu/repository/bitstream/JRC85709/final%20version%20online%20ipts%20jrc%2085709.pdf>, Accessed, 29 June 2020.

European Union. (2005). Summary of notifications received by the Commission until 31 December 2004 pursuant to Article 5 of Regulation (EC) No 258/97 of the European Parliament and of the Council. OJ, C 208/2, 12.6.2003.

European Union. (2009). Commission Decision (2009/777/EC) concerning the extension of uses of algal oil from the micro-algae *Ulkenia* sp. as a novel food ingredient under Regulation (EC) No 258/97 of the European Parliament and of the Council. OJ, L278/54, 23.10.2009.

European Union. (2015a). COMMISSION IMPLEMENTING DECISION (EU) 2015/545 of 31 March 2015 authorizing the placing on the market of oil from the micro-algae *Schizochytrium* sp. (ATCC PTA-9695) as a novel food ingredient under Regulation (EC) No 258/97 of the European Parliament and of the Council. OJ, L 90/7, 2.04.2015.

European Union. (2015b). COMMISSION IMPLEMENTING DECISION (EU) 2015/546 of 31 March 2015 authorizing an extension of use of DHA and EPA-rich oil from the micro-algae Schizochytrium sp. as a novel food ingredient under Regulation (EC) No 258/97 of the European Parliament and of the Council. OJ, L 90/11, 2.04.2015.

European Union. (2015c). Regulation (EU) 2015/2283 of the European Parliament and of the Council of 25 November 2015 on novel foods. Official Journal of the European Union, L 327/1 of 11 December 201, pp 1–22. http://eur-lex.europa.eu/legal-content/EN/TXT/PDF/?uri = CELEX:32015R2283&from = EN.

Fernández, I., Acién, F. G., Fernández, J. M., Guzmán, J. L., Magán, J. J., & Berenguel, M. (2012). Dynamic model of microalgal production in tubular photobioreactors. *Bioresource Technology, 126*, 172.

Figueroa, F. L., Jerez, C. G., & Korbee, N. (2013). Use of *in vivo* chlorophyll fluorescence to estimate photosynthetic activity and biomass productivity in microalgae grown in different culture systems. *Latin American Journal of Aquatic Research, 41*, 801–819.

Forehead, H., & O'Kelly, C. (2013). Small doses, big troubles: modelling growth dynamics of organisms affecting microalgal production cultures in closed photobioreactors. *Bioresource Technology, 129*, 329–334.

FSANZ. (2016). Novel food-record of views. Record of views formed by FSANZ. Novel Food Group or Advisory Committee on Noel Food, March 2016.

Giannelli, L., Scoma, A., & Torzillo, G. (2009). Interplay between light intensity, chlorophyll concentration and culture mixing on the hydrogen production in sulfur-deprived *Chlamydomonas reinhardtii* cultures grown in laboratory photobioreactors. *Biotechnology and Bioengineering*, *104*(1), 76–90.

Grivalský, T., Ranglová, K., Câmara Manoel, J. A., Lakatos, G. E., Lhotský, R., & Masojídek, J. (2019). Development of thin-layer cascades for microalgae cultivation: milestones (review). *Folia Microbiologica*, *64*, 603–614.

Grobbelaar, J. U. (1989). Do light/dark cycles of medium frequency enhance phytoplankton productivity? *Journal of Applied Phycology*, *1*, 333–340.

Grobbelaar, J. U. (2009). Upper limits of photosynthesis productivity and problems of scaling. *Journal of Applied Phycology*, *21*, 519–522.

Grobbelaar, J. U. (2010). Microalgal biomass production: Challenges and realities. *Photosynthesis Research*, *106*, 201–206.

Grobbelaar, J. U., Ladislav, N., & Tichy, V. (1996). Influence of high frequency light/dark fluctuations on photosynthetic characteristics of microalgae photoacclimated to different light intensities and applications for mass algal cultivation. *Journal of Applied Phycology*, *8*, 335–343.

Grobbelaar, J. U., Nedbal, L., Tichy, L., & Setlik, I. (1995). Variation in some photosynthetic characteristics of microalgae cultured in outdoor thin-layered sloping reactors. *Journal of Applied Phycology*, *7*, 175–184.

Guiry, M. (2012). How many species of algae are there? *Journal of Phycology*, *48*, 1057–1063.

Hamed, I. (2016). The evolution and versatility of microalgal biotechnology: A review. *Comprehensive Reviews in Food Science and Food Safety*, *15*, 1104–1123.

Hartig, P., Grobbelaar, J. U., Soeder, C. J., & Groeneveg, J. (1980). On the mass culture of microalgae. Areal density as an important factor for achieving maximal productivity. *Biomass*, *15*, 211–221.

Havelková-Doušová, H., Prašil, O., & Behrenfeld, M. J. (2004). Photoacclimation of *Dunaliella tertiolecta* (Chlorophyceae) under fluctuating irradiance. *Photosynthetica*, *42*(2), 273–281.

Hidasi, N., & Belay, A. (2018). Diurnal variation of various culture and biochemical parameters of *Arthrospira platensis* in large scale outdoor raceway ponds. *Algal Research*, *29*, 121–129.

Hoffman, Y., Aflalo, C., Zarka, A., Gutman, J., James, T., & Boussiba, S. (2008). Isolation and characterization of a novel chytrid species (phylum Blastocladiomycota), parasitic on the green alga *Haematococcus*. *Mycological Research*, *112*, 70–81.

Holen, D. A., & Boraas, M. E. (1995). Mixotrophy in chrysophytes. In C. D. Sandgreen, J. P. Smol, & J. Kristiansen (Eds.), *Crysophytae algae: Ecology, phylogeny, and development* (pp. 119–140). Cambridge University Press.

Hue, N. T., Vandamme, D., & Muylaert, K. (2020). Potential of garlic oil to control biological contamination of *Chlamydomonas* cultures by the ciliate *Oxytricha*. *Journal of Applied Phycology*, *32*, 831–837.

Johnsen, G., & Sakshaug, E. (2007). Biooptical characteristics of PSII and PSI in 33 species (13 pigment groups) of marine phytoplankton, and the relevance for pulse-amplitude-modulated and fast-repetition-rate fluorometry. *Journal of Phycology*, *43*(6), 1236–1251.

Kazbar, A., Cogne, G., Urbain, B., Marec, H., Le-Gouic, B., Tallec, J., ... Pruvost, J. (2019). Effect of dissolved oxygen concentration on microalgal culture in photobioreactors. *Algal Research*, *39*, 101432.

Kliphuis, A. M. J., Martens, D. E., Janssen, M., & Wijffels, R. H. (2011). Effect of O_2:CO_2 ratio on the primary metabolism of *Chlamydomonas reinhardtii*. *Biotechnology and Bioengineering*, *108*, 2390–2402.

Knoppová, J., Masojídek, J., & Pokorný, J. (1993). Chlorophyll fluorescence quenching caused by inorganic carbon depletion in the green alga *Scenedesmus quadricauda*. *Photosynthetica*, *28*, 541–547.

Kok, B. (1953). Experiments on photosynthesis by chlorella in flashing light. In S. J. Burlew (Ed.), *Algal culture from laboratory to pilot plant* (pp. 63–77). Washington, Washington DC: Carnegie Institution.

Kromkamp, J. C., Beardall, J., Sukenik, A., Kopecký, J., Masojídek, J., van Bergeijk, S., ... Yamshon, A. (2009). Short term variation in photosynthetic parameters of *Nannochloropsis* grown in two different types of outdoor mass cultures. *Aquatic Microbial Ecology*, *56*, 309–322.

Law, E. A., & Berning, J. L. (1991). A study of the energetics and economics of microalgae mass culture with marine chlorophyte *Tetraselmis suecica*: implications for use of power plant stack gases. *Biotechnology and Bioengineering*, *37*, 936–947.

Law, E. A., Terry, K. L., Wickman, J., & Chalup, M. S. (1983). A simple algal production system designed to utilize the flashing light effect. *Biotechnology and Bioengineering*, *25*, 2319–2335.

Leeper, D. A., & Porter, K. G. (1995). Toxicity of the mixotrophic chrysophyte Poterioochromonas malhamensis to the cladoceran *Daphnia ambigua*. *Archiv für Hydrobiologie*, *134*, 207–222.

Ma, M., Yuan, D., Yue, H., Park, M., & Gong, Y. (2017). Effective control of *Poterioochromonas malhamensis* in pilot-scale culture of *Chlorella sorokiniana* GT-1 by maintaining CO2-mediated low culture pH. *Algal Research*, *26*, 436–444.

Malapascua, J. R. F., Jerez, C. G., Sergejevová, M., Figueroa, F. L., & Masojídek, J. (2014). Photosynthesis monitoring to optimize growth of microalgae mass cultures: Application of chlorophyll fluorescence techniques. *Aquatic Biology*, *22*, 123–140.

Masojidek, Jiri, Masojídek, J., M. Sergejevová , J. R., Malapascua, J., Kopecký (2015).. (2015). *Thin-layer systems for mass cultivation of microalgae: flat panels and sloping cascades. In Prokop, A., Rakesh, K., Zoppi, M.E. (Eds.)., Algal Biorefineries, Vol. 2: Products and refinery design (pp. 237-262).* (2, pp. 237–262). Switzerland 2015: Springer International Publishing.

Masojídek, J., Kopecký, J., Giannelli, L., & Torzillo, G. (2011a). Productivity correlated to photobiochemical performance of *Chlorella* mass cultures grown outdoors in thin-layer cascades. *Journal of Industrial Microbiology & Biotechnology*, *38*, 307–317.

Masojídek, J., Torzillo, G., Koblížek, M., Kopecký, J., Bernardini, P., Sacchi, A., & Komenda, J. (1999). Photoadaptation of two members of the Chlorophyta (*Scenedesmus* and *Chlorella*) in laboratory and outdoor cultures: changes of chlorophyll fluorescence quenching and the xanthophyll cycle. *Planta*, *209*, 126–135.

Masojídek, J., Torzillo, G., Kopecký, J., Koblížek, M., Nidiaci, L., Komenda, J., ... Sacchi, A. (2000). Changes in chlorophyll fluorescence quenching and pigment composition in the green alga *Chlorococcum* sp. grown under nitrogen deficiency and salinity stress. *Journal of Applied Phycology*, *12*, 417–426.

Masojídek, J., Vonshak, A., & Torzillo, G. (2011b). Chlorophyll fluorescence applications in microalgae mass cultures. In D. J. Suggett, O. Prášil, & M. A. Borowitzka (Eds.), *Chlorophyll a fluorescence in aquatic sciences: Methods and applications* (pp. 277–292). Dordrecht: Springer.

Mathur, S., Agrawal, D., & Jajoo, A. (2014). Photosynthesis: Response to high temperature stress. *Journal of Photochemistry and Photobiology B: Biology*, *137*, 116–126.

Maxwell, K., & Johnson, G. N. (2000). Chlorophyll fluorescence—A practical guide. *Journal of Experimental Botany*, *51*, 659–668.

Mendoza, J. L., Granados, M. R., de Godos, I., Aciér, F. G., Molina, E., Heaven, S., & Banks, C. J. (2013). Oxygen transfer and evolution in microalgal culture in open raceways. *Bioresource Technology*, *137*, 188–195.

Merrett, M., & Armitage, T. L. (1982). The effect of oxygen concentration on photosynthetic biomass production by algae. *Planta*, *155*, 95–96.

Milledge, J. (2012). Microalgae—Commercial potential for fuel, food and feed. *Food Science and Technology*, *26*(1), 26–28.

Moheimani, N. R., & Borowitska, M. A. (2006). Limits to productivity of the alga *Pleurochrysis carteae* (Haptophyta) grown in outdoor raceway ponds. *Biotechnology and Bioengineering*, *96*, 27–36.

Molina Grima, E., Acién Fernández, F. G., Camacho, G., & Chisti, Y. (1999). Photobioreactors: Light regime, mass transfer, and scale-up. *Journal of Biotechnology*, *70*, 231–247.

Niccolai, A., Bigagli, E., Biondi, N., Rodolfi, L., Cinci, L., Luceri, C., & Tredici, M. R. (2016). In vitro toxicity of microalgal and cyanobacterial strains of interest as food source. *Journal of Applied Phycology*, *29*, 199–209.

Niccolai, A., Chini Zittelli, G., Rodolfi, L., Biondi, N., & Tredici, M. R. (2019). Microalgae of interest as food source: Biochemical composition and digestibility. *Algal Research*, *42*, 101617.

Nishiyama, Y., Allakhverdiev, S. I., Yamamoto, I., Hayashio, H., & Murata, N. (2004). Singlet oxygen inhibits the repair of photosystem II by suppressing the translation elongation of the D1 protein in *Synechocystis sp*. PCC 6803. *Biochemistry*, *43*, 11321–11330.

Nishiyama, Y., & Murata, N. (2014). Revised scheme for the mechanism of photoinhibition and its application to enhance the abiotic stress tolerance of the photosynthetic machinery. *Applied Microbiology and Biotechnology*, *98*, 8777–8796.

Nota Mipaaf. (2012) 03 12 - n. 5391 Alga Spirulina.

Oliveira, M. A. C. L., De Monteiro, M. P. C., Robbs, P. G., & Leite, S. G. F. (1999). Growth and chemical composition of *Spirulina maxima* and *Spirulina platensis* biomass at different temperatures. *Aquaculture International*, *7*, 261–275.

Oswald, J. A. (1988). Large-scale algal culture systems (engineering aspects). In L. J. Borowitzka, & M. A. Borowitzka (Eds.), *Microalgal biotechnology* (pp. 357–395). Cambridge: Cambridge University Press, 1988.

Payer, H. D., Chiemvichak, Y., Hosakul, K., Kongpanichkul, C., Kraidej, L., Nguitragul, M., ... Buri, P. (1980). Temperature as an important climatic factor during mass production of microscopic algae. In G. Shelef, & C. J. Soeder (Eds.), *Algae biomass. Production and use* (pp. 389–399). Amsterdam: Elsiever/North Holland Biomedical Press.

Pirt, S. J., Kee, Y. K., Walach, M. R., Pirt, M. W., Balyuzi, H. H. M., & Bazin, M. L. (1983). A tubular bioreactor for photosynthetic production of biomass from carbon dioxide: Design and performance. *Journal of Chemical Technology and Biotechnology*, *33B*, 35–58, Oxford, Oxfordshire: 1986.

Pulz, O., & Gross, W. (2004). Valuable products from biotechnology of microalgae. *Applied Microbiology and Biotechnology*, *65*, 635–648.

Raja, R., Hemaiswarya, S., Kumar, N. A., Sridhar, S., & Rengasamy, R. (2008). A perspective on the biotechnological potential of microalgae. *Critical Review in Microbiology*, *34*(2), 77–88.

Ralph, P. J., & Gademann, R. (2005). Rapid light curves: A powerful tool to assess photosynthetic activity. *Aquatic Botany*, *82*, 222–237.

Ranglová, K., Lakatos, G. E., Câmara Manoel, J. A., Grivalský, T., & Masojídek, J. (2019). Rapid screening test to estimate temperature optima for microalgae growth using photosynthesis activity measurements. *Folia Microbiologica*, *64*, 615–625.

Ras, M., Steyer, J. P., & Bernard, O. (2013). Temperature effect on microalgae: A crucial factor for outdoor production. *Reviews in Environmental Science and Biotechnology*, *12*, 153–164.

Raso, S., van Genugten, B., Vermuë, M., & Wijffels, R. H. (2012). Effect of oxygen concentration on the growth of *Nannochloropsis* sp. at low light intensity. *Journal of Applied Phycology*, *24*, 863–871.

Rego, A., Redondo, L. M., Geraldes, V., Costa, L., Navalho, J., & Pereira, M. T. (2015). Control of predators in industrial scale microalgae cultures with pulsed electric fields. *Bioelectrochemistry*, *103*, 60–64, Amsterdam, Netherlands.

Regulation (EC) No 178/2002 of the European parliament and of the Council of 28 January 2002 laying down the general principles and requirements of food law, establishing the European food safety authority and laying down procedures in matters of food safety. Official Journal of the European Communities, L, 31, 1-24.

Regulation (EC) No 258/97 of the European Parliament and of the Council of 27 January 1997 concerning novel foods and novel food ingredients. Official Journal of the European Union, L 43/1 of 14 February 1997, pp 1–6. http://eur-lex.europa.eu/legal-content/EN/TXT/PDF/?uri = CELEX:31997R0258&from = EN\\ Accessed, 29 June 2020.

Regulation (EC) No 889/2008 of 5 September 2008 laying down detailed rules for the implementation of Council Regulation (EC) No 834/2007 on organic production and labelling of organic products with regard to organic production, labelling and control.

Richmond, A. (2013). Biological principles of mass cultivation of photoautotrophic microalgae. In A. Richmond, & Q. Hu (Eds.), *Handbook of microalgal culture: Applied phycology and biotechnology* (pp. 171–204). Chichester: Wiley.

Richmond, A., & Becker, E. W. (1986). In A. Richmond (Ed.), *Technological aspects of mass cultivation, a general outline* (pp. 245–263). Boca Raton, Florida: CRC Press, Inc..

Ritchie, R. J., & Larkum, A. W. D. (2012). Modelling photosynthesis in shallow algal production ponds. *Photosynthetica*, *50*, 481–500.

Schreiber, U. (2004). Pulse-amplitude-modulation (PAM) fluorometry and saturation pulse method: an overview. In G. C. Papageorgiou, & Govindjee (Eds.), *Chlorophyll* a *fluorescence: A signature of photosynthesis. Advances in photosynthesis and respiration* (19, pp. 279–319). Dordrecht: Springer.

Schreiber, U., Endo, T., Mi, H., & Asada, K. (1995). Quenching analysis of chlorophyll fluorescence by the saturation pulse method: Particular aspects relating to the study of eukaryotic algae and cyanobacteria. *Plant & Cell Physiology*, *36*, 873–882.

Schreiber, U., Schliwa, U., & Bilger, W. (1986). Continuous recording of photochemical and nonphotochemical fluorescence quenching with a new type of modulation fluorometer. *Photosynthesis Research*, *10*, 51–62.

Setlik, I., Sust, V., & Malek, I. (1970). Dual purpose circulation units for large scale culture of algae in temperate zones. I. Basic design considerations and scheme of pilot plant. *Algological Studies*, *11*, 111–164.

Sforza, E., Simionato, D., Giacometti, G. M., Bertucco, A., & Morosinotto, T. (2012). Adjusted light and dark cycles can optimize photosynthetic efficiency in algae growing in photobioreactors. *PLoS One*, *7*(6), e38975.

Silva Benavides, A. M., Ranglova, K., Malapascua, J. R., Masojidek, J., & Torzillo, G. (2017). Diurnal changes of photosynthesis and growth of *Arthrospira platensis* cultured in a thin-layer cascade and an open pond. *Algal Research*, *28*, 48–56.

Soeder, C. J. (1980). Massive cultivation of microalgae: Results and prospects. *Hydrobiologia, 72*, 197–209.

Sousa, C., de Winter, L., Janssen, M., Vermuë, M. H., & Wijffels, R. H. (2012). Growth of the microalgae *Neochloris oleabundans* at high partial oxygen pressure and sub-saturating light intensity. *Bioresource Technology, 104*, 565–570.

Spolaore, P., Joannis-cassan, C., Duran, E., & Isambert, A. (2006). Commercial applications of microalgae. *Journal of Bioscience and Bioengineering, 101*, 87–96.

Strasser, R. J., Srivastava, A., & Govindjee. (1995). Polyphasic chlorophyll *a* fluorescence transient in plants and cyanobacteria. *Photochemistry and Photobiology, 61*, 33–42.

Strasser, R. J., Tsimili-Michael, M., & Srivastava, A. (2004). Analysis of the chlorophyll a fluorescence transient. In G. C. Papageorgiou, & Govindjee (Eds.), Chlorophyll a fluorescence: A signature of photosynthesis. *Advances in photosynthesis and respiration* (19, pp. 321–362). The Netherlands: Springer.

Szabó, M., Parker, K., Guruprasad, S., Kuzhiumparambil, U., Lilley, R. M., Tamburic, B., … Ralph, P. J. (2014). Photosynthetic acclimation of *Nannochloropsis oculata* investigated by multi-wavelength chlorophyll fluorescence analysis. *Bioresource Technology, 167*, 521–529.

Takahashi, S., & Murata, N. (2008). How do environmental stresses accelerate photoinhibition? *Trends in Plant Science, 13*, 178–182.

Tomaselli, L., Giovannetti, L., Sacchi, A., & Bocci, F. (1988). Effects of temperature on growth and biochemical composition in *Spirulina platensis* strain M2. In T. Stadler, J. Mellion, M. C. Verdus, Y. Karamanos, H. Morvan, & D. Christiaen (Eds.), *Algal biotechnology* (pp. 303–314). London: Elsevier Applied Science.

Tomaselli, L., Giovannetti, L., & Torzillo, G. (1993). Physiology stress response in Spirulina spp. In F. Doumerge, H. Duran-Chastel, & A. Toulemont (Eds.), *Algae of life* (pp. 65–75). Monaco: Bullettin de l'Institut océanographique, Special 12.

Tomaselli, L., Torzillo, G., Giovannetti, L., Pushparaj, B., Bocci, F., Tredici, M., … Materassi, R. (1987). Recent research on Spirulina in Italy. *Hydrobiologia, 151/152*, 79–82.

Torzillo, G., 1980. Two years of mass culture of Spirulina maxima in tubular system (in Italian). In: Materassi, R. (Ed.), Proceedings of conference "Perspectives of mass cultivation of Spirulina in Italy". Florence, Accademia dei Georgofili, 20 – 21 November, 1980.

Torzillo, G., Accolla, P., Pinzani, E., & Masojidek, J. (1996). In situ monitoring of chlorophyll fluorescence to assess the synergistic effect of low temperature and high irradiance stresses in *Spirulina* cultures grown outdoors in photobioreactors. *Journal of Applied Phycology, 8*, 283–291.

Torzillo, G., Bernardini, P., & Masojidek, J. (1998). On-line monitoring of chlorophyll fluorescence to assess the extent of photoinhibition of photosynthesis induced by high oxygen concentration and low temperature and its effect on the productivity of outdoor cultures of *Spirulina platensis* (Cyanobacteria). *Journal of Phycology, 34*, 504–510.

Torzillo, G., & Chini Zittelli, G. (2015). Tubular photobioreactors. In A. Prokop, R. K. Bajpai, & M. E. Zappi (Eds.), *Algal biorefineries volume 2: Products and refinery design* (pp. 187–212). Switzerland: Springer International Publishing.

Torzillo, G., Giovannetti, L., Bocci, F., & Materassi, R. (1984). Effect of oxygen concentration on the protein content of *Arthrospira* biomass. *Biotechnology and Bioengineering, 26*, 1134–1135.

Torzillo, G., Pushparaj, B., Bocci, F., Balloni, W., Materassi, R., & Florenzano, G. (1986). Production of *Spirulina* biomass in closed photobioreactors. *Biomass, 11*, 61–74.

Torzillo, G., Sacchi, A., Materassi, R., & Richmond, A. (1991). Effect of temperature on yield and biomass loss in Spirulina platensis grown outdoors in tubulr photobioreactors. *Journal of Applied Phycology, 3*, 103–109.

Torzillo, G., & Vonshak, A. (1994). Effect of light and temperature on the photosynthetic activity of the cyanobacterium *Spirulina platensis*. *Biomass and Bioenergy*, *6*, 399–403.

Touloupakis, E., Cicchi, B., Silva Benavides, A. M., & Torzillo, G. (2016). Effect of high pH on growth of *Synechocystis* sp. PCC 6803 cultures and their contamination by golden algae (*Poterioochromonas* sp.). *Applied Microbiology and Biotechnology*, *100*, 1333–1341.

Touloupakis, E., Cicchi, B., & Torzillo, G. (2015). A bioenergetic assessment of photosynthetic growth of *Synechocystis* sp. PCC 6803 in continuous cultures. *Biotechnology for Biofuels*, *8*, 133–144.

Ugwu, C. U., Aoyagi, H., & Uchiyama, H. (2007). Influence of irradiance, dissolved oxygen concentration, and temperature on the growth of *Chlorella sorokiniana*. *Photosynthetica*, *45*, 309–311.

Van der Spiegel, M., Noordam, M. Y., & van der Fels-Klerx, H. J. (2013). Safety of novel protein sources (insects, microalgae, seaweed, duckweed, and rapeseed) and legislative aspects for their application in food and feed production. *Comprehensive Reviews in Food Science and Food Safety*, *12*, 662–678.

Venancio, H. C., Cella, H., Garcia Lopes, R., & Bianchini Derner, R. (2020). Surface-to-volume ratio influence on the growth of *Scenedesmus obliquus* in a thin-layer cascade system. *Journal of Applied Phycology*, *32*, 821–829.

Vigani, M., Parisi, C., Rodríguez-Cerezo, E., Barbosa, M., Sijtsma, L., Ploeg, M., & Enzing, C. (2015). Food and feed products from micro-algae: Market opportunities and challenges for the EU. *Trends in Food Science and Technology*, *42*, 81–92.

Vonshak, A., & Novoplansky, N. (2008). Acclimation to low temperature of two *Arthrospira platensis* (cyanobacteria) strains involves downregulation of PSII and improved resistance to photoinhibition. *Journal of Phycology*, *4*, 1071–1079.

Vonshak, A., Novoplansky, N., Silva Benavides, A. M., Torzillo, G., Beardall, J., & Palacios, Y. M. (2020). Photosynthetic characterization of two *Nannochloropsis* species and its relevance to outdoor cultivation. *Journal of Applied Phycology*, *32*, 909–922.

Vonshak, A., & Richmond, A. (1988). Mass production of the blue-green alga Spirulina: An overview. *Biomass*, *15*(4), 233–247.

Vonshak, A., Torzillo, G., Masojídek, J., & Boussiba, S. (2001). Sub-optimal morning temperature induces photoinhibition in dense outdoor cultures of the alga *Monodus subterraneus* (Eustigmatophyta). *Plant, Cell & Environment*, *24*, 1113–1118.

Vonshak, A., Torzillo, G., & Tomaselli, L. (1994). Use of chlorophyll fluorescence to estimate the effect of photoinhibition in outdoor cultures of *Spirulina platensis*. *Journal of Applied Phycology*, *6*, 31–34.

Wells, M. L., Potin, P., Craigie, J. S., Raven, J. A., Merchant, S. S., Helliwell, K. E., ... Brawley, S. H. (2017). Algae as nutritional and functional food sources: revisiting our understanding. *Journal of Applied Phycology*, *29*, 949–982.

Wijffels, R., & Barbosa, M. J. (2010). An outlook on microalgal biofuels. *Science (New York, N.Y.)*, *329*(5993), 796–799.

Wilson, A., Ajlani, G., Verbavatz, J. M., Vass, I., Kerfeld, C. A., & Kirilowsky, D. (2006). A soluble carotenoid protein involved in phycobilisome-related energy dissipation in cyanobacteria. *The Plant Cell*, *18*, 992–1007.

Zhu, X.-G., Long, S. P., & Ort, D. R. (2008). What is the maximum efficiency with which photosynthesis can convert energy into biomass? *Current Opinion in Biotechnology*, *19*, 153–159.

Chapter 2

Unconventional microalgae species and potential for their use in the food industry

Juan Roberto Benavente-Valdés[1], Alejandro Méndez-Zavala[1], Israel Hernández-López[2], Berinka Araceli Carreón-González[2], María Elena Velázquez-Arellano[3], Lourdes Morales-Oyervides[1] and Julio César Montañez-Saénz[1]

[1]*Department of Chemical Engineering, Faculty of Chemical Sciences, Autonomous University of Coahuila, Saltillo, México,* [2]*Faculty of Biological Sciences, Autonomous University of Coahuila, Torreón, México,* [3]*Department of Animal Nutrition, The Antonio Narro Agrarian Autonomous University, Saltillo, México*

2.1 Introduction

In the early 1950s, microalgae were considered to be a good supplement and fortification in diets for malnourished children and adults. Commercial production of microalgae for human nutrition has become a reality all over the world. Microalgae are an enormous biological resource, representing one of the most promising sources of new products, ingredients, and applications. They can be used to enhance the nutritional value of food and feed due to their well-balanced chemical composition (Costa & de Morais, 2013, p. 405). Numerous combinations of microalgae or mixtures with foods can be found in the market as nutritional supplements in the form of tablets, powders, capsules, pastilles, and liquids. They can also be incorporated into food products (e.g., different kinds of pasta, biscuits, bread, snack foods, candies, yogurts, or soft drinks), providing health benefits (Matos, 2017, pp. 1333–1350; Vigani et al., 2015, pp. 81–92). Microalgae are being considered as an ingredient with high nutritional content, that is, innovative and easily accessible to consumers. The amount of enriched food products with microalgal biomass available in the market is increasing, occupying ever increasing space in the global market as healthy foods (Lafarga, 2019, pp. 167–174).

Despite the many advantages of microalgal biomass in food, only a few species are used commercially for the production of food products, with only

Cultured Microalgae for the Food Industry. DOI: https://doi.org/10.1016/B978-0-12-821080-2.00010-1

six recognized as safe for human consumption, of which the *Spirulina* and *Chlorella* species are the microalgae with the higher market shares. However, there are a large number of other diverse species that offer attractive phytochemical compositions for applications in food that could improve the nutritional quality of food. In addition, a wide range of colors can be found for use as natural dyes to improve the appearance of food. This chapter describes microalgae that have been studied for nonfood purposes, which have nutritional quality and functional properties that make them attractive for new food additives development. In addition, such microalgae have been classified according to safe consumption and their frequency of utilization for food purposes is described.

2.2 Microalgae as a food source

Algae have been used by humanity for food since ancient times. Biomass from microalgae can be used directly as food to increase the nutrient content of prepared foods, and as a probiotic and supplement agent with a positive effect on human and animal health. Microalgae are photosynthetic microorganisms that can produce and accumulate biomass, which is used to produce food, feeds, cosmetics, pharmaceuticals, and fuels. Microalgae are an important source of high-valuable molecules, including lipids, proteins, carbohydrates, and a variety of pigments, and represent a promising ecological and economical option for applications in various industries. Therefore increasing interest in their production has been evident lately. Microalgae can grow almost everywhere if there is water, sunlight, a carbon source, and other micro- and macronutrients. Nowadays, worldwide annual production of microalgal biomass is around 35,000 t, with *Arthrospira* (*Spirulina*), the most produced and commercialized microalgae (Barsanti & Gualtieri, 2018, pp. 107–115; Lozano-Garcia et al., 2019, p. 101459). The utilization of these microorganisms as food and food supplements is not new, as microalgae have been used for thousands of years. They were utilized by ancient cultures, mostly in North America, Asia, and Africa. Jacob-Lopes et al. (2019, pp. 1–7) stated that, since the 1950s, microalgae have been considered as a potential source of proteins useful for human consumption. However, it was not until the 1960s (*Chlorella*) and 1970s (*Spirulina*) that these microalgae were commercially cultured for their protein content. Recent studies have revealed the high amount of different valuable molecules in microalgae which are of particular interest: polyunsaturated fatty acids (PUFAs), carbohydrates, proteins, pigments, lipids, vitamins A, B1, B2, C and E, and some minerals including Fe, Mg, and Ca. Microalgae composition is fundamental to meeting the need for a well-balanced and nutritional diet worldwide that can counteract the fast lifestyle and cheap fast food, which have led to a high amount of health problems, including obesity, heart disease, and diabetes. Such health issues have increased the interest in healthy food consumption, introducing the opportunity of microalgae evaluation as "superfoods". According to Koyande et al. (2019), within the microalgae-derived products, dried *Spirulina* production is the highest worldwide, with 12,000 t per year, followed by *Chlorella* sp., *Dunaliella salina*, *Aphanizomenon flosaquae*, and *Haematococcus pluvialis*, with 5000, 3000, 1500, and 700 t produced annually, respectively, with an estimated market value of US$44.6 billion by 2023. *Spirulina* has the most promising estimated market value of US$2.0 billion by 2026 and *Chlorella* US$700 million by 2022.

2.3 Microalgae currently used in food

Microalgal biomass and microalgae-derived products have earned an important place in the food market. Several companies have started to develop microalgae-containing foods. For example, OHi (OHi Foods, United States) is a gluten-free bar, made from grains and soy that contains *Spirulina* and *Chlorella* biomass; the famous M&M's chocolates (Mars Wrigley Confectionery, Ireland) use *Spirulina* as natural colorant; Gullón Vitalday (Galletas Gullón, Spain) are oat and rice cakes with 1% *Spirulina*. Frecious Slow Juice (Frecious, the Netherlands) contains 2.4% *Chlorella*. Vegetable juice (Happy Planet Foods, Canada) provides a green fruit smoothie with both *Spirulina* and *Chlorella*. Also, there are breads and pastas enriched with *Dunaliella* sp. (Koyande et al., 2019; Lafarga, 2019, pp. 167–174).

All the mentioned products are commercialized mainly in Europe, Asia, and North America, fulfilling an organic, ecological, and vegan consumer market. Besides the "superfood" classification, thanks to their nutrimental properties, microalgae contribute to the prevention of diverse health issues such as obesity, diabetes, Alzheimer disease, some types of cancer, and heart diseases, among others (Sathasivam, Radhakrishnan, Hashem, & Abd_Allah, 2019, pp. 709–722).

Table 2.1 shows different kinds of foods that have already been supplemented with complete microalgal biomass and with some microalgae-derived ingredients. Also listed are the microalgae species used, the nutrimental and functional properties obtained from microalgae addition, and the benefits obtained from their consumption. It is noted that the benefits resulting from the addition of microalgal biomass to food are recurrent among studies, such as the high antioxidant and anticancer activities, the nutrimental enhancement with higher protein content, and the capacity to prevent cardiac diseases. Thanks to the high antioxidant activity of microalgae, which is related to the pigments content, different kinds of diseases can be prevented.

The nutrimental contents of the most used worldwide microalgae species are listed in Table 2.2. According to the literature reviewed, some of the most valuable metabolites are proteins, carbohydrates, and lipids, among others. The one with the highest protein content is *Spirulina maxima*,

TABLE 2.1 Microalgae used in food products and their functional and nutritional properties.

Type of product	Microalgae species	Nutrimental properties	Benefits	References
Bread	*Arthrospira platensis*	Proteins, fiber, and mineral content	Antioxidant	(Pan-utai & Iamtham, 2019; Vieira et al., 2020)
	Arthrospira sp.		Antioxidant	Costa and de Morais (2013)
	Dunaliella spp.		Anti-Parkinson disease	(Abd El-baky, El Baz, & El-Baroty, 2004)
	Chlorella vulgaris		Anticancer; antiinflammatory activity	(Abd El-Hack et al., 2019)
Cookies	*Haematococcus pluvialis*	Proteins and pigments	Antioxidant activity	(Chu, 2012)
	C. vulgaris		Anticancer activity	(Sathasivam, Radhakrishnan, Hashem, & Abd_Allah, 2019)
Pasta	*A. platensis*	Proteins, EPA, DHA, and antioxidant compounds	Protein enrichment	(Chu, 2012)
	Dunaliella salina		Anticancer activity	(Capelli & Cysewski, 2010)
	Chlorella sorokiniana		Night blindness prevention/antiaging activity	(Matsukawa, Hotta, Masuda, Chihara, & Karube, 2000)
Yogurt	*Arthrospira* sp.	Proteins, PUFAs, and antioxidant compounds	Protein content enhancement/linoleic acid	(Khan, Shin, & Kim, 2018)
	Chlorella sp.		Antioxidative activity, PUFA enhancement	(Bhosale & Bernstein, 2005)

DHA, Docosahexaenoic acid; *EPA*, eicosapentaenoic acid; *PUFA*, polyunsaturated fatty acid.

TABLE 2.2 Some microalgae species and their nutrimental content (values in % dry matter).

Microalgae	Proteins	Lipids	Carbohydrates	References
Aphanizomenon flosaquae	62	3	23	(Campana et al., 2017)
Chlorella vulgaris	51–58	14–22	12–17	(Yadavalli et al., 2020)
Chlorella pyranoidosa	57	2	26	
Dunaliella salina	57	6	32	(Milledge, 2012)
Dunaliella bioculata	49	8	4	Caporgno and Mathys (2018)
Haematococcus pluvialis	48	15	27	(Bleakley & Hayes, 2017)
Porphyridium cruentum	28–39	09–14	40–57	Caporgno and Mathys (2018)
Spirulina maxima	60–71	6–7	13–16	(Bleakley & Hayes, 2017)
Spirulina platensis	46–63	4–9	8–14	

followed by *A. flosaquae*, and *Spirulina platensis* in third place. *H. pluvialis* holds first place in lipids content, while *Porphyridium cruentum* is in the first place for its carbohydrate content.

2.4 Unconventional microalgae in food

A limited number of microalgae species are used for the formulation of food products due to the scarce information available about the safety of these microorganisms when added into a food matrix. According to the Food and Drug Administration (FDA) and the European Food Safety Authority (EFSA), algal biomass is classified as an "other dietary supplement." *Spirulina*, *Chlorella*, *Dunaliella*, *Haematococcus*, *Schizochytrium*, *P. cruentum*, and *Crypthecodinium cohnii* are classified in the generally recognized as safe category (García, de Vicente, & Galán, 2017, pp. 1017–1024). Moreover, according to a report made by Enzing, Ploeg, Barbosa, and Sijtsma (2014, pp. 19–37), 19 species are classified as nontoxic. These microalgae are *Synechococcus* sp., *Tetraselmis* sp., *Phaeodactylum tricornutum*, *Chlamydomonas reinhardtii*, *Thalassiosira pseudonana*, *H. pluvialis*, *Odontella aurita*, *Parietochloris incisa*, *Skeletonema* sp., *Chlorococcum* sp., *Monodus subterraneus*, *Scenedesmus*, *Navicula* sp., *Synechococcus* sp., *Nitzschia dissipata*, *Nannochloropsis* sp., *Desmodesmus* sp., and *Isochrysis* sp. In addition, other microalgae such as *Tetraselmis*, *Isochrysis*, *Pavlova*, *Phaeodactylum*, *Chaetoceros*, *Nannochloropsis*, *Skeletonema*, and

Thalassiosira are also used as feeds in aquaculture (Sousa, Gouveia, Batista, Raymundo, & Bandarra, 2008, pp. 75–112).

The development of foods enriched with microalgae is dominated by the *Chlorella* and *Spirulina* species (Lafarga et al., 2019, pp. 167–174). However, numerous species have been studied for nonfood purposes, such as the production of biofuels, wastewater treatment, or obtention of high value-added compounds. Such compounds represent a potential option for their utilization in food due to their nutritional attributes and functional properties, let alone their controlled production, and high yields and recoveries (Chen et al., 2018, pp. 336–346; Molinuevo-Salces, Riaño, Hernández, & Cruz García-González, 2019, pp. 505–533). Microalgae are divided into different groups, including cynaobacteria (blue-green algae), rhodophytes (red algae), chlorophytes (green algae), and chromophytes (all other algae). Each group contains hundreds of species and each of these species has thousands of different strains. Only a small number of these varieties have been studied for food purposes. Table 2.3 shows the microalgae that have been rarely or never used in food processing.

2.5 Potential uses of unconventional microalgae species in food: nutrimental and physical properties

Several studies support the available information on the growth and characteristics of microalgae species that are not used commercially in food or applied in research for the development of new food products. The selection of microalgae species with balanced nutritional profiles is essential for the success of the development of new foods, therefore a detailed physicochemical and functional characterization of the microalgae is a crucial stage that will determine which is the most suitable for different applications and purposes. In addition, the determination of toxic agents in the cell is essential for its potential use as a food additive. This section describes the potential uses of unconventional microalgal biomass according to the main components of interest for use in food.

2.5.1 Proteins

Microalgal biomass is generally rich in protein, competing in quality and quantity against conventional food proteins. Additionally, when it comes to amino acid composition, microalgae have been shown to have a high nutritional value of proteins, comparing favorably to eggs, soybean, and fish. The substitution of meat, milk, and egg proteins with microalgal proteins is a promising and sustainable option. An advantage of using microalgae to obtain proteins is that the land requirements are low ($<2.5\ m^2\ kg^{-1}$ of protein) compared to animal protein (up to $250\ m^2\ kg^{-1}$ of protein) (Caporgno & Mathys, 2018, p. 58). The protein content range for microalgal biomass is

TABLE 2.3 Main species of microalgae studied with safety aspects and uses in food.

Microorganism	Species	Safety aspect	Use in food
Cyanobacteria	*Anabaena* sp.	ND	UC
	Phormidium sp.	ND	UC
	Nostoc sp.	NT	C
	Spirulina/Arthrospira sp.	GRAS	C
	Synechococcus sp.	NT	UC
Chlorophyta	*Botryococcus* sp.	ND	UC
	Chlamydomonas sp.	NT	NF
	Chlorella sp.	GRAS	C
	Chlorococcum sp.	ND	UC
	Desmodesmus sp.	NT	UC
	Dunaliella sp.	GRAS	C
	Haematococcus sp.	GRAS	C
	Muriellopsis sp.	ND	NF
	Oocystis sp.	ND	UC
	Parachlorella sp.	NT	UC
	Parietochloris sp.	ND	UC
	Scenedesmus sp.	NT	NF
	Trentepohlia sp.	ND	UC
	Tetraselmis sp.	NT	NF
Dynophyta	*Crypthecodinium cohnii*	GRAS	C
Heterokonta	*Schizochytrium* sp.	GRAS	C
Bacillariophyta	*Chaetoceros* sp.	ND	UC
	Nitzschia sp.	NT	UC
	Odontella sp.	NT	UC
	Phaeodactylum sp.	NT	UC
	Skeletonema sp.	NT	UC
	Thalassiosira sp.	NT	UC
Ochrophyta	*Monodus* sp.	NT	UC
	Nannochloropsis sp.	NT	NF

(*Continued*)

TABLE 2.3 (Continued)

Microorganism	Species	Safety aspect	Use in food
Haptophyta	*Diacronema* sp.	NT	NF
	Isochrysis sp.	NT	NF
	Pavlova sp.	NT	NF
Rodophyta	*Galdiera* sp.	ND	UC
	Porphyridium cruentum	GRAS	C

C, Conventional use; *GRAS*, generally recognized as safe; *ND*, undefined; *NF*, nonfrequent use; *NT*, nontoxic; *UC*, unconventional.

generally 30–70% (de Carvalho et al., 2020, p. 122719; Loveday, 2019, pp. 311–339).

C. reinhardtii (48%), *Scenedesmus obliquus* (50%–56%), and *Synechococcus* sp. (63%) are among the unconventional microalgae with higher biomass protein content (% of dry matter) (Becker, 2007, pp. 207–210). This estimate usually comes from the total nitrogen content of the bulk biomass and not necessarily from the actual protein content, depending on the method of analysis used. The cyanobacteria *Anabaena variabilis*, is a promising source of protein, with the ability to accumulate protein by up to 50.1% (Han et al., 2016, pp. 16–22). Likewise, *Phormidium fragile* produces 80 mg g^{-1} of protein under high light intensity exposure (Ganesan, Jailani, Mani, Paramasivam, & Kareem, 2019, pp. 4115–4118). The accumulation of proteins by these cyanobacteria is below that reported in *Spirulina* species (60%–70% w/w); however, they represent a viable protein source.

The biochemical composition of the microalgal biomass varies greatly depending on the type of microorganism and its production system. Teuling, Wierenga, Schrama, and Gruppen (2017, pp. 7989–8002), showed these differences in classical microalgae used in foods (such as *Arthrospira* sp.), as well as in unconventional microalgae proposed as potential food sources (such as *Nannochloropsis* sp., *Tetraselmis* sp., and *Scenedesmus* sp.) with variations (% w/w, DW) from 26% to 67% for proteins, 6% to 36% for carbohydrates, and 7% to 25% for lipids. According to Teuling et al. (2017, pp. 7989–8002), after *Arthrospira maxima*, the strains with the highest protein content are *Nannochloropsis gaditana*, *Tetraselmis impellucida*, and *Scenedesmus dimorphus* (45%, 35%, and 27% of protein, respectively), based on the analysis of total amino acids. Moreover, the strain *Parachlorella kessleri* GB1 could be a candidate as a source of unicellular protein. Parul and Thiyam (2019, p. 108205) showed that culture systems can modify the biochemical composition of microalgae and their antioxidant properties, the main component being protein with values of up to 45% in mixotrophic cultures supplemented with glucose. Similar results were shown by Piasecka, Krzeminska, and Tys (2017, pp. 1735–1743), who reported that it was possible to increase the protein content in *P. kessleri* when using beet molasses as culture medium in heterotrophic and mixotrophic cultures.

A problem associated with the use of dry microalgal biomass as a source of protein is its low digestibility and bioavailability (Khanra et al., 2018, pp. 60–84) due to the presence of cell wall polysaccharides (de Carvalho et al., 2020, p. 122719), which implies the use of protein separation processes to improve their bioavailability (Becker, 2007, pp. 207–210). Therefore it is crucial to evaluate the methods of isolation, recovery, concentration, and purification of protein fractions so that they can be used in food formulations (Grossmann, Ebert, Hinrichs, & Weiss, 2018, pp. 67–76).

The use of unconventional microalgae as a source of protein in food entails the possibility of introducing risks related to production or extraction techniques. Those risks include allergens, toxins, pathogens, heavy metals, and pesticides. Therefore their commercialization and potential mass production should be subjected to the implementation of quality control and food safety programs, such as Good Manufacturing Practices (GMP), Standard Sanitation Operation Procedures (SSOP), and Hazard Analysis and Critical Control Points (HACCP) (Van der Spiegel, Noordam, & van der Fels-Klerx, 2013, pp. 662–678). For that reason, the bioactive components of the microalgae for food application could be obtained using methods that allow the generation of food-grade ingredients. Thus, lately, research into innovative technologies such as supercritical fluid extraction, microwave-assisted extraction, ultrasound-assisted extraction, pulsed electric field, and other methods have been evaluated for the recovery of microalgal compounds with an ecological approach (Vieria Costa, Bastos Freitas, Moraes, Zaparoli, & Greques Morais, 2020, p. 122727).

2.5.2 Lipids

Lipids have three functions in microalgae: as an energy store, for signaling activity, and as a structural part of the cell membrane. Lipid content in microalgae may vary from 1% to 40% w/w. Other reports suggested that the lipid content of microalgal biomass is between 5% and 30% (de Carvalho et al., 2020, p. 122719). Microalgae are a source of high added-value lipids such as PUFAs. This group includes eicosapentaenoic acid (EPA, ω-3), docosahexaenoic acid (DHA, ω-3), arachidonic acid (ω-6), and linolenic acid (LNA, ω-6). PUFAs are an essential source of bioactive compounds intended to combat nutritional problems associated with chronic degenerative diseases (diabetes and hypertension, among others) (Balakrishnan et al., 2019). A potential industrial interest of microalgal lipids in the food industry could be in cooking oils or mixtures of oils with other sources due to their properties and proportion of omega oils.

Some unconventional microalgae with high production of PUFAs for promising applications in the food industry include *Isochrysis galbana, Aurantiochytrium* sp., *Ulkenia, C. cohnii, Thraustochytrids, Nannochloropsis, Cadophora fracta, Parietochloris incise, P. tricornutum*, and *Nitzschia* sp. (Katiyar & Arora, 2020, p. 101800). In this sense, it has been reported that the cyanobacteria *A. variabilis* was able to accumulate lipids by up to 46.9% when exposed to an ultrasound treatment, improving the content of PUFAs (Han et al., 2016, pp. 16–22).

On the other hand, chlorophytes stand out for their high lipids and fatty acids contents, when compared to cyanobacteria. *Chlorococcum* sp. has been described as a promising lipid producer. Montero, Olguín, De Philippis, and Reverchon (2018, pp. 2847–2857) reported a lipid content of 98.7 mg g^{-1} when this microalga is cultivated using digestate from pig manure as the culture medium. The FAME profile showed that palmitic acid and linoleic acid are the most abundant fatty acids, representing more than 60% of the total lipids. Similarly, Correia (2018) reported a total lipid content of 18.35%, where 74.70% were PUFAs when *Chlorococcum* sp. was cultivated in a pilot-scale PBR under autotrophic conditions. Another potential strain for use as a lipid source is *Oocystis* sp., with a high content of fatty acids such as hexadecanoic and linoleic. However, this strain has only been studied for bioenergetic purposes and has not been evaluated in animal or human nutrition (Anthony & Stuart, 2015; Chavez, 2018). *Trentepohlia arborum*, a filamentous microalga, demonstrated the presence of large and copious lipid bodies in the cytoplasm, reaching a productivity of 21% (w/w) when grown under high concentrations of ammonia and light intensities (Chen, Zhang, & Liu, 2016, pp. 567–573). Similar lipid contents have been observed in *Tetraselmis* and *O. aurita* species, with levels ranging between 22% and 28% (w/w) (Niccolai, Zittelli, Rodolfi, Biondi, & Tredici, 2019, p. 101617; Upadhyay, Lawrence, Shekhar, & Anantharaman, 2017, pp. 1259–1269).

Other unconventional microalga with high lipid production and, thus, with potential applications in the food sector include *P. kessleri*. You, Zhang, Peng, and Miao (2019, pp. 510–526) showed that *P. kessleri* responds to salt stress in the culture medium through the accumulation of lipids (C18: 1 and C18: 2 of total fatty acids). Similar behavior has been observed in *P. kessleri* TY02 under nitrogen stress (depletion): the microalgae accumulated higher lipid content (up to 45%) when compared to the system without nitrogen limitations (24%) (Gao et al., 2019, p. 1188). Nitrogen stress has also increased the lipid content in *Scenedesmus acuminatus* (54% DW) with C16 and C18 fatty acids (Zhang, Wu, Sun, Peng, & Li, 2018, pp. 73–102).

Furthermore, EPA and DHA are essential fatty acids for proper infantile brain development, thus they are essential in pregnancy and lactation. In addition, these compounds are not produced by the human body, so their incorporation into the diet is necessary. Some unconventional microalgae have the capacity to accumulate these essential fatty acids. For example, *P. tricornutum*, a saltwater microalgae specie, reported a total lipid content of 25% w/w and a high concentration of PUFAs such as EPA (Ryckebosch, Muylaert, Eeckhout, Ruyssen, & Foubert, 2011, pp. 11063–11069). Also, the haptophyte *Pavlova lutheri*, a marine microalga, is also rich in long-chain PUFAs and is capable of producing large amounts of ω-3 fatty acids such as EPA and DHA (Guihéneuf & Stengel, 2017, pp. 113–125; Ljubic & Jacobsen, 2018). Also, *T. pseudonana*, commonly used in the feed for white shrimp, has been reported as a good source of fatty acids, with 42.4% of PUFAs, of which 16.4% are EPA and 1.7% DHA (Van Nguyen, 2018, p. 138). Likewise, *M. subterraneus* and *Skeletonema menzelii* showed high levels of EPA and DHA (Hu et al., 2019, pp. 140–149; Jiang, Han, Gao, & Gao, 2016, pp. 723–732).

Biomasses of *I. galbana and Diacronema vlkianum* have already been used in food processing, providing a significant amount of EPA and DHA when the dry biomass was added in the preparation of pasta, maintaining their content after the cooking process (Fradique et al., 2013, pp. 312–319). Similarly, the addition of *I. galbana* biomass to cookies showed an increase in EPA (Gouveia et al., 2008, pp. 891–896). Moreover, the incorporation of biomass in both foods imparted color to the final products.

The above indicates that many species of microalgae can be used as a source of healthy fatty acids by incorporating biomass into food products.

2.5.3 Carbohydrates

Microalgae have a high carbohydrate content, representing about 50% of their dry weight. Microalgae can store carbohydrates such as starch; additionally, the cell wall acts as a reservoir. The potential use of nonconventional microalgal biomass to obtain, first polysaccharides, and then individual sugars opens the opportunity to obtain sugar-rich extracts for the formulation of candies, jams, beverages, and other food products (de Carvalho et al., 2020, p. 122719). The carbohydrate content in microalgal biomass ranges from 25% to 60% DW. The wall cell of the microalgae is a complex matrix of polysaccharides and its composition and concentration depend on its phylogeny. Microalgae do not contain lignin, which facilitates the process of hydrolyzing the wall cell (de Carvalho et al., 2020, p. 122719). Based on its biochemical composition, it is possible to obtain different concentrations of simple sugars such as glucose, xylose, or cellobiose by means of enzymatic hydrolysis.

Chlorophytes accumulate more carbohydrates than cyanobacteria. For example, *Trentepohlia* sp. is capable of producing erythritol, which is a sugar alcohol used as a food additive and sugar substitute (Elshobary et al., 2016, pp. 646–656). *Oocystis minuta* has been studied for carbohydrate obtention for bioenergetic purposes, reaching about 45% w/w under autotrophy conditions and high agitation rate (Ajala & Alexander, 2020, pp. 1–11). In the case of *Scenedesmus raciborskii* WZKMT, the sugar-rich biomass extract has been used to produce ethanol (Alam et al., 2019, pp. 122–219). Ueno et al. (2017, pp. 400–406) state that from the biomass of *Parachlorella* sp. and with macrophage-stimulating activity, a complex polysaccharide (approximately 400 kDa) composed mainly of galactose, glucose, galacturonic acid, and mannose can be isolated. Likewise, from *Scenedesmus* sp., an exopolysaccharide containing mainly glucose, glucosamine, and D-mannose has been isolated and their antitumoral activity has been demonstrated on human colon cancer cell lines HCT116 (19.2% cell viability) and HCT8 (38.6% cell viability) (Zhang, Liu, Ren, & Chen, 2019, pp. 761–767). This suggests the promising use and application of these microalgae as sources of functional carbohydrates.

Furthermore, the red microalgae *Galdieria sulphuraria* can accumulate mostly polysaccharides and proteins, depending on the trophic culture conditions. Graziani et al. (2013, pp. 144–152) evaluated 43 strains of *G. sulphuraria* in autotrophic and heterotrophic cultures and in both systems, the main component of the biomass was polysaccharides. This characteristic positions *G. sulphuraria* as a plausible candidate for the preparation of healthy foods. As highlighted previously, the biochemical composition of the biomass can be modified by varying the carbon source in heterotrophy. For example, the cultivation of *G. sulphuraria* in spent cherry-brine liquid increased the carbohydrate content to levels higher than those of proteins and lipids (Massa et al., 2019, pp. 9–15). Likewise, *G. sulphuraria* can accumulate glycogen in a natural way that represents an advantage for its production and utilization in food products compared to starch. Glycogen isolated from microalgae is a prospective alternative for the formulation of energy drinks and other food products due to its solubilization and rheologic properties (Martínez-García, Kormpa, & van der Maarel, 2017, pp. 75–82). In addition to this carbohydrate, the production and isolation of floridoside (2-*O*-α-D-galactopyranosylglycerol) from *G. sulphuraria* has been reported (Martinez-Garcia & van der Maarel, 2016, p. 71); the authors reported an accumulation of this carbohydrate depending on the culture conditions, suggesting a change in the metabolic behavior as an adaptation mechanism to the environment.

Botryococcus braunii can also accumulate a high concentration of carbohydrates (more than 50% DW), with high glucose and galactose content, as well as fucose in some strains (Gouveia et al., 2017, pp. 77–86). However, more studies are required to characterize such materials for future applications. It has also been proved that this microorganism could be a source of polyhydroxybutyrate. According to Kavitha, Kurinjimalar, Sivakumar, Palani, and Rengasamy (2016, pp. 700–706), *B. braunii* can accumulate up to 20% DW of the polymer. This type of polymers can be used as an immobilizing or encapsulating agent of bioactive compounds as antioxidants.

Another approach to microalgal biomass may be its use in the formulation of biodegradable composites, making it possible to reduce the number of plasticizer materials and replace them with microalgal biomass. Fabra, Martínez-Sanz, Gómez-Mascaraque, Gavara, and López-Rubio (2018, pp. 184–191) showed that the use of biomass from *Nannochloropsis* sp., can be used for the formulation of cornstarch-based composites, with positive effects on the structural and permeation properties, and projected them as an alternative for the manufacture of biodegradable packaging for food.

Various polysaccharides have shown biological activities; among them, chrysolaminarin is recognized as a scavenging reactive oxygen species that can be used to treat or prevent serious diseases, including cancer (Xia et al., 2014, pp. 4883–4897). This polysaccharide is a major energy storage carbohydrate in diatoms as *Odontella* sp. with a maximum content of 60.3% DW (Xia, Gao, Fu, Xiong, & Zhang, 2018, pp. 723–729).

2.5.4 Pigments

Natural pigments found in microalgae play an essential role in their photosynthetic metabolism, growth, and coloration. Nevertheless, they also have profoundly beneficial activities biologically speaking, as antioxidants, anticancer, anti-inflammatories, among others. Microalgal pigments are known as a promising source of eco-friendly colorants, nutraceuticals, and antioxidants. Carotenoids, phycobilins, and chlorophylls are the principal pigments found in microalgal cells.

The cyanobacteria *Anabaena* has recently been recognized as a potential source of carotenoids and the biosynthesis pathway in *Anabaena* 7120 has been extensively studied (Prasanna, Kumar, Sood, Prasanna, & Singh, 2006, pp. 187–202). This microalga is capable of accumulating 0.396 μg of carotenoids and 0.822 μg of chlorophyll *a* per mg of protein at autotrophy conditions at 23°C. It also accumulates β-carotene when grown at a low temperature. In contrast, the increase in temperature caused ketocarotenoids such as echinenone, canthaxanthin, and keto-myxoxanthophyll to increase in number (Kłodawska et al., 2019, pp. 34–40). Likewise, the cyanobacteria *P. fragile* produces 15 mg g^{-1} of chlorophyll under high light intensity exposure (Ganesan et al., 2019, pp. 4115–4118).

On the other hand, charophyte species are a popular source of pigments. Some unconventional microalgae that produce pigments (like carotenoids and phycobiliproteins) are *Scenedesmus almeriensis*, *Muriellopsis* sp., *Tetraselmis* sp., and *Chlamydomonas* sp., among others. *T. arborum* shows an attractive pigment profile when cells are green or orange. In the green state, chlorophyll *a* is the dominant pigment, while in the orange state, β-carotene is the pigment with the highest content. This strain also can accumulate neoxanthin, lutein, zeaxanthin, and chlorophyll *b* (Chen et al., 2016, pp. 567–573). The accumulation of different pigment profiles makes

T. arborum biomass a potential dye when added to food, providing green and orange colors.

The indigenous diatom *Odontella* sp. has shown the ability to produce fucoxanthin, a carotenoid with potent antioxidant, anticancer, and antiinflammatory activities. Xia et al. (2018, pp. 723–729) reported a maximum fucoxanthin content of 2.5% w/w when different nitrogen regimens were evaluated. In contrast, Heo, Cho, Kim, and Kim (2018, pp. 26–33) showed a lower yield of this carotenoid with values of 0.33% w/w.

The xanthophyll carotenoid fucoxanthinol (the deacetylated form of fucoxanthin) has been isolated from the diatom *Nitzschia laevis*, showing neuroprotective activity, allowing its possible application in pharmaceutical products or food supplements (Li et al., 2020, p. 116). In addition, bioaccessibility studies have been carried out with the aim of evaluating the possible application of fucoxanthin rich extracts of *N. laevis* in food systems (Guo et al., 2020, pp. 1844–1850). Mok, Lee, Kim, Pan, and Kim (2018, pp. 79–86) showed that fucoxanthin from *P. tricornutum* increases its stability and improves its bioavailability when added to dairy products. Fucoxanthin is a microalgal pigment now commercialized by an Israel-based company Algatech (www.algatech.com) that has an all-natural 3% fucoxanthin oleoresin patented product with the brand name Fucovital of the diatom *P. tricornutum* (Guo et al., 2020, pp. 1844–1850) and another astaxanthin-rich pigment product from *H. pluvialis* (Astapure). Among the microalgae that can produce fucoxanthin under different culture conditions *Navicula* sp. and *Isochrysis* sp. are included (Sun, Wang, & Liu, 2019, p. 101545; Telussa, Rusnadi, & Naruchman, 2019, p. 101530).

Botryococcus sp. is a microalga with high carotenoid production potential, particularly lutein. In addition, some strains are producers of nonconventional carotenoids such as canthaxanthin, violaxanthin, loroxanthin, botryoxanthin-A, and B, α-botryoxanthin-A, etc. The wide range of compounds produced as a function of the environmental conditions allows this microalga to be promoted as a commercial source for various foods/ingredients or nutritional supplements (Ambati et al., 2018, pp. 541–558). In this same sense, the recovery of lutein from *Scenedesmus* sp. has been demonstrated with the use of microwave-assisted binary phase solvent extraction (Low, Idris, & Yusof, 2020, p. 125631).

In some food products such as ice cream, biomass of *Nannochloropsis oculata* and *D. vlkianum* (as spray-dried biomass) has been added as an ingredient to obtain a green color due to their content of pigments such as carotenoids, anthocyanins, and chlorophylls (Durmaz et al., 2020, p. 101811). Apart from color, biomass components like proteins and lipids provide functional properties, which also modify physicochemical properties such as viscosity. However, it is necessary to improve the sensory properties, for instance, the "marine" taste and odor. If a specific techno-functional property in the food is required, then it will require the use of a particular type of microalgae or its bioactive compounds (Caporgno & Mathys, 2018, p. 58).

In another case, the addition of *I. galbana* biomass at different concentrations (1% and 3% w/w) modified the color of biscuits, according to Gouveia et al. (2008, pp. 891–896). In this study, stability studies during storage were evaluated using the CIE Lab color system; the total color difference (ΔE^*) had a high value (from 19 to 17) during the first days of formulation as a result of the oxidation and instability of chlorophyll, however, after approximately 2 weeks a $\Delta E < 5$ implied color stability. That behavior indicated the need for encapsulation or stabilization of the microalgal biomass to reduce losses in quality aspects during cooking process conditions. This approach has been analyzed for the stability of lutein from different sources through encapsulation technology such as liposomes and emulsions. A detailed review of those systems was performed by Steiner, McClements, and Davidov-Pardo (2018, pp. 71–81). In particular, the microalgal biomass encapsulation approach has been evaluated in *Tetraselmis chuii* to improve the stability of the carotenoids. According to Bonilla-Ahumada, Khandual, and Lugo-Cervantes (2018, pp. 229–238), the microencapsulation of *T. chuii* biomass can preserve the β-carotene content in 80%–92% after 3 months of storage. Similarly, the use of spray-dried microalgal biomass (*I. galbana* and *N. oculata*) has also been added to a chewing gum formulation to provide a green or golden-brown color to the product due to the chlorophyll *a* and carotenoid content in microalgal biomass. The addition of these microalgae allows the use of natural pigments without affecting textural properties; however, the odor can affect the sensory perception of the consumer; in addition, shelf-life studies must be carried out to guarantee the color stability during storage (Palabiyik et al., 2018, pp. 1031–1039). Likewise, the addition of microalgae in wheat bread modified the color of the product without altering other physicochemical properties such as texture; García-Segovia, Pagán-Moreno, Lara, and Martínez-Monzó (2017, pp. 437–447) added four microalgal biomasses to bread, resulting in yellow-green intensities in both the crust and the crumb, this effect was more noticeable with the use of *S. almeriensis* in comparison to the control (no addition).

G. sulphuraria is an unconventional microalga producer of phycobiliproteins, and is thus considered as a promising candidate for food applications. The culture system and conditions can regulate the composition of the pigment type; for example, it has been shown that in the autotrophic culture the main pigment produced is allophycocyanin (799 g kg^{-1}), while in the heterotrophic culture system the primary pigment is phycoerythrin (Graziani et al., 2013, pp. 144–152). The increase in phycocyanin concentration has been presented by Wan et al. (2016, pp. 272–278), achieving up to 13.8% phycocyanin in dry weight of the biomass through a sequence culture strategy that included a heterotrophic culture of high cell density, a cell dilution, and subsequently a culture in high light radiation as a photo-inductor of the pigment.

The use of this technique allowed the production of a high concentration of phycobiliproteins for industrial scaling up.

The pigment content of unconventional species has drawn attention for their utilization as biomass when using the powder as a dye, or the use of an extract that will allow obtaining the benefits of the pigment of interest without the risk of any toxin present in the biomass.

2.5.5 Others

Microalgae are capable of synthesizing a great diversity of compounds such as antioxidants, vitamins, and minerals (Barsanti & Gualtieri, 2018, pp. 107–115; Sathasivam & Ki, 2018, p. 26). For example, vitamins can be also found in microalgal cells. Vitamin E comprises a group of lipid-soluble compounds from which α-tocopherol is the most abundant and has the highest antioxidant activity. *Euglena gracilis*, *N. oculata*, and *Tetraselmis suecica* are species that have reported the presence of this vitamin, being higher in *E. gracilis* (Carballo-Cárdenas, Tuan, Janssen, & Wijffels, 2003, pp. 139–147; Durmaz, 2007, pp. 717–722). Vitamin K1, a potential nutrient for the prevention of chronic diseases, can be produced by *Anabaena cylindrica* where the consumption of 1 g of biomass provides two to three times the daily adult intake (Tarento et al., 2018, pp. 77–87).

Microalgae also may serve as a continuous and reliable source of natural products, including antioxidants. In this area, *Trentepohlia* species (*T. abietina*, *T. arborum*, *T. diffracta*, and *T. umbrina*) have been widely studied, demonstrating a high content of phenolic and flavonoid compounds, with the latter mostly being responsible for antioxidant activity (Kharkongor & Ramanujam, 2017, p. 00270). *Tetraselmis* species have also shown antioxidant activity. Amna, Yoo, and Park (2018) demonstrated that extracellular polysaccharides produced by *Tetraselmis tertiolecta* showed antioxidant activity depending on their molecular weight. Likewise, antioxidant activity has been reported for the species of *Tetraselmis tetrathele* and *T. chuii* (Correia, Barraca, Marques, & Leandro, 2016; Farahin, Yusoff, Nagao, Basri, & Shariff, 2016, p. 631). Lyophilized *T. chuii* biomass is already marketed for animal feed under the TetraSOD brand and was studied by Mantecón, Moyano, Cameán, and Jos (2019, p. 110810). The authors demonstrated through a study with rats, that it is nontoxic and does not cause adverse effects on the organism, which potentiates the use of *T. chuii* biomass for use as an additive in human nutrition. Indeed, the dried biomass of *T. chuii* has been recently authorized as a novel food in accordance with Regulation (EC) No 258/97 in the EU.

Biomass from microalgae also can be a source of a large variety of minerals and micronutrients. For example, *T. pseudonana* showed I, Fe, Cu, Mn, Mg, Ca, and Co content (Van Nguyen, 2018, p. 138).

Some microalgae have shown a therapeutic effect in favor of the immune system. In this sense, the mutants of the cyanobacteria *Synechococcus* have shown effects on the immune enzymatic activity of shrimp, helping them against common infections in these types of animals. The evaluation of such species as immune system boosters in animals and humans is worthy of study.

2.6 Final remarks

Microalgae are a unique resource that can produce different metabolites for the elaboration of a great variety of high-value products. Among all the microalgae species identified, only a limited number of species are known for their chemical composition and metabolites produced, and also produced at an industrial scale for applications in the food sector. However, there is a wide variety of unconventional microalgae with the potential to be applied in food products. For these unconventional species, most of the literature reported is focused on the production of biofuels, chemicals, or simply in the treatment of effluents. Also, those unconventional microalgae are being cultivated for the production of added-value compounds for the nonfood sector. Thus it is necessary to extrapolate the benefits of other species of microalgae by assessing their toxicity and digestibility in order to use them as food additives.

References

Abd El-baky., El Baz., & El-Baroty, G. S. (2004). Production of antioxidant by the green alga *Dunaliella salina. International Journal of Agriculture and Biology*, *6*, 49–57.

Abd El-Hack., Abdelnour, S., Alagawany, M., Abdo, M., Sakr, M. A., ... Gebriel, M. G. (2019). Microalgae in modern cancer therapy: Current knowledge. *Biomedicine & Pharmacotherapy*, *111*, 42–50.

Ajala, S. O., & Alexander, M. L. (2020). Evaluating the effects of agitation by shaking, stirring and air sparging on growth and accumulation of biochemical compounds in microalgae cells. *Biofuels*, 1–11.

Alam, M. A., Yuan, T., Xiong, W., Zhang, B., Lv, Y., & Xu, J. (2019). Process optimization for the production of high-concentration ethanol with *Scenedesmus raciborskii* biomass. *Bioresource Technology*, *294*, 122–219.

Ambati, R. R., Gogisetty, D., Gokare, R. A., Ravi, S., Bikkina, P. N., Su, Y., & Lei, B. (2018). *Botryococcus* as an alternative source of carotenoids and its possible applications – An overview. *Critical Reviews in Biotechnology*, *38*(4), 541–558.

Amna, S., Yoo, A. Y., & Park, J. K. (2018). Antioxidant activity of extracellular polysaccharides (EPS) from marine microalga *Tetraselmis tertiolecta* depending upon their molecular weight. *Korean Journal of Chitosan*, *23*(3), 176–181.

Anthony, R., & Stuart, B. (2015). Solvent extraction and characterization of neutral lipids in *Oocystis* sp. *Frontiers in Energy Research*, *2*, 64.

Balakrishnan, J., Dhavamani, S., Sadasivam, S. G., Arumugam, M., Vellaikumar, S., Ramalingam, J., & Shanmugam, K. (2019). Omega-3-rich *Isochrysis* sp. biomass enhances brain docosahexaenoic acid levels and improves serum lipid profile and antioxidant status in Wistar rats. *Journal of the Science of Food and Agriculture*, *99*(13), 6066–6075.

Barsanti, L., & Gualtieri, P. (2018). Is exploitation of microalgae economically and energetically sustainable? *Algal Research*, *31*, 107–115.

Becker, E. W. (2007). Micro-algae as a source of protein. *Biotechnology Advances*, *25*, 207–210.

Bhosale, P., & Bernstein, P. S. (2005). Microbial xanthophylls. *Applied Microbiology and Biotechnology*, *68*(4), 445–455.

Bleakley, S., & Hayes, M. (2017). Algal proteins: extraction, application, and challenges concerning production. *Foods*, *6*(5), 33.

Bonilla-Ahumada, F. J., Khandual, S., & Lugo-Cervantes, E. C. (2018). Microencapsulation of algal biomass (*Tetraselmis chuii*) by spray-drying using different encapsulation materials for better preservation of beta-carotene and antioxidant compound. *Algal Research*, *36*, 229–238.

Campana, R., Martinelli, V., Scoglio, S., Colombo, E., Benedetti, S., & Baffone, W. (2017). Influence of *Aphanizomenon flos-aquae* and two of its extracts on growth ability and antimicrobial properties of *Lactobacillus acidophilus* DDS-1. *LWT—Food Science and Technology*, *81*, 291–298.

Capelli, B., & Cysewski, G. R. (2010). Potential health benefits of spirulina microalgae. *Nutrafoods*, *9*(2), 19–26.

Caporgno, M. P., & Mathys, A. (2018). Trends in microalgae incorporation into innovative food products with potential health benefits. *Frontiers in Nutrition*, *5*, 58.

Carballo-Cárdenas, E. C., Tuan, P. M., Janssen, M., & Wijffels, R. H. (2003). Vitamin E (α-tocopherol) production by the marine microalgae *Dunaliella tertiolecta* and *Tetraselmis suecica* in batch cultivation. *Biomolecular Engineering*, *20*(4–6), 139–147.

Chavez, L.M. (2018). Nannochloris *sp. and* Oocystis *sp. algae exposed to osmotic stress for increased lipid production for use in biofuels* (Doctoral dissertation). The University of Texas Rio Grande Valley.

Chen, J., Li, J., Dong, W., Zhang, X., Tyagi, R. D., Drogui, P., & Surampalli, R. Y. (2018). The potential of microalgae in biodiesel production. *Renewable and Sustainable Energy Reviews*, *90*, 336–346.

Chen, L., Zhang, L., & Liu, T. (2016). Concurrent production of carotenoids and lipid by a filamentous microalga *Trentepohlia arborum*. *Bioresource Technology*, *214*, 567–573.

Chu, W. L. (2012). Biotechnological applications of microalgae. *International e-Journal of Science, Medicine & Education*, *6*(1), S24–S37.

Correia, C. R., Barraca, C., Marques, D. F., & Leandro, S. M. (2016). Water recycling influence in growth and antioxidant activity of Rhodomonas lens and *Tetraselmis chuii*. *Frontiers in Marine Science*, *3*.

Correia, N. S. G. (2018). *Characterization and analysis of the biotechnological potential of a* Chlorococcum *strain isolated at the microalgae production unit*—ALGAFARM (Master's thesis).

Costa, J. A. V., & de Morais, M. G. (2013). *16 Microalgae for food production* (p. 405) Boca Raton, FL: CRC Press, Taylor & Francis Group.

de Carvalho, J. C., Magalhaes, A. I., Jr, de Melo Pereira, G. V., Pedrioni Medeiros, A. B., Bittencourt Sydney, E., Rodrigues, C., … Soccol, C. R. (2020). Microalgal biomass pretreatment for integrated processing into biofuels, food, and feed. *Bioresource Technology*, *300*, 122719.

Durmaz, Y. (2007). Vitamin E (α-tocopherol) production by the marine microalgae *Nannochloropsis oculata* (Eustigmatophyceae) in nitrogen limitation. *Aquaculture (Amsterdam, Netherlands)*, *272*(1–4), 717–722.

Durmaz, Y., Kilici, M., Toker, O. S., Konar, N., Palabiyik, I., & Tamtürk, F. (2020). Using spray-dried microalgae in ice cream formulation as a natural colorant: Effect on physico-chemical and functional properties. *Algal Research*, *47*, 101811.

Elshobary, M. E., Osman, M. E., Abo-Shady, A. M., Komatsu, E., Perreault, H., Sorensen, J., & Piercey-Normore, M. D. (2016). Algal carbohydrates affect polyketide synthesis of the lichen-forming fungus *Cladonia rangiferina*. *Mycologia*, *108*(4), 646–656.

Enzing, C., Ploeg, M., Barbosa, M., & Sijtsma, L. (2014). Microalgae-based products for the food and feed sector: An outlook for Europe. *JRC Scientific and Policy Reports*, 19–37.

Fabra, M. J., Martínez-Sanz, M., Gómez-Mascaraque, L. G., Gavara, R., & López-Rubio, A. (2018). Structural and physicochemical characterization of thermoplastic corn starch films containing microalgae. *Carbohydrate Polymers*, *186*, 184–191.

Farahin, A. W., Yusoff, F. M., Nagao, N., Basri, M., & Shariff, M. (2016). Phenolic content and antioxidant activity of *Tetraselmis tetrathele* (West) Butcher 1959 cultured in annular photo-bioreactor. *Journal of Environmental Biology*, *37*(4), 631.

Fradique, M., Batista, A. P., Nunes, M. C., Gouveia, L., Bandarra, N. M., & Raymundo, A. (2013). *Isochrysis galbana* and *Diacronema vlkianum* biomass incorporation in pasta products as PUFA's source. *LWT-Food Science and Technology*, *50*(1), 312–319.

Ganesan, J., Jailani, M. H., Mani, V., Paramasivam, M., & Kareem, K. A. A. (2019). Characterization of growth of marine cyanobacterium *Phormidium fragile* JN 112 under different regimes of pH, light intensity and media composition. *Journal of Pharmacognosy and Phytochemistry*, *8*(3), 4115–4118.

Gao, Y., Feng, J., Lv, J., Liu, Q., Nan, F., Liu, X., & Xie, S. (2019). Physiological changes of *Parachlorella kessleri* TY02 in lipid accumulation under nitrogen stress. *International Journal of Environmental Research and Public Health*, *16*, 1188.

García, J. L., de Vicente, M., & Galán, B. (2017). Microalgae, old sustainable food and fashion nutraceuticals. *Microbial Biotechnology*, *10*(5), 1017–1024.

García-Segovia, P., Pagán-Moreno, M. J., Lara, I. F., & Martínez-Monzó, J. (2017). Effect of microalgae incorporation on physicochemical and textural properties in wheat bread formulation. *Food Science and Technology International*, *23*(5), 437–447.

Gouveia, J. D., Ruiz, J., van der Broek, L. A. M., Hesselink, T., Peters, S., Kleinegris, D. M. M., ... Wijffels, R. H. (2017). *Botryococcus braunii* strains compared for biomass productivity, hydrocarbon and carbohydrate content. *Journal of Biotechnology*, *248*, 77–86.

Gouveia, L., Coutinho, C., Mendonça, E., Batista, A. P., Sousa, I., Bandarra, N. M., & Raymundo, A. (2008). Functional biscuits with PUFA-ω3 from *Isochrysis galbana*. *Journal of the Science of Food and Agriculture*, *88*(5), 891–896.

Graziani, G., Schiavo, S., Nicolai, M. A., Buono, S., Fogliano, V., Pinto, G., & Pollio, A. (2013). Microalgae as human food: Chemical and nutritional characteristics of the thermo-acidophilic microalga *Galdieria sulphuraria*. *Food and Function*, *4*, 144–152.

Grossmann, L., Ebert, S., Hinrichs, J., & Weiss, J. (2018). Production of protein-rich extract from disrupted microalgae cells: Impact of solvent treatment and lyophilization. *Algal Research*, *36*, 67–76.

Guihéneuf, F., & Stengel, D. B. (2017). Interactive effects of light and temperature on pigments and n-3 LC-PUFA-enriched oil accumulation in batch-cultivated *Pavlova lutheri* using high-bicarbonate supply. *Algal Research*, *23*, 113–125.

Guo, B., Oliviero, T., Fogliano, V., Ma, Y., Chen, F., & Capuano, E. (2020). Gastrointestinal bioaccessibility and colonic fermentation of fucoxanthin from extract of the microalga *Nitzschia laevis*. *Journal of Agricultural and Food Chemistry*, *68*, 1844–1850.

Han, F., Pei, H., Hu, W., Jiang, L., Cheng, J., & Zhang, L. (2016). Beneficial changes in biomass and lipid of microalgae *Anabaena variabilis* facing the ultrasonic stress environment. *Bioresource Technology*, *209*, 16–22.

Heo, J., Cho, D. H., Kim, U., & Kim, H. S. (2018). Isolation and characterization of indigenous diatom, *Odontella* sp. BS-003 as potential fucoxanthin and omega-3 fatty acid producer. *Journal of Marine Bioscience and Biotechnology*, *10*(1), 26–33.

Hu, H., Li, J. Y., Pan, X. R., Zhang, F., Ma, L. L., Wang, H. J., & Zeng, R. J. (2019). Different DHA or EPA production responses to nutrient stress in the marine microalga *Tisochrysis lutea* and the freshwater microalga *Monodus subterraneus*. *Science of the Total Environment*, *656*, 140–149.

Jacob-Lopes, E., Maroneze, M. M., Deprá, M. C., Sartori, R. B., Dias, R. R., & Zepka, L. Q. (2019). Bioactive food compounds from microalgae: An innovative framework on industrial biorefineries. *Current Opinion in Food Science*, *25*, 1–7.

Jiang, X., Han, Q., Gao, X., & Gao, G. (2016). Conditions optimising on the yield of biomass, total lipid, and valuable fatty acids in two strains of *Skeletonema menzelii*. *Food Chemistry*, *194*, 723–732.

Katiyar, R., & Arora, A. (2020). Health promoting functional lipids from microalgae pool: A review. *Algal Research*, *46*, 101800.

Kavitha, G., Kurinjimalar, C., Sivakumar, K., Palani, P., & Rengasamy, R. (2016). Biosynthesis, purification and characterization of polyhydroxybutyrate from *Botryococcus braunii* kütz. *International Journal of Biological Macromolecules*, *89*, 700–706.

Khanra, S., Mondal, M., Halder, G., Tiwari, O. N., Gayen, K., & Bhowmick, T. K. (2018). Downstream processing of microalgae for pigments, protein, and carbohydrate in industrial application: A review. *Food and Bioproducts Processing*, *110*, 60–84.

Khan, M. I., Shin, J. H., & Kim, J. D. (2018). The promising future of microalgae: Current status, challenges, and optimization of a sustainable and renewable industry for biofuels, feed, and other products. *Microbial Cell Factories*, *17*(1), 1–21.

Kharkongor, D., & Ramanujam, P. (2017). Antioxidant activities of four dominant species of *Trentepohlia* (Trentepohliales, Chlorophyta). *International Journal of Complementary & Alternative Medicine*, *8*, 00270.

Kłodawska, K., Bujas, A., Turos-Cabal, M., Żbik, P., Fu, P., & Malec, P. (2019). Effect of growth temperature on biosynthesis and accumulation of carotenoids in cyanobacterium *Anabaena* sp. PCC 7120 under diazotrophic conditions. *Microbiological Research*, *226*, 34–40.

Koyande, A. K., Chew, K. W., Rambabu, K., Tao, Y., Chu, D. T., & Show, P. L. (2019). Microalgae: A potential alternative to health supplementation for humans. *Food Science and Human Wellness*, *8*.

Lafarga, T. (2019). Effect of microalgal biomass incorporation into foods: Nutritional and sensorial attributes of the end products. *Algal Research*, *41*(101566), 1–10.

Lafarga, T., Acién-Fernández, F. G., Castellari, M., Villaró, S., Bobo, G., & Aguiló-Aguayo, I. (2019). Effect of microalgae incorporation on the physicochemical, nutritional, and sensorial properties of an innovative broccoli soup. *LWT*, *111*, 167–174.

Li, Y., Liu, L., Sun, P., Zhang, Y., Wu, T., Sun, H., … Chen, F. (2020). Fucoxanthinol from diatom *Nitzschia leavis* ameliorates neuroinflammatory responses in lipopolysaccharides-stimulated BV-2 microglia. *Marine Drugs*, *18*, 116.

Ljubic, A., & Jacobsen, C. (2018). Interactive effects of salinity and light on growth of EPA and DHA rich-microalgae *Pavlova lutheri*. In: *AlgaEurope 2017*.

Loveday, S. M. (2019). Food proteins: Technological, nutritional, and sustainability attributes of traditional and emerging proteins. *Annual Review of Food Science and Technology*, *10*, 311–339.

Low, K. L., Idris, A., & Yusof, N. M. (2020). Novel protocol optimized for microalgae lutein used as food additives. *Food Chemistry*, *307*, 125631.

Lozano-Garcia, D. F., Cuellar-Bermudez, S. P., del Rio-Hinojosa, E., Betancourt, F., Aleman-Nava, G. S., & Parra-Saldivar, R. (2019). Potential land microalgae cultivation in Mexico: From food production to biofuels. *Algal Research*, *39*, 101459.

Mantecón, L., Moyano, R., Cameán, A. M., & Jos, A. (2019). Safety assessment of a lyophilized biomass of *Tetraselmis chuii* (TetraSOD®) in a 90 day feeding study. *Food and Chemical Toxicology*, *133*, 110810.

Martínez-García, M., Kormpa, A., & van der Maarel, M. J. E. C. (2017). The glycogen of *Galdieria sulphuraria* as alternative to starch for the production of slowly digestible and resistant glucose polymers. *Carbohydrate Polymer*, *169*, 75–82.

Martinez-Garcia, M., & van der Maarel, M. J. E. C. (2016). Floridoside production by the red microalga *Gladieria sulphuraria* under different conditions of growth and osmotic stress. *AMB Express*, *6*, 71.

Massa, M., Buono, S., Langellotti, A. L., Martello, A., Russo, G. L., Troise, D. A., ... Fogliano, V. (2019). Biochemical composition and *in vitro* digestibility of *Galdieria sulphuraria* grown on spent cherry-brine liquid. *New Biotechnology*, *53*, 9–15.

Matos, Â. P. (2017). The impact of microalgae in food science and technology. *Journal of the American Oil Chemists' Society*, *94*(11), 1333–1350.

Matsukawa, R., Hotta, M., Masuda, Y., Chihara, M., & Karube, I. (2000). Antioxidants from carbon dioxide fixing *Chlorella sorokiniana. Journal of Applied Phycology*, *12*(3), 263–267.

Milledge, J. J. (2012). Microalgae-commercial potential for fuel, food and feed. *Food Science & Technology*, *26*(1), 28–30.

Mok, I. K., Lee, J. K., Kim, J. H., Pan, C. H., & Kim, S. M. (2018). Fucoxanthin bioavailability from fucoxanthin-fortified milk: *In vivo* and *in vitro* study. *Food Chemistry* (258), 79–86.

Molinuevo-Salces, B., Riaño, B., Hernández, D., & Cruz García-González, M. (2019). Microalgae and wastewater treatment: Advantages and disadvantages. In M. Alam, & Z. Wang (Eds.), *Microalgae Biotechnology for Development of Biofuel and Wastewater Treatment* (pp. 505–533). Singapore: Springer.

Montero, E., Olguín, E. J., De Philippis, R., & Reverchon, F. (2018). Mixotrophic cultivation of *Chlorococcum* sp. under non-controlled conditions using a digestate from pig manure within a biorefinery. *Journal of Applied Phycology*, *30*(5), 2847–2857.

Niccolai, A., Zittelli, G. C., Rodolfi, L., Biondi, N., & Tredici, M. R. (2019). Microalgae of interest as food source: Biochemical composition and digestibility. *Algal Research*, *42*, 101617.

Palabiyik, I., Durmaz, Y., Öner, B., Toker, O. S., Coksari, G., Konar, N., & Tamtürk, F. (2018). Using spray-dried microalgae as a natural coloring agent in chewing gum: Effects on color, sensory, and textural properties. *Journal of Applied Phycology*, *30*, 1031–1039.

Parul, S. A. K., & Thiyam, G. (2019). Variation of both chemical composition and antioxidant properties of newly isolated *Parachlorella kessleri* GB1, by growing in different culture conditions. *LWT-Food Science and Technology*, *112*, 108205.

Pan-Utai, W., & Iamtham, S. (2019). Extraction, purification and antioxidant activity of phycobiliprotein from *Arthrospira platensis*. *Process Biochemistry*, *82*, 189–198.

Piasecka, A., Krzeminska, I., & Tys, J. (2017). Enrichment of *Parachlorella kessleri* biomass with bioproducts: Oil and protein by utilization of beet molasses. *Journal of Applied Phycology*, *29*, 1735–1743.

Prasanna, R., Kumar, R., Sood, A., Prasanna, B. M., & Singh, P. K. (2006). Morphological, physiochemical and molecular characterization of Anabaena strains. *Microbiological Research*, *161*(3), 187–202.

Ryckebosch, E., Muylaert, K., Eeckhout, M., Ruyssen, T., & Foubert, I. (2011). Influence of drying and storage on lipid and carotenoid stability of the microalga *Phaeodactylum tricornutum*. *Journal of Agricultural and Food Chemistry*, *59*(20), 11063–11069.

Sathasivam, R., & Ki, J. S. (2018). A review of the biological activities of microalgal carotenoids and their potential use in healthcare and cosmetic industries. *Marine Drugs*, *16*(1), 26.

Sathasivam, R., Radhakrishnan, R., Hashem, A., & Abd_Allah, E. F. (2019). Microalgae metabolites: A rich source for food and medicine. *Saudi Journal of Biological Sciences*, *26*(4), 709–722.

Sousa, I., Gouveia, L., Batista, A. P., Raymundo, A., & Bandarra, N. M. (2008). Microalgae in novel food products. *Food Chemistry Research Developments*, 75–112.

Steiner, B. M., McClements, D. J., & Davidov-Pardo, G. (2018). Encapsulation systems for lutein: A review. *Trends in Food Science and Technology*, *82*, 71–81.

Sun, Z., Wang, X., & Liu, J. (2019). Screening of *Isochrysis* strains for simultaneous production of docosahexaenoic acid and fucoxanthin. *Algal Research*, *41*, 101545.

Tarento, T. D., McClure, D. D., Vasiljevski, E., Schindeler, A., Dehghani, F., & Kavanagh, J. M. (2018). Microalgae as a source of vitamin K1. *Algal Research*, *36*, 77–87.

Telussa, I., Rusnadi., & Naruchman, Z. (2019). Dynamics of b-carotene and fucoxanthin of tropical marine *Navicula* sp. as a response to light stress conditions. *Algal Research*, *41*, 101530.

Teuling, E., Wierenga, P. A., Schrama, J. W., & Gruppen, H. (2017). Comparison of protein extracts from various unicellular green sources. *Journal of Agriculture and Food Chemistry*, *65*, 7989–8002.

Ueno, M., Cho, K., Hirata, N., Yamashita, K., Yamaguchi, K., Kim, D., & Oda, T. (2017). Macrophage-stimulating activities of newly isolated complex polysaccharides from *Parachlorella kessleri* strain KNK-A001. *International Journal of Biological Macromolecules*, *104*, 400–406.

Upadhyay, S., Lawrence, K., Shekhar, S. K., & Anantharaman, P. (2017). Lipid screening, extraction and production of biodiesel from marine diatom *Odontella aurita*. *International Journal of Current Microbiology and Applied Sciences*, *6*(3), 1259–1269.

Van der Spiegel, M., Noordam, M. Y., & van der Fels-Klerx, H. J. (2013). Safety of novel protein sources (insects, microalgae, seaweed, duckweed, and rapeseed) and legislative aspects for their application in food and feed production. *Comprehensive Reviews in Food Science and Food Safety*, *12*, 662–678.

Van Nguyen, C. (2018). Fatty acid profile and nutrition values of microalga (*Thalassiosira pseudonana*) commonly used in white shrimp (*Litopenaeus vannamei*) culturing. *Vietnam Journal of Science and Technology*, *56*(4A), 138.

Vieira, M. V., Oliveira, S. M., Amado, I. R., Fasolin, L. H., Vicente, A. A., Pastrana, L. M., & Fuciños, P. (2020). 3D printed functional cookies fortified with *Arthrospira platensis*: Evaluation of its antioxidant potential and physical-chemical characterization. *Food Hydrocolloids*, *107*, 105893.

Vieria Costa, J. A., Bastos Freitas, B. C., Moraes, L., Zaparoli, M., & Greques Morais, M. (2020). Progress in the physicochemical treatment of microalgae biomass for value-added product recovery. *Bioresource Technology*, *301*, 122727.

Vigani, M., Parisi, C., Rodríguez-Cerezo, E., Barbosa, M. J., Sijtsma, L., Ploeg, M., & Enzing, C. (2015). Food and feed products from micro-algae: Market opportunities and challenges for the EU. *Trends in Food Science & Technology*, *42*(1), 81–92.

Wan, M., Wang, Z., Zhang, Z., Wang, J., Li, S., Yu, A., & Li, Y. (2016). A novel paradigm for the high-efficient production of phycocyanin from *Galderia sulphuraria*. *Bioresource Technology*, *218*, 272–278.

Xia, S., Gao, B., Fu, J., Xiong, J., & Zhang, C. (2018). Production of fucoxanthin, chrysolaminarin, and eicosapentaenoic acid by *Odontella aurita* under different nitrogen supply regimes. *Journal of Bioscience and Bioengineering*, *126*(6), 723–729.

Xia, S., Gao, B., Li, A., Xiong, J., Ao, Z., & Zhang, C. (2014). Preliminary characterization, antioxidant properties and production of chrysolaminarin from marine diatom *Odontella aurita*. *Marine Drugs*, *12*(9), 4883–4897.

Yadavalli, R., Ratnapuram, H., Motamarry, S., Reddy, C. N., Ashokkumar, V., & Kuppam, C. (2020). Simultaneous production of flavonoids and lipids from *Chlorella vulgaris* and *Chlorella pyrenoidosa*. *Biomass Conversion and Biorefinery*, 1–9.

You, Z., Zhang, Q., Peng, Z., & Miao, X. (2019). Lipid droplets mediate salt stress tolerance in *Parachlorella kesskeri*. *Plant Physiology*, *181*(2), 510–526.

Zhang, J., Liu, L., Ren, Y., & Chen, F. (2019). Characterization of exopolysaccharides produced by microalgae with antitumor activity on human colon cancer cells. *International Journal of Biological Macromolecules*, *128*, 761–767.

Zhang, Y., Wu, H., Sun, M., Peng, Q., & Li, A. (2018). Photosynthetic physiological performance and proteomic profiling of the oleaginous algae *Scenedesmus acuminatus* reveal the mechanism of lipid accumulation under low and high nitrogen supplies. *Photosynthesis Research*, *138*, 73–102.

Chapter 3

Conventional and novel approaches to extract food ingredients and nutraceuticals from microalgae

Bhagya Yatipanthalawa and Gregory Martin
Algal Processing Group, Department of Chemical and Biomolecular Engineering, The University of Melbourne, Parkville, VIC, Australia

3.1 Introduction

With the increasing global population, there is a growing demand for novel food products that are sustainable. Microalgal biomass is a potential feedstock for the production of food/feed and nutraceuticals due to the presence of lipids, proteins, carbohydrates, and high-value metabolites such as vitamins, carotenoids, and polyunsaturated fatty acids (PUFAs). Fast growth rates, flexibility in location, and high CO_2 uptake efficiency are some of the main advantages of using microalgae as a feedstock for the aforementioned applications (Sirajunnisa & Surendhiran, 2016). The accumulation of these high value components will depend on the microalgae species and the growth conditions used (Olmstead et al., 2013; Postma et al., 2016). Microalgae as a food and nutraceutical product have been gaining interest owing to their nutritional and functional properties.

To date, microalgae have mostly been considered as a whole-food ingredient. An alternative is the extraction of food ingredients and nutraceutical compounds from microalgae cells to enable better preservation and more targeted utilization of their nutritional and functional properties. There are different components that can be targeted for extraction, and many different unit operations that can be applied. The process of extraction must be cost-effective, able to recover high yields of the targeted components, and avoid degradation or contamination of the final products. This chapter introduces readers to the many conventional and novel processing methods that can be applied in extracting food ingredients and nutraceuticals from microalgae to help in the selection and development of appropriate processes for different end products.

Cultured Microalgae for the Food Industry. DOI: https://doi.org/10.1016/B978-0-12-821080-2.00009-5

3.2 Algae harvesting and concentration

Prior to extraction of food ingredients, the microalgae biomass needs to be harvested. It is also usually desirable for the algae to be concentrated, and often for the cells to be ruptured, and even heat-treated for food applications. This section summarizes the options for pre-processing of microalgae in relation to microalgae food ingredients.

Due to shelf-shading from the available light, the concentration of biomass that can be reached in photoautotrophic cultivation is very low—in the order of $1\ g\ L^{-1}$ (0.1% wt./wt. solids). To reduce the size of processing equipment, the culture must be concentrated [e.g., to ca. 5% wt./wt. solids by flocculation (Şirin et al., 2013)] and dewatered [e.g., to ca. 25% wt./wt. solids by centrifugation (Vandamme et al., 2013)]. However, further water removal requires thermal drying, which is highly energy demanding (Cooney, Young, & Pate, 2011; Lardon, Hélias, Sialve, Steyer, & Bernard, 2009) and can damage sensitive biological components.

Different methods are available for concentrating algae biomass, including flocculation, filtration, and centrifugation. Flocculation and filtration are typically best used as a preconcentration step, with centrifugation able to be subsequently applied with reduced throughput and energy demand to produce a concentrated slurry.

3.2.1 Preconcentration

Microalgal cells are small and do not settle well without being aggregated. Therefore various pre-concentration techniques such as chemical flocculation, bioflocculation, electro-coagulation, gravity sedimentation, and floatation are used in order to pre-concentrate microalgae cells which can then be easily concentrated by filtration or centrifugation.

Chemical flocculants include metal salts such as alum and ferric chloride, synthetic polymers, and biopolymers such as chitosan. Although metal salts are widely used in industries such as wastewater treatment, their use is not recommended in microalgae flocculation due to possible issues associated with contaminating the biomass with these chemicals (Vandamme et al., 2013). However, in some cases, the addition of chemical flocculants may be beneficial for food applications, for instance, the use of $FeCl_3$ as a flocculant was shown to improve the lipolytic stability of *Tisochrysis* during storage of wet cell slurries (Balduyck et al., 2019). Biopolymers such as chitosan are suitable flocculants for harvesting microalgae for food applications, as they are nontoxic. However, these are relatively expensive, and their use for harvesting microalgae is still emerging. Bacteria-derived bioflocculants such as *Actinobacterium* bioflocculant from *Streptomyces* sp., have been used to effectively recover *Nannochloropsis* sp. cells (Sivasankar et al., 2020). These are harmless to microalgae, allowing them to be recycled and they can be produced at a low cost (Sivasankar et al., 2020). Self-flocculation is another potential low-cost method for harvesting microalgae. Only a few species of microalgae have been seen to have the ability to self-flocculate, which has led to more efforts being directed toward coculturing of microalgae with other microorganisms that induce flocculation (Zhao et al., 2019). However, this novel area of bioflocculation needs to be further studied in order to identify self-flocculating microalgae with relevance to food ingredients, and the effect of exogenous organisms on food safety and nutritional and functional properties when incorporating into food and pharmaceutical products. Further, the presence of flocculants can affect the downstream extraction processes due to the alteration of rheological properties in flocculated and concentrated biomass (Yap et al., 2016).

3.2.2 Dewatering

Dewatering techniques such as centrifugation and filtration are often used to concentrate microalgae into pastes. However, filtration techniques are often not preferred for dewatering as the deposition of microalgae leads to fouling and reduced flux, which requires chemical cleaning to recover, leading to high operational costs (Barros et al., 2015) and potential chemical contamination of the product.

Although centrifugation incurs high capital costs, it is highly effective at concentrating microalgae and can be used in a continuous operation without requiring the addition of any chemicals. It is most energy efficient to apply centrifugation to pre-concentrated streams, therefore centrifugation is often used as the final step in algae harvesting to obtain a concentrated paste.

3.2.3 Drying

So far, the commercial production of microalgae for food applications has mainly focused on producing powdered microalgae to be consumed as whole-cell proteins. Dry powders of whole microalgae are relatively stable and easy to transport (Balduyck et al., 2016; De Farias Neves, Demarco, & Tribuzi, 2019). Drying of algae can also be applied prior to extraction of food ingredients and nutraceuticals, for instance to facilitate solvent extraction of microalgal lipids.

Producing a dry powder requires thermal drying to remove intracellular water remaining at concentrations beyond the close-packing limit of the cells, which is dependent on the species of microalgae, and typically about 15%–40% (wt./wt.). Solar drying, spray drying, drum drying, and lyophilization can all be used for drying microalgae. Solar drying and drum drying are lower cost and more energy efficient than the other methods, but are also harsher on sensitive food and nutraceutical components. Lyophilization, or freeze-drying, is the gentlest method, but also the most costly. As a reasonably gentle yet efficient method, spray drying is the most widely used for food applications (Cooney et al., 2011; De Farias Neves et al., 2019).

As drying is highly energy intensive compared to physical dewatering, processes to extract intracellular components from wet microalgae slurries are being developed. While drying can prevent potential degradation of high-value components during wet storage, a short heat treatment can extend the wet storage of microalgae by inactivating lipolytic enzymes (Balduyck et al., 2019). Further, it has been observed that spray drying microalgae can also degrade the cell structure, resulting in the subsequent degradation of valuable components when stored under unfavorable conditions (De Farias Neves et al., 2019).

3.3 Cell rupture

3.3.1 Need for cell rupture

Rather than just producing whole microalgae cells, it is possible to add value to microalgae biomass by extracting high-value products for food and nutraceutical applications. Prior cell rupture is usually required for the extraction of high-value components from microalgae cells due to the presence of rigid cell walls that hinder the accessibility of these components. Cell rupture can also improve the bioavailability of microalgal cell components, even if they are not extracted (Bernaerts et al., 2020). For instance, it has been shown that the proteins and lipids inside intact *Nannochloropsis* sp. cells could not be accessed by mammalian digestive enzymes, whereas cell rupture increased their accessibility (Cavonius et al., 2016).

The presence of lipids, proteins, polysaccharides, and cell debris from ruptured microalgae can also affect the overall functional properties of the slurry (Law et al., 2018). Many cellular components are surface active and can enhance emulsion stability (Law et al., 2018), suggesting that rupturing microalgae can improve the functional properties of microalgae for some food applications in addition to their nutritional value.

However, wet storage of ruptured biomass can increase the free fatty acid (FFA) content compared with that of intact cells due to the release of lipolytic enzymes, which is disadvantageous for food applications (Balduyck et al., 2017). Thus it may be preferable to deactivate the lipolytic enzymes, for instance, using a mild heat treatment (Balduyck et al., 2019). Further, lipid stability has been found to be higher in food systems formulated with the inclusion of intact or disrupted *Nannochloropsis* biomass than with extracted oils (Gheysen et al., 2019).

Different cell disruption techniques that have been widely studied in the literature are summarized below. The selection of the cell disruption method needs to be made considering the feasibility for process scale-up, energy requirements, and the efficiency of the particular method for a given species.

3.3.2 High-pressure homogenization

High-pressure homogenization (HPH) has widely been used for the disruption of microalgae cells where it has been found to be efficient for a wide range of species (Spiden et al., 2013; Yap et al., 2014; Martin 2016). The rupture efficiency of a microalgae slurry during HPH depends solely on the homogenization pressure and the species used, thus, high solids concentrations can be used to minimize energy consumption (Yap et al., 2015). The robustness of microalgae cells in relation to mechanical rupture by HPH varies considerably across different species (Spiden et al., 2013). In particular, some industrially relevant microalgae species such as *Nannochloropsis* and *Chlorella* have strong cell walls which require the use of several passes at 100 MPa pressure, whereas algae with larger, weaker cells like *Tetraselmis* are susceptible to cell rupture at much lower pressures (Spiden et al., 2013). It has been shown that the strong cell walls of *Nannochloropsis* sp. can be weakened by incubating the cells, thereby reducing the required HPH pressure and energy demands (Olmstead et al., 2013). Ultra-high-pressure homogenization (UHPH) at 250 MPa can effectively rupture *Nannochloropsis* cells, reducing the number of passes required for cell disruption (Bernaerts et al., 2019). However, aggregate formation of released intracellular material due to extensive heating has been observed at ultra-high pressures, representing possible degradation of the heat-sensitive components (Bernaerts et al., 2019). Nevertheless, UHPH can be used for spore inactivation or sterilization of the system in food production (Sevenich & Mathys, 2018).

HPH is considered to be the most industrially scalable and energy-efficient method to rupture microalgae with strong cell walls (Yap et al., 2015). However, it can result in the formation of FFAs, due to the release of compartmentalized lipases resulting from the loss of cell integrity (Balduyck et al., 2018). The potential impact of this on downstream extraction processes and the quality of microalgae lipid products needs to be considered for food applications.

3.3.3 Enzymes and autolysis

Biological processes for the degradation of cell walls of microalgae are attractive due to the high energy costs associated with mechanical cell disruption methods. Autolysis and enzymatic hydrolysis are two established methods for breaking down microbial biomass (Demuez et al., 2015; Middelberg, 1995) and producing food extracts, for example from yeast and microalgae (Mitsuda and Shikanai, 1959). Microalgae cell walls mainly consist of polysaccharides and glycoproteins (Scholz et al., 2014), which can be broken down by selective enzymes under mild reaction conditions.

As enzymes can be expensive and their recovery and reuse is difficult (Sierra et al., 2017), autolysis is an alternative approach that utilizes the cells' own hydrolytic enzymes to degrade the cell (Demuez et al., 2015; Kightlinger et al., 2014; Middelberg, 1995). Autolysis can be induced at low cost, simply by incubating the cells at elevated temperatures (e.g., 30°C–50°C) (Halim et al., 2019; Halim et al., 2016; Kightlinger et al., 2014; Middelberg, 1995). This method has minimal energy and equipment requirements and could therefore be beneficially incorporated into industrial-scale microalgal biorefineries. The use of in situ enzymes generated under stress conditions that degrade cell wall material can also be used to weaken the cells prior to mechanical cell rupture (Halim et al., 2016) or for cell permeabilization (Sierra et al., 2017). However, the effect of autolysis on proteins, lipids, and polysaccharides of interest needs to be further explored, since it is likely that the autolytic enzymes will also break down these biomolecules (e.g., via proteolysis or lipolysis). Further, the response to stress conditions will vary between different microalgae species, suggesting the requirement for species-specific techniques for autolysis.

3.3.4 Ultrasound

Ultrasonication is another technology that has been studied for the rupture of microalgal cells. While ultrasonication is not energy efficient if applied to dilute slurries, it can be used for the disruption of microalgae cells at high solid concentrations (Yao et al., 2018). However, the effect of ultrasound attenuates as the viscosity of the slurry increases significantly at increasing solids concentrations, meaning it may not be sufficiently energy efficient for the rupture of microalgae with tough cell walls (Yao et al., 2018).

Apart from using ultrasound to rupture the cells, ultrasonic-assisted extraction has been used to accelerate the mass transfer of the components of interest such as lipids and pigments to the solvent phase (Li et al., 2019). Partial denaturation of proteins has been observed during ultrasonication, whereas it has been suggested that ultrasonication itself does not hydrolyze the lipids (Li et al., 2019; Yao et al., 2018). Ultrasonic processing units can be used at an industrial scale and can be operated in a continuous mode, however, currently HPH appears more feasible for large-scale rupture of microalgae.

3.3.5 Pulsed electric fields

The use of pulsed electric fields (PEFs) is a novel method for permeabilizing microalgae cells, rather than completely rupturing them, by increasing the transmembrane voltage (Frey et al., 2017). This technology has been used to permeabilize the cells of freshwater microalgae such as *Chlorella vulgaris* (Buchmann et al., 2019) and cyanobacteria (*Arthrospira*) (Buchmann et al., 2018), enabling the release of soluble proteins. Continuous extraction of proteins from viable microalgae cell cultures of freshwater *Chlorella vulgaris* using PEF has also been achieved (Buchmann et al., 2019). Selective release of proteins could be achieved from a cell wall mutant of *Chlamydomonas reinhardtii* using PEF, which permeabilized the microalgae cells instead of completely fragmenting them ('T Lam et al., 2017). However, the presence of a rigid cell wall can adversely affect the release of intercellular proteins using PEF, with comparatively high amounts of carbohydrates and small ionic solutes being released instead (Postma et al., 2016). Incubation of microalgae cells after PEF treatment has also improved subsequent protein and lipid extraction efficiencies (Silve et al., 2018). Thus PEF can be used as a permeabilization technique to selectively release water-soluble components that would make the product recovery easier. One limitation of PEF is that the energy consumption is impractically high if the conductivity is too high. This limits its application to microalgae in low-salinity media.

3.3.6 Other methods

Bead milling can also be used to rupture microalgae (Postma et al., 2015), with the efficiency of the process depending on the bead type, microalgae species, biomass concentration, bead filling ratio, and the flow rate of the suspension (Postma et al., 2017). However, the energy consumption of bead milling is relatively high (Doucha & Livansky, 2008). For instance, the reported minimum energy required to rupture *Tetraselmis suecica*, a weak microalga, was around 1.7 MJ kg^{-1} dry algae (Postma et al., 2017), considerably higher than that required to rupture the same species using HPH (Yap et al., 2015).

Dilute acid hydrolysis is a thermochemical process used to release fermentable sugars from lignocellulosic biomass via acid-catalyzed hydrolysis. This method has also been applied to microalgae biomass at temperatures between 120°C and 170°C (Dong et al., 2016). However, as these temperatures will denature the proteins and likely degrade other nutritional compounds, for instance via the Maillard reaction (Van Boekel, 2001), this method may not be suitable for the purpose of extracting food additives and nutraceuticals.

3.4 Lipid and pigment extraction

The lipid composition of microalgae is highly dependent on the growth conditions and species (Olmstead et al., 2013). Microalgal lipids include nonpolar lipids, polar phospholipids and glycolipids, and pigments. Among these, the non-polar triacylglyceride (TAG) content is of interest as a food oil, whereas polar membrane lipids, which can be rich in long-chain polyunsaturated fatty acids (LC-PUFAs) such as eicosapentaenoic acid (EPA) (Martin et al., 2014), are preferred as a high-value food additive or nutraceuticals.

There are two broad strategies for the extraction of lipids from microalgae: extraction from dried biomass or wet biomass. Extracting lipids from wet algal biomass is more technically challenging than from dried biomass, however wet lipid extraction may be preferred due to the high energy cost associated with obtaining a dried biomass (Cooney et al., 2011; Dong et al., 2016). Lipids can be extracted from dried biomass using organic solvents (Jiménez Callejón et al., 2020; Mahmood et al., 2017) or supercritical CO_2 (Halim et al., 2011). Lipid extraction from wet algae biomass can be performed using organic solvents (Jiménez Callejón et al., 2014; Olmstead et al., 2013) or with so-called switchable solvents whose polarity can be changed to facilitate recovery (Du et al., 2013; Samorì et al., 2013).

Organic solvents have been widely reported for the extraction of lipids from microalgae. The choice of solvent depends on the polarity of the lipids of interest, the mode of extraction, as well as various safety and cost factors. Both monophasic or biphasic systems can be used to extract lipids from wet microalgae (Yap et al., 2014). Monophasic systems involve organic solvents that are miscible with water and often are comprised of polar and non-polar solvent mixtures. These solvents are able to penetrate whole algae cells due to their miscibility with water and can recover both polar and nonpolar lipids (Yap et al., 2014). These systems are often used to extract the lipids from microalgae for analytical purposes and for the recovery of high-value polar lipids containing long-chain omega-3 fatty acids. However, if water-miscible solvents are used to extract lipids from wet algae biomass, solvent recovery will require energy-intensive distillation, which can potentially be avoided if water-immiscible solvents are used that can be physically recovered by centrifugation (Olmstead et al., 2013). However, cell rupture has been found to be essential for biphasic lipid extraction in most microalgae species due to the inability of the water-immiscible solvents to penetrate the cell walls (Yap et al., 2014). Also, the solvent phase forms a stable emulsion in the ruptured algae slurry, which can make centrifugal separation of the lipid-rich solvent phase problematic (Law et al., 2017).

The following section describes techniques applicable for the extraction and separation of TAGs, PUFAs, and pigment components that are of interest to the food industry.

3.4.1 Recovery of triacylglycerides

TAGs are located as droplets within microalgae cells, where they function as energy and carbon storage reservoirs (Olmstead et al., 2013). TAGs in microalgae are similar to vegetable oils that are widely used in food applications, however they have their own specific fatty acid profile dependent on the species and growth conditions. TAGs are neutral, saponifiable lipids that will be preferentially extracted over polar membrane lipids by nonpolar solvents. Biphasic lipid extraction using nonpolar solvents like hexane can extract the TAGs effectively from wet slurries of ruptured microalgal biomass (Jiménez Callejón et al., 2014; Yap et al., 2014). A key requirement for energy efficiency is to maximize the concentration of the slurry and to reduce the amount of solvent used in relation to the amount of TAG recovered (Martin, 2016). Nonpolar solvents such as hexane can also be used to directly extract TAG from dried algal biomass (Dos Santos et al., 2015).

Although hexane has long been used as a solvent for food applications, such as extraction of soybean oil, it is increasingly seen as problematic due to its toxicity, flammability, and potential to cause environmental damage (Campbell et al., 2011). The use of renewable bio-based alternative solvents such as ethyl acetate, 2-methyltetrahydrofuran, cyclophentyl methyl ether, and ethyl lactate is an emerging research area for the extraction of lipids (Mahmood et al., 2017). These solvents can potentially replace hexane, having been shown to be effective for selective extraction of neutral lipids (Mahmood et al., 2017) and requiring similar amounts of energy for evaporative solvent recovery (Breil et al., 2016). Nonetheless, these extraction processes need to be further optimized to minimize the volume of solvent used for extracting TAGs from wet microalgae, especially given the cost of these solvents in comparison to hexane. In addition, although these solvents may be claimed to be more environmentally benign than hexane, they are still toxic and flammable, presenting similar issues for food applications.

Supercritical fluid extraction is another potential method for extracting TAGs from dried microalgal biomass. This process, which is widely used for high-value food applications such as decaffeination, uses a supercritical fluid that is held at a pressure and temperature above its critical point. These supercritical fluids are effective for lipid extraction from biomass, as they have gas-like properties of high diffusivity and low viscosity, and liquid-like density and absorption capacity. Carbon dioxide has been the most widely studied fluid for supercritical fluid extraction as it is non-flammable and non-toxic, making it highly suitable for use in the food industry. Lipid extraction using SC-CO_2 can be highly effective and selective toward TAGs (Halim et al., 2011). However, the need for dried biomass and the high pressures associated with this method lead to very high operational and capital costs due to the use of pressure vessels. In addition, ionic liquids (Di Salvo et al., 2013) have also been researched as substitutes for conventional organic solvents for the extraction of microalgal lipids, however the applicability of these solvents in industrial-scale applications is not yet clear.

In addition to extracting the lipids into the solvent, efficient separation of the lipid-rich solvent from the de-lipidated biomass is also important. During the extraction from dried biomass the lipid-rich solvent permeats out of the biomass and can be easily recovered. Wet lipid extraction relies on physically recovering the lipid-rich solvent from the biomass, which can be a significant challenge due to the stability of the emulsion formed during extraction. It has been shown that the separation process can be limited by barriers to coalescence presented by the strongly emulsifying properties of the ruptured algae cells (Law et al., 2017). Following this, the solvent must be recovered from the lipid to be recycled. For biphasic lipid extraction systems using low solvent-to-biomass ratios, this process can be relatively energy efficient as only the solvent has to be evaporated (Martin, 2016). However, water-miscible solvents must be distilled out of the aqueous phase, meaning energy is required to vaporize the water, obviating the energetic benefit of avoiding drying the biomass. There is the potential for further improvements if novel technologies such as the use of switchable solvents (Du et al., 2013; Samorì et al., 2013) or membrane-based solvent–lipid separation (De Melo et al., 2015) can be successfully developed.

3.4.2 Recovery of polyunsaturated fatty acids

The ω-3 LC-PUFAs, mainly EPA and docosahexaenoic acid (DHA), in certain microalgae are one of their most valuable nutritional components. In microalgae such as *Nannochloropsis*, the ω-3 LC-PUFAs are predominantly associated with the polar membranes lipids, rather than the TAG oils (Martin et al., 2014; Simionato et al., 2013). As such, ω-3 LC-PUFAs can be extracted from microalgae using a combination of water-immiscible and water-miscible solvents, such as hexane/IPA and dichloromethane/ethanol (Ryckebosch et al., 2013), which are able to extract the polar membrane lipids. Such extraction also recovers pigments which have antioxidative properties that can protect the chemical integrity and nutritive value of the ω-3 LC-PUFAs during storage and processing in food applications (Ryckebosch et al., 2013). As an alternative to adding a blend of polar and non-polar solvents, such pairs of solvents can be added sequentially. For instance, a recent study initially used hexane to extract the neutral lipids from dried algae biomass, and then extracted the polar lipids in a second extraction using ethanol (Jiménez Callejón et al., 2020). This method requires two extraction and solvent separation steps but has the benefit of producing two relatively selective lipid streams. This could facilitate subsequent purification of the PUFAs using methods such as crystallization and chromatography (Molina Grima et al., 1999).

As described in the previous section, the use of water-immiscible solvents that form biphasic systems is preferred over water-miscible solvent systems from an energy perspective due to the avoidance of distillation provided by the physical separation of the solvent and water. However, since water-immiscible solvents preferentially extract the neutral lipids, they may not effectively extract the polar lipid-associated PUFAs (Olmstead et al., 2013). One approach is to esterify the ω-3 LC-PUFAs and then recover the low-polarity fatty acid methyl esters (FAME) using a biphasic extraction system. For instance, lipase-assisted in situ transesterification of lipids allowed conversion of saponifiable lipids including EPA to FAME, which could be recovered from wet algal biomass using a biphasic extraction system (Law et al., 2018). In another study, direct saponification of PUFAs was achieved using potassium hydroxide in ethanol followed by precipitation in an antisolvent such as acetone or acetonitrile to extract and concentrate PUFAs from thraustochytrids (nonphotosynthetic protists) (Mirmehrabi & Brar, 2020). The PUFAs can be administered as FFAs or triacylglycerols, both being similarly absorbed for medical and dietetic purposes (Medina et al., 1998). Thus, enzymatic reactions can be carried out to produce PUFA-enriched triglycerides from the extracted polar lipids or FAMEs, to enhance the nutritional value of the microalgal oil (Medina et al., 1998).

3.4.3 Pigment extraction

Microalgae-based pigments can be used as food coloring, and perhaps more importantly, can provide antioxidant, anticarcinogenic, and neuroprotective properties (Cuellar-Bermudez et al., 2015). As such, these pigments can be used as functional food additives or as nutraceuticals. The pigments found in microalgae can be grouped into carotenoids (e.g., β-carotene, lutein, and astaxanthin), chlorophylls, and phycobiliproteins (e.g., phycocyanin, phycoerythrin, and allophycocyanin) (Cuellar-Bermudez et al., 2015), with the composition varying depending on the species and growth conditions.

Pigments are usually co-extracted with lipids during solvent extraction. Combinations of water-immiscible and water-miscible solvents such as hexane and isopropanol can extract pigments such as carotenoids (Ryckebosch et al., 2013). Alkaline saponification can increase the efficiency of monophasic extraction of carotenoids (Cerón-García et al., 2018). However, the treatment conditions need to be selected based on the microalgae species and the carotenoid composition. Simultaneous extraction and production of β-carotene from *Dunaliella salina* was demonstrated by so-called "milking," using dodecane, a biocompatible organic solvent which could be applied without leading to cell death (Hejazi et al., 2002). The ability to milk *Dunaliella salina* is mainly due to the absence of a rigid cell wall, which allows the exocytosis of the β-carotene globules in the presence of organic biocompatible solvents (Hejazi et al., 2004). This has allowed the selective extraction of β-carotene, minimizing the coextraction of chlorophylls.

The use of SC-CO_2 can also be favorable for the extraction of pigments as it can achieve high recovery rates and involves a non-toxic solvent that assures food safety (Molino et al., 2020). The extraction of the compound of interest can be optimized by varying the temperature and pressure of the system, by changing the density and solvation power of supercritical CO_2 (Molino et al., 2018). For instance, SC-CO_2 extraction can also be used for the extraction of astaxanthin, an antioxidant, from *H. pluvialis*. However, due to the low solubility of astaxanthin in CO_2, relatively high pressures are required for the efficient recovery (Kim et al., 2016). The extraction efficiency of astaxanthin increases with the temperature until a maximum value, beyond which denaturation of the pigment becomes problematic (Molino et al., 2018). Astaxanthin can also be selectively extracted using organic solvents, for example, dodecane and methanol (Kang & Sim, 2007), although less toxic organic solvents such as acetone, ethyl acetate, and ethanol are preferable for food applications.

Phycocyanin is a blue-colored pigment and is a phycobiliprotein found mostly in cyanobacteria such as Spirulina (Viskari & Colyer, 2003). These are water soluble and therefore can be separated from the cell debris after cell rupture. The techniques used for cell rupture and extraction need to be mild, as C-phycocyanin and allophycocyanin degrade with thermal and acidic treatment (Böcker et al., 2019).

The varying polarity, solubility, and chemical stability of the pigments present in microalgae require different selective extraction techniques. The selection of an extraction method will also depend on the nutritional and functional properties that are required in the extracted pigments. After extraction, separation and purification steps are also required to achieve high purity of these metabolites. This can be achieved by evaporation (Gouveia et al., 2007), precipitation, chromatographic techniques (Chan et al., 2013; Molino et al., 2018), or membrane filtration (Lawrenz, Fedewa, & Richardson, 2011), depending on the metabolite of interest.

In food applications, the toxicity of the solvent used for extraction is a major consideration due to increasingly strict regulations. Therefore technologies such as SC-CO_2 are of increasing interest despite the high costs compared with more conventional organic solvent extraction methods. The stability of the lipids and pigments is also very important for their quality as food ingredients, and it has been shown that this can be enhanced by the co-extraction of the pigments and lipids (Ryckebosch et al., 2013). Further research is required to help develop low-cost extraction processes that can ensure minimal damage to the nutritional and functional properties of pigments, while avoiding the use of toxic organic solvents.

3.5 Protein recovery

Microalgal proteins have long been considered to have potential as a sustainable source of dietary protein for human consumption (Hedenskog et al., 1969). In addition to possessing desirable amino acid profiles, some proteins extracted from microalgae have shown antioxidative and antihypertensive properties which increase their health benefits (Medina et al., 2015; Norzagaray-Valenzuela et al., 2017). Microalgal proteins can be utilized as whole-cell proteins, protein concentrates, isolates, hydrolysates, or bioactive peptides (Soto-Sierra et al., 2018). Despite the high nutritional value of the proteins in microalgae, they are not yet commonly used as protein ingredients in food due to the presence of components like chlorophyll and algal lipids that alter their color, odor, and appearance (Patinier, 2019). The extraction and purification of algal proteins offer a means of producing higher value products for more targeted food applications.

Microalgal proteins can be broadly considered to be either free, water-soluble proteins that reside in the cytosol and organelle interiors, or bound, amphiphilic proteins associated with the membranes and cell walls. Following cell rupture, water-soluble proteins will be released into the surrounding aqueous phase and can then be separated from the cell debris by centrifugation (Safi et al., 2017; Schwenzfeier et al., 2011). As protein release relies on cell rupture, it is necessary to maximize this in order to obtain high yields of the soluble proteins.

Further increasing protein yields requires accessing the membrane- and wall-bound protein fraction. Alkaline conditions have been shown to improve the solubility of microalgal proteins, although the functional properties of the proteins can be altered (Gerde et al., 2013; Parimi et al., 2015). For example, Ursu et al. (2014) reported that the emulsifying capacity of proteins solubilized at alkaline conditions was lower than that at neutral pH. Thus, the effect of pH during alkaline extraction on the extent of denaturation in microalgal proteins and their effects on the nutritional and functional properties require investigation. Another option is the use of enzymes such as Alcalase, which can increase the solubilization of proteins, presumably due to hydrolysis of the proteins (Safi et al., 2017). Thus, this method of solubilizing proteins is most suited to producing a protein hydrolysate.

After separating the protein-rich supernatant from the cell debris, the proteins can be concentrated and purified using different methods such as membrane filtration (Safi et al., 2017), adsorption using ion exchange columns (Schwenzfeier et al., 2011), protein precipitation (Parimi et al., 2015) and liquid–liquid extraction methods such as aqueous two-phase systems (ATPSs) and three-phase partitioning (TPP) (Suarez Ruiz et al., 2018; Waghmare et al., 2016). Schwenzfeier et al. (2011) developed a method to isolate proteins from centrifugal supernatants of ruptured *Tetraselmis* sp. using ion exchange chromatography. Liquid–liquid extraction methods for protein separation and concentration from microalgae biomass are emerging technologies. One added advantage of such processes is that different molecules such as lipids, proteins, and pigments will migrate into different phases and it becomes easier to fractionate them. In TPP, proteins are separated from the cell debris and concentrated using ammonium sulfate and *t*-butanol. The mixture results in the formation of three phases, with non-polar components partitioning into an upper solvent phase, the polar components concentrating in an aqueous lower phase, and the proteins isolated in the middle phase (Waghmare et al., 2016). Further, microwaves (Chew et al., 2019) and ultrasonic waves (Chia et al., 2019) have been used to improve protein recovery in the TPP process.

ATPSs consist of two immiscible aqueous phases formed by two polymers, two salts, or a salt and a polymer, which separate into distinct phases when they are present above a certain critical concentration (Desai et al., 2014; Suarez Ruiz et al., 2018). ATPSs have been demonstrated to be successful in fractionating biomolecules including proteins from ruptured microalgae under mild conditions (Suarez Ruiz et al., 2018). Nonetheless, Desai et al. (2014) observed Rubisco protein to form aggregates at high ionic concentrations during ionic liquid-based ATPS. Therefore it is necessary to select the ATPSs carefully to minimize the effect on protein quality while maximizing the recovery of the proteins. Application of this method to the food industry requires consideration on the use and reuse of concentrated salts, polymers, and toxic chemicals like *t*-butanol.

Precipitation techniques such as isoelectric point precipitation and precipitation using solvents such as acetone and trichloroacetic acid have been used to separate plant and algae proteins (Niu et al., 2018; Ursu et al., 2014). Such precipitates can be enriched with both proteins and fatty acids, and can be used as a functional food ingredient (Cavonius et al., 2015). Protein concentrates can also be obtained by membrane filtration. Ultrafiltration is appropriate for concentrating solubilized proteins, for example in a supernatant, with diafiltration able to be used to reduce the concentration of smaller organic molecules and salts (Safi et al., 2017; Ursu et al., 2014). Protein concentration by membrane filtration can preserve the functional properties of the proteins due to the mild conditions used (Ursu et al., 2014). Overall, the choice of protein separation and purification will depend on the desired quality and price of the targeted protein product and the specific food application. In this regard, Soto-Sierra et al. (2018) provide a detailed review of the extraction and fractionation of different protein products from microalgae.

3.6 Carbohydrate recovery

The carbohydrates in microalgae are mostly either storage polysaccharides or structural carbohydrates (Bernaerts et al., 2019). Storage carbohydrates are located in the intercellular compartments, typically as starch granules that serve as sugar reserves. Along with the proteins, starch granules can be released from the interior of the microalgae cells once the cells are ruptured. In comparison, structural carbohydrates are compositional components of the cell wall, forming a large part of the cell debris upon rupture. The cell walls of most microalgae are composed of polysaccharides and are covered by a layer of external polymers, known as extracellular polymeric substances or exopolysaccharides (EPS) (Bernaerts et al., 2019). The EPS content varies among the species and growth conditions, and typically includes a considerable carbohydrate component (Baroni et al., 2020; Henderson et al., 2008). Cell wall polysaccharides are composed of heteropolysaccharides with a substantial amount of uronic acids and sulfate groups (Bernaerts et al., 2018). These polysaccharides have the ability to act as functional food material, as gelling agents, thickeners, and even as antioxidants (Bernaerts et al., 2019; Zhang et al., 2010). Despite this, the extraction and fractionation of polysaccharides from microalgae has not been widely studied to date.

The existing literature on carbohydrate extraction from microalgae has mainly been focused on bioethanol production, where the hydrolysis and saccharification of the carbohydrates are also required. Thus harsh treatment methods such as acid and alkali hydrolysis (Castro, Ellis, Miller, & Sims, 2015; Harun, Jason, Cherrington, & Danquah, 2011) and enzymatic treatment using amylases and cellulases (Velazquez-Lucio et al., 2018) have been used. However, these treatments result in the degradation of the carbohydrates, proteins, and lipids, which is not preferred when it comes to food applications for which the nutritional and functional properties need to be preserved.

Cell wall carbohydrates can be recovered from mechanically ruptured cells via ethanol precipitation following removal of lipids and proteins (Bernaerts et al., 2018). Polysaccharides released as EPSs from the microalga *Porphyridium cruentum* have been extracted with hot water and concentrated using ultrafiltration using a patented method by Barnier et al. (1990). In addition, novel techniques such as microwave- (Yuan & Macquarrie, 2015), ultrasonic- (Rahimi, Tabarsa, & Rezaei, 2016), and enzyme-assisted (Rodrigues et al., 2015) extractions have been explored for recovering microalgal polysaccharides. However, the techniques used are costly and energy intensive compared with conventional hot water extraction. Further, one cannot guarantee the extraction of a pure polysaccharide fraction, as other water-soluble macromolecules such as proteins can be coextracted with the polysaccharides. Schwenzfeier et al. (2014) and Patel et al. (2013) found that diafiltration using a membrane with a 300 kDa cut-off was more efficient than solvent precipitation and dialysis. Flocculation of polysaccharides using non-toxic flocculants followed by centrifugation to separate the flocculated polysaccharides is another potential technique for polysaccharide separation (Liu et al., 2019).

3.7 Concluding discussion

The recovery of food-grade ingredients from microalgae requires several processing steps. Once the algae are produced, they need to be harvested and concentrated to reduce the processing volume. From an energy-efficiency perspective, wet algae biomass is preferred over dry algae powders due to the high amounts of energy required to thermally remove water. However, metabolic and microbial degradation of high-value components during wet storage will need to be considered as an important factor in food and nutraceutical production (Balduyck et al., 2016). In order to inhibit the activity of endogenous enzymes, cells can be subjected to a short heat treatment (Balduyck et al., 2019). For example, heating *Tisochrysis lutea* at 80°C for 8 min improved lipid stability during wet storage by avoiding the formation of FFA, without compromising pigment or lipid stability (Balduyck et al., 2019). Heat treatment may also be useful in mitigating any food safety issues associated with the presence of bacteria in algae cultures. It is likely that species- and product-specific pasteurization methods will be needed to optimize enzyme and bacterial inactivation for food and nutraceutical applications.

Except for solvent extraction from dried biomass, cell rupture is necessary to increase the availability of intracellular products. Selection of a cell disruption technique needs to consider the energy requirements, process scalability, effectiveness on different species, and any negative effects on the high-value components. Novel techniques such as PEF are emerging for the selective release of components from microalgae, that can then be recovered from whole cells. Much of the proteins and carbohydrates can be directly recovered after cell disruption as they are released into the surrounding medium. However, recovery of membrane- and wall-bound proteins and carbohydrates requires alkaline or hot water extraction.

For lipid and pigment extraction, novel extraction methods such as supercritical CO_2 extraction, bio-based solvents, and ionic liquids are emerging due to challenges associated with the use of conventional organic solvents. In particular, supercritical CO_2 is a safe alternative to the highly toxic organic solvents. However, the high costs associated with such extractions limit their application. The challenge is to find lipid extraction processes that require only non-toxic additives and are also cheap to implement at large scale for food and nutraceutical applications. Further research is therefore needed to develop and optimize greener and low-cost extraction processes for the extraction of food and nutraceutical components from microalgae.

Extraction of multiple components from microalgae can enhance the economics of the process. Depending on the product/products of interest, a cascading process that allows the extraction of multiple components, with minimal effect on their quality, is therefore desirable. As an example, the

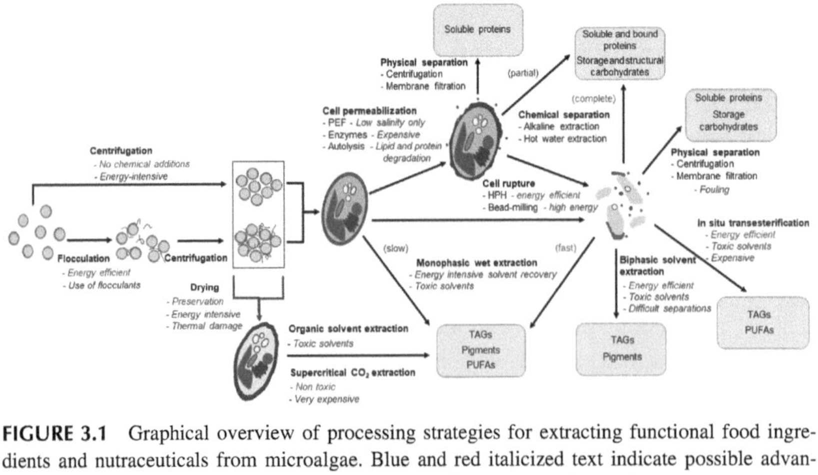

FIGURE 3.1 Graphical overview of processing strategies for extracting functional food ingredients and nutraceuticals from microalgae. Blue and red italicized text indicate possible advantages and limitations, respectively, associated with different processing options.

extraction of proteins prior to lipid extraction will preserve their nutritional and functional properties (Gerde et al., 2013).

Further, the coextraction of compounds from microalgae can have added advantages in incorporating them into food matrices. As an example, the coextraction of polysaccharides with proteins improves their functional properties while the coextraction of pigments such as carotenoids with lipids improves their stability (Bernaerts et al., 2020; Schwenzfeier et al., 2014). Therefore when selecting the extraction methods, the effect on the stability and functional properties of the final product will also need to be considered. Fig. 3.1 provides a summary of how existing novel and conventional technologies can be used to obtain protein, lipid, carbohydrate, and pigment fractions from microalgae for food and nutraceutical applications.

Acknowledgments

This work was supported by funding from the Australian Research Council (ARC) through the Discovery Projects Grant Scheme (DP170103791). Bhagya Yatipanthalawa is also grateful to the University of Melbourne for support provided by a Melbourne Research Scholarship (MRS).

The authors are affiliated with the Future Food Hallmark Research Initiative of The University of Melbourne.

References

Balduyck, L., Bijttebier, S., Bruneel, C., Jacobs, G., Voorspoels, S., Van Durme, J., . . . Foubert, I. (2016). Lipolysis in T-Isochrysis lutea during wet storage at different temperatures. *Algal Research*, *18*, 281–287.

Balduyck, L., Bruneel, C., Goiris, K., Dejonghe, C., & Foubert, I. (2018). Influence of high pressure homogenization on free fatty acid formation in Nannochloropsis sp. *European Journal of Lipid Science and Technology*, *120*.

Balduyck, L., Dejonghe, C., Foubert, I. & Vandamme, D. (2019). Simultaneous harvesting and inhibition of lipolytic reactions in T-Isochrysis by FeCl3 flocculation.

Balduyck, L., Dejonghe, C., Goos, P., Jooken, E., Muylaert, K., & Foubert, I. (2019). Inhibition of lipolytic reactions during wet storage of T-Isochrysis lutea biomass by heat treatment. *Algal Research*, *38*, 101388.

Balduyck, L., Stock, T., Bijttebier, S., Bruneel, C., Jacobs, G., Voorspoels, S., . . . Foubert, I. (2017). Integrity of the microalgal cell plays a major role in the lipolytic stability during wet storage. *Algal Research*, *25*, 516–524.

Barnier, H., Dos Santos, P.F.F., Gudin, C. & Thepenier, C. (1990). Process for the production and extraction of polysaccharides from a porphyridium cruentum culture and apparatus for performing the process. Google Patents.

Baroni, E., Cao, B., Webley, P. A., Scales, P. J., & Martin, G. J. O. (2020). Nitrogen availability and the nature of extracellular organic matter of microalgae. *Industrial & Engineering Chemistry Research*, *59*, 6795–6805.

Barros, A. I., Gonçalves, A. L., Simões, M., & Pires, J. C. (2015). Harvesting techniques applied to microalgae: A review. *Renewable and Sustainable Energy Reviews*, *41*, 1489–1500.

Bernaerts, T. M., Gheysen, L., Foubert, I., Hendrickx, M. E., & Van Loey, A. M. (2019). The potential of microalgae and their biopolymers as structuring ingredients in food: A review. *Biotechnology Advances*, 107419.

Bernaerts, T. M. M., Gheysen, L., Foubert, I., Hendrickx, M. E., & Van Loey, A. M. (2019). Evaluating microalgal cell disruption upon ultra high pressure homogenization. *Algal Research*, *42*.

Bernaerts, T. M. M., Gheysen, L., Kyomugasho, C., Jamsazzadeh Kermani, Z., Vandionant, S., Foubert, I., ... Van Loey, A. M. (2018). Comparison of microalgal biomasses as functional food ingredients: Focus on the composition of cell wall related polysaccharides. *Algal Research*, *32*, 150–161.

Bernaerts, T. M. M., Verstreken, H., Dejonghe, C., Gheysen, L., Foubert, I., Grauwet, T., & Van Loey, A. M. (2020). Cell disruption of Nannochloropsis sp. improves in vitro bioaccessibility of carotenoids and ω3-LC-PUFA. *Journal of Functional Foods*, *65*, 103770.

Böcker, L., Ortmann, S., Surber, J., Leeb, E., Reineke, K., & Mathys, A. (2019). Biphasic short time heat degradation of the blue microalgae protein phycocyanin from Arthrospira platensis. *Innovative Food Science & Emerging Technologies*, *52*, 116–121.

Breil, C., Meullemiestre, A., Vian, M., & Chemat, F. (2016). Bio-based solvents for green extraction of lipids from oleaginous yeast biomass for sustainable aviation biofuel. *Molecules (Basel, Switzerland)*, *21*, 196.

Buchmann, L., Bloch, R., & Mathys, A. (2018). Comprehensive pulsed electric field (PEF) system analysis for microalgae processing. *Bioresource Technology*, *265*, 268–274.

Buchmann, L., Brändle, I., Haberkorn, I., Hiestand, M., & Mathys, A. (2019). Pulsed electric field based cyclic protein extraction of microalgae towards closed-loop biorefinery concepts. *Bioresource Technology*, *291*.

Campbell, K. A., Glatz, C. E., Johnson, L. A., Jung, S., De Moura, J. M. N., Kapchie, V., & Murphy, P. (2011). Advances in aqueous extraction processing of soybeans. *Journal of the American Oil Chemists Society*, *88*, 449–465.

Castro, Y. A., Ellis, J. T., Miller, C. D., & Sims, R. C. (2015). Optimization of wastewater microalgae saccharification using dilute acid hydrolysis for acetone, butanol, and ethanol fermentation. *Applied Energy*, *140*, 14–19.

Cavonius, L. R., Albers, E., & Undeland, I. (2015). pH-shift processing of Nannochloropsis oculata microalgal biomass to obtain a protein-enriched food or feed ingredient. *Algal Research*, *11*, 95–102.

Cavonius, L. R., Albers, E., & Undeland, I. (2016). In vitro bioaccessibility of proteins and lipids of pH-shift processed Nannochloropsis oculata microalga. *Food & Function*, *7*.

Cerón-García, M., González-López, C., Camacho-Rodríguez, J., López-Rosales, L., García-Camacho, F., & Molina-Grima, E. (2018). Maximizing carotenoid extraction from microalgae used as food additives and determined by liquid chromatography (HPLC). *Food Chemistry*, *257*, 316–324.

Chan, M.-C., Ho, S.-H., Lee, D.-J., Chen, C.-Y., Huang, C.-C., & Chang, J.-S. (2013). Characterization, extraction and purification of lutein produced by an indigenous microalga Scenedesmus obliquus CNW-N. *Biochemical Engineering Journal*, *78*, 24–31.

Chew, K. W., Chia, S. R., Lee, S. Y., Zhu, L., & Show, P. L. (2019). Enhanced microalgal protein extraction and purification using sustainable microwave-assisted multiphase partitioning technique. *Chemical Engineering Journal*, *367*, 1–8.

Chia, S. R., Chew, K. W., Zaid, M., Fatimah, H., Chu, D.-T., Tao, Y., & Show, P. L. (2019). Microalgal protein extraction from Chlorella vulgaris FSP-E using triphasic partitioning technique with sonication. *Frontiers in Bioengineering and Biotechnology*, *7*, 396.

Cooney, M. J., Young, G., & Pate, R. (2011). Bio-oil from photosynthetic microalgae: Case study. *Bioresource Technology*, *102*, 166–177.

Cuellar-Bermudez, S. P., Aguilar-Hernandez, I., Cardenas-Chavez, D. L., Ornelas-Soto, N., Romero-Ogawa, M. A., & Parra-Saldivar, R. (2015). Extraction and purification of high-value metabolites from microalgae: essential lipids, astaxanthin and phycobiliproteins. *Microbial Biotechnology*, *8*, 190–209.

De Farias Neves, F., Demarco, M., & Tribuzi, G. (2019). *Drying and quality of microalgal powders for human alimentation. Microalgae-From Physiology to Application*. IntechOpen.

De Melo, J. R. M., Tres, M. V., Steffens, J., Vladimir Oliveira, J., & Di Luccio, M. (2015). Desolventizing organic solvent-soybean oil miscella using ultrafiltration ceramic membranes. *Journal of Membrane Science*, *475*, 357–366.

Demuez, M., Mahdy, A., Tomas-Pejo, E., Gonzalez-Fernandez, C., & Ballesteros, M. (2015). Enzymatic cell disruption of microalgae biomass in biorefinery process. *Biotechnology and Bioengineering*, *112*, 1955–1966.

Desai, R. K., Streefland, M., Wijffels, R. H., & Eppink, M. H. (2014). Extraction and stability of selected proteins in ionic liquid based aqueous two phase systems. *Green Chemistry*, *16*, 2670–2679.

Di Salvo, R., Reich, A., Dykes, H.W.H. & Teixeira, R. (2013). Lipid extraction from microalgae using a single ionic liquid. Google Patents.

Dong, T., Knoshaug, E. P., Pienkos, P. T., & Laurens, L. M. L. (2016). Lipid recovery from wet oleaginous microbial biomass for biofuel production: A critical review. *Applied Energy*, *177*, 879–895.

Dong, T., Van Wychen, S., Nagle, N., Pienkos, P. T., & Laurens, L. M. L. (2016). Impact of biochemical composition on susceptibility of algal biomass to acid-catalyzed pretreatment for sugar and lipid recovery. *Algal Research*, *18*, 69–77.

Dos Santos, R. R., Moreira, D. M., Kunigami, C. N., Aranda, D. A. G., & Teixeira, C. M. L. L. (2015). Comparison between several methods of total lipid extraction from Chlorella vulgaris biomass. *Ultrasonics Sonochemistry*, *22*, 95–99.

Doucha, J., & Livansky, K. (2008). Influence of processing parameters on disintegration of *Chlorella* cells in various types of homogenizers. *Applied Microbiology and Biotechnology*, *81*, 431–440.

Du, Y., Schuur, B., Samorì, C., Tagliavini, E., & Brilman, D. W. F. (2013). Secondary amines as switchable solvents for lipid extraction from non-broken microalgae. *Bioresource Technology*, *149*, 253–260.

Frey, W., Gusbeth, C., Sakugawa, T., Sack, M., Mueller, G., Sigler, J., … Raso, J. (2017). *Environmental applications, food and biomass processing by pulsed electric fields. Bioelectrics*. Springer.

Gerde, J. A., Wang, T., Yao, L., Jung, S., Johnson, L. A., & Lamsal, B. (2013). Optimizing protein isolation from defatted and non-defatted Nannochloropsis microalgae biomass. *Algal Research*, *2*, 145–153.

Gheysen, L., Lagae, N., Devaere, J., Goiris, K., Goos, P., Bernaerts, T., … Foubert, I. (2019). Impact of Nannochloropsis sp. dosage form on the oxidative stability of n-3 LC-PUFA enriched tomato purees. *Food Chemistry*, *279*, 389–400.

Gouveia, L., Nobre, B., Marcelo, F., Mrejen, S., Cardoso, M., Palavra, A., & Mendes, R. (2007). Functional food oil coloured by pigments extracted from microalgae with supercritical CO2. *Food Chemistry*, *101*, 717–723.

Halim, R., Gladman, B., Danquah, M. K., & Webley, P. A. (2011). Oil extraction from microalgae for biodiesel production. *Bioresource Technology*, *102*, 178–185.

Halim, R., Hill, D. R. A., Hanssen, E., Webley, P. A., Blackburn, S., Grossman, A. R., ... Martin, G. J. O. (2019). Towards sustainable microalgal biomass processing: anaerobic induction of autolytic cell-wall self-ingestion in lipid-rich *Nannochloropsis* slurries. *Green Chemistry*, *21*, 2967–2982.

Halim, R., Webley, P. A., & Martin, G. J. (2016). The CIDES process: fractionation of concentrated microalgal paste for co-production of biofuel, nutraceuticals, and high-grade protein feed. *Algal Research*, *19*, 299–306.

Harun, R., Jason, W., Cherrington, T., & Danquah, M. K. (2011). Exploring alkaline pre-treatment of microalgal biomass for bioethanol production. *Applied Energy*, *88*, 3464–3467.

Hedenskog, G., Enebo, L., Vendlova, J., & Prokes, B. (1969). Investigation of some methods for increasing the digestibility *in vitro* of microalgae. *Biotechnology and Bioengineering*, *11*, 37–51.

Hejazi, M., De Lamarliere, C., Rocha, J., Vermue, M., Tramper, J., & Wijffels, R. (2002). Selective extraction of carotenoids from the microalga Dunaliella salina with retention of viability. *Biotechnology and Bioengineering*, *79*, 29–36.

Hejazi, M., Kleinegris, D., & Wijffels, R. (2004). Mechanism of extraction of β-carotene from microalga Dunaliellea salina in two-phase bioreactors. *Biotechnology and Bioengineering*, *88*, 593–600.

Henderson, R. K., Baker, A., Parsons, S. A., & Jefferson, B. (2008). Characterisation of algogenic organic matter extracted from cyanobacteria, green algae and diatoms. *Water Research*, *42*, 3435–3445.

Jiménez Callejón, M. J., Robles Medina, A., González Moreno, P. A., Esteban Cerdán, L., Orta Guillén, S., & Molina Grima, E. (2020). Simultaneous extraction and fractionation of lipids from the microalga Nannochloropsis sp. for the production of EPA-rich polar lipid concentrates. *Journal of Applied Phycology*, *32*, 1117–1128.

Jiménez Callejón, M. J., Robles Medina, A., Macías Sánchez, M. D., Hita Peña, E., Esteban Cerdán, L., González Moreno, P. A., & Molina Grima, E. (2014). Extraction of saponifiable lipids from wet microalgal biomass for biodiesel production. *Bioresource Technology*, *169*, 198–205.

Kang, C. D., & Sim, S. J. (2007). Selective extraction of free astaxanthin from Haematococcus culture using a tandem organic solvent system. *Biotechnology Progress*, *23*, 866–871.

Kightlinger, W., Chen, K., Pourmir, A., Crunkleton, D. W., Price, G. L., & Johannes, T. W. (2014). Production and characterization of algae extract from *Chlamydomonas reinhardtii*. *Electric Journal of Biotechnology*, *17*, 14–18.

Kim, D.-Y., Vijayan, D., Praveenkumar, R., Han, J.-I., Lee, K., Park, J.-Y., ... Oh, Y.-K. (2016). Cell-wall disruption and lipid/astaxanthin extraction from microalgae: Chlorella and Haematococcus. *Bioresource Technology*, *199*, 300–310.

Lardon, L., Hélias, A., Sialve, B., Steyer, J.-P., & Bernard, O. (2009). Life-cycle assessment of biodiesel production from microalgae. *Environmental Science & Technology*, *43*, 6475–6481.

Law, S. Q., Mettu, S., Ashokkumar, M., Scales, P. J., & Martin, G. J. (2018). Emulsifying properties of ruptured microalgae cells: Barriers to lipid extraction or promising biosurfactants? *Colloids and Surfaces B: Biointerfaces*, *170*, 438–446.

Law, S. Q. K., Chen, B., Scales, P. J., & Martin, G. J. O. (2017). Centrifugal recovery of solvent after biphasic wet extraction of lipids from a concentrated slurry of *Nannochloropsis* sp. biomass. *Algal Research*, *24*, 299–308.

Law, S. Q. K., Halim, R., Scales, P. J., & Martin, G. J. O. (2018). Conversion and recovery of saponifiable lipids from microalgae using a nonpolar solvent via lipase-assisted extraction. *Bioresource Technology*, *260*, 338–347.

Lawrenz, E., Fedewa, E. J., & Richardson, T. L. (2011). Extraction protocols for the quantification of phycobilins in aqueous phytoplankton extracts. *Journal of Applied Phycology*, *23*, 865–871.

Li, W., Gamlath, C.J., Pathak, R., Martin, G.J. & Ashokkumar, M. (2019). Ultrasound–The Physical and Chemical Effects Integral to Food Processing.

Liu, Q., Yao, C., Sun, Y., Chen, W., Tan, H., Cao, X., ... Yin, H. (2019). Production and structural characterization of a new type of polysaccharide from nitrogen-limited Arthrospira platensis cultivated in outdoor industrial-scale open raceway ponds. *Biotechnology for Biofuels*, *12*, 131.

Mahmood, W. M. A. W., Theodoropoulos, C., & Gonzalez-Miquel, M. (2017). Enhanced microalgal lipid extraction using bio-based solvents for sustainable biofuel production. *Green Chemistry*, *19*, 5723–5733.

Martin, G. J. O., Hill, D. R. A., Olmstead, I. L. D., Bergamin, A., Shears, M. J., Dias, D. A., ... Callahan, D. L. (2014). Lipid profile remodeling in response to nitrogen deprivation in the microalgae *Chlorella* sp. (Trebouxiophyceae) and *Nannochloropsis* sp. (Eustigmatophyceae). *PLoS ONE*, *9*, e103389.

Martin, G. J. O. (2016). Energy requirements for wet solvent extraction of lipids from microalgal biomass. *Bioresource Technology*, *205*, 40–47.

Medina, A. R., Grima, E. M., Giménez, A. G., & González, M. I. (1998). Downstream processing of algal polyunsaturated fatty acids. *Biotechnology Advances*, *16*, 517–580.

Medina, C., Rubilar, M., Shene, C., Torres, S., & Verdugo, M. (2015). Protein fractions with techno-functional and antioxidant properties from Nannochloropsis gaditana microalgal biomass. *Journal of Biobased Materials and Bioenergy*, *9*, 417–425.

Middelberg, A. P. J. (1995). Process-scale disruption of microorganisms. *Biotechnology Advances*, *13*, 491–551.

Mirmehrabi, M., & Brar, S. K. (2020). A novel process for isolation and purification of polyunsaturated fatty acids from a thraustochytrid. *Algal Research*, *46*, 101806.

Mitsuda, H., & Shikanai, T. (1959). Studies on the utilization of Chlorella for food. *Journal of Japanese Society of Food Nutrition*, *12*, 40–58.

Molina Grima, E., Robles Medina, A., & Gimenez Gimenez, A. (1999). Recovery of algal PUFAs. In Z. Cohen (Ed.), *Chemicals from Microalgae*. London: Taylor & Francis Ltd.

Molino, A., Mehariya, S., Di Sanzo, G., Larocca, V., Martino, M., Leone, G. P., ... Musmarra, D. (2020). Recent developments in supercritical fluid extraction of bioactive compounds from microalgae: Role of key parameters, technological achievements and challenges. *Journal of CO2 Utilization*, *36*, 196–209.

Molino, A., Rimauro, J., Casella, P., Cerbone, A., Larocca, V., Chianese, S., ... Hristoforou, E. (2018). Extraction of astaxanthin from microalga Haematococcus pluvialis in red phase by using generally recognized as safe solvents and accelerated extraction. *Journal of Biotechnology*, *283*, 51–61.

Niu, L., Zhang, H., Wu, Z., Wang, Y., Liu, H., Wu, X., & Wang, W. (2018). Modified TCA/acetone precipitation of plant proteins for proteomic analysis. *PLoS One*, *13*.

Norzagaray-Valenzuela, C. D., Valdez-Ortiz, A., Shelton, L. M., Jiménez-Edeza, M., Rivera-López, J., Valdez-Flores, M. A., & Germán-Báez, L. J. (2017). Residual biomasses and protein hydrolysates of three green microalgae species exhibit antioxidant and anti-aging activity. *Journal of Applied Phycology*, *29*, 189–198.

Olmstead, I. L., Hill, D. R., Dias, D. A., Jayasinghe, N. S., Callahan, D. L., Kentish, S. E., ... Martin, G. J. (2013). A quantitative analysis of microalgal lipids for optimization of biodiesel and omega-3 production. *Biotechnology and Bioengineering*, *110*, 2096–2104.

Olmstead, I. L. D., Kentish, S. E., Scales, P. J., & Martin, G. J. O. (2013). Low solvent, low temperature method for extracting biodiesel lipids from concentrated microalgal biomass. *Bioresource Technology*, *148*, 615–619.

Parimi, N. S., Singh, M., Kastner, J. R., Das, K. C., Forsberg, L. S., & Azadi, P. (2015). Optimization of protein extraction from Spirulina platensis to generate a potential co-product and a biofuel feedstock with reduced nitrogen content. *Frontiers in Energy Research*, *3*, 30.

Patel, A. K., Laroche, C., Marcati, A., Ursu, A. V., Jubeau, S., Marchal, L., ... Michaud, P. (2013). Separation and fractionation of exopolysaccharides from Porphyridium cruentum. *Bioresource Technology*, *145*, 345–350.

Patinier, S. (2019). Method for extracting soluble proteins from microalgal biomass. Google Patents.

Postma, P., Barbosa, M., Wijffels, R., Eppink, M. & Olivieri, G. (2016). Microalgal biorefinery for bulk and high-value products: Product extraction within cell disintegration. *Handbook of electroporation*.

Postma, P., Miron, T., Olivieri, G., Barbosa, M., Wijffels, R., & Eppink, M. (2015). Mild disintegration of the green microalgae Chlorella vulgaris using bead milling. *Bioresource Technology*, *184*, 297–304.

Postma, P., Pataro, G., Capitoli, M., Barbosa, M., Wijffels, R. H., Eppink, M., ... Ferrari, G. (2016). Selective extraction of intracellular components from the microalga Chlorella vulgaris by combined pulsed electric field–temperature treatment. *Bioresource Technology*, *203*, 80–88.

Postma, P., Suarez-Garcia, E., Safi, C., Yonathan, K., Olivieri, G., Barbosa, M., ... Eppink, M. (2017). Energy efficient bead milling of microalgae: effect of bead size on disintegration and release of proteins and carbohydrates. *Bioresource Technology*, *224*, 670–679.

Rahimi, F., Tabarsa, M., & Rezaei, M. (2016). Ulvan from green algae Ulva intestinalis: optimization of ultrasound-assisted extraction and antioxidant activity. *Journal of Applied Phycology*, *28*, 2979–2990.

Rodrigues, D., Sousa, S. R., Silva, A., Amorim, M., Pereira, L., Rocha-Santos, T. A., ... Freitas, A. C. (2015). Impact of enzyme-and ultrasound-assisted extraction methods on biological properties of red, brown, and green seaweeds from the central west coast of Portugal. *Journal of Agricultural and Food Chemistry*, *63*, 3177–3188.

Ryckebosch, E., Bruneel, C., Termote-Verhalle, R., Lemahieu, C., Muylaert, K., Van Durme, J., ... Foubert, I. (2013). Stability of omega-3 LC-PUFA-rich photoautotrophic microalgal oils compared to commercially available omega-3 LC-PUFA oils. *Journal of Agricultural and Food Chemistry*, *61*, 10145–10155.

Safi, C., Olivieri, G., Campos, R. P., Engelen-Smit, N., Mulder, W., Van Den Broek, L., & Sijtsma, L. (2017). Biorefinery of microalgal soluble proteins by sequential processing and membrane filtration. *Bioresource Technology*, *225*, 151–158.

Samorì, C., Barreiro, D. L., Vet, R., Pezzolesi, L., Brilman, D. W., Galletti, P., & Tagliavini, E. (2013). Effective lipid extraction from algae cultures using switchable solvents. *Green chemistry*, *15*, 353–356.

Scholz, M. J., Weiss, T. L., Jinkerson, R. E., Jing, J., Roth, R., Goodenough, U., ... Gerken, H. G. (2014). Ultrastructure and composition of the Nannochloropsis gaditana cell wall. *Eukaryotic Cell*, *13*, 1450–1464.

Schwenzfeier, A., Wierenga, P. A., Eppink, M. H., & Gruppen, H. (2014). Effect of charged polysaccharides on the techno-functional properties of fractions obtained from algae soluble protein isolate. *Food Hydrocolloids*, *35*, 9–18.

Schwenzfeier, A., Wierenga, P. A., & Gruppen, H. (2011). Isolation and characterization of soluble protein from the green microalgae Tetraselmis sp. *Bioresource Technology*, *102*, 9121–9127.

Sevenich, R., & Mathys, A. (2018). Continuous versus discontinuous ultra-high-pressure systems for food sterilization with focus on ultra-high-pressure homogenization and high-pressure thermal sterilization: A review. *Comprehensive Reviews in Food Science and Food Safety*, *17*, 646–662.

Sirajunnisa, A. R., & Surendhiran, D. (2016). Algae – A quintessential and positive resource of bioethanol production: A comprehensive review. *Renewable and Sustainable Energy Reviews*, *66*, 248–267.

Sierra, L. S., Dixon, C. K., & Wilken, L. R. (2017). Enzymatic cell disruption of the microalgae Chlamydomonas reinhardtii for lipid and protein extraction. *Algal Research*, *25*, 149–159.

Silve, A., Kian, C. B., Papachristou, I., Kubisch, C., Nazarova, N., Wüstner, R., ... Frey, W. (2018). Incubation time after pulsed electric field treatment of microalgae enhances the efficiency of extraction processes and enables the reduction of specific treatment energy. *Bioresource Technology*, *269*, 179–187.

Simionato, D., Block, M. A., La Rocca, N., Jouhet, J., Maréchal, E., Finazzi, G., & Morosinotto, T. (2013). The response of Nannochloropsis gaditana to nitrogen starvation includes de novo biosynthesis of triacylglycerols, a decrease of chloroplast galactolipids, and reorganization of the photosynthetic apparatus. *Eukaryotic Cell*, *12*, 665–676.

Şirin, S., Clavero, E., & Salvadó, J. (2013). Potential pre-concentration methods for Nannochloropsis gaditana and a comparative study of pre-concentrated sample properties. *Bioresource Technology*, *132*, 293–304.

Sivasankar, P., Poongodi, S., Lobo, A. O., & Pugazhendhi, A. (2020). Characterization of a novel polymeric bioflocculant from marine actinobacterium Streptomyces sp. and its application in recovery of microalgae. *International Biodeterioration & Biodegradation*, *148*, 104883.

Soto-Sierra, L., Stoykova, P., & Nikolov, Z. L. (2018). Extraction and fractionation of microalgae-based protein products. *Algal Research*, *36*, 175–192.

Spiden, E. M., Yap, B. H. J., Hill, D. R. A., Kentish, S. E., Scales, P. J., & Martin, G. J. O. (2013). Quantitative evaluation of the ease of rupture of industrially promising microalgae by high pressure homogenization. *Bioresource Technology*, *140*, 165–171.

Suarez Ruiz, C. A., Emmery, D. P., Wijffels, R. H., Eppink, M. H., & Van Den Berg, C. (2018). Selective and mild fractionation of microalgal proteins and pigments using aqueous two-phase systems. *Journal of Chemical Technology & Biotechnology*, *93*, 2774–2783.

'T Lam, G. P., Van Der Kolk, J. A., Chordia, A., Vermuë, M. H., Olivieri, G., Eppink, M. H., & Wijffels, R. H. (2017). Mild and selective protein release of cell wall deficient microalgae with pulsed electric field. *ACS Sustainable Chemistry & Engineering*, *5*, 6046–6053.

Ursu, A.-V., Marcati, A., Sayd, T., Sante-Lhoutellier, V., Djelveh, G., & Michaud, P. (2014). Extraction, fractionation and functional properties of proteins from the microalgae Chlorella vulgaris. *Bioresource Technology*, *157*, 134–139.

Van Boekel, M. A. J. S. (2001). Kinetic aspects of the Maillard reaction: A critical review. *Food/Nahrung*, *45*, 150–159.

Vandamme, D., Foubert, I., & Muylaert, K. (2013). Flocculation as a low-cost method for harvesting microalgae for bulk biomass production. *Trends in Biotechnology*, *31*, 233–239.

Velazquez-Lucio, J., Rodríguez-Jasso, R. M., Colla, L. M., Sáenz-Galindo, A., Cervantes-Cisneros, D. E., Aguilar, C. N., ... Ruiz, H. A. (2018). Microalgal biomass pretreatment for bioethanol production: a review. *Biofuel Research Journal*, *5*, 780–791.

Viskari, P. J., & Colyer, C. L. (2003). Rapid extraction of phycobiliproteins from cultured cyanobacteria samples. *Analytical Biochemistry*, *319*, 263–271.

Waghmare, A. G., Salve, M. K., Leblanc, J. G., & Arya, S. S. (2016). Concentration and characterization of microalgae proteins from Chlorella pyrenoidosa. *Bioresources and Bioprocessing*, *3*, 16.

Yao, S., Mettu, S., Law, S. Q. K., Ashokkumar, M., & Martin, G. J. O. (2018). The effect of high-intensity ultrasound on cell disruption and lipid extraction from high-solids viscous slurries of Nannochloropsis sp. biomass. *Algal Research*, *35*, 341–348.

Yap, B. H., Crawford, S. A., Dumsday, G. J., Scales, P. J., & Martin, G. J. (2014). A mechanistic study of algal cell disruption and its effect on lipid recovery by solvent extraction. *Algal Research*, *5*, 112–120.

Yap, B. H., Dumsday, G. J., Scales, P. J., & Martin, G. J. O. (2015). Energy evaluation of algal cell disruption by high pressure homogenization. *Bioresource Technology*, *184*, 280–285.

Yap, B. H., Martin, G. J., & Scales, P. J. (2016). Rheological manipulation of flocculated algal slurries to achieve high solids processing. *Algal Research*, *14*, 1–8.

Yuan, Y., & Macquarrie, D. (2015). Microwave assisted extraction of sulfated polysaccharides (fucoidan) from Ascophyllum nodosum and its antioxidant activity. *Carbohydrate Polymers*, *129*, 101–107.

Zhang, Z., Wang, F., Wang, X., Liu, X., Hou, Y., & Zhang, Q. (2010). Extraction of the polysaccharides from five algae and their potential antioxidant activity in vitro. *Carbohydrate Polymers*, *82*, 118–121.

Zhao, F., Xiao, J., Ding, W., Cui, N., Yu, X., Xu, J.-W., ... Zhao, P. (2019). An effective method for harvesting of microalga: Coculture-induced self-flocculation. *Journal of the Taiwan Institute of Chemical Engineers*, *100*, 117–126.

Chapter 4

Potential applications of microalgae-derived proteins and peptides in the food industry

Caleb Acquah[1], Flora-Glad Ekezie[1,2] and Chibuike C. Udenigwe[1,3]

[1]School of Nutrition Sciences, Faculty of Health Sciences, University of Ottawa, Ottawa, ON, Canada, [2]School of Food Science and Engineering, South China University of Technology, Guangzhou, P.R. China, [3]Department of Chemistry and Biomolecular Sciences, University of Ottawa, Ottawa, ON, Canada

4.1 Introduction

An increase in the global population from 7.7 billion in 2019 to 9.7 billion in 2050, as projected by the United Nations, implies that the demand for food is expected to also rise (FAO, 2009). Concomitantly, the global dietary protein demand is anticipated to increase from 202 million tonnes in 2015 to between 360 and 1250 million tonnes by 2050 (Soto-Sierra, Stoykova, Nikolov, & Nikolov, 2018). This will require raising global food production by over 70% using sustainable and affordable resources (FAO, 2009). Microalgae have garnered considerable attention as new model organisms in the search for healthy and sustainable biomasses for commercial food protein applications (Chiong, Acquah, Lau, Khor, & Danquah, 2016; Ejike et al., 2017; Koyande et al., 2019). Notably, the consumption of microalgae has been in existence for more than 1000 years and they have been demonstrated to be rich sources of proteins, lipids, pigments, polyphenols, and bioactive molecules that can boost physiological processes in the body (Katiyar & Arora, 2020). In addition to their nutritional and biopharmaceutical applications, microalgae can be valorized for other purposes such as biofuel production, wastewater treatment, and cosmetics production (Artan, Acquah, Danquah, & Ongkudon, 2015; Edeseyi et al., 2015).

Microalgae are genetically diverse unicellular organisms with about 200,000 species and can be broadly categorized into 11 major phyla, namely Chlorophyta, Charophyta, Chlorarachniophyta, Cryptophyta, Cyanophyta, Dinophyta, Euglenophyta, Glaucophyta, Haptophyta, Heterokontophyta, and

Cultured Microalgae for the Food Industry. DOI: https://doi.org/10.1016/B978-0-12-821080-2.00011-3

Rhodophyta (Norton, Melkonian, & Andersen, 1996). Cultivation of microalgae does not require the same amounts of arable land as terrestrial sources of food proteins. Microalgae can grow in a vast array of aquatic environments from freshwater to extremely saline water and are tolerant to variation in temperatures, pH, and light intensities (Barsanti et al., 2008). For instance, most microalgae species grow in extreme saline environments, whereas green algae grow only in fresh or saline water (Gladyshev & Gubelit, 2019). Similarly, there have been microalgae blooms in hot deserts and others in Antarctic snow (Alam et al., 2019; Soto, Fuentes, Huovinen, & Gómez, 2020), thus indicating their resilience and essence in the biosphere.

The protein content of microalgae is estimated to be 30%–55% of the dry weight (González López et al., 2010). Bleakley and Hayes (2017) reported that the yield of protein in microalgae is estimated to be in the region of 4–15 tonnes/ha/year, which is higher than values for the most commonly consumed protein sources such as soybeans (0.6–1.2 tonnes ha^{-1} per year), wheat (1.1 tonnes ha^{-1} per year), and pulses (1–2 tonnes ha^{-1} per year). Notably, *Chlorella vulgaris* (51%–58% protein content dry weight) is reportedly the most utilized microalgae species for several industrial applications, whereas *Arthrospira platensis* is one of the microalgae species with the highest protein content (55%–74% dry weight) (Becker, 2007; Coelho et al., 2020; Markou et al., 2019). Proteins derived from microalgae can be processed like conventional proteins into several forms to produce protein flours, animal feed, food additives, functional foods, and nutraceuticals.

Bioactive peptides can be generated from microalgae proteins through microbial fermentation, in vitro and in vivo enzymatic hydrolysis, with subsequent downstream processing to remove insoluble materials (Acquah, Chan, Pan, Agyei, & Udenigwe, 2019; Sun, Acquah, Aluko, & Udenigwe, 2020). Bioactive peptides generated from microalgae proteins have been shown to have anticancer, antihypertensive, anticoagulant, antioxidative, antineoplastic, hypolipidemic, hepatoprotective, and immunomodulatory activities (Suttisuwan et al., 2019). These properties enhance the prospects of microalgae peptides as drug candidates, nutraceuticals, and in the formulation of functional and novel food products.

Despite the prolonged existence of microalgae as sustainable sources of food, there are limited studies on the potential of maximizing the commercial benefits of microalgae proteins relative to those from conventional terrestrial plant and animal sources. The most utilized techniques in microalgae protein extraction are alkaline solubilization-isoelectric precipitation and aqueous extraction. This results in a low protein extraction yield due to the inherent rigid cell wall structure of microalgae (Safi, Liu, et al., 2014). Moreover, commercial cultivation and application of microalgae proteins are impeded by factors such as seasonal changes, biogeography, extraction technologies, and regulations on the cultivation and use of microalgae. This chapter discusses the commercial use of microalgae proteins as alternative sources of proteins in the food industry, and various techniques used in enhancing the extraction and technofunctional properties of microalgae proteins.

4.2 Microalgae protein extraction

Microalgae are sustainable sources of biomass with high protein content and health-promoting molecules. This has heightened interest in the extraction and commercial applications of microalgae proteins (Grossmann, Ebert, Hinrichs, & Weiss, 2018). However, raw or unprocessed microalgae have low protein digestibility. Conventional methods for the extraction of microalgae proteins include aqueous, acidic, and alkaline extraction techniques, which are followed by a series of recovery steps such as centrifugation, ultrafiltration, precipitation, or chromatography. The efficiency of chemical-based approaches involving acid or alkali treatments for extracting proteins from microalgae were established in several species including *Porphyridium* spp., *Scenedesmus* spp., and *Chlorella* spp. (Kose, Ozen, Elibol, & Oncel, 2017; Lorenzo-Hernando, Ruiz-Vegas, Vega-Alegre, & Bolado-Rodríguez, 2019; Stack et al., 2018), among others. Owing to the presence of high-viscosity and anionic cell-wall polysaccharides that contribute to high cell rigidity of microalgae, the protein compositions are entrapped in the trilayered structures, thereby impeding their release during extraction. Therefore implementing a cell disruption pretreatment step prior to the actual extraction is crucial to facilitate the efficient extraction of microalgal proteins (Chiong et al., 2016). Cell disruption methods are categorized into two main groups: (1) mechanical and (2) nonmechanical methods, based on the working mechanism for cellular disintegration.

4.2.1 Microalgal cell disruption methods

4.2.1.1 Mechanical methods

The cell wall of microalgae can be disintegrated using mechanical forces such as solid or liquid-shear forces (e.g., grinding using bead mill, homogenizers, and pulverization). The high throughput, large sample-loading capacity, effective temperature control, and easy upscaling for industrial implementation makes bead milling a fascinating approach for cell disintegration. The most popular design of a bead mill comprises of a shaft in a fixed vessel that has agitators that transfer kinetic energy to the beads in the chamber resulting in many collisions. The cells are disintegrated in the bead collision zones by shear forces, as energy is transferred from the beads to the microalgal cells. By using a bead mill, cell disintegration can be effectively achieved but a significant amount of energy is required during upscaling. However, at a biomass concentration of 100–200 g L^{-1}, low energy consumption can be attained (Greenwell, Laurens, Shields, Lovitt, & Flynn, 2009). An earlier study focused on bead milling for mild cell disruption and promotion of protein recovery reported that at a temperature over 35°C, 32%–42% release and yield of water-soluble proteins was possible (Postma et al., 2015). *Chlorella vulgaris* was the microalgae investigated and the authors established that energy consumption reduced to levels below 2.5 $kWh\,kg^{-1}$ dry weight (Postma et al., 2015). Alternatively, a homogenizer can be used for mechanical disintegration of the microalgal cell wall as reported for *Chlorella pyrenoidosa* and *Nannochloropsis salina* (Ljubic, Safafar, & Jacobsen, 2019). Traditional mortars are still employed for cell wall disintegration of microalgae. However, an aqueous media of *Haematococcus pluvialis*, *Nannochloropsis oculata*, *Porphyridium cruentum*, and *Arthrospira platensis*, hand-ground by a mortar for 5 min was shown to have lower protein concentration compared to other disruption strategies such as high-pressure cell disruption, sonication, or chemical pretreatments (Safi, Ursu, et al., 2014).

4.2.1.2 Nonmechanical methods

Nonmechanical approaches may include the conventional use of a variety of solvents, enzymatic hydrolysis, osmotic stress, blanching, and other physical methods. In enzymatic hydrolysis, enzymes are applied as a pretreatment procedure for disintegration of cells before protein extraction to enhance the protein yield. A typical illustration was demonstrated in the preparation anti-inflammatory peptides from *Synechococcus* sp. VDW using trypsin digestion followed by further purification by ultrafiltration (Suttisuwan et al., 2019). Other gastrointestinal endopeptidases, that is, α-chymotrypsin and pepsin, have been used to recover potential antiinflammatory peptides from enzymatic hydrolysate of the edible microalgae *Spirulina maxima* (Vo, Ryu, & Kim, 2013). In addition, methods involving the application of osmotic stress have also been used to enhance the yield of water-soluble proteins from microalgae, with better efficiency than a high shear force exerted by Ultra-turrax® T25 Basic tool (Harnedy & FitzGerald, 2013). Alternatively, merging different extraction methods can present additional advantages for microalgal protein extraction (Shokrkar, Ebrahimi, & Zamani, 2017).

When choosing a method for cell disruption, care must be taken to prevent denaturation of the proteins. In addition, microalgae tissues contain proteases which can deter the process of extraction and purification of proteins (Wijesinghe & Jeon, 2012). From an economic viewpoint, it is vital to optimize the disintegration processes used to access the biomolecules in microalgae. The current processes are energy intensive, costly, and sometimes require large amounts of organic solvents with limited extraction efficiency and environmental implications. To enhance the process economics and product commercialization, extraction processes should be energy efficient, eco-friendly, and be able to preserve the product quality. Nowadays, it is preferable to use green and advanced cell disruption and extraction techniques that do not involve the use of toxic solvents and retain product quality.

Some emerging methods for protein extraction from microalgae are discussed below.

4.2.2 Nonconventional methods for microalgal protein extraction

4.2.2.1 Ultrasound-assisted extraction

Ultrasound-assisted extraction (UAE) is gaining popularity in the food industry for several applications including improving the liberation of microalgal proteins, which is induced by the creation, multiplication, and collapse of bubbles formed by acoustic cavitation (Ashokkumar et al., 2008). The intense collapse of these bubbles forms micropockets with high pressure and temperature, leading to excitation and particle rupture. These perturbations caused by ultrasound can destroy the cell walls, enhance solvent penetration into the matrices, and accelerate diffusion. UAE processes are known to be suitable for compounds that are sensitive to high temperature. Its other major advantages include quick processing time, minimal solvent consumption, and the higher purity of the final product compared to conventional extraction methods. In a recent study, pretreatment using ultrasonication coupled with subcritical water extraction (SWE) increased protein extraction from *Spirulina platensis*, when compared to each of the treatments alone (Fan, Hu, Wang, Yang, & Zhang, 2020). The enhanced extraction was likely caused by the eroding effect of the acoustic waves on the tough microalgal cell wall. A high-power intensity ($0.4\ \text{kWh}\ \text{L}^{-1}$) optimized protein fractionation from microalgae belonging to the Chlorococcales order during wastewater treatment (Keris-Sen, Sen, Soydemir, & Gurol, 2014). To further improve protein extraction, two alternating countercurrent frequencies can be employed during UAE (Qu, Ma, Wang, & Zheng, 2012). Although ultrasonication is a simple and easy method to implement, limitations such as high-power consumption and possible heat generation during prolonged usage could lead to protein degradation. In addition, the probe may require frequent replacement due to its distinct short lifetime and durability, thereby limiting its suitability for industrial application.

4.2.2.2 Microwave-assisted extraction

The microwave-assisted extraction (MAE) technique involves selective interactions of ionic and polar components with high-frequency (300 MHz to 300 GHz) nonionizing electromagnetic radiation for the extraction of target compounds. The principle of MAE is based on heating of the materials via two mechanisms, that is, dipole rotation or ionic conduction upon application of microwave power. This causes cleavage of the hydrogen bonds and movement of dissolved ions, leading to enhanced penetration of solvents. In essence, the moisture in the biomass evaporates, forming bubbles with high pressure that collapse to disrupt the cell contents and facilitate the extraction of the target compounds (Zhang et al., 2018). Higher yields of soluble proteins were reportedly extracted from a biomass of microalgae comprising of species of *Stigeoclonium*, *Monoraphidium*, *Nitzschia*, and *Navicula*, that were subjected to microwave pretreatment compared to those exposed to ultrasonication (Passos, Carretero, & Ferrer, 2015). MAE has garnered considerable attention as a promising method for the extraction of compounds from the food matrix due to its advantages over conventional methods, including higher extraction rate and yield, reduced solvent consumption, and lower operating cost. However, the use of MAE in microalgal protein extraction may be limited by temperature sensitivity, thus requiring further research to optimize its application.

4.2.2.3 Pulsed electric field

The pulsed electric field (PEF) extraction technique involves perforating cell membranes by the action of high electric fields (100 to 300 $kV\ cm^{-1}$), leading to temporary or permanent electroporation and formation of pores within the membranes (Vanthoor-Koopmans, Wijffels, Barbosa, & Eppink, 2013). Electroporation allows the introduction of various foreign components to cells, including proteins, by the application of a short duration of electric pulses (Fox et al., 2006). The irreversible electroporation caused my PEF aids the liberation of intracellular proteins and other high-value components. Some of the earliest works on PEF-assisted extraction of algal components reported an increase in the yield of various high-value compounds such as carbohydrates, lipids, chlorophyll, and carotenoids from microalgae (Lai, Parameswaran, Li, Baez, & Rittmann, 2014; Luengo et al., 2015; Postma et al., 2016; Zbinden et al., 2013). Since then, PEF has been used to increase the protein yield from species of *Spirulina* and *Chlorella* by 13% and 27%, respectively, under PEF conditions of 15 $kV\ cm^{-1}$ and 100 $kJ\ kg^{-1}$ (Coustets, Al-Karablieh, Thomsen, & Teissié, 2013). Another study reported a significant increase in protein yield from *Chorella vulgaris* and *Nannochloropsis salina* during PEF-assisted extraction, thereby permitting the release of intracellular cytosolic proteins (Coustets et al., 2013). Notwithstanding, factors such as conductivity and electrode gap need to be taken into consideration as they could limit the possibility of upscaling of this technology for industrial extraction purposes.

4.2.2.4 Other cell wall disruption and extraction techniques

Several other techniques can be used for algal protein extraction. For instance, subcritical and supercritical fluid extraction have become increasingly popular in the past two decades. In subcritical water extraction (SWE), hot water (100°C–374°C) is kept in the liquid state under high pressure (~10 bars). In supercritical fluid extraction (SFE), a fluid is heated above its critical point. Both SWE and SFE require expensive equipment and are commonly used for the extraction of lipids. Other methods include various physical methods such as steam explosion, hydrothermal liquefaction, freeze-drying, and the use of high-voltage electrical discharges (Bleakley & Hayes, 2017). In addition, membrane technologies, for example, microfiltration, ultrafiltration, nanofiltration, and reverse osmosis, are also used as enrichment methods to further enhance protein recovery from microalgal sources, in conjunction with a cell disintegration technique. These techniques have the advantages of being environmentally friendly and not requiring the application of heat (Kumar et al., 2013). They are also well-suited for maximizing the valorization of all the components of microalgae during biorefinery processes. Ultrafiltration has been used as a secondary step to high-pressure homogenization during the extraction of multiple components from *Tetraselmis suecica* (Safi, Liu, et al., 2014). Regardless of the method used, extraction efficiency varies with a range of factors such as the characteristics (cell size, shape, and structure) of the microalga, composition of extraction media (type of solvent, biomass concentration), and processing conditions (pressure, temperature, stirring effect, electric fields). A summary of the nonconventional techniques in microalgal protein extraction and their limitations is presented in Table 4.1.

4.3 Technological approaches for microalgae peptide production

The main techniques used for producing bioactive peptides from algal sources include enzymatic hydrolysis during gastrointestinal digestion or with the use of external proteolytic processes mediated by exogenous proteases, and microbial fermentation using enzymes secreted by the microorganisms (Perez Espitia et al., 2012). Chemical methods involving alkaline and acid hydrolysis are also utilized in some industries due to their economic advantage. In addition to these conventional methods, a trending approach is using membrane bioreactor systems or immobilized enzymes to enhance the product yield and decrease the quantity of enzymes used in microalgae peptide production (Jaradat, 2017). These methods are described below.

4.3.1 Peptide production by enzymatic hydrolysis

Several approaches based on using protein substrates have been developed to produce peptides from microalgae. These methods could be either proteolysis with gastrointestinal enzymes (in vivo), proteolysis with commercial enzymes (in vitro), or proteolysis via microbial fermentation (in situ). For enzymatic hydrolysis, the reaction is typically carried out at pH 4–8 and a temperature of 40°C–60°C depending on the optimum activity of the hydrolyzing proteases. These enzymes can act on microalgae proteins, forming fragments with distinct functional and biological activities (Hajfathalian, Ghelichi, García-Moreno, Moltke Sørensen, & Jacobsen, 2017; Ogino et al., 1999). In addition to optimum pH and temperature, the concentration of the enzyme and protein substrate need to be taken into consideration. To achieve high stability of the protein hydrolysate, antioxidants should be added before hydrolysis, or lipids and

TABLE 4.1 Benefits and limitations of the nonconventional processing techniques used to recover proteins from microalgae.

Technique	Mechanism	Efficiency	Benefits	Limitations
Ultrasound	Cavitation and free radical formation	+++	Economical Eco-friendly High cell disruption No toxic solvents Scalable	Longer treatment time Poor selectivity Difficulties with purification
Microwave	Temperature increase, molecular energy increase	+++	Short treatment duration	Thermal degradation of proteins High energy consumption
Pulsed electric field	Proliferation due to electricity	++	Eco-friendly Short treatment duration Less heating effect Energy savings Selective extraction Avoids toxic solvents	High cost
High-voltage electrical discharge	Proliferation due to electricity	+++	Eco-friendly Less heating effect No toxic solvents	Nonselective extraction Possible degradation of proteins
High-pressure homogenization	Cavitation and shear stress	+++	Eco-friendly High cell disruption High extraction yields	Pumpable solutions Nonselective extraction High cost
Subcritical fluid extraction	Substrate interaction	+++	Eco-friendly No toxic solvents	Low extraction yields High temperatures required for high extraction yields

(*Continued*)

TABLE 4.1 (Continued)

Technique	Mechanism	Efficiency	Benefits	Limitations
Supercritical fluid extraction	Substrate interaction	+ + +	Highly reduced toxic organic solventsCan be directly coupled with analytical techniques such as chromatography	Long extraction times Possible use of modifiers Can be relatively simple or highly complex, thus increasing the cost

" + ," slightly efficient; "++," efficient; "++ + ," highly efficient.

heme-containing proteins should be removed. Upon completion of hydrolysis, the reaction is stopped by adjusting the pH, or increasing the temperature, to render the proteases inactive. Subsequently, the mixture containing both the liberated peptides and unhydrolyzed proteins is subjected to filtration or centrifugation (Boeriu et al., 2010; Machado, Liria, Proti, Remuzgo, & Miranda, 2004). The resulting filtrate or supernatant is concentrated by drying. The specific biological activity possessed by the hydrolysate varies with the type of proteases used for the reaction, raw materials, and hydrolysis conditions. The peptides derived from this process also retain their nutritional value. Enzymatic hydrolysis is nontoxic, and gives higher yield and purity of peptides compared to solvent extraction or chemical approaches (Perez Espitia et al., 2012).

In proteolytic fermentation, microorganisms such as lactic acid bacteria, which produce proteases, act on the protein substrate to release different polypeptides (Elfahri, Donkor, & Vasiljevic, 2014; Sanjukta & Rai, 2016). In this case, it is important to optimize the growth parameters and the proteolytic activity exerted by the microorganisms. Additionally, to further improve the bioactivity of peptides, different proteases can be used in a multistep hydrolysis process (Hajfathalian et al., 2017). Moreover, a multistep recycling coupled with an ultrafiltration membrane system can be used to achieve a series of enzymatic digestion and isolate bioactive peptides from the proteins (Hajfathalian et al., 2017; Phong et al., 2018; Rizzello et al., 2016).

4.3.2 Peptide production by chemical methods

Due to low yields of purified peptides, chemical synthesis can be used to scale up the production of microalgal peptides identified after the downstream processing in Section 4.3.1. The production of peptides by chemical methods (i.e., liquid- and solid-phase synthesis) involves a condensation reaction between the carboxylic group (C-terminus) of one amino acid to the amino group (N-terminus) of another, in order to yield organic molecules with distinct bioactivities (Jaradat, 2017). In liquid-phase synthesis, all the reagents and reaction products in the medium are dissolved. In solid-phase synthesis, amino acids are added to create the peptide on the polymeric support, which remains insoluble in the medium (Blackburn, 1998; Vaino & Janda, 2000). This polymeric support consists of an insoluble resin that has a functional group located on its surface (linker) and permits the formation of stable bonds in the peptide sequence. In solid-phase synthesis, two steps are involved (Jaradat, 2017). First, acylation of an amino acid using the insoluble polymeric support occurs, followed by removal of the N-terminal group (unprotecting phase) and attachment of the next amino acid to the "peptide–linker–resin" complex. These two steps are repeated continuously until the target peptide arrangement is attained. Finally, a cleaving reagent is used to separate the complex. Generally, there are two distinct α-amino protecting groups: *tert*-butyloxycarbonyl (Boc) or 9-fluorenylmethyl carbamates (Fmoc) that are utilized for peptide synthesis, depending on the type of protector of the reactive group of the amino acids (N-terminal) participating in the synthesis (Hernández-Ledesma, del Mar Contreras, & Recio, 2011). In the first protocol, which is based on differences in sensitivity to acid, Boc serves as the N-amino group protector, while trifluoroacetic acid (TFA) is used to detach the Boc group. However, in cases where the protecting groups are lateral chains of ester, ether, and urethrane by-products, hydrofluoric acid is used. In the second protocol, Fmoc serves as the N-amino group protector and Fmoc group can be removed with ease using alkaline solutions like piperidine. The Fmoc protocol gives better yield and high-quality peptides, when compared to *t*-Boc protocol, due to its higher degree of chemoselectivity without modification of the acid-sensitive lateral chains (Perez Espitia et al., 2012).

Furthermore, regardless of the type of method (enzymatic hydrolysis or chemical approaches) used to produce microalgal peptides, a high degree of purity, that is, 95% or higher, is required before their molecular and biological activities can be maximized (Ovando et al., 2016). Hence, several processes are being developed for the separation of peptides from microalgae to obtain high-purity products. The choice of separation and purification approach depends solely on the distinct characteristics of the peptide, such as molecular weight, polarity, and solubility (Agyei, Pan, Acquah, Bekhit, & Danquah, 2019; Agyei, Pan, Acquah, & Danquah, 2017). Examples of peptide purification techniques widely explored include electrodialysis with ultrafiltration, capillary electrophoresis, and chromatography using reverse-phase, affinity, size exclusion, ion-exchange, and gel-permeation platforms (Acquah et al., 2019; Hajfathalian et al., 2017). In choosing a purification technique, special consideration is often given to the structural features of the target peptides. It is important to pay attention to the possibility of upscaling, the percentage yield, and the target end-market.

In addition, due to the possible loss of uniformity and distortion of structures, an additional step is needed to confirm the peptide purity and structural integrity. Techniques currently used for characterizing the peptides include fast atom bombardment mass spectrometry, electrospray ionization, and matrix-assisted laser desorption/ionization mass spectrometry (Perez Espitia et al., 2012). These methods can accurately provide information regarding the chemical structure and molecular mass–charge ratio of the peptide.

4.4 Technofunctional properties of microalgae proteins and peptides

The technofunctionality of proteins refers to those distinct properties that proteins exhibit in food formulations, aside from their regular nutritional functions. These physicochemical properties influence their technological roles during food processing, and contribute significantly to the organoleptic attributes (including texture, viscosity, palatability, and mouthfeel) of the processed food products. The technofunctional properties can be classified into: surface properties (solubility, emulsification, foaming, fat- and flavor-binding) and hydrodynamic properties (gelation, viscosity, thickening, and texturization) (Chen et al., 2019). The search for novel sources of functional ingredients remains an important objective of the food industry. Nowadays, there is a growing interest in microalgae proteins as sustainable and economical alternatives to the conventional animal- or plant-based proteins. As shown in Table 4.2 several studies have reported the technofunctional properties of microalgae proteins, especially those derived from species of *Spirulina* and *Tetraselmis*. When incorporating microalgae proteins into traditional or novel food formulations, it is important to consider factors such as the type of processing, food matrix (such as emulsion, gel, or foam systems), and potential interactions of the proteins with other food components.

4.4.1 Emulsifying properties

The emulsifying properties of proteins are dependent on their surface properties and are influenced by processes which modify the molecule size, charge, and distribution of hydrophilic and hydrophobic residues (Acquah, Di Stefano, & Udenigwe, 2018). Like proteins and peptides from animal and plant sources, the microalgae proteins are surface-active due to the presence of hydrophilic and hydrophobic functional groups, giving them the ability to adsorb to an interface and act as emulsifiers. An earlier study on *Tetraselmis* sp. demonstrated that microalgae proteins have good emulsifying properties and therefore can be used as functional ingredients in food formulations

TABLE 4.2 Studies on the technofunctional properties of crude protein extracts from microalgae.

Microalgae	Property	Remarks	References
Haematococcus pluvialis	Emulsification	High emulsion stability at pH 7 (94%) compared to pH 5.7 (84%)	Ba, Ursu, Laroche, & Djelveh (2016)
Chlorella minutissima *Isochrysis galbana* *Picochlorum* sp.	Water-holding capacity Oil-binding capacity	Maintained texture of final fish products by improving their functional properties	Ben Atitallah et al. (2019)
Arthrospira platensis	Emulsification Water-holding capacity Oil-binding capacity Foam formation Gelation	Functional properties significantly affected by pH; the emulsifying and foaming capacities were positively correlated to the protein solubility	Benelhadj et al. (2016)
Arthrospira platensis	Water-holding capacity Oil-binding capacity	Ultrasound pretreatment improved the protein content, water-holding capacity, and oil-binding capacity	Boukhari, Doumandji, Sabrine Ait chaouche, & Ferradji (2018)
Chlorella sorokiniana *Phaeodactylum tricornutum*	Emulsification	Emulsions prepared with both microalgae extracts were able to withstand very high ionic strengths	Ebert et al. (2019)
Tetraselmis suecica	Foaming Emulsification Gelation	Good functionality was obtained with crude samples, even after minimal separation steps	Garcia et al. (2018)
Porphyridium cruentum *Nannochloropsis* spp. *Phaeodactylum tricornutum*	Water-holding capacity Oil-binding capacity Emulsification Viscosity	*P. cruentum* and *Ph. tricornutum* biomass had functional properties comparable to those of soybean flour	Guil-Guerrero, Navarro-Juárez, López-Martínez, Campra-Madrid, & Rebolloso-Fuentes (2004)

(*Continued*)

TABLE 4.2 (Continued)

Microalgae	Property	Remarks	References
Nannochloropsis gaditana	Protein solubility Emulsification Foaming	High protein solubility observed to positively correlate with the emulsifying and foaming properties	Medina, Rubilar, Shene, Torres, & Verdugo (2015)
Tetraselmis sp.	Foam formation	Favors the production of stable foams by the selective adsorption of the protein fraction to the air–water interface	Schwenzfeier, Lech, Wierenga, Eppink, & Gruppen (2013)
Tetraselmis sp.	Foam formation Emulsification	Fractionation led to improved emulsion stability and foaming properties	Schwenzfeier et al. (2014)
Tetraselmis sp.	Gelation Foam formation Emulsification	Displayed better surface activity and gelation behavior compared to whey protein isolate	Suarez Garcia et al. (2018)
Arthrospira platensis	Foam formation	Better foaming properties compared to whey protein isolate	Buchmann et al. (2019)
Chlorella pyrenoidosa *Arthospira platensis* *Nannochloropsis oceanica*	Protein solubility Gelation Foam formation Emulsification Water-holding capacity Oil-binding capacity	Reduced functionality was observed near the isoelectric point. Functional properties were comparable to those of some commercial products	Chen et al. (2019)
Nannochloropsis gaditana *Tetraselmis impellucida* *Arthrospira maxima*	Emulsification	No correlation between interfacial and molecular properties (adsorption rate constant and zeta-potential) and the emulsion behavior	Teuling, Schrama, Gruppen, & Wierenga (2019)

(Schwenzfeier, Helbig, Wierenga, & Gruppen, 2013). More recent results have demonstrated a substantial degree of interfacial activity at the oil–water interface for proteins derived from *Phaeodactylum tricornutum* and *Chlorella sorokiniana* (Ebert, Grossmann, Hinrichs, & Weiss, 2019). Even after 7 days of storage, emulsions formed by the microalgae proteins were stable. At higher extract concentrations, the swift adsorption and increased viscosity confirmed that the reduction in interfacial tension exhibited by the microalgae was caused by their protein fractions and not polysaccharides, similar to an earlier report (Schwenzfeier, Wierenga, Eppink, & Gruppen, 2014).

In addition, emulsions from microalgae proteins show higher stability toward changes in pH ($pH \geq 5$) and can withstand high salt concentrations. This high tolerance to saline environments may be attributed to the beneficial protective effects of the charged groups at the interface, leading to denser emulsions with high stability. *Chlorella sorokiniana* was found to be well suited for stabilizing emulsions (oil/water) from pH 5–7, with a particle size (d_{43}) of $<1\ \mu m$, even during storage (Ebert et al., 2019). This supersedes the emulsifying property of many conventional food proteins, which typically show low emulsifying properties at pH 4–6, close to their isoelectric points. Other microalgae protein-fractions have also been confirmed to have high emulsion stability at various acidic pH values (Grossmann, Hinrichs, & Weiss, 2019; McClements & Gumus, 2016; Schwenzfeier, Helbig, et al., 2013) due to their high solubility. For example, proteins extracted from *Tetraselmis* sp. formed stable emulsions at pH 5–7 (Schwenzfeier, Wierenga, & Gruppen, 2011) and proteins from *Chlorella protothecoides* had high emulsification capacity at pH 2–12 (Grossmann et al., 2019; Schwenzfeier, Helbig, et al., 2013). These studies support the superior property of microalgal proteins in forming stable emulsions at a wide pH range compared to many conventional food proteins. Many factors influence the emulsifying properties of microalgae proteins, including the protein purity, molecular weight, charge distribution, and extent of glycosylation by carbohydrate moieties.

4.4.2 Foaming properties

The foaming capacity of a protein refers to the extent of interfacial area that can be formed by the protein, while foam stability is the ability of protein to retain the air bubbles formed. Several studies have demonstrated the superior foam formation and stability of microalgal proteins (Table 4.2). Some factors such as the pH are important in determining the foaming properties of proteins. For instance, the foaming capacity of proteins isolated from *Spirulina platensis* increased at pH values above 3.0 and over 250% at alkaline condition, pH 10 (Benelhadj, Gharsallaoui, Degraeve, Attia, & Ghorbel, 2016). The high foaming capacity at alkaline conditions is due to the higher net charge of the proteins, which increases the protein structural flexibility. This permits the protein to diffuse faster, encapsulate air bubbles, and increase the foam formation (Aluko & Yada, 1995). Therefore the foaming capacity of microalgal proteins can be maximized by exposing them to alkaline pH. Another study also confirmed that crude microalgal protein fractions from *Tetraselmis suecica* presented a higher surface activity than whey protein isolate for foaming and emulsification (Garcia et al., 2018). Even after chemical treatment, defatted cyanophyceae powder demonstrated a superior foaming capacity (Mahajan & Ahluwalia, 2010; Nirmala, Prakash, & Venkataraman, 1992). Furthermore, Devi and Venkataraman (1984) reported that *Arthrospira platensis* protein concentrate exhibited foaming property similar to that of soybean meal.

For foam stability, protein concentrates and isolates from *Spirulina* sp. LEB 18 were found to retain this foaming property for a longer time when compared to the original biomass, with 80% of the foam remaining after 60 min (Pereira, Lisboa, & Costa, 2018). Protein concentration is the major factor that influences the foaming properties. The higher the protein concentration, the higher the strength of air bubbles formed, and this is because of the reduced size of the bubbles and increase in the viscosity of the solution. A potential application of the high foam stability of microalgal protein concentrates and isolates is in the production of protein bars, where milk proteins are conventionally used. One problem encountered in the formulation of this type of product is the increase in hardness over time (Loveday, Hindmarsh, Creamer, & Singh, 2009). The high foam stability of *Spirulina* protein extracts can be used to maintain the product texture over longer durations. In addition, charged polysaccharides affect the functional properties of microalgal protein concentrates and isolates because they can modify the interfacial composition by coadsorption at the air–water interface (Schwenzfeier, Lech, et al., 2013). Smaller amounts of charged polysaccharides formed the most stable foams in *Tetraselmis* sp. protein isolate fractions (Schwenzfeier et al., 2014). In general, factors that contribute to the foaming properties of proteins include the source, composition, concentration, extraction, processing, and solubility of the protein, as well as pH and temperature of the medium, and the presence of carbohydrates and lipids. There is a dearth of detailed information on the influence of these factors on the foaming properties of microalgae proteins.

4.4.3 Gelling properties

Gelation is an important technofunctional property of microalgae proteins. Gelation is more favorable for proteins with high molecular weights due to their ability to form extensively cross-linked networks (Jeewanthi, Lee, & Paik, 2015). When proteins are partially unfolded, the exposed polypeptide segments interact at specific reactive groups to form the cross-linked networks. The native proteins must be denatured, or their conformation altered, followed by aggregation, before the formation of the gel matrix. Noncovalent bonds such as electrostatic and hydrophobic interactions are the main forces for gelation because they promote aggregation and subsequent gel settings (Jeewanthi et al., 2015), thereby reinforcing the rigidity of the gel network. In addition, intermolecular sulfhydryl and disulfide bonds are formed during gelation and contribute significantly to the elasticity of the gel structures. Extracts of *Tetraselmis suecica* containing 10% proteins were reported to show a rapid increase in gel rigidity (G′) within 1 min of heating compared to the extract containing 5% protein. It appears that the proteins denatured quickly to form the gels; this also demonstrates the effect of protein concentration on the formation of stable gels. However, Chronakis (2001) reported that the gelation capacity of *Spirulina platensis* proteins can be enhanced at a low protein concentration (1.5–2.5 wt.%) by using a stabilizing agent such as $CaCl_2$. Other factors that affect protein gelation include the presence of reducing agents and denaturants, and the temperature, pH, and ionic strength of the solution (Hajfathalian et al., 2017).

4.4.4 Water-holding capacity

Water-holding capacity is the ability of a protein to hold its own and added water during physical processing or when force is applied. It is an important functionality of proteins in various food formulations because it contributes significantly to food texture, especially in baked goods and comminuted meat products. Examples of microalgal species studied so far for their water-holding capacity include *Chlorella* sp., *Spirulina* sp., and *Tetraselmis* sp. (Lafarga et al., 2019). These microalgae could be potentially useful for reducing moisture loss and maintaining freshness and moist mouthfeel of food products, especially baked goods. *Tetraselmis* sp. was reported to show the least water-holding capacity values of the three microalgae, possibly due to the dissolution of salts from the biomass into the aqueous media during the assay (Lafarga et al., 2019). Although the proteins are thought to have these effects, other components of the microalgae such as the polysaccharides are expected to also contribute significantly. High water-holding capacity values of the microalgae proteins would also be favorable in food products such as sauces, soups, or custards, where they can give a thickening and viscous effect. Intrinsic factors that determine the water-holding capacity of food proteins include the conformation, amino acid composition, and surface hydrophobicity of the proteins (Garcia et al., 2018).

4.4.5 Fat absorption

Fat absorption capacity is an essential functional property in food products such as meat extenders or replacements, mainly because it promotes flavor retention and mouthfeel. The fat absorption capacity of *Arthrospira platensis* protein isolate (up to 252.7 ± 9.9 g of oil per 100 g) was found to be higher than that of soy protein (Kinsella, 1979). This high oil-binding ability could be because of the exposed nonpolar side chain residues associated with the proteins. In addition, physical pretreatment can contribute to improving the fat absorption capacity of the proteins. For instance, the fat absorption capacity value obtained for *Spirulina* powder (2.3 ml H_2O g^{-1}) was slightly higher than the value obtained by Devi and Venkataraman (1984) due to the influence of ultrasound treatment. The process may have broken noncovalent interactions, thereby exposing the hydrophobic regions of the proteins for enhanced fat binding. Detailed studies on isolated or purified microalgae proteins are needed to assess their potential use as functional ingredients in food products.

4.5 Application of microalgae proteins and peptides

Unprocessed microalgae have low protein digestibility due to the high viscosity in water and cell wall structures of the organisms. Therefore extraction of the proteins is expected to enhance the protein quality, digestibility, and bioavailability. Microalgae proteins can be utilized for several applications such as human nutrition, animal feed, biopharmaceutical production, and other industrial applications because of their diverse compositions, amino acid sequences, and structures resulting from their adaptation to complex habitats (Apone, Barbulova, & Colucci, 2019). Examples of companies that are involved in commercial production of microalgae biomass, food, and medicinal products are listed in Table 4.3.

4.5.1 Human nutrition and health benefits

Adequate food protein intake is crucial in maintaining normal metabolism and preventing protein-energy malnutrition and wasting. The presence of high amounts of proteins and other essential nutrients, such as carotenoids, vitamins, fatty acids, and minerals, in microalgae makes them sustainable alternatives to many conventional food sources. In addition, microalgae have been used as a source of human nutrition around the world for many centuries (Wells et al., 2017), although they remain largely underutilized in the food industry. Only a few microalgae species have been commercialized as functional foods or nutrient-dense supplements. The initial microalgae species to be commercialized as "health food" were *Chlorella* and *Spirulina* in Japan, Mexico, and Taiwan (Borowitzka, 2013). This has been followed by species such as *Dunaliella salina*, *Haematococcus pluvialis*, and *Crypthecodinium cohnii* (Borowitzka, 2013).

Microalgae proteins can be consumed as a part of the biomass or in other forms such as extracted powders, food additives, and functional food products. The proteins consist of the essential amino acids, including isoleucine, leucine, lysine, tryptophan, tyrosine, and valine, which make up ~35% of

TABLE 4.3 Commercial utilization of microalgae species for nutritional and medicinal purposes.

Commercial producer	Microalgae product	Country
Algae Tec	Nutraceuticals and aquafeed	Australia
Algomed	Sports nutrition, functional foods, and nutraceuticals	Germany
Algosource	Nutraceuticals and functional foods	France
Algatechnologies	Food and beverages, nutrition, and cosmetics	Israel
Aurora Algae	Pharmaceuticals, food, and nutrition	United States
Algaeon	Animal feed, food, and nutrition	United States
Aquafana	AlgaMac ProteinPlus	United States
Algae Technology Solution	Sustainable algae-based foods	Australia
Allmicroalgae	Food supplements and food ingredients	Portugal
BlueBioTech	Algae powder, nutraceuticals, and nutrition	Germany
Canadian Pacific Algae	Algae biomass and algae microbeads	Canada
Hortifarm Corporation	Algae cultures and bioreactors	India
Phytobloom	Aqua and animal feed, pharmaceutical products, and functional foods	Portugal
Oilgae	Functional foods and beverages, and bioenergy	India
Parry Nutraceuticals	Astaxanthin, food supplements, and nutraceuticals	India
Phytolutions	Food supplements	Germany
Subitec	Algae powder	Germany
Solazyme, Inc.	Biofuels, cosmetics, nutrition	United States
Sun Chlorella Corporation	Health products	Japan
Taiwan Chlorella Manufacturing	Assorted algae food products (including chocolates and juices) and powders	Taiwan

the total amino acids (Waghmare, Salve, LeBlanc, & Arya, 2016). *Spirulina*, an edible cyanobacteria, is the most commonly consumed microalgae and is considered to be a "superfood" because to its nutrient density, including its high protein content of 60%–70% dry weight (Koyande et al., 2019). The commercial production of *Spirulina* and *Chlorella* spp. is ~7500 tonnes of dry biomass annually, which is an important contribution toward combating the protein gap (Andrade, 2018). Human studies with *Spirulina platensis* as food supplements demonstrated an enhanced nutritional status in undernourished HIV-infected and non-HIV-infected children, particularly in ameliorating anemia and weight loss in Burkina Faso (Simpore et al., 2006, 2005). Despite this promise, there is the potential issue of consumer acceptability of microalgae-based food products, especially in populations that do not normally consume microalgae as food. Nonetheless, the incorporation of 0.25%–4% of microalgae biomass into cookies, pasta, and bread did not negatively alter the texture and rheological properties of the products (Batista et al., 2017; Graça, Fradinho, Sousa, & Raymundo, 2018; Mostolizadeh, Moradi, Mortazavi, Motallebi, & Ghaeni, 2020). More so, the use of mild cell wall disruption techniques was reported to enhance the release of proteins from the microalgae, the formation of macromolecular networks, and the bioactivity of wheat bread (Nunes et al., 2020). The findings suggest the potential dual role of microalgae proteins in food structuring and product formulation, and as a bioactive agent in functional food development.

Functional foods are foods that when consumed have the potential to improve physiological functions and minimize the risk of diseases in addition to their basic nutritional benefits. The most promising use of microalgae proteins for health promotion is through the bioactive peptides released by enzymatic hydrolysis or during digestion in the body (Ejike et al., 2017). For instance, Suttisuwan et al. (2019) reported the identification of antioxidative peptides (AILQSYSAGKTK, ALNKTHLIQTK, LLVHAPVK, IPDAHPVK, and VVVLRDGAVQQLGTPR) from *Synechococcus* sp. VDW proteins after enzymatic hydrolysis using trypsin. Similarly, Sedighi, Jalili, Ranaei-Siadat, and Amrane (2016) reported an increase in the antimicrobial activity after pepsin hydrolysis of *Chlorella vulgaris* proteins compared to the microalgae biomass or intact proteins. Nonetheless, the stability of the microalgae peptides during food product development and in the gastrointestinal tract is not clear. These factors are crucial, as the native structure of the peptides needs to remain intact within the food matrix in order to maintain their bioaccessibility, bioavailability, and physiological bioactivity (Sun & Udenigwe, 2020; Udenigwe & Fogliano, 2017). Because bioactive peptides are susceptible to protease degradation during digestion in the body, it is imperative to discover microalgae peptides without the scissile bonds. This will prevent the loss of bioactivity when the peptides are used in formulating new food products; see Sun et al. (2020) for a detailed discussion. To circumvent the structural stability issue, Wang and Zhang (2013) demonstrated that the molecular integrity of bioactive peptides generated from microalgae proteins can be preserved through micro-/nanoencapsulation.

4.5.2 Animal feed

Aquaculture and livestock farming are heavily reliant on using protein-rich meals as feeds, with soybean meal and fish meal being the primary sources (Mo, Man, & Wong, 2020; OECD/FAO, 2018). With 40%–70% protein content, microalgae are considered more environmentally sustainable and affordable sources of protein relative to soybean meals (45%–57% protein content). Microalgae also have good protein quality and protein yields (per hectare per year) that are 20 and 40 times higher than those of soybean and corn, respectively (Spiller et al., 2020). This makes microalgae more suitable as the source of proteins for animal feed. The biochemical composition of microalgae species can be optimized to have improved protein yield and nutritional quality for livestock feeding by altering the mode of culture conditions (Hulatt, Wijffels, Bolla, & Kiron, 2017; Jia et al., 2015; Li et al., 2020). Mo et al. (2020) also reported the lowering of the cost of culturing gold-lined sea bream by almost half by replacing 60% of the fish meal with digested soybean dreg with microalgae supplement. In our opinion, the formulated feed can be further improved by replacing the entire soybean component with microalgae because of the high amount of antinutritional factors in legumes that inhibit protein digestibility. Previous studies have shown that incorporating 2.5%–20% of microalgae into the fish diet enhanced the growth performance, lipid metabolism, and feed efficiency of fish (Norambuena et al., 2015; Yadav, Meena, Sahoo, Das, & Sen, 2020). Similar findings were also reported in poultry that were fed with microalgae-supplemented diets (Dineshbabu, Goswami, Kumar, Sinha, & Das, 2019; Gadiev et al., 2019). Despite the cost effectiveness, the quality, especially palatability, of the microalgae-supplemented animal feed should be taken into consideration.

4.5.3 Other commercial applications

Besides food and feed, microalgae proteins can be utilized in many other medicinal and cosmetic applications due to their biophysicochemical properties. Bioactive peptides derived from microalgae proteins have been demonstrated to have potential cosmeceutical applications in skincare (Chen, Liou, Chen, & Shih, 2011; Shih & Cherng, 2012). For example, *Chlorella*-derived peptides were found to reduce the UVB-induced matrix metalloproteinase-1 and cysteine-rich 61 (CYR61) mRNA expression, and monocyte chemoattractant protein-1 production in UV-B irradiated fibroblasts (Chen et al., 2011). Despite the promising in vitro results, there is a dearth of clinical studies on the potential of microalgae peptides in formulating cosmetic products. Some microalgae species such as cyanobacteria and rhodophyta contain phycobiliproteins, which are intensely fluorescent, water-soluble proteins that are covalently bound, via the cysteine residues, to chromophores known as phycobilins. Examples of phycobiliproteins include allophycocyanin, phycocyanin, and phycoerythrin. Industrial applications of phycobilin proteins include the production of fluorescent tags, development of antioxidants in cosmetics, formulation of functional foods, and as photosensitizers in photodynamic therapies (Borowitzka, 2013). Presently, companies such as Earthrise Nutritionals, Parry Nutraceuticals, and Fuqing King Dirmsa Spirulina Co., Ltd. commercially produce "Generally Regarded as Safe" (GRAS) pigments from microalgae phycobiliproteins (Costa et al., 2019).

4.6 Conclusion

Beyond the biofuel buzz, microalgae cultivation can also be a sustainable means of producing high-quality proteins for food and feed. Compared to terrestrial sources of food proteins, microalgae can be cultivated under various environmental conditions and with minimal requirement for arable lands. Microalgae also have high protein content and balanced essential amino acid compositions, with the potential to diversify the protein profile in response to diverse and harsh environmental stimuli. Furthermore, the physicochemical and technofunctional properties (e.g., foaming, emulsion, gelation, and fat absorption capacity) of microalgae proteins compare favorably to the properties of traditional food proteins. Therefore microalgae proteins can be used to formulate new food and feed products. Microalgae proteins can be processed through enzymatic hydrolysis or microbial fermentation to generate bioactive peptides that can be used in the production of functional foods, nutraceuticals, cosmeceuticals, and drugs. Nonetheless, microalgae have rigid cell walls that limit the bioavailability and digestibility of proteins embedded within the cells. Pretreatments of microalgae using mechanical and nonmechanical processes can disrupt the cell walls and liberate the proteins. Mechanical pretreatment using a bead mill is one of the most commonly utilized techniques. It is beneficial to use a cascade of nonmechanical techniques that preserve the molecular integrity of the extracted microalgae proteins. Future research and development should focus on the upscaling and application of microalgae proteins in the food industry.

References

Acquah, C., Chan, Y. W., Pan, S., Agyei, D., & Udenigwe, C. C. (2019). Structure-informed separation of bioactive peptides. *Journal of Food Biochemistry*, *43*(1), e12765. Available from https://doi.org/10.1111/jfbc.12765.

Acquah, C., Di Stefano, E., & Udenigwe, C. C. (2018). Role of hydrophobicity in food peptide functionality and bioactivity. *Journal of Food Bioactives*, *4*, 88–98. Available from https://doi.org/10.31665/jfb.2018.4164.

Agyei, D., Pan, S., Acquah, C., Bekhit, A. E.-D. A., & Danquah, M. K. (2019). Structure-informed detection and quantification of peptides in food and biological fluids. *Journal of Food Biochemistry*, *43*(1), e12482. Available from https://doi.org/10.1111/jfbc.12482.

Agyei, D., Pan, S., Acquah, C., & Danquah, M. K. (2017). Bioactivity profiling of peptides from food proteins. In A. M. Grumezescu, & A. M. Holban (Eds.), *Soft Chemistry and Food Fermentation* (pp. 49–77). Academic Press/Elsevier. Available from https://doi.org/10.1016/b978-0-12-811412-4.00003-5.

Alam, M. M., Mumtaz, A. S., Hallenbeck, P. C., Shah, S. A., Abidin, S. Z. U., & Inam, F. (2019). Morphological characterization, growth appraisal, and probing biofuels potential of newly isolated Scenedesmus sp. from desert Cholistan. *Microscopy Research and Technique*, *82*(12), 2079–2088. Available from https://doi.org/10.1002/jemt.23380.

Aluko, R. E., & Yada, R. Y. (1995). Structure-function relationships of cowpea (Vigna unguiculata) globulin isolate: influence of pH and NaCl on physicochemical and functional properties. *Food Chemistry*, *53*(3), 259–265. Available from https://doi.org/10.1016/0308-8146(95)93931-g.

Andrade, L. M. (2018). Chlorella and spirulina microalgae as sources of functional foods, nutraceuticals, and food supplements; an overview. *MOJ Food Processing & Technology*, *6*(1), 45–58. Available from https://doi.org/10.15406/mojfpt.2018.06.00144.

Apone, F., Barbulova, A., & Colucci, M. G. (2019). Plant and microalgae derived peptides are advantageously employed as bioactive compounds in cosmetics. *Frontiers in Plant Science*, *10*, 756. Available from https://doi.org/10.3389/fpls.2019.00756.

Artan, A., Acquah, C., Danquah, M. K., & Ongkudon, C. M. (2015). Process analysis of microalgae biomass thermal disruption for biofuel production. In P. Ravindra (Ed.), *Advances in Bioprocess Technology* (pp. 113–131). Cham: Springer. Available from https://doi.org/10.1007/978-3-319-17915-5_7.

Ashokkumar, M., Sunartio, D., Kentish, S., Mawson, R., Simons, L., Vilkhu, K., & Versteeg, C. (2008). Modification of food ingredients by ultrasound to improve functionality: a preliminary study on a model system. *Innovative Food Science & Emerging Technologies*, *9*(2), 155–160. Available from https://doi.org/10.1016/j.ifset.2007.05.005.

Ba, F., Ursu, A. V., Laroche, C., & Djelveh, G. (2016). Haematococcus pluvialis soluble proteins: extraction, characterization, concentration/fractionation and emulsifying properties. *Bioresource Technology*, *200*, 147–152. Available from https://doi.org/10.1016/j.biortech.2015.10.012.

Barsanti, L., Primo, C., Valtere, E., Frassanito, A. M., Passarelli, V., Vesentini, N., & Gualtieri, P. (2008). Oddities and curiosities in the algal world. In V. Evangelista, L. Barsanti, A. M. Frassanito, V. Passarelli, & P. Gualtieri (Eds.), *Algal Toxins: Nature, Occurrence, Effect and Detection. NATO Science for Peace and Security Series A: Chemistry and Biology* (pp. 353–391). Dordrecht: Springer. Available from https://doi.org/10.1007/978-1-4020-8480-5_17.

Batista, A. P., Niccolai, A., Fradinho, P., Fragoso, S., Bursic, I., Rodolfi, L., ... Raymundo, A. (2017). Microalgae biomass as an alternative ingredient in cookies: sensory, physical and chemical properties, antioxidant activity and in vitro digestibility. *Algal Research*, *26*, 161–171. Available from https://doi.org/10.1016/j.algal.2017.07.017.

Becker, E. W. (2007). Micro-algae as a source of protein. *Biotechnology Advances*, *25*(2), 207–210. Available from https://doi.org/10.1016/J.BIOTECHADV.2006.11.002.

Ben Atitallah, A., Hentati, F., Dammak, M., Hadrich, B., Fendri, I., Ayadi, M.-A., ... Barkallah, M. (2019). Effect of microalgae incorporation on quality characteristics and functional and antioxidant capacities of ready-to-eat fish burgers made from common carp (*Cyprinus carpio*). *Applied Sciences*, *9*(9), 1830. Available from https://doi.org/10.3390/app9091830.

Benelhadj, S., Gharsallaoui, A., Degraeve, P., Attia, H., & Ghorbel, D. (2016). Effect of pH on the functional properties of Arthrospira (Spirulina) platensis protein isolate. *Food Chemistry*, *194*, 1056–1063. Available from https://doi.org/10.1016/j.foodchem.2015.08.133.

Blackburn, C. (1998). Polymer supports for solid-phase organic synthesis. *Peptide Science*, *47*(5), 311–351, https://doi.org/10.1002/(sici)1097-0282(1998)47:5 < 311::aid-bip2 > 3.0.co;2-v.

Bleakley, S., & Hayes, M. (2017). Algal proteins: extraction, application, and challenges concerning production. *Foods*, *6*(5), 33. Available from https://doi.org/10.3390/foods6050033.

Boeriu, C. G., Frissen, A. E., Boer, E., van Kekem, K., van Zoelen, D.-J., & Eggen, I. F. (2010). Optimized enzymatic synthesis of C-terminal peptide amides using subtilisin A from Bacillus licheniformis. *Journal of Molecular Catalysis B: Enzymatic*, *66*(1–2), 33–42. Available from https://doi.org/10.1016/j.molcatb.2010.03.010.

Borowitzka, M. A. (2013). High-value products from microalgae-their development and commercialisation. *Journal of Applied Phycology*, *25*, 743–756. Available from https://doi.org/10.1007/s10811-013-9983-9.

Boukhari, N., Doumandji, A., Sabrine Ait chaouche, F., & Ferradji, A. (2018). Effect of ultrasound treatment on protein content and functional properties of Spirulina powder grown in Algeria. *Mediterranean Journal of Nutrition and Metabolism*, *11*(3), 235–249. Available from https://doi.org/10.3233/mnm-180220.

Buchmann, L., Bertsch, P., Böcker, L., Krähenmann, U., Fischer, P., & Mathys, A. (2019). Adsorption kinetics and foaming properties of soluble microalgae fractions at the air/water interface. *Food Hydrocolloids*, *97*, 105182. Available from https://doi.org/10.1016/j.foodhyd.2019.105182.

Chen, C.-L., Liou, S.-F., Chen, S.-J., & Shih, M.-F. (2011). Protective effects of Chlorella-derived peptide on UVB-induced production of MMP-1 and degradation of procollagen genes in human skin fibroblasts. *Regulatory Toxicology and Pharmacology*, *60*(1), 112–119. Available from https://doi.org/10.1016/J.YRTPH.2011.03.001.

Chen, Y., Chen, J., Chang, C., Chen, J., Cao, F., Zhao, J., … Zhu, J. (2019). Physicochemical and functional properties of proteins extracted from three microalgal species. *Food Hydrocolloids*, *96*, 510–517. Available from https://doi.org/10.1016/j.foodhyd.2019.05.025.

Chiong, T., Acquah, C., Lau, S. Y., Khor, E. H., & Danquah, M. K. (2016). Microalgal-based protein by-products: extraction, purification, and applications. In G. S. Dhillon (Ed.), *Protein Byproducts Transformation from Environmental Burden Into Value-Added Products* (pp. 213–234). Elsevier/Academic Press. Available from https://doi.org/10.1016/B978-0-12-802391-4.00012-4.

Chronakis, I. S. (2001). Gelation of edible blue-green algae protein isolate (Spirulina platensis strain Pacifica): thermal transitions, rheological properties, and molecular forces involved. *Journal of Agricultural and Food Chemistry*, *49*(2), 888–898. Available from https://doi.org/10.1021/jf0005059.

Coelho, D., Lopes, P. A., Cardoso, V., Ponte, P., Brás, J., Madeira, M. S., … Prates, J. A. M. (2020). A two-enzyme constituted mixture to improve the degradation of Arthrospira platensis microalga cell wall for monogastric diets. *Journal of Animal Physiology and Animal Nutrition*, *104*(1), 310–321. Available from https://doi.org/10.1111/jpn.13239.

Costa, J. A. V., Freitas, B. C. B., Rosa, G. M., Moraes, L., Morais, M. G., & Mitchell, B. G. (2019). Operational and economic aspects of Spirulina-based biorefinery. *Bioresource Technology*, *292*, 121946. Available from https://doi.org/10.1016/j.biortech.2019.121946.

Coustets, M., Al-Karablieh, N., Thomsen, C., & Teissié, J. (2013). Flow process for electroextraction of total proteins from microalgae. *The Journal of Membrane Biology*, *246*(10), 751–760. Available from https://doi.org/10.1007/s00232-013-9542-y.

Devi, M. A., & Venkataraman, L. V. (1984). Functional properties of protein products of mass cultivated blue-green alga Spirulina platensia. *Journal of Food Science*, *49*(1), 24–27. Available from https://doi.org/10.1111/j.1365-2621.1984.tb13660.x.

Dineshbabu, G., Goswami, G., Kumar, R., Sinha, A., & Das, D. (2019). Microalgae–nutritious, sustainable aqua- and animal feed source. *Journal of Functional Foods*, *62*, 103545. Available from https://doi.org/10.1016/j.jff.2019.103545.

Ebert, S., Grossmann, L., Hinrichs, J., & Weiss, J. (2019). Emulsifying properties of water-soluble proteins extracted from the microalgae *Chlorella sorokiniana* and *Phaeodactylum tricornutum*. *Food & Function*, *10*(2), 754–764. Available from https://doi.org/10.1039/c8fo02197j.

Edeseyi, M. E., Kaita, A. Y., Harun, R., Danquah, M. K., Acquah, C., & Sia, J. K. M. (2015). Rethinking sustainable biofuel marketing to titivate commercial interests. *Renewable and Sustainable Energy Reviews*, *52*, 781–792. Available from https://doi.org/10.1016/j.rser.2015.07.117.

Ejike, C. E. C. C., Collins, S. A., Balasuriya, N., Swanson, A. K., Mason, B., & Udenigwe, C. C. (2017). Prospects of microalgae proteins in producing peptide-based functional foods for promoting cardiovascular health. *Trends in Food Science & Technology*, *59*, 30–36. Available from https://doi.org/10.1016/j.tifs.2016.10.026.

Elfahri, K. R., Donkor, O. N., & Vasiljevic, T. (2014). Potential of novel Lactobacillus helveticus strains and their cell wall bound proteases to release physiologically active peptides from milk proteins. *International Dairy Journal*, *38*(1), 37–46. Available from https://doi.org/10.1016/j.idairyj.2014.03.010.

Fan, X., Hu, S., Wang, K., Yang, R., & Zhang, X. (2020). Coupling of ultrasound and subcritical water for peptides production from Spirulina platensis. *Food and Bioproducts Processing*, *121*, 105–112. Available from https://doi.org/10.1016/j.fbp.2020.01.012.

FAO. (2009). Global agriculture towards 2050. *High Level Expert Forum-How to Feed the World 2050*, pp. 1–4. Retrieved from http://www.fao.org/fileadmin/templates/wsfs/docs/Issues_papers/HLEF2050_Global_Agriculture.pdf

Fox, M. B., Esveld, D. C., Valero, A., Luttge, R., Mastwijk, H. C., Bartels, P. V., ... Boom, R. M. (2006). Electroporation of cells in microfluidic devices: a review. *Analytical and Bioanalytical Chemistry*, *385*(3), 474–485. Available from https://doi.org/10.1007/s00216-006-0327-3.

Gadiev, R. R., Khaziev, D. D., Galina, C. R., Farrakhov, A. R., Farhutdinov, K. D., Dolmatova, I. Y., ... Latypova, G. F. (2019). The use of chlorella in goose breeding. *AIMS Agriculture and Food*, *4*(2), 349–361. Available from https://doi.org/10.3934/agrfood.2019.2.349.

Garcia, E. S., van Leeuwen, J. J. A., Safi, C., Sijtsma, L., van den Broek, L. A. M., Eppink, M. H. M., ... van den Berg, C. (2018). Techno-functional properties of crude extracts from the green microalga Tetraselmis suecica. *Journal of Agricultural and Food Chemistry*, *66* (29), 7831–7838. Available from https://doi.org/10.1021/acs.jafc.8b01884.

Gladyshev, M. I., & Gubelit, Y. I. (2019). Green tides: new consequences of the eutrophication of natural waters. *Contemporary Problems of Ecology*, *12*, 109–125. Available from https://doi.org/10.1134/S1995425519020057.

González López, C. V., García, M., del, C. C., Fernández, F. G. A., Bustos, C. S., Chisti, Y., & Sevilla, J. M. F. (2010). Protein measurements of microalgal and cyanobacterial biomass. *Bioresource Technology*, *101*(19), 7587–7591. Available from https://doi.org/10.1016/j.biortech.2010.04.077.

Graça, C., Fradinho, P., Sousa, I., & Raymundo, A. (2018). Impact of Chlorella vulgaris on the rheology of wheat flour dough and bread texture. *LWT - Food Science and Technology*, *89*, 466–474. Available from https://doi.org/10.1016/j.lwt.2017.11.024.

Greenwell, H. C., Laurens, L. M. L., Shields, R. J., Lovitt, R. W., & Flynn, K. J. (2009). Placing microalgae on the biofuels priority list: a review of the technological challenges. *Journal of The Royal Society Interface*, *7*(46), 703–726. Available from https://doi.org/10.1098/rsif.2009.0322.

Grossmann, L., Ebert, S., Hinrichs, J., & Weiss, J. (2018). Effect of precipitation, lyophilization, and organic solvent extraction on preparation of protein-rich powders from the microalgae Chlorella protothecoides. *Algal Research*, *29*, 266–276. Available from https://doi.org/10.1016/j.algal.2017.11.019.

Grossmann, L., Hinrichs, J., & Weiss, J. (2019). Solubility and aggregation behavior of protein fractions from the heterotrophically cultivated microalga Chlorella protothecoides. *Food Research International*, *116*, 283–290. Available from https://doi.org/10.1016/j.foodres.2018.08.037.

Guil-Guerrero, J. L., Navarro-Juárez, R., López-Martínez, J. C., Campra-Madrid, P., & Rebolloso-Fuentes, M. (2004). Functional properties of the biomass of three microalgal species. *Journal of Food Engineering*, *65*(4), 511–517. Available from https://doi.org/10.1016/j.jfoodeng.2004.02.014.

Hajfathalian, M., Ghelichi, S., García-Moreno, P. J., Moltke Sørensen, A.-D., & Jacobsen, C. (2017). Peptides: production, bioactivity, functionality, and applications. *Critical Reviews in Food Science and Nutrition*, *58*(18), 3097–3129. Available from https://doi.org/10.1080/10408398.2017.1352564.

Harnedy, P. A., & FitzGerald, R. J. (2013). Extraction of protein from the macroalga Palmaria palmata. *LWT - Food Science and Technology*, *51*(1), 375–382. Available from https://doi.org/10.1016/j.lwt.2012.09.023.

Hernández-Ledesma, B., del Mar Contreras, M., & Recio, I. (2011). Antihypertensive peptides: production, bioavailability and incorporation into foods. *Advances in Colloid and Interface Science*, *165*(1), 23–35. Available from https://doi.org/10.1016/j.cis.2010.11.001.

Hulatt, C. J., Wijffels, R. H., Bolla, S., & Kiron, V. (2017). Production of fatty acids and protein by nannochloropsis in flat-plate photobioreactors. *PLoS One*, *12*(1), e0170440. Available from https://doi.org/10.1371/journal.pone.0170440.

Jaradat, D. M. M. (2017). Thirteen decades of peptide synthesis: key developments in solid phase peptide synthesis and amide bond formation utilized in peptide ligation. *Amino Acids*, *50*(1), 39–68. Available from https://doi.org/10.1007/s00726-017-2516-0.

Jeewanthi, R. K. C., Lee, N.-K., & Paik, H.-D. (2015). Improved functional characteristics of whey protein hydrolysates in food industry. *Korean Journal for Food Science of Animal Resources*, *35*(3), 350–359. Available from https://doi.org/10.5851/kosfa.2015.35.3.350.

Jia, J., Han, D., Gerken, H. G., Li, Y., Sommerfeld, M., Hu, Q., & Xu, J. (2015). Molecular mechanisms for photosynthetic carbon partitioning into storage neutral lipids in Nannochloropsis oceanica under nitrogen-depletion conditions. *Algal Research*, *7*, 66–77. Available from https://doi.org/10.1016/j.algal.2014.11.005.

Katiyar, R., & Arora, A. (2020). Health promoting functional lipids from microalgae pool: a review. *Algal Research*, *46*, 101800. Available from https://doi.org/10.1016/j.algal.2020.101800.

Keris-Sen, U. D., Sen, U., Soydemir, G., & Gurol, M. D. (2014). An investigation of ultrasound effect on microalgal cell integrity and lipid extraction efficiency. *Bioresource Technology*, *152*, 407–413. Available from https://doi.org/10.1016/j.biortech.2013.11.018.

Kinsella, J. E. (1979). Functional properties of soy proteins. *Journal of the American Oil Chemists' Society*, *56*(3Part1), 242–258. Available from https://doi.org/10.1007/bf02671468.

Kose, A., Ozen, M. O., Elibol, M., & Oncel, S. S. (2017). Investigation of in vitro digestibility of dietary microalga Chlorella vulgaris and cyanobacterium Spirulina platensis as a nutritional supplement. *3 Biotech*, *7*(3), 170. Available from https://doi.org/10.1007/s13205-017-0832-4.

Koyande, A. K., Chew, K. W., Rambabu, K., Tao, Y., Chu, D.-T., & Show, P.-L. (2019). Microalgae: a potential alternative to health supplementation for humans. *Food Science and Human Wellness, 8*(1), 16–24. Available from https://doi.org/10.1016/J.FSHW.2019.03.001.

Kumar, P., Sharma, N., Ranjan, R., Kumar, S., Bhat, Z. F., & Jeong, D. K. (2013). Perspective of membrane technology in dairy industry: a review. *Asian-Australasian Journal of Animal Sciences, 26*(9), 1347–1358. Available from https://doi.org/10.5713/ajas.2013.13082.

Lafarga, T., Acién-Fernández, F. G., Castellari, M., Villaró, S., Bobo, G., & Aguiló-Aguayo, I. (2019). Effect of microalgae incorporation on the physicochemical, nutritional, and sensorial properties of an innovative broccoli soup. *LWT, 111*, 167–174. Available from https://doi.org/10.1016/j.lwt.2019.05.037.

Lai, Y. S., Parameswaran, P., Li, A., Baez, M., & Rittmann, B. E. (2014). Effects of pulsed electric field treatment on enhancing lipid recovery from the microalga, Scenedesmus. *Bioresource Technology, 173*, 457–461. Available from https://doi.org/10.1016/j.biortech.2014.09.124.

Li, T., Chen, Z., Wu, J., Wu, H., Yang, B., Dai, L., ... Xiang, W. (2020). The potential productivity of the microalga, Nannochloropsis oceanica SCS-1981, in a solar powered outdoor open pond as an aquaculture feed. *Algal Research, 46*, 101793. Available from https://doi.org/10.1016/j.algal.2020.101793.

Ljubic, A., Safafar, H., & Jacobsen, C. (2019). Recovery of microalgal biomass and metabolites from homogenized, swirl flash-dried microalgae. *Journal of Applied Phycology, 31*(4), 2355–2363. Available from https://doi.org/10.1007/s10811-019-1733-1.

Lorenzo-Hernando, A., Ruiz-Vegas, J., Vega-Alegre, M., & Bolado-Rodríguez, S. (2019). Recovery of proteins from biomass grown in pig manure microalgae-based treatment plants by alkaline hydrolysis and acidic precipitation. *Bioresource Technology, 273*, 599–607. Available from https://doi.org/10.1016/j.biortech.2018.11.068.

Loveday, S. M., Hindmarsh, J. P., Creamer, L. K., & Singh, H. (2009). Physicochemical changes in a model protein bar during storage. *Food Research International, 42*(7), 798–806. Available from https://doi.org/10.1016/j.foodres.2009.03.002.

Luengo, E., Martínez, J. M., Coustets, M., Álvarez, I., Teissié, J., Rols, M.-P. P., & Raso, J. (2015). A comparative study on the effects of millisecond- and microsecond-pulsed electric field treatments on the permeabilization and extraction of pigments from Chlorella vulgaris. *The Journal of Membrane Biology, 248*(5), 883–891. Available from https://doi.org/10.1007/s00232-015-9796-7.

Machado, A., Liria, C. W., Proti, P. B., Remuzgo, C., & Miranda, M. T. M. (2004). Sínteses química e enzimática de peptídeos: princípios básicos e aplicações. *Química Nova, 27*(5), 781–789. Available from https://doi.org/10.1590/s0100-40422004000500018.

Mahajan, A., & Ahluwalia, A. S. (2010). Effect of processing on functional properties of spirulina protein preparations. *African Journal of Microbiology Research, 4*(1), 55–060.

Markou, G., Arapoglou, D., Eliopoulos, C., Balafoutis, A., Taddeo, R., Panara, A., & Thomaidis, N. (2019). Cultivation and safety aspects of Arthrospira platensis (Spirulina) grown with struvite recovered from anaerobic digestion plant as phosphorus source. *Algal Research, 44*, 101716. Available from https://doi.org/10.1016/j.algal.2019.101716.

McClements, D. J., & Gumus, C. E. (2016). Natural emulsifiers—biosurfactants, phospholipids, biopolymers, and colloidal particles: molecular and physicochemical basis of functional performance. *Advances in Colloid and Interface Science, 234*, 3–26. Available from https://doi.org/10.1016/j.cis.2016.03.002.

Medina, C., Rubilar, M., Shene, C., Torres, S., & Verdugo, M. (2015). Protein fractions with techno-functional and antioxidant properties from Nannochloropsis gaditana microalgal biomass. *Journal of Biobased Materials and Bioenergy, 9*(4), 417–425. Available from https://doi.org/10.1166/jbmb.2015.1534.

Mo, W. Y., Man, Y. B., & Wong, M. H. (2020). Soybean dreg pre-digested by enzymes can effectively replace part of the fishmeal included in feed pellets for rearing gold-lined seabream. *Science of the Total Environment*, *704*, 135266. Available from https://doi.org/10.1016/j.scitotenv.2019.135266.

Mostolizadeh, S., Moradi, Y., Mortazavi, M. S., Motallebi, A. A., & Ghaeni, M. (2020). Effects of incorporation Spirulina platensis (Gomont, 1892) powder in wheat flour on chemical, microbial and sensory properties of pasta. *Iranian Journal of Fisheries Sciences*, *19*(1), 410–420. Available from https://doi.org/10.22092/ijfs.2019.119107.

Nirmala, C., Prakash, V., & Venkataraman, L. V. (1992). Physico-chemical and functional properties of proteins from spray dried algae (Spirulina platensis). *Food / Nahrung*, *36*(6), 569–577. Available from https://doi.org/10.1002/food.19920360608.

Norambuena, F., Hermon, K., Skrzypczyk, V., Emery, J. A., Sharon, Y., Beard, A., & Turchini, G. M. (2015). Algae in fish feed: performances and fatty acid metabolism in juvenile atlantic salmon. *PLoS One*, *10*(4), e0124042. Available from https://doi.org/10.1371/journal.pone.0124042.

Norton, T. A., Melkonian, M., & Andersen, R. A. (1996). Algal biodiversity. *Phycologia*, *35*(4), 308–326. Available from https://doi.org/10.2216/i0031-8884-35-4-308.1.

Nunes, M. C., Graça, C., Vlaisavljević, S., Tenreiro, A., Sousa, I., & Raymundo, A. (2020). Microalgal cell disruption: effect on the bioactivity and rheology of wheat bread. *Algal Research*, *45*, 101749. Available from https://doi.org/10.1016/j.algal.2019.101749.

OECD/FAO. (2018). Oilseeds and oilseed products. Retrieved March 4, 2020, from https://www.oecd-ilibrary.org/oilseeds-and-oilseed-products_5j8qpj1dg2lx.pdf?itemId = %2Fcontent%2Fcomponent%2Fagr_outlook-2018-7-en&mimeType = pdf

Ogino, H., Yamada, M., Watanabe, F., Ichinose, H., Yasuda, M., & Ishikawa, H. (1999). Peptide synthesis catalyzed by organic solvent-stable protease from Pseudomonas aeruginosa PST-01 in monophasic aqueous-organic solvent systems. *Journal of Bioscience and Bioengineering*, *88*(5), 513–518. Available from https://doi.org/10.1016/s1389-1723(00)87668-9.

Ovando, C. A., de Carvalho, J. C., Vinícius de Melo Pereira, G., Jacques, P., Soccol, V. T., & Soccol, C. R. (2016). Functional properties and health benefits of bioactive peptides derived from Spirulina: a review. *Food Reviews International*, *34*(1), 34–51. Available from https://doi.org/10.1080/87559129.2016.1210632.

Passos, F., Carretero, J., & Ferrer, I. (2015). Comparing pretreatment methods for improving microalgae anaerobic digestion: thermal, hydrothermal, microwave and ultrasound. *Chemical Engineering Journal*, *279*, 667–672. Available from https://doi.org/10.1016/j.cej.2015.05.065.

Pereira, A. M., Lisboa, C. R., & Costa, J. A. V. (2018). High protein ingredients of microalgal origin: obtainment and functional properties. *Innovative Food Science & Emerging Technologies*, *47*, 187–194. Available from https://doi.org/10.1016/j.ifset.2018.02.015.

Perez Espitia, P. J., de Fátima Ferreira Soares, N., dos Reis Coimbra, J. S., de Andrade, N. J., Souza Cruz, R., & Alves Medeiros, E. A. (2012). Bioactive peptides: synthesis, properties, and applications in the packaging and preservation of food. *Comprehensive Reviews in Food Science and Food Safety*, *11*(2), 187–204. Available from https://doi.org/10.1111/j.1541-4337.2011.00179.x.

Phong, W. N., Show, P. L., Ling, T. C., Juan, J. C., Ng, E.-P., & Chang, J.-S. (2018). Mild cell disruption methods for bio-functional proteins recovery from microalgae—recent developments and future perspectives. *Algal Research*, *31*, 506–516. Available from https://doi.org/10.1016/j.algal.2017.04.005.

Postma, P. R., Miron, T. L., Olivieri, G., Barbosa, M. J., Wijffels, R. H., & Eppink, M. H. M. (2015). Mild disintegration of the green microalgae Chlorella vulgaris using bead milling. *Bioresource Technology*, *184*, 297–304. Available from https://doi.org/10.1016/j.biortech.2014.09.033.

Postma, P. R., Pataro, G., Capitoli, M., Barbosa, M. J., Wijffels, R. H., Eppink, M. H. M., ... Ferrari, G. (2016). Selective extraction of intracellular components from the microalga Chlorella vulgaris by combined pulsed electric field–temperature treatment. *Bioresource Technology*, *203*, 80–88. Available from https://doi.org/10.1016/j.biortech.2015.12.012.

Qu, W., Ma, H., Wang, T., & Zheng, H. (2012). Alternating two-frequency countercurrent ultrasonic-assisted extraction of protein and polysaccharide from Porphyra yezoensis. *Transactions of the Chinese Society of Agricultural Engineering*, *29*(1), 285–292.

Rizzello, C. G., Tagliazucchi, D., Babini, E., Sefora Rutella, G., Taneyo Saa, D. L., & Gianotti, A. (2016). Bioactive peptides from vegetable food matrices: research trends and novel biotechnologies for synthesis and recovery. *Journal of Functional Foods*, *27*, 549–569. Available from https://doi.org/10.1016/j.jff.2016.09.023.

Safi, C., Liu, D. Z., Yap, B. H. J., Martin, G. J. O., Vaca-Garcia, C., & Pontalier, P.-Y. (2014). A two-stage ultrafiltration process for separating multiple components of Tetraselmis suecica after cell disruption. *Journal of Applied Phycology*, *26*(6), 2379–2387. Available from https://doi.org/10.1007/s10811-014-0271-0.

Safi, C., Ursu, A. V., Laroche, C., Zebib, B., Merah, O., Pontalier, P.-Y., & Vaca-Garcia, C. (2014). Aqueous extraction of proteins from microalgae: effect of different cell disruption methods. *Algal Research*, *3*, 61–65. Available from https://doi.org/10.1016/j.algal.2013.12.004.

Sanjukta, S., & Rai, A. K. (2016). Production of bioactive peptides during soybean fermentation and their potential health benefits. *Trends in Food Science & Technology*, *50*, 1–10. Available from https://doi.org/10.1016/j.tifs.2016.01.010.

Schwenzfeier, A., Helbig, A., Wierenga, P. A., & Gruppen, H. (2013). Emulsion properties of algae soluble protein isolate from Tetraselmis sp. *Food Hydrocolloids*, *30*(1), 258–263. Available from https://doi.org/10.1016/j.foodhyd.2012.06.002.

Schwenzfeier, A., Lech, F., Wierenga, P. A., Eppink, M. H. M., & Gruppen, H. (2013). Foam properties of algae soluble protein isolate: effect of pH and ionic strength. *Food Hydrocolloids*, *33*(1), 111–117. Available from https://doi.org/10.1016/j.foodhyd.2013.03.002.

Schwenzfeier, A., Wierenga, P. A., Eppink, M. H. M., & Gruppen, H. (2014). Effect of charged polysaccharides on the techno-functional properties of fractions obtained from algae soluble protein isolate. *Food Hydrocolloids*, *35*, 9–18. Available from https://doi.org/10.1016/j.foodhyd.2013.07.019.

Schwenzfeier, A., Wierenga, P. A., & Gruppen, H. (2011). Isolation and characterization of soluble protein from the green microalgae Tetraselmis sp. *Bioresource Technology*, *102*(19), 9121–9127. Available from https://doi.org/10.1016/j.biortech.2011.07.046.

Sedighi, M., Jalili, H., Ranaei-Siadat, S. O., & Amrane, A. (2016). Potential health effects of enzymatic protein hydrolysates from Chlorella vulgaris. *Applied Food Biotechnology*, *3*(3), 160–169. Available from https://doi.org/10.22037/afb.v3i3.11306.

Shih, M. F., & Cherng, J. Y. (2012). Protective effects of chlorella-derived peptide against UVC-induced cytotoxicity through inhibition of Caspase-3 activity and reduction of the expression of phosphorylated FADD and cleaved PARP-1 in skin fibroblasts. *Molecules (Basel, Switzerland)*, *17*(8), 9116–9128. Available from https://doi.org/10.3390/molecules17089116.

Shokrkar, H., Ebrahimi, S., & Zamani, M. (2017). Bioethanol production from acidic and enzymatic hydrolysates of mixed microalgae culture. *Fuel*, *200*, 380–386. Available from https://doi.org/10.1016/j.fuel.2017.03.090.

Simpore, J., Kabore, F., Zongo, F., Dansou, D., Bere, A., Pignatelli, S., ... Musumeci, S. (2006). Nutrition rehabilitation of undernourished children utilizing Spiruline and Misola. *Nutrition Journal*, *5*(1), 3. Available from https://doi.org/10.1186/1475-2891-5-3.

Simpore, J., Zongo, F., Kabore, F., Dansou, D., Bere, A., Nikiema, J. B., ... Musumeci, S. (2005). Nutrition rehabilitation of HIV-infected and HIV-negative undernourished children utilizing spirulina. *Annals of Nutrition and Metabolism*, *49*(6), 373–380. Available from https://doi.org/10.1159/000088889.

Soto, D. F., Fuentes, R., Huovinen, P., & Gómez, I. (2020). Microbial composition and photosynthesis in Antarctic snow algae communities: integrating metabarcoding and pulse amplitude modulation fluorometry. *Algal Research*, *45*, 101738. Available from https://doi.org/10.1016/j.algal.2019.101738.

Soto-Sierra, L., Stoykova, P., Nikolov, Z. L., & Nikolov, Z. L. (2018). Extraction and fractionation of microalgae-based protein products. *Algal Research*, *36*, 175–192. Available from https://doi.org/10.1016/j.algal.2018.10.023.

Spiller, M., Muys, M., Papini, G., Sakarika, M., Buyle, M., & Vlaeminck, S. E. (2020). Environmental impact of microbial protein from potato wastewater as feed ingredient: comparative consequential life cycle assessment of three production systems and soybean meal. *Water Research*, *171*, 115406. Available from https://doi.org/10.1016/j.watres.2019.115406.

Stack, J., Le Gouic, A. V., Tobin, P. R., Guihéneuf, F., Stengel, D. B., & FitzGerald, R. J. (2018). Protein extraction and bioactive hydrolysate generation from two microalgae, *Porphyridium purpureum* and *Phaeodactylum tricornutum*. *Journal of Food Bioactives*, *1*, 153–165. Available from https://doi.org/10.31665/JFB.2018.1134.

Suarez Garcia, E., van Leeuwen, J., Safi, C., Sijtsma, L., Eppink, M. H. M., Wijffels, R. H., & van den Berg, C. (2018). Selective and energy efficient extraction of functional proteins from microalgae for food applications. *Bioresource Technology*, *268*, 197–203. Available from https://doi.org/10.1016/j.biortech.2018.07.131.

Sun, X., Acquah, C., Aluko, R. E., & Udenigwe, C. C. (2020). Considering food matrix and gastrointestinal effects in enhancing bioactive peptide absorption and bioavailability. *Journal of Functional Foods*, *64*, 103680. Available from https://doi.org/10.1016/j.jff.2019.103680.

Sun, X., & Udenigwe, C. C. (2020). Chemistry and biofunctional significance of bioactive peptide interactions with food and gut components. *Journal of Agricultural and Food Chemistry*. Available from https://doi.org/10.1021/acs.jafc.9b07559.

Suttisuwan, R., Phunpruch, S., Saisavoey, T., Sangtanoo, P., Thongchul, N., & Karnchanatat, A. (2019). Isolation and characterization of anti-inflammatory peptides derived from trypsin hydrolysis of microalgae protein (*Synechococcus* sp. VDW). *Food Biotechnology*, *33*(4), 303–324. Available from https://doi.org/10.1080/08905436.2019.1673171.

Teuling, E., Schrama, J. W., Gruppen, H., & Wierenga, P. A. (2019). Characterizing emulsion properties of microalgal and cyanobacterial protein isolates. *Algal Research*, *39*, 101471. Available from https://doi.org/10.1016/j.algal.2019.101471.

Udenigwe, C. C., & Fogliano, V. (2017). Food matrix interaction and bioavailability of bioactive peptides: two faces of the same coin? *Journal of Functional Foods*, *35*, 9–12. Available from https://doi.org/10.1016/j.jff.2017.05.029.

Vaino, A. R., & Janda, K. D. (2000). Solid-phase organic synthesis: a critical understanding of the resin. *Journal of Combinatorial Chemistry*, *2*(6), 579–596. Available from https://doi.org/10.1021/cc000046o.

Vanthoor-Koopmans, M., Wijffels, R. H., Barbosa, M. J., & Eppink, M. H. M. (2013). Biorefinery of microalgae for food and fuel. *Bioresource Technology*, *135*, 142–149. Available from https://doi.org/10.1016/j.biortech.2012.10.135.

Vo, T.-S., Ryu, B., & Kim, S.-K. (2013). Purification of novel anti-inflammatory peptides from enzymatic hydrolysate of the edible microalgal Spirulina maxima. *Journal of Functional Foods*, *5*(3), 1336–1346. Available from https://doi.org/10.1016/j.jff.2013.05.001.

Waghmare, A. G., Salve, M. K., LeBlanc, J. G., & Arya, S. S. (2016). Concentration and characterization of microalgae proteins from Chlorella pyrenoidosa. *Bioresources and Bioprocessing, 3* (1), 16. Available from https://doi.org/10.1186/s40643-016-0094-8.

Wang, X., & Zhang, X. (2013). Separation, antitumor activities, and encapsulation of polypeptide from Chlorella pyrenoidosa. *Biotechnology Progress, 29*(3), 681–687. Available from https://doi.org/10.1002/btpr.1725.

Wells, M. L., Potin, P., Craigie, J. S., Raven, J. A., Merchant, S. S., Helliwell, K. E., ... Brawley, S. H. (2017). Algae as nutritional and functional food sources: revisiting our understanding. *Journal of Applied Phycology, 29*(2), 949–982. Available from https://doi.org/10.1007/s10811-016-0974-5.

Wijesinghe, W. A. J. P., & Jeon, Y.-J. (2012). Enzyme-assistant extraction (EAE) of bioactive components: A useful approach for recovery of industrially important metabolites from seaweeds: A review. *Fitoterapia, 83*(1), 6–12. Available from https://doi.org/10.1016/j.fitote.2011.10.016.

Yadav, G., Meena, D. K., Sahoo, A. K., Das, B. K., & Sen, R. (2020). Effective valorization of microalgal biomass for the production of nutritional fish-feed supplements. *Journal of Cleaner Production, 243*, 118697. Available from https://doi.org/10.1016/j.jclepro.2019.118697.

Zbinden, M. D. A., Sturm, B. S. M., Nord, R. D., Carey, W. J., Moore, D., Shinogle, H., & Stagg-Williams, S. M. (2013). Pulsed electric field (PEF) as an intensification pretreatment for greener solvent lipid extraction from microalgae. *Biotechnology and Bioengineering, 110* (6), 1605–1615. Available from https://doi.org/10.1002/bit.24829.

Zhang, R., Parniakov, O., Grimi, N., Lebovka, N., Marchal, L., & Vorobiev, E. (2018). Emerging techniques for cell disruption and extraction of valuable bio-molecules of microalgae Nannochloropsis sp. *Bioprocess and Biosystems Engineering, 42*(2), 173–186. Available from https://doi.org/10.1007/s00449-018-2038-5.

Chapter 5

Carbohydrates derived from microalgae in the food industry

Rajeev Ravindran and Gaurav Rajauria
School of Agriculture and Food Science, University College Dublin, Dublin, Ireland

5.1 Introduction

Feeding an ever-growing population with depleted natural resources and increasing environmental problems is one of the biggest challenges that we face in the 21st century (Torres-Tiji, Fields, & Mayfield, 2020). Meanwhile, the growing importance of functional foods and ingredients has fueled consumer demand for minimally processed foods with clean and high biological value components from natural origin. The increasing consumer knowledge pertaining to the positive impact of healthy eating and good food choices on human health has ushered in a new era of novel functional foods and nutraceutical products. Microalgae have gained enormous attention and have shown immense potential to address food security, food for health, alternative energy resources, and global warming challenges (Salla et al., 2016). They are considered one of the most promising natural sources of functional ingredients for new functional food products (Matos, Cardoso, Bandarra, & Afonso, 2017). They have excellent potential to not only enhance the nutritional value of foods due to their well-balanced nutrient composition, their biologically active compounds also have immense capacity to minimize the risk of cancer, inflammation, and metabolic disorders (Batista, Gouveia, Bandarra, Franco, & Raymundo, 2013). Moreover, microalgae do not compete with human or animal food crops for cultivation on land as they are easy to grow, can be cultivated with relatively low nutrient requirements in nonarable land, and do not depend on the climate or season for harvesting (Costa & de Morais, 2011; de Jesus Raposo, de Morais, & de Morais, 2014). Within the mariculture industry, microalgae continue to be cultivated to ensure sustainability of the environment and product quality, but only a few species are being cultivated commercially and explored for their potential industrial applications. The total global algae market is very large and contains a diverse set of products which are generally sold at a small scale, which is thus difficult to estimate (Torres-Tiji

Cultured Microalgae for the Food Industry. DOI: https://doi.org/10.1016/B978-0-12-821080-2.00007-1

et al., 2020). Microalgal dry biomass as a food ingredient alone has a market value well above US$1.25 billion per year, which does not include food supplements and food additive products that are obtained from this biomass (Pulz & Gross, 2004). The global market share of microalgae-based product is expected to attain a market value of US$3.2 billion by the end of 2030 (Transparency Market Research, 2020).

Microalgae are some of the oldest, most economically promising photosynthetic (prokaryotic or eukaryotic) plants on the planet (Pittman, Dean, & Osundeko, 2011). They are unicellular microscopic organisms which grow individually, aggregated, or in a filamentous form in marine, freshwater, wastewater, or terrestrial environments (Ozkurt, 2009; Mata et al., 2018). Over 50,000 microalgae species are disseminated globally, performing more than 50% of the primary photosynthetic productivity and producing approximately half of the atmospheric oxygen on Earth (Chisti, 2007; Rajkumar, Yaakob, & Takriff, 2014). However, only 50 species of microalgae (mainly *Spirulina*, *Chlorella*, *Porphyry*, *Nannochloropsis*, *Haematococcaceae*, and *Dunaliella*) have been investigated for physiological, biochemical, and potential commercial applications (Adarme-Vega et al., 2012; Anbuchezhian, Karuppiah, & Li, 2015). Based on the microalgal species/strain and growth conditions, they produce a wide variety of different valuable compounds, such as carbohydrates, proteins with essential amino acids, lipids, polyunsaturated fatty acids, phycobiliproteins, carotenoid pigments, vitamins, sterols, antioxidants, and dietary fiber (de Jesus Raposo, de Morais, & de Morais, 2013). They are economical, viable, and sustainable resources that are therefore potentially exploited as a promising feedstock for functional foods, nutraceutical, pharmaceuticals, cosmetics, biofuels, and other high-value industrial products (Shahid, Khan, Ahmad, Farooq, & Mehmood, 2020; Skorupskaite, Makareviciene, & Levisauskas, 2015).

Microalgal biomass contains different amounts of the primary essential organic nutrient groups including carbohydrates, proteins, and lipids. However, their growth and biomass composition can be manipulated by modifying the algal strain through genetic engineering or cultivation techniques and environmental factors including nutrients, light, temperature, etc. (Markou, Angelidaki, & Georgakakis, 2012). Hence, the functional properties and physicochemical characteristics of the bioactive they produce can also be manipulated or maintained (de Jesus Raposo, de Morais, & de Morais, 2014). Nowadays, microalgae are mainly cultivated and manipulated for lipid production, which has huge potential in biofuel production. However, algal carbohydrates, which possess the lowest energy content per unit biomass, have substantial applications in the food industry and are also the main substrate for several significant biofuel productions by biotechnological conversion technologies (Markou et al., 2012). A schematic diagram for the cost-effective and efficient production of microalgal carbohydrate-based products is illustrated in Fig. 5.1.

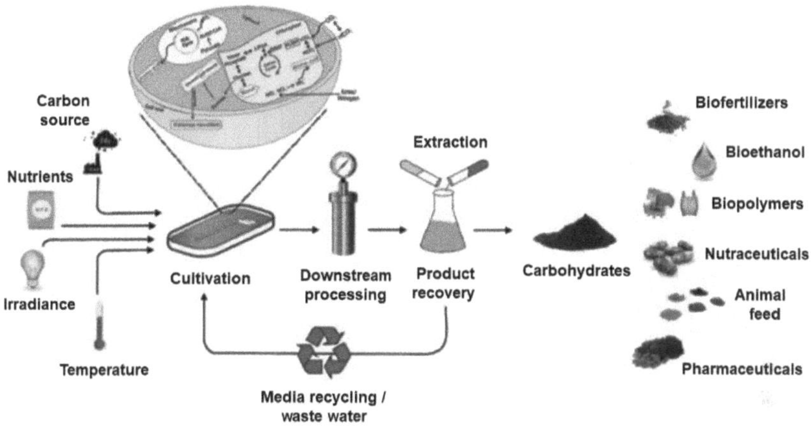

FIGURE 5.1 Schematic diagram for the cost-effective and efficient production of microalgal carbohydrate-based products. *Adapted and modified from Shahid et al. (2020), Microalgal carbohydrates and proteins: synthesis, extraction, applications, and challenges. In Microalgae Biotechnology for Food, Health and High Value Products. Springer, Singapore. pp. 433–468.*

Structurally, microalgal cells mainly consist of an inner and outer cell wall layer. The carbohydrate is produced from photosynthesis and carbon fixation metabolism (Ho, Chen, Lee, & Chang, 2011). The carbohydrate in microalgae is mainly deposited in the form of starch in plastids, and cellulose, glycogen, and soluble polysaccharides in the cell wall layers (Khan, Shin, & Kim, 2018). The outer cell wall layer, which varies from species to species, is mainly composed of specific polysaccharides, such as pectin, agar, and alginate, while the inner cell wall layer primarily consists of cellulose and hemicellulose (Yamada & Sakaguchi, 1982). The carbohydrate content in microalgae accounted for approximately 15%–75% of dry biomass, which may vary according to the microalgal species (Chen et al., 2013; Pina-Pérez, Brück, Brück, & Beyrer, 2019). *Chlorella* sp., which is one of the most cultivated green microalgae, was reported to accumulate a higher amount of carbohydrate content, at up to 37%–55% of its dry biomass (Chen et al., 2013). Similarly, *Spirogyra* sp. and *Porphyridium cruentum* are comprised of 33%–64% and 40%–57% carbohydrate, respectively, while *Dunaliella salina* was reported to accumulate as much as 85.58% carbohydrate (Halder & Azad, 2019). Table 5.1 summarizes the carbohydrate or starch content of green microalgal species. While, the cellular carbohydrates provide structural support, intracellular carbohydrates are used as a source of energy for metabolic functions or provide protection for survival against environmental stress conditions (Markou et al., 2012). These algal polysaccharides are natural biopolymers which are easily modified, safe, stable, nontoxic, and biodegradable, and therefore have many applications in food, agriculture, biofuel, biomedical, beauty care products, and pharmaceutical

TABLE 5.1 Carbohydrate or starch content of green microalgal species.

Microalgal species	Carbohydrate or starch content (%)	References
Chlorella vulgaris IAM C-534	37.0 (starch)	Hirano, Ueda, Hirayama, and Ogushi (1997)
C. vulgaris CCAP 211/11B	55.0	Illman, Scragg, and Shales (2000)
C. vulgaris P12	41.0 (starch)	Dragone, Fernandes, Abreu, Vicente, and Teixeira (2011)
C. vulgaris	55.0 (starch)	Brennan and Owende (2010)
Chlamydomonas reinhardtii UTEX 90	60.0	Hirano et al. (1997)
C. reinhardtii IAM C-238	55.0 (starch)	Kim et al. (2006)
Chlorococcum sp.	32.5	Ike, Toda, Tsuji, Hirata, and Miyamoto (1997)
Chlorococcum sp. TISTR8583	26.0 (starch)	Harun and Danquah (2011)
Scenedesmus acutiformis TISTR 8495	16.4 (starch)	Spolaore et al. (2006)
S. obliquus CNW-N	51.8	Ho, Chen, and Chang (2012)
Tetraselmis sp.CS-362	26.0	Brown, McCausland, and Kowalski (1998)

Source: Adapted from Chen et al., 2013 Microalgae-based carbohydrates for biofuel production. Biochemical Engineering Journal, 78, 1–10.

industries. Since the overall digestibility of algal-derived carbohydrate is high, they can be directly incorporated as dried whole microalgae in food items (Becker, 2004). In the food industry, they have significant applications as emulsifiers, thickeners, and stabilizers in beverages, animal feed, and human food products (Bixler & Porse, 2011). They also possess a wide range of biological properties including antioxidant, antitumor, antivirus, anticoagulant, and antimicrobial, and therefore are being explored in drug delivery and functional food products development (de Jesus Raposo, de Morais, & de Morais, 2014; Pina-Pérez et al., 2019).

5.2 Microalgal carbohydrates

Microalgae synthesize carbohydrates to store energy and to support the cell and cell wall structures. Cellulose and starch are the most abundant polysaccharides found in microalgae. However, they have also been reported to synthesize other forms of carbohydrates. These are usually in the form of β-glucans and other related glucans. While the storage carbohydrates are either laminarins or paramylon, structural polysaccharides are comprised of callose, cellulose, chitins, etc. The nature of the polysaccharides found in microalgae such as protozoans and chromistans is more or less in correlation with the taxonomy. Most carbohydrates found in microalgae are heterogeneous in nature with the major constituents being glucose, galactose, and xylose. However, rhamnose, fucose, fructose, and methylated sugars have also been found in microalgae. Table 5.2 demonstrates the neutral sugar contents of selected microalgal species. These sugars are not only linked to each other with glycosidic linkage, they are also covalently connected to other cell molecules such as amino acids, peptides, proteins, and lipids (Pina-Pérez et al., 2019). Detailed classifications of polysaccharides produced by microalgae are explained in the sections below (Fig. 5.1).

5.2.1 Storage polysaccharides

Microalgae and cyanobacteria are photosynthetic in nature. While microalgae produce and store starch, cyanobacteria synthesize glycogen (Liu et al., 2019). Starch and glycogen are the most abundant forms of storage polysaccharides found in microalgal species. Both polysaccharides are homopolymers of repeating units of glucose with structural differences owing to the bonds that link individual monomer units. Starch is formed by a combination of amylose and amylopectin. Amylose is a linear chain of glucose-linked α-1,4-glycosidic linkages, while amylopectin has branched structures formed by repeating glucose units with α-1,4- and α-1,6-glycosidic bonds between them. While starch is formed by repeating units of glucose units linked by α-1,4-glycosidic bonds, glycogen comprises of the same glucose units lined through α-1,6-glycosidic linkages. On the other hand, floridean starch is different from starch found in terrestrial plants due to the absence of amylose structures. They are found to be accumulated in the cytoplasm of Glaucophyta which is a class of microalgae. It is more commonly found in red macroalgae.

Laminarin-type polysaccharides are formed of glucose subunits that are linked together with β-(1→3)- and β-(1→6)-glycosidic bonds. Two examples of this type of storage polysaccharide found in microalgae are chrysolaminarin and paramylon. Chrysolaminarin is generally found in phytoplankton species such as *Bacillariophyta* and plays the role of food reserves. They are formed by a cluster of glucose units linked by β-1,3- and β-1,6-glycosidic linkages. They are soluble in water and stored in the cytoplasm of the organism in vacuoles. Microalgae belonging to the class Chrysophyceae have also been reported to contain chrysolaminarin in their cytoplasm. Paramylon is a linear β-1,3-glucan present exclusively in

TABLE 5.2 Carbohydrates of selected microalgal species.

	Microalgae					
Sugar[a]	*Chlorococcum* sp.	*Spirulina platensis*	*Chlamydomonas reinhardtii*	*Nitzchia ciosterium*	*Phaeodactylum tricornutum*	*Dunaliella tertiolecta*
Xylose (%)	27	7.0	–	7.0	7.5	1.0
Mannose (%)	15	9.3	2.3	16.8	45.9	4.5
Glucose (%)	47	54.4	74.9	32.6	21.0	85.3
Galactose (%)	9	–	4.5	18.4	8.9	1.1
Rhamnose (%)	–	22.3	1.5	7.7	8.6	5.5
References	Harun and Danquah (2011)	Sekharam, Venkataraman, and Salimath (1989)	Choi, Nguyen, and Sim (2010)	Brown (1991)	Brown (1991)	Brown (1991)

[a]*Sugars are expressed as percentage of total carbohydrates.*
Source: Adapted from Markou et al., 2012 Microalgal carbohydrates: an overview of the factors influencing carbohydrates production, and of main bioconversion technologies for production of biofuels. Applied Microbiology and Biotechnology, 96(3), 631–645.

euglenoids. It is insoluble in nature, has a high molecular mass, and is naturally crystalline in nature. In protists such as *Euglena gracilis* it is found as membrane-bound granules in the cytoplasm.

5.2.2 Cell wall-related polysaccharides

The microalgal cell wall is mainly composed of polysaccharides. However, other biopolymers such as proteins, resistant polymers, and calcified and silicified structures also form an integral part of the cell wall. Interestingly, a majority of microalgal species have a layer of polymers covering the cell. The nature and proliferation of this layer depend on the growth conditions and environmental conditions. These polymers can either remain in contact with the cells or spread into the surrounding environment. These polysaccharides are wide in variety and little literature exists that comprehensively characterizes them. Hence these polymers are addressed using various terms such as exopolysaccharides, extracellular polysaccharides, released polysaccharides, dissolved organic matter, and algogenic organic matter (De Philippis & Vincenzini, 1998; Henderson, Baker, Parsons, & Jefferson, 2008; Mühlenbruch, Grossart, Eigemann, & Voss, 2018; Zhou, Hu, Zhang, Yang, & Lin, 2016). Exopolysaccharides contribute to almost 10% of the total dry biomass (Bernaerts et al., 2018). Some of the most common microalgal organisms that synthesize exopolysaccharides include cyanobacteria, diatoms, green microalgae, and red microalgae. It is believed that exocellular polysaccharides protect the microalgal cell from environmental fluctuations and predators. The identification of exopolysaccharide produced by microalgae is rather challenging due to the adherence of bacterial polysaccharides to microalgal varieties (Sekar, Venugopalan, Satpathy, Nair, & Rao, 2004). Additionally, a clear distinction between cell wall polysaccharides and exopolysaccharides is at times nonexistent such as in the case of several red microalgal species (Arad & Levy-Ontman, 2010).

Depending upon the microalgal species, both storage and cell wall-related polysaccharides are produced either during the stationary phase or during the exponential growth phase, and have exhibited significant biological properties in both in vitro and in vivo model systems (Pina-Pérez et al., 2019). However, the bioactive potential of these polysaccharides is influenced by factors such as microalgal species, monomer compositions and distribution, type of linkage and interaction with other compounds, a lack of repeating monomer units, molecular weight, and the presence of nonsugar compounds (Pina-Pérez et al., 2019). Bernaerts et al. (2018) revealed that the occurrence of sulfate groups and uronic acids in cell wall polysaccharides of some of the algal species contributes toward the biological properties of these molecules, thereby the polysaccharide content is mostly characterized by the existence of uronic acids.

5.3 Biochemical composition and physical properties of microalgal carbohydrates

5.3.1 Structure

The most common exopolysaccharide produced by microalgae is usually sulfated and consists of uronic acid in various proportions. This gives the exopolysaccharide an overall negative charge, making them excellent anionic polymers. Sulfated exopolysaccharides may be homogeneous or heterogeneous in nature. *Gyrodinium impudicum* is an example of a sulfated homogeneous polysaccharide producer with repeating units of galactose along with uronic acid (Lee et al., 2008; Yim et al., 2004). Heterogeneous sulfated exopolysaccharides are commonly comprised of galactose, glucose, xylose, and glucuronic acid (Gloaguen et al., 2004). Cyanobacterial species produce anionic exopolysaccharides that are complex in nature. They contain a variety of monosaccharides, uronic acids, and various functional substituents such as methyl, acetate, pyruvate, sulfate groups, and proteins. The flexibility imparted by the macromolecular chain, the nature of the carbon skeleton, molecular weight, distribution of mass, the presence of functional groups, their position on the macromolecular chain, and the presence of other biomolecules such as proteins all contribute to the complexity of the exopolysaccharide (Pierre et al., 2019). Trabelsi, M'sakni, Ben Ouada, Bacha, and Roudesli (2009) partially characterized the exopolysaccharide produced by *Arthrospira platensis*, a cyanobacterial species, and revealed that 55% of the heteropolymer was comprised of proteins.

5.3.2 Rheology

The differences in component monosaccharides determines the difference in properties in microalgal polysaccharides. Combined with these differences, the percentage of uronic acid and sulfate govern the rheological properties of polysaccharides. This can lead to the aqueous solutions of such polysaccharides being highly viscous. A pictorial representation of how the viscosity of different algal polysaccharides compares with each other has been provided in Fig. 5.2. The majority of studies on the rheological properties of microalgal exopolysaccharides have been performed on *Porphyridium* sp., a red microalgae. Arad et al. (2006) performed an extensive study on the rheological properties of the exopolysaccharides from *Porphyridium* sp. They reported that the viscosity of the polysaccharide was stable over a wide range of temperatures (25°C–70°C), pH values, and salinities. *Porphyridium* exopolysaccharide have been characterized as exhibiting strong shear-thinning behavior similar to xanthan solutions. The strong shear-thinning property has been attributed to multiple interactions between polysaccharide chains that break under shear (Geresh, Adin, Yarmolinsky, & Karpasas, 2002). Dilute concentrations of

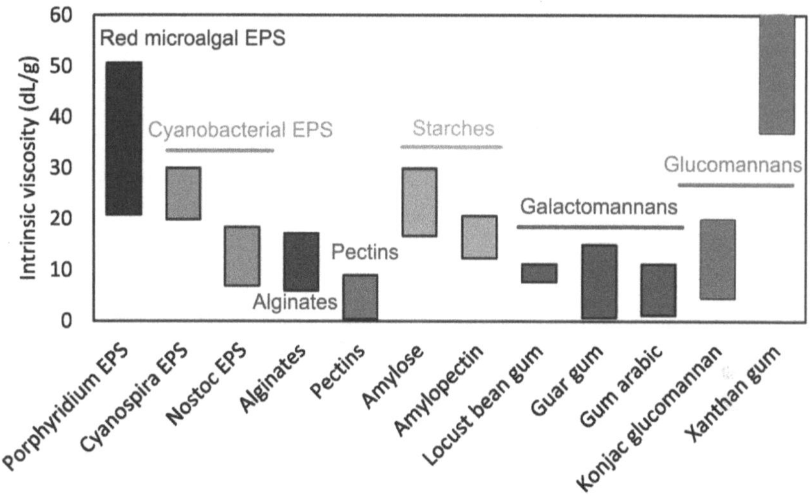

FIGURE 5.2 Intrinsic viscosity of microalgal and cyanobacterial extracellular polysaccharides (EPS) in comparison with common thickening and gelling agents. Intrinsic viscosities of xanthan gum are not completely shown in the graph, but values up to 183 dL/g have been reported. *Adapted from Bernaerts, Gheysen, Foubert, Hendrickx, and Van Loey (2019), The potential of microalgae and their biopolymers as structuring ingredients in food: a review. Biotechnology Advances, 37(8), 107419.*

microalgae with low viscosities exhibit the properties of a Newtonian fluid; whereas beyond a critical concentration the properties of flow resembles a non-Newtonian fluid. The value of this critical concentration is dependent on the species of microalgae. For example, *Chlorella vulgaris* at a concentration of 4% exhibits Newtonian flow, while sheer thinning behavior increased at higher concentrations indicated by flow behavior indices in the range of 0.62–0.78 (Yatirajula, Shrivastava, Saxena, & Kodavaty, 2019).

5.4 Food applications of carbohydrates derived from microalgae

Several studies have reported the incorporation of microalgae-derived ingredients for improving the nutritional profile of foods. The technofunctional as well as antioxidant properties of foods have been reportedly improved by the addition of microalgae-based products. Although microalgal biopolymers can be used as thickening agents in the food industry, the idea of using whole microalgal cells is gaining popularity. Microalgae have been introduced in dairy products to deliver bioactive compounds. Microalgal cells tend to produce exopolysaccharides which stimulate the growth of bacteria, thus exhibiting prebiotic properties. For example, *Chlorella* has successfully been incorporated in cheeses and yogurts (Beheshtipour, Mortazavian, Haratian, &

Darani, 2012; Jeon, 2006). The addition of microalgae to cookies has been reported to increase their taste, versatility, and general appeal. In a study, *Chlorella vulgaris* was incorporated into cookies and resulted in improvement of overall firmness and imparted a greenish tone (Gouveia, Batista, Miranda, Empis, & Raymundo, 2007).

Albeit being a cheap and innovative source of starch, microalgal starch is not commonly exploited as a replacement for starch obtained from terrestrial plants. The freshwater algae, *Chlorella* is the most common source of microalgal species that is exploited for starch (Brányiková et al., 2011). Meanwhile, due to its low gelatinization and pasting temperatures, floridean starch finds wide applications in the preparation of instant noodles and deep-frozen foods (Yu et al., 2002). This microalgal polysaccharide exhibits low gelatinization temperature, low viscosity, high clarity, and little or no retrogradation upon repetitive freeze–thaw cycles, as studied by differential scanning calorimetry and rapid viscosity analysis. On the other hand, chrysolaminarin is a proven antioxidant agent and has been exploited as a food ingredient along with being a potent drug-delivery agent and is used in the synthesis of biomaterials (Xia et al., 2014). Paramylon also exhibits antioxidant properties and has successfully prevented the oxidative damage caused to the liver by carbon tetrachloride. Mass cultivation of *Euglena gracilis* has enabled reliable and economical supply of paramylon. Paramylon can generally be used for any application that requires β-glucans (Nakashima et al., 2018).

Due to the rich nutrient compositions, microalgae are rapidly being incorporated in food and feed preparations. Some species of microalgae have been widely explored, commercialized, and utilized as nutritional supplements for human foods and as animal feeds. They already have found many applications as food additives in food products, such as pasta, cookies, breads, noodles, bean curd, ice cream, candies, and many other common foodstuffs, or in the form of tablets or pills, as nutraceuticals (de Jesus Raposo, de Morais, & de Morais, 2015; Liang Liu, Chen, & Chen, 2004). The incorporation of microalgae in food formulations or the preparation of functional foods/nutraceuticals is a progressive step for the development of tailor-made healthy products in the near future. Fig. 5.3 demonstrates a schematic representation of the different approaches to introduce microalgal biopolymers as structuring agents in food products. Species such as *Chlorella vulgaris*, *Arthrospira platensis*, *Euglena gracilis*, *Chlamydomonas reinhardtii*, *Auxenochlorella protothecoides*, and *Dunaliella bardawil* are even listed as Generally Recognized As Safe (GRAS) for human consumption by the FDA (Rasala & Mayfield, 2015; Torres-Tiji et al., 2020). However, the GRAS status is limited to the United States, European countries and other countries such as Canada, China, India, and Japan have different regulations for algal-derived products. For instance, astaxanthin, a color additive from *H. pluvialis* has been approved for salmon feed and as a supplement for food fortification for human consumption in several European countries

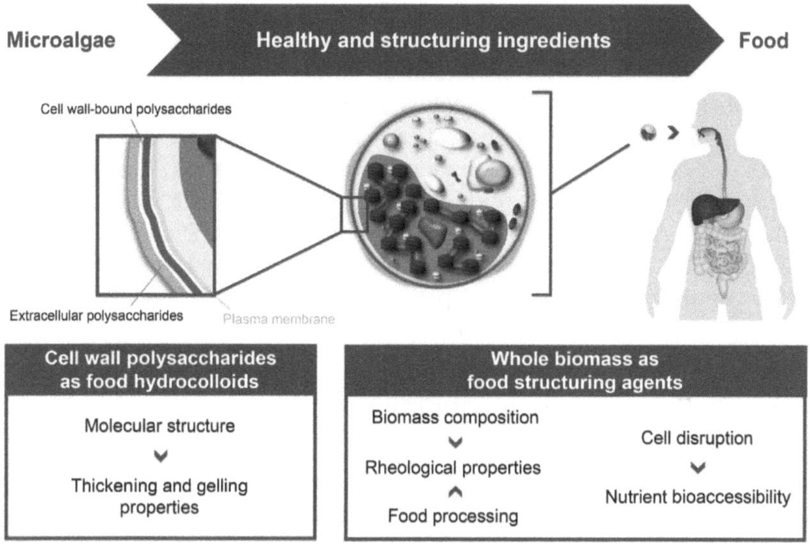

FIGURE 5.3 Schematic representation of the different approaches to introduce microalgal biopolymers as structuring agents in food products. *Adapted from Bernaerts et al. (2019), The potential of microalgae and their biopolymers as structuring ingredients in food: a review. Biotechnology Advances, 37(8), 107419.*

(maximum intake of 8 mg/day), the United States, and Japan (Yuan, Peng, Yin, & Wang, 2011). Meanwhile the same compound, extracted using cleaner supercritical CO_2 extraction technology has been registered as a Novel Food by the UK Foods Standard Agency (FSA) (Shah, Liang, Cheng, & Daroch, 2016). Similarly, *A. platensis* microalga is recognized to be safe to consume in the EU, Canada, China, India, and Japan, while *E. gracilis* is authorized in China, Canada, and Japan (EFSA, 2020; FSSAI, 2016; NHFPC, 2016; NNHPD, 2019).

Algal carbohydrate, especially polysaccharides, have found significant applications as thickeners, emulsifiers, flocculants, stabilizers, and gelling and hydrating agents in the food industry (Bernal & Llamas, 2012). The emulsifying characteristics of these polysaccharides are because of the presence of deoxy sugars such as rhamnose and fucose and protein/peptide moieties granting them a significant hydrophobic behavior (Flaibani, Olsen, & Painter, 1989; Shepherd, Rockey, Sutherland, & Roller, 1995). They show great ability to be used in edible films and food coatings, alone or in combination with other compounds (Marceliano, 2009), and have the great possibility to be used as fat replacers in low-fat or fat-free food products such as mayonnaises, salad dressings, and food emulsions (Franco, Raymundo, Sousa, & Gallegos, 1998; Raymundo, Franco, Gallegos, Empis, & Sousa, 1998; Raymundo, Gouveia, Batista, Empis, & Sousa, 2005). The sulfated polysaccharides from marine microalgae, due to their fiber content, have shown a good prebiotic characteristic. Properties such as their acid-binding and cation exchange abilities and fecal bulking capability, make them an ideal candidate for novel nutraceutical development (Ciferri, 1983). For instance, β-glucans, a type of polysaccharides typically found in the cell walls of many green algae, have been shown to decrease bad cholesterol (LDL, low-density lipoprotein) and lower the risk of cardiovascular diseases when food products are fortified with such algae (El Khoury, Cuda, Luhovyy, & Anderson, 2012; Spolaore, Joannis-Cassan, Duran, & Isambert, 2006). Table 5.3 demonstrates some strains of microalgae/blue-green algae producing polysaccharides, their main neutral sugars, and associated biological properties.

5.5 Challenges and concluding remarks

Microalgal-derived nutrients have excellent potential in the food industry as functional food ingredients due to their antioxidant, anticancer, and antiinflammatory properties. The selection of microalgal species or strains with balanced nutritional profiles is fundamental for successful novel food development. The safety of microalgal species, their nutritional and toxicological relevance, and their use as food supplements or as a source of functional ingredients need to be addressed. In addition, due to the strict food safety regulations, the modification of microalgal species for human consumption, by applying biotechnological approaches, is restricted to a very few algal species.

Algal carbohydrates, not only because of their wide range of biological activities but also due to their properties such as thickening, emulsifying, stabilizing, gelling, and hydrating characteristics, are gaining popularity in functional food preparations. Despite their potential as a technological food ingredient, molecular structure, composition, and biological characteristics of microalgal polysaccharides are poorly understood. The bioactivity of polysaccharides is related to their precise molecular structure, therefore molecular characterization, a detailed physicochemical characterization, and targeted functionalization of polysaccharide polymer from individual microalgal strains is important for successfully utilizing them as novel food hydrocolloids in food products. In addition, carbohydrates represent a large portion of the algal biomass but the impact of cultivation conditions on the carbohydrate composition relates wide variability in polysaccharide composition, even within a species, or a strain for microalgae. Subsequently, it generates the opportunity to modify cultivation parameters as an efficient strategy for enriching microalgal biomass with carbohydrate content, which will help in determining the best suited species/strain for different applications and purposes. Little knowledge in this domain has been gathered so far because microalgae have enormous complexity in terms of carbohydrate composition and molecular structural organization. Only a limited number of microalgae species have been investigated in this research area, thus more clarity is

TABLE 5.3 Species of microalgae/blue-green algae producing polysaccharide and their main neutral sugars.

Microalgae	Polysaccharide	Main neutral sugars	Biological activity	References
Chlorella stigmatophora	Sulfated polysaccharides	Glucose, xylose, fucose	Antiinflammatory, immunomodulator	Guzman, Gato, Lamela, Freire-Garabal, and Calleja (2003)
Chlorella vulgaris	β-(1,3)-glucan	Glucose	Antitumor, infection preventive agent	Ogawa, Ikeda, and Kondo (1999)
	polysaccharides	Rhamnose, galactose		
Dunaliella salina	Exo-polysaccharides	Xylose, galactose, glucose, fructose		Mishra, Kavita, and Jha (2011)
Tetraselmis sp.	Sulfated polysaccharides		Antiadhesive	Guzman-Murillo and Ascencio (2000)
Porphyridium sp.	Sulfated polysaccharides	Xylose, galactose, glucose	Antiinflammatory, immunomodulator, prevention of tumor cell growth, antiadhesive, antiviral, biolubricant	Geresh and Arad (1991), Matsui, Muizzuddin, Arad, and Marenus (2003), Arad et al. (2006), Talyshinsky, Souprun, and Huleihel (2002)
Porphyridium cruentum	Sulfated polysaccharides	Xylose, galactose, glucose, Glucuronic acid 3-O-methyl-xyl.	Antioxidant and free radical scavenging, immunomodulator, antiviral, antibacterial, antilipidemic, antiglycemic	de Jesus Raposo, de Morais, and de Morais (2014), Huang, Chen, and You (2001), Sun, Wang, and Zhou (2012)
Rhodella reticulata	Sulfated polysaccharides	Xylose, rhamnose, 3-*O*-methyl-rham, 4-*O*-methyl-gal	Antiviral, antilipidemic, antiglycemic, prevention of tumor cell growth	Arad and Levy-Ontman (2010), Geresh and Arad (1991)
Cyanobacteria				
Arthrospira platensis	Exo-polysaccharides s-spirulan	Galactose, fructose, xylose glucose Rhamnose, fucose, glucose, 3-*O*-methyl-rham	Antiviral, antibacterial, prevention of tumor cell growth Antiproliferative, antiadhesive, antimetastatic	Challouf et al. (2011), Hayashi, Hayashi, Maeda, and Kojima (1996), Kaji et al. (2004)

Source: Adapted from Chanda, Merghoub, & Arroussi, 2019 Microalgae polysaccharides: the new sustainable bioactive products for the development of plant bio-stimulants? World Journal of Microbiology and Biotechnology, 35(11), 177.

needed in terms of the carbohydrate accumulation pattern within (cell wall bound) or outside the cells (extracellular).

However, the incorporation of whole microalgal biomass into food products is more sustainable (to reduce waste generation) and economical (to overcome extraction and purification steps), incorporation of isolated pure functional compounds/extracts (such as polysaccharides) creates high-quality food products with nutritional benefits. However, challenges associated with the extraction and purification of specific polysaccharides ought to be undertaken. Fundamental knowledge on the polysaccharide composition may predict the functionality of polysaccharide-rich extracts, but bio-accessibility and rheological properties of these polysaccharides may change when the food products undergo processing. Therefore food processing operations and the impact of processing on foods should be measured and examined in more detail to exploit the structuring capacity of these compounds. Additionally, to preserve the functionality and physiochemical properties of these polymers, novel food process operations/conditions such as pressure levels for high-pressure homogenization and time–temperature combinations for thermal processing should also be explored. Another challenge associated with microalgal polysaccharides is their competencies against plant-based and macroalgae- or seaweed-based conventional polysaccharides which have low production costs and limited regulatory issues. Therefore there is an urgent need for greater insight into more economic commercial cultivation techniques, product safety, market demand, commercial factors, and specific processing and preparation requirements for microalgal polysaccharides, and for their potential to be realized, they must be cultivated with the desired characteristics that conventional agriculture crops possess.

References

Adarme-Vega, T. C., Lim, D. K., Timmins, M., Vernen, F., Li, Y., & Schenk, P. M. (2012). Microalgal biofactories: a promising approach towards sustainable omega-3 fatty acid production. *Microbial Cell Factories, 11*(1), 96.

Anbuchezhian, R., Karuppiah, V., & Li, Z. (2015). *Prospect of marine algae for production of industrially important chemicals. Algal biorefinery: An integrated approach* (pp. 195–217). Cham: Springer.

Arad, S., Rapoport, L., Moshkovich, A., van Moppes, D., Karpasas, M., Golan, R., & Golan, Y. (2006). Superior biolubricant from a species of red microalga. *Langmuir: the ACS Journal of Surfaces and Colloids, 22*(17), 7313–7317.

Arad, S. M., & Levy-Ontman, O. (2010). Red microalgal cell-wall polysaccharides: biotechnological aspects. *Current Opinion in Biotechnology, 21*(3), 358–364.

Batista, A. P., Gouveia, L., Bandarra, N. M., Franco, J. M., & Raymundo, A. (2013). Comparison of microalgal biomass profiles as novel functional ingredient for food products. *Algal Research, 2*(2), 164–173.

Becker, W. (2004). *18 microalgae in human and animal nutrition.* In: *Handbook of microalgal culture: biotechnology and applied phycology* (Vol. 312). Wiley Online Library.

Beheshtipour, H., Mortazavian, A. M., Haratian, P., & Darani, K. K. (2012). Effects of *Chlorella vulgaris* and Arthrospira platensis addition on viability of probiotic bacteria in yogurt and its biochemical properties. *European Food Research and Technology, 235*(4), 719–728.

Bernaerts, T. M., Gheysen, L., Foubert, I., Hendrickx, M. E., & Van Loey, A. M. (2019). The potential of microalgae and their biopolymers as structuring ingredients in food: a review. *Biotechnology Advances, 37*(8), 107419.

Bernaerts, T. M. M., Gheysen, L., Kyomugasho, C., Jamsazzadeh Kermani, Z., Vandionant, S., Foubert, I., … Van Loey, A. M. (2018). Comparison of microalgal biomasses as functional food ingredients: focus on the composition of cell wall related polysaccharides. *Algal Research, 32*, 150–161.

Bernal, P., & Llamas, M. A. (2012). Promising biotechnological applications of antibiofilm exopolysaccharides. *Microbial Biotechnology, 5*(6), 670.

Bixler, H. J., & Porse, H. (2011). A decade of change in the seaweed hydrocolloids industry. *Journal of Applied Phycology, 23*(3), 321–335.

Brányiková, I., Maršálková, B., Doucha, J., Brányik, T., Bisova, K., Zachleder, V., & Vitova, M. (2011). Microalgae-novel highly efficient starch producers. *Biotechnology and Bioengineering, 108*, 766–776.

Brennan, L., & Owende, P. (2010). Biofuels from microalgae—a review of technologies for production, processing, and extractions of biofuels and co-products. *Renewable and Sustainable Energy Reviews, 14*(2), 557–577.

Brown, M. R. (1991). The amino-acid and sugar composition of 16 species of microalgae used in mariculture. *Journal of Experimental Marine Biology and Ecology, 145*(1), 79–99.

Brown, M. R., McCausland, M. A., & Kowalski, K. (1998). The nutritional value of four Australian microalgal strains fed to Pacific oyster Crassostrea gigas spat. *Aquaculture (Amsterdam, Netherlands), 165*(3-4), 281–293.

Challouf, R., Trabelsi, L., Ben Dhieb, R., El Abed, O., Yahia, A., Ghozzi, K., … Ben Ouada, H. (2011). Evaluation of cytotoxicity and biological activities in extracellular polysaccharides released by cyanobacterium Arthrospira platensis. *Brazilian Archives of Biology and Technology, 54*(4), 831–838.

Chanda, M. J., Merghoub, N., & Arroussi, H. E. (2019). Microalgae polysaccharides: the new sustainable bioactive products for the development of plant bio-stimulants? *World Journal of Microbiology and Biotechnology, 35*(11), 177.

Chen, C.-Y., Zhao, X.-Q., Yen, H.-W., Ho, S.-H., Cheng, C.-L., Lee, D.-J., … Chang, J.-S. (2013). Microalgae-based carbohydrates for biofuel production. *Biochemical Engineering Journal, 78*, 1–10.

Choi, S. P., Nguyen, M. T., & Sim, S. J. (2010). Enzymatic pretreatment of Chlamydomonas reinhardtii biomass for ethanol production. *Bioresource Technology, 101*(14), 5330–5336.

Ciferri, O. (1983). Spirulina, the edible microorganism. *Microbiological Reviews, 47*(4), 551–578.

Chisti, Y. (2007). Biodiesel from microalgae. *Biotechnology Advances, 25*(3), 294–306.

Costa, J. A. V., & de Morais, M. G. (2011). The role of biochemical engineering in the production of biofuels from microalgae. *Bioresource Technology, 102*(1), 2–9.

de Jesus Raposo, M. F., de Morais, A. M. B., & de Morais, R. M. S. C. (2015). Marine polysaccharides from algae with potential biomedical applications. *Marine Drugs, 13*(5), 2967–3028.

de Jesus Raposo, M. F., de Morais, A. M. M. B., & de Morais, R. M. S. C. (2014). Influence of sulphate on the composition and antibacterial and antiviral properties of the exopolysaccharide from Porphyridium cruentum. *Life Sciences, 101*(1-2), 56–63.

de Jesus Raposo, M. F., de Morais, A. M. M. B., & de Morais, R. M. S. C. (2014). Bioactivity and applications of polysaccharides from marine microalgae. In K. Ramawat, & J. M. Mérillon (Eds.), *Polysaccharides: Bioactivity and Biotechnology* (pp. 1–38). Cham, Switzerland: Springer.

de Jesus Raposo, M. F., de Morais, R. M. S. C., & de Morais, A. M. M. B. (2013). Bioactivity and applications of sulphated polysaccharides from marine microalgae. *Marine Drugs, 11*(1), 233–252.

De Philippis, R., & Vincenzini, M. (1998). Exocellular polysaccharides from cyanobacteria and their possible applications. *FEMS Microbiology Reviews*, *22*(3), 151–175.

Dragone, G., Fernandes, B. D., Abreu, A. P., Vicente, A. A., & Teixeira, J. A. (2011). Nutrient limitation as a strategy for increasing starch accumulation in microalgae. *Applied Energy*, *88*(10), 3331–3335.

EFSA, 2020. EU Novel Food Catalogue (v.1.1). European Food Safety Authority.

El Khoury, D., Cuda, C., Luhovyy, B. L., & Anderson, G. H. (2012). Beta glucan: health benefits in obesity and metabolic syndrome. *Journal of Nutrition and Metabolism*, 2012.

Flaibani, A., Olsen, Y., & Painter, T. J. (1989). Polysaccharides in desert reclamation: compositions of exocellular proteoglycan complexes produced by filamentous blue-green and unicellular green edaphic algae. *Carbohydrate Research*, *190*(2), 235–248.

Franco, J. M., Raymundo, A., Sousa, I., & Gallegos, C. (1998). Influence of processing variables on the rheological and textural properties of lupin protein-stabilized emulsions. *Journal of Agricultural and Food Chemistry*, *46*(8), 3109–3115.

FSSAI (2016). The Gazette of India, Food Safety and Standards Authority of India.

Geresh, S., Adin, I., Yarmolinsky, E., & Karpasas, M. (2002). Characterization of the extracellular polysaccharide of Porphyridium sp.: molecular weight determination and rheological properties. *Carbohydrate Polymers*, *50*(2), 183–189.

Geresh, S., & Arad, S. (1991). The extracellular polysaccharides of the red microalgae: chemistry and rheology. *Bioresource Technology*, *38*(2-3), 195–201.

Gloaguen, V., Ruiz, G., Morvan, H., Mouradi-Givernaud, A., Maes, E., Krausz, P., & Strecker, G. (2004). The extracellular polysaccharide of Porphyridium sp.: an NMR study of lithium-resistant oligosaccharidic fragments. *Carbohydrate Research*, *339*(1), 97–103.

Gouveia, L., Batista, A. P., Miranda, A., Empis, J., & Raymundo, A. (2007). Chlorella vulgaris biomass used as colouring source in traditional butter cookies. *Innovative Food Science & Emerging Technologies*, *8*(3), 433–436.

Guzman, S., Gato, A., Lamela, M., Freire-Garabal, M., & Calleja, J. M. (2003). Anti-inflammatory and immunomodulatory activities of polysaccharide from *Chlorella stigmatophora* and *Phaeodactylum tricornutum. Phytotherapy Research*, *17*(6), 665–670.

Guzman-Murillo, M. A., & Ascencio, F. (2000). Anti-adhesive activity of sulphated exopolysaccharides of microalgae on attachment of red sore disease-associated bacteria and Helicobacter pylori to tissue culture cells. *Letters in Applied Microbiology*, *30*(6), 473–478.

Halder, P., & Azad, A. K. (2019). Recent trends and challenges of algal biofuel conversion technologies. In A. K. Azad, & M. Rasul (Eds.), *Advanced Biofuels: Applications, Technologies and Environmental Sustainability* (pp. 167–179). UK: Woodhead Publishing.

Harun, R., & Danquah, M. K. (2011). Enzymatic hydrolysis of microalgal biomass for bioethanol production. *Chemical Engineering Journal*, *168*(3), 1079–1084.

Hayashi, T., Hayashi, K., Maeda, M., & Kojima, I. (1996). Calcium spirulan, an inhibitor of enveloped virus replication, from a blue-green alga Spirulina platensis. *Journal of Natural Products*, *59*(1), 83–87.

Henderson, R. K., Baker, A., Parsons, S. A., & Jefferson, B. (2008). Characterisation of algogenic organic matter extracted from cyanobacteria, green algae and diatoms. *Water Research, 42*(13), 3435–3445.

Hirano, A., Ueda, R., Hirayama, S., & Ogushi, Y. (1997). CO2 fixation and ethanol production with microalgal photosynthesis and intracellular anaerobic fermentation. *Energy, 22*(2-3), 137–142.

Ho, S. H., Chen, C. Y., & Chang, J. S. (2012). Effect of light intensity and nitrogen starvation on CO2 fixation and lipid/carbohydrate production of an indigenous microalga Scenedesmus obliquus CNW-N. *Bioresource Technology, 113*, 244–252.

Ho, S. H., Chen, C. Y., Lee, D. J., & Chang, J. S. (2011). Perspectives on microalgal CO_2-emission mitigation systems—a review. *Biotechnology Advances, 29*(2), 189–198.

Huang, J., Chen, B., & You, W. (2001). Studies on separation of extracellular polysaccharide from *Porphyridium cruentum* and its anti-HBV activity in vitro. *Chinese Journal of Marine Drugs*. (05).

Ike, A., Toda, N., Tsuji, N., Hirata, K., & Miyamoto, K. (1997). Hydrogen photoproduction from CO2-fixing microalgal biomass: application of halotolerant photosynthetic bacteria. *Journal of Fermentation and Bioengineering, 84*(6), 606–609.

Illman, A. M., Scragg, A. H., & Shales, S. W. (2000). Increase in Chlorella strains calorific values when grown in low nitrogen medium. *Enzyme and Microbial Technology, 27*(8), 631–635.

Jeon, J.-K. (2006). Effect of chlorella addition on the quality of processed cheese. *Journal of the Korean Society of Food Science and Nutrition, 35*, 373–377.

Kaji, T., Okabe, M., Shimada, S., Yamamoto, C., Fujiwara, Y., Lee, J. B., & Hayashi, T. (2004). Sodium spirulan as a potent inhibitor of arterial smooth muscle cell proliferation in vitro. *Life Sciences, 74*(19), 2431–2439.

Khan, M. I., Shin, J. H., & Kim, J. D. (2018). The promising future of microalgae: current status, challenges, and optimization of a sustainable and renewable industry for biofuels, feed, and other products. *Microbial Cell Factories, 17*(1), 36.

Kim, M. S., Baek, J. S., Yun, Y. S., Sim, S. J., Park, S., & Kim, S. C. (2006). Hydrogen production from Chlamydomonas reinhardtii biomass using a two-step conversion process: anaerobic conversion and photosynthetic fermentation. *International Journal of Hydrogen Energy, 31*(6), 812–816.

Lee, C.-K., Kim, H. S., Nam, J. R., Lee, M.-J., Yim, J.-H., & Lee, H. K. (2008). Antiviral effects of sulfated exopolysaccharide from the marine microalga Gyrodinium impudicum strain KG03. *Antiviral Research, 78*(2), A59.

Liang, S., Liu, X., Chen, F., & Chen, Z. (2004). *Current microalgal health food R & D activities in China. Asian pacific phycology in the 21st century: Prospects and challenges* (pp. 45–48). Dordrecht: Springer.

Liu, Q., Yao, C., Sun, Y., Chen, W., Tan, H., Cao, X., … Yin, H. (2019). Production and structural characterization of a new type of polysaccharide from nitrogen-limited Arthrospira platensis cultivated in outdoor industrial-scale open raceway ponds. *Biotechnology for Biofuels, 12*(1), 131.

Marceliano, M. B. (2009). Structure and function of polysaccharide gum-based edible films and coatings. In M. E. Embuscado, & K. C. Huber (Eds.), *Edible films and coatings for food applications*. Dordrecht: Springer.

Markou, G., Angelidaki, I., & Georgakakis, D. (2012). Microalgal carbohydrates: an overview of the factors influencing carbohydrates production, and of main bioconversion technologies for production of biofuels. *Applied Microbiology and Biotechnology, 96*(3), 631–645.

Mata, T. M., Cameira, M., Marques, F., Santos, E., Badenes, S., Costa, L., ... Martins, A. A. (2018). Carbon footprint of microalgae production in photobioreactor. *Energy Procedia, 153*, 432–437.

Matos, J., Cardoso, C., Bandarra, N. M., & Afonso, C. (2017). Microalgae as healthy ingredients for functional food: a review. *Food & Function, 8*(8), 2672–2685.

Matsui, M. S., Muizzuddin, N., Arad, S., & Marenus, K. (2003). Sulfated polysaccharides from red microalgae have antiinflammatory properties in vitro and in vivo. *Applied Biochemistry and Biotechnology, 104*(1), 13–22.

Mishra, A., Kavita, K., & Jha, B. (2011). Characterization of extracellular polymeric substances produced by micro-algae Dunaliella salina. *Carbohydrate Polymers, 83*(2), 852–857.

Mühlenbruch, M., Grossart, H.-P., Eigemann, F., & Voss, M. (2018). Mini-review: phytoplankton-derived polysaccharides in the marine environment and their interactions with heterotrophic bacteria. *Environmental Microbiology, 20*(8), 2671–2685.

Nakashima, A., Yamada, K., Iwata, O., Sugimoto, R., Atsuji, K., Ogawa, T., ... Suzuki, K. (2018). β-Glucan in foods and its physiological functions. *Journal of Nutritional Science and Vitaminology, 64*(1), 8–17.

NHFPC (2016). List of New Food Ingredients and General Food, National Health and Family Planning Commission, People Republic of China.

NNHPD (2019). Natural Health Products Ingredients Database, Natural and Non-prescription Health Products Directorate, Canada.

Ogawa, K., Ikeda, Y., & Kondo, S. (1999). A new trisaccharide, α-d-glucopyranuronosyl-(1→ 3)-α-l-rhamnopyranosyl-(1→ 2)-α-l-rhamnopyranose from *Chlorella vulgaris*. *Carbohydrate Research, 321*(1-2), 128–131.

Ozkurt, I. (2009). Qualifying of safflower and algae for energy. *Energy Education Science and Technology Part A-Energy Science and Research, 23*(1-2), 145–151.

Pierre, G., Delattre, C., Dubessay, P., Jubeau, S., Vialleix, C., Cadoret, J.-P., ... Michaud, P. (2019). What is in store for EPS microalgae in the next decade? *Molecules (Basel, Switzerland), 24*(23), 4296.

Pina-Pérez, M.C., Brück, W.M., Brück, T., & Beyrer, M. (2019). Microalgae as healthy ingredients for functional foods. The Role of Alternative and Innovative Food Ingredients and Products in Consumer Wellness, 103–137.

Pittman, J. K., Dean, A. P., & Osundeko, O. (2011). The potential of sustainable algal biofuel production using wastewater resources. *Bioresource Technology, 102*, 17–25.

Pulz, O., & Gross, W. (2004). Valuable products from biotechnology of microalgae. *Applied Microbiology and Biotechnology, 65*, 635–648.

Rajkumar, R., Yaakob, Z., & Takriff, M. S. (2014). Potential of micro and macro algae for biofuel production: a brief review. *Bioresources, 9*(1), 1606–1633.

Rasala, B. A., & Mayfield, S. P. (2015). Photosynthetic biomanufacturing in green algae; production of recombinant proteins for industrial, nutritional, and medical uses. *Photosynthesis Research, 123*(3), 227–239.

Raymundo, A., Franco, J., Gallegos, C., Empis, J., & Sousa, I. (1998). Effect of thermal denaturation of lupin protein on its emulsifying properties. *Food/Nahrung, 42*(03-04), 220–224.

Raymundo, A., Gouveia, L., Batista, A. P., Empis, J., & Sousa, I. (2005). Fat mimetic capacity of Chlorella vulgaris biomass in oil-in-water food emulsions stabilized by pea protein. *Food Research International, 38*(8-9), 961–965.

Salla, A. C. V., Margarites, A. C., Seibel, F. I., Holz, L. C., Brião, V. B., Bertolin, T. E., & Costa, J. A. V. (2016). Increase in the carbohydrate content of the microalgae Spirulina in culture by nutrient starvation and the addition of residues of whey protein concentrate. *Bioresource Technology*, *209*, 133–141.

Sekar, R., Venugopalan, V., Satpathy, K., Nair, K., & Rao, V. (2004). *Laboratory studies on adhesion of microalgae to hard substrates*. In: *Asian Pacific Phycology in the 21st Century: Prospects and Challenges* (pp. 109–116). Springer.

Sekharam, K. M., Venkataraman, L. V., & Salimath, P. V. (1989). Structural studies of a glucan isolated from blue-green alga Spirulina platensis. *Food Chemistry*, *31*(2), 85–91.

Shah, M. M. R., Liang, Y., Cheng, J. J., & Daroch, M. (2016). Astaxanthin-producing green microalga *Haematococcus pluvialis*: from single cell to high value commercial products. *Frontiers in Plant Science*, *7*, 531.

Shahid, A., Khan, F., Ahmad, N., Farooq, M., & Mehmood, M. A. (2020). *Microalgal carbohydrates and proteins: synthesis, extraction, applications, and challenges*. In: *Microalgae Biotechnology for Food, Health and High Value Products* (pp. 433–468). Singapore: Springer.

Shepherd, R., Rockey, J., Sutherland, I. W., & Roller, S. (1995). Novel bioemulsifiers from microorganisms for use in foods. *Journal of Biotechnology*, *40*(3), 207–217.

Skorupskaite, V., Makareviciene, V., & Levisauskas, D. (2015). Optimization of mixotrophic cultivation of microalgae *Chlorella* sp. for biofuel production using response surface methodology. *Algal Research*, *7*, 45–50.

Spolaore, P., Joannis-Cassan, C., Duran, E., & Isambert, A. (2006). Commercial applications of microalgae. *Journal of Bioscience and Bioengineering*, *101*(2), 87–96.

Sun, L., Wang, L., & Zhou, Y. (2012). Immunomodulation and antitumor activities of different-molecular-weight polysaccharides from *Porphyridium cruentum*. *Carbohydrate Polymers*, *87* (2), 1206–1210.

Talyshinsky, M. M., Souprun, Y. Y., & Huleihel, M. M. (2002). Anti-viral activity of red microalgal polysaccharides against retroviruses. *Cancer Cell International*, *2*(1), 8.

Torres-Tiji, Y., Fields, F. J., & Mayfield, S. P. (2020). Microalgae as a future food source. *Biotechnology Advances*, *41*, 107536.

Trabelsi, L., M'sakni, N. H., Ben Ouada, H., Bacha, H., & Roudesli, S. (2009). Partial characterization of extracellular polysaccharides produced by cyanobacterium Arthrospira platensis. *Biotechnology and Bioprocess Engineering*, *14*(1), 27–31.

Transparency Market Research (2020). Microalgae-based Products Market (Product: Spirulina, Chlorella, Astaxanthin, Beta Carotene, PUFA Fatty Acid, Lutein, and Others; and Application: Food & Beverages, Personal Care, Nutraceuticals & Dietary Supplements, Pharmaceuticals, Chemicals, Fuel, Animal Feed, and Others) - Global Industry Analysis, Size, Share, Growth, Trends, and Forecast, 2020 – 2030. <https://www.transparencymarketresearch.com/microalgae-based-products-market.html> (Access 25.09.2020).

Xia, S., Gao, B., Li, A., Xiong, J., Ao, Z., & Zhang, C. (2014). Preliminary characterization, antioxidant properties and production of chrysolaminarin from marine diatom Odontella aurita. *Marine Drugs*, *12*, 4883–4897.

Yamada, T., & Sakaguchi, K. (1982). Comparative studies on Chlorella cell walls: induction of protoplast formation. *Archives of Microbiology*, *132*(1), 10–13.

Yatirajula, S. K., Shrivastava, A., Saxena, V. K., & Kodavaty, J. (2019). Flow behavior analysis of Chlorella vulgaris microalgal biomass. *Heliyon*, *5*(6), e01845.

Yim, J. H., Kim, S. J., Ahn, S. H., Lee, C. K., Rhie, K. T., & Lee, H. K. (2004). Antiviral effects of sulfated exopolysaccharide from the marine microalga Gyrodinium impudicum strain KG03. *Marine Biotechnology (NY)*, *6*(1), 17–25.

Yu, S., Blennow, A., Bojko, M., Madsen, F., Olsen, C. E., & Engelsen, S. B. (2002). Physico-chemical characterization of Floridean starch of red algae. *Starch - Stärke*, *54*(2), 66–74.

Yuan, J.-P., Peng, J., Yin, K., & Wang, J.-H. (2011). Potential health-promoting effects of astaxanthin: a high-value carotenoid mostly from microalgae. *Molecular Nutrition & Food Research*, *55*, 150–165.

Zhou, K., Hu, Y., Zhang, L., Yang, K., & Lin, D. (2016). The role of exopolymeric substances in the bioaccumulation and toxicity of Ag nanoparticles to algae. *Scientific Reports*, *6*(1), 32998.

Chapter 6

Microalgae as source of edible lipids

Etiele G. Morais[1], Lisa Schüler[1], Hugo Pereira[1], Inês Maia[1], Katkam N. Gangadhar[1], Jorge A.V. Costa[2], Michele G. Morais[3], João Varela[1] and Luísa Barreira[1]

[1]CCMAR-Centre of Marine Sciences, University of the Algarve, Gambelas Campus, Faro, Portugal, [2]College of Chemistry and Food Engineering, Federal University of Rio Grande, Laboratory of Biochemical Engineering, Rio Grande, RS, Brazil, [3]College of Chemistry and Food Engineering, Federal University of Rio Grande, Laboratory of Microbiology and Biochemistry, Rio Grande, RS, Brazil

6.1 Introduction

Polyunsaturated fatty acids (PUFAs) are essential to human health, as they are not naturally produced by the human body and need to be provided through the diet. These compounds are responsible for the correct functioning of a series of biological processes and a source of energy for metabolism. The consumption of PUFAs can benefit human health, as they reduce inflammation and prevent neurological and cardiovascular diseases (Adarme-Vega et al., 2012). In addition, insufficient consumption of PUFAs can lead to the development of diseases such as obesity and osteoporosis (Kelly, Gilman, Kim, & Ilich, 2013). Increased public awareness of the health and clinical benefits of PUFAs has improved consumer acceptance and preference for food products enriched with these compounds (Kumar, Deviram, Mathimani, Duc, & Pugazhendhi, 2019).

Microalgal biomass is already widely used as a food supplement due to its composition, being rich in bioactive compounds such as polysaccharides, phenolic compounds, pigments, and PUFAs. Some species such as *Spirulina*, *Chlorella*, *Dunaliella*, and *Hematococcus* are Generally Recognized as Safe (GRAS) certified by the Food and Drug Administration, which allows the application of microalgal biomass in food products or as a supplement for humans (García, Vicente, & Galán, 2017). In the EU, *Arthrospira* (*Spirulina*) *platensis* and *Chlorella vulgaris* are approved for human consumption due to their history of safe use and are well established in the market, while

Cultured Microalgae for the Food Industry. DOI: https://doi.org/10.1016/B978-0-12-821080-2.00015-0

Odontella aurita and *Tetraselmis chui* were recently approved as novel foods by the European Food Safety Authority (EU 2017/2470).

Although the application of microalgal lipids in the production of biodiesel has moved microalgal research forward, it soon became evident that the production of microalgae as a source of PUFAs is much more profitable. For this reason, microalgae producers have focused on PUFA-rich strains, whose biomass is more valuable than that produced for low-value commodities such as biofuels (Sayeda, Gamila, & Farouk, 2015).

Microalgae can accumulate from 20% to 50% of their dry weight in lipids; additionally, these microorganisms have the ability to change their intracellular composition, including lipid concentration and profile, as a response to changes in environmental or cultivation conditions such as temperature, luminosity, pH, and nutrient concentration (Katiyar & Arora, 2020). The composition of the biomass is also strain-specific; while some species such as *A. platensis* are recognized for their protein content, others such as *Phaeodactylum tricornutum* or *Tisochrysis lutea* are known to produce high amounts of PUFAs. In general, microalgae have high PUFA contents and are considered to be an alternative to fish oil in the provision of these functional nutrients (Woods & Fearon, 2009). Microalgae, however, are photosynthetic organisms and the main PUFA producers in the oceans. They have the additional advantages of rapid growth coupled with high lipid content, being able to grow in harsh environments, and some strains, can be cultivated throughout the year. This chapter addresses the main topics of microalgae cultivation from optimum cultivation conditions to obtain optimal lipid profiles to their application as a nutritional supplement or new foods including products already commercialized and the main challenges and future prospects.

6.2 Microalgal lipid composition

Microalgae represent a diverse group of organisms that are able to convert carbon dioxide into oxygen and other metabolites, which include proteins, carbohydrates, and lipids (Hamed, 2016; Pereira et al., 2011). Over the years, there has been increasing interest in microalgae-produced compounds such as lipids, which are used in several applications, such as the food, chemical, and pharmaceutical industries (Mimouni et al., 2012). Lipids are water-insoluble biomolecules that have many important biological roles such as energy storage, signal molecules, and as structural components of cell membranes (Fahy, Cotter, Sud, & Subramaniam, 2011). Lipids are, however, stored in different ways, depending on their function, microalgal species or strain, microalgal growth phase, and environmental or cultivation conditions (Sajjadi, Chen, Raman, & Ibrahim, 2018). Microalgal species that can accumulate a considerable amount of lipids are considered oleaginous strains. Lipids from microalgae are mainly divided into two categories: polar and neutral (Zhu, Li, & Hiltunen, 2016). Microalgae are also able to generate other types of lipids such as waxes, hydrocarbons, sterols, prenyl derivatives like carotenoids, and pyrrole derivatives such as chlorophylls (Sajjadi et al., 2018). Neutral lipids, such as triacylglycerol (TAG), sterols, free fatty acids (FFAs), and pigments are considered to be energy reservoirs and are used for storage, having no structural function (Huerlimann, de Nys, & Heimann, 2010). Storage lipids are primarily found in the form of TAGs and have a high saturated fatty acid content and also some unsaturated fatty acids. When cultivating conditions are growth-limiting for microalgae, they tend to accumulate carbohydrates, carotenoids, and neutral lipids (Schüler et al., 2017). Under these conditions, the synthesis of lipids is switched from structural lipids, necessary for cellular division, diverting the photosynthetic energy to the accumulation of TAGs (Huerlimann et al., 2010). As these TAGs cannot be used for membrane biosynthesis and assembly, they are stored, being easily catabolized as metabolic energy (Pal et al., 2019). They are mainly biosynthesized when microalgae are exposed to light, and accumulated in lipid bodies, being later used for the synthesis of polar lipids in the absence of light (Sharma, Schuhmann, & Schenk, 2012). This process allows cells to survive and adapt to extreme environmental factors.

Polar lipids, such as phospholipids and glycolipids, which are essential components of cell membranes, including those of the thylakoids in the chloroplast, are considered structural lipids. Sterols and phospholipids act as a permeable barrier, protecting microalgal cells (Sajjadi et al., 2018) and aiding in the separation of intracellular organelles (Gopalakrishnan & Ramamurthy, 2014). Other polar lipids behave as intermediates in signaling pathways and contribute to the cellular response to environmental factors such as temperature and light (Chen, Wang, Qiu, & Ge, 2018).

Both neutral and polar lipids can be comprised of different types of fatty acids. Polar lipids are mainly comprised of PUFAs, which are fatty acids with more than one double bound, whereas monounsaturated fatty acids (MUFAs) and saturated fatty acids (SFAs) are predominantly present in neutral lipids (Olofsson et al., 2012). SFAs and MUFAs are mainly produced and accumulated when microalgae cells are under unfavorable growth conditions, as a survival mechanism (Paliwal et al., 2017). PUFAs, however, given their higher proportion in polar lipids, are mainly produced when microalgae are growing under optimal growth conditions, as they are essential for maintaining membrane function (Paliwal et al., 2017) and fluidity upon a shift in a given abiotic factor (Schüler et al., 2017). Microalgae have the ability to synthesize significant amounts of PUFAs, as they are the primary source of these compounds (Bellou et al., 2014). They are known for having antioxidant (Plaza & Herrero, 2009) and antiinflammatory (Wall, Ross, Fitzgerald, & Stanton, 2010) activities, preventing conditions like Alzheimer's (Hooper, Barreto, Pahor, Weiner, & Vellas, 2018), fatty liver (Jump, Lytle, Depner, & Tripathy, 2018), and cardiovascular diseases (Rimm et al., 2018).

TABLE 6.1 Fatty acids (% of total fatty acids) composition of several strains of microalgae.

Microalgal strains	C16:0	C18:0	∑ SFA	C16:1	C18:1	∑ MUFA	C18:2	C18:3*n* − 3	C18:4*n* − 3	C20:5*n* − 3	C22:6*n* − 3	∑ PUFA	References
							LA	ALA		EPA	DHA		
Arthrospira platensis	47.7	12.5	60.1	–	6.3	6.3	10.4	–	–	–	–	10.4	Kim et al. (2013)
Botryococcus braunii	16.5	8.2	24.7	15.5	34.2	49.7	12.3	–	–	–	–	12.3	Ranga Rao, Ravishankar, and Sarada (2012)
Chlamydomonas mexicana	35.1	1.1	36.2	1.0	6.7	7.7	40.9	14.3	–	–	–	55.2	Salama et al. (2013)
Chlorella vulgaris	20.2	3.0	23.2	0.7	12.0	12.7	–	18.3	–	0.9	0.5	19.6	Khoeyi, Seyfabadi, and Ramezanpour (2012)
Dunaliella salina	9.8	3.1	12.9	14.7	8.0	22.7	4.9	12.9	–	–	15.4	33.1	Fakhry and Maghraby (2013)
Emiliania huxleyi	10.3	10.8	21.1	–	42.2	42.2	–	–	8.7	–	9.2	17.9	Lang, Hodac, Friedl, and Feussner (2011)
Nannochloropsis oculata	18.4	–	18.4	26.1	3.0	29.1	6.2	–	–	34.0	–	40.2	Gangadhar et al. (2016a)

Nannochloropsis salina	32.0	3.2	35.2	29.9	9.4	39.3	2.6	0.9	–	12.7	–	16.2	Ma, Wang, Yu, Yin, and Zhou (2014)
Odontella aurita	13.7	2.2	15.9	22.5	10.1	32.6	0.7	0.8	1.3	25.6	3.8	32.2	Guihéneuf, Mimouni, Ulmann, and Tremblin (2009)
Pavlova lutheri	16.3	3.0	19.3	14.4	3.9	18.3	1.1	0.5	8.3	23.4	13.5	46.8	Guihéneuf et al. (2009)
Phaeodactylum tricornutum	7.1	0.3	7.4	9.4	9.4	18.8	1.9	–	–	26.1	1.7	29.7	Quelhas et al. (2019)
Porphyridium cruentum	26.8	–	26.8	0.3	0.6	0.9	3.9	–	–	25.4	0.1	29.4	Durmaz, Monteiro, Bandarra, Gökpinar, and Işik (2007)
Rhodomonas sp.	12.4	2.0	14.4	3.9	3.8	7.7	1.3	19.8	28.1	7.7	3.8	60.7	Renaud, Thinh, Van, Lambrinidis, and Parry (2002)
Scenedesmus sp.	20.5	0.8	21.2	19.0	0.9	19.9	10.3	39.3	–	–	–	49.6	Custódio et al. (2014)
Skeletonema sp.	6.0	0.1	6.1	21.9	0.3	22.2	0.8	0.1	3.2	10.8	2.2	17.1	Cardoso et al. (2020)

(Continued)

TABLE 6.1 (Continued)

Microalgal strains	C16:0	C18:0	∑ SFA	C16:1	C18:1	∑ MUFA	C18:2	C18:3*n* − 3	C18:4*n* − 3	C20:5*n* − 3	C22:6*n* − 3	∑ PUFA	References
							LA	ALA		EPA	DHA		
Synechococcus sp.	43.0	8.8	51.8	23.0	8.1	31.1	–	–	–	–	–	–	Blažina et al. (2019)
Tetraselmis sp.	24.9	0.4	25.3	7.9	27.1	34.9	21.3	–	–	9.4	–	30.7	Custódio et al. (2014)
Tetraselmis sp. CTP4	15.3	0.5	15.8	1.9	11.8	13.7	5.3	18.8	4.9	3.7	–	32.7	Cardoso et al. (2020)
Tetraselmis sp. IMP3	27.1	0.2	27.3	1.0	18.7	19.7	6.2	13.6	2.0	6.4	–	28.2	Cardoso et al. (2020)
Tisochrysis lutea	14.4	0.3	14.7	20.5	15.2	35.7	12.1	–	–	2.8	12.7	27.5	Custódio et al. (2014)

ALA, α-Linolenic; *DHA*, docosahexaenoic; *EPA*, eicosapentaenoic; *LA*, linoleic; *MUFA*, monounsaturated fatty acid; *PUFA*, polyunsaturated fatty acid; *SFA*, saturated fatty acid.

Vertebrates can synthesize most essential PUFAs, except for α-linolenic (ALA) and linoleic (LA) acids. These fatty acids have to be obtained through the diet, in both feed and food, as they are the pathway precursors for the biosynthesis of *n*-3 and *n*-6 PUFAs (Pereira et al., 2012). Because of limited elongation of ALA into docosahexaenoic (DHA) and eicosapentaenoic (EPA) acids in humans, which is the precursor of DHA, these FAs must also be incorporated into the diet. DHA has an important role in the central nervous system, as it has been defined as a structural component (Guesnet & Alessandri, 2011) and EPA has an impact on different dysfunctions (El-Mowafy, Abdel-Dayem, Abdel-Aziz, El-Azab, & Said, 2011).

Different species of microalgae can be a source of these different types of fatty acids (Table 6.1), which can be accumulated under the right growth conditions. These different types of fatty acids can be extracted and applied to different areas. Species like *Botryococcus braunii*, *Emiliania huxleyi*, *Nannochloropsis salina*, and the cyanobacteria *Synechococcus* spp. show high SFA and MUFA contents (Table 6.1). Higher amounts of PUFAs are found in species like *Chlamydomonas mexicana*, *Pavlova lutheri*, *Rhodomonas* sp., *Scenedesmus* sp., and *Tetraselmis* sp. CTP4. However, species like *O. aurita*, *P. lutheri*, *P. tricornutum*, *Skeletonema* sp., and *T. lutea* demonstrate huge potential for food purposes, as they accumulate both EPA and DHA, which are *n*-3 PUFAs in great demand.

Variation of lipid content is, normally, between 20% and 50% of dry biomass (Bellou et al., 2014; Chen et al., 2018) in genera like *Dunaliella*, *Isochrysis*, *Tetraselmis*, and *Phaeodactylum*, with some species being able to achieve higher lipid content. Species like *B. braunii*, *C. vulgaris*, and *Nannochloropsis* sp. have the potential to reach up to 62% in lipids of their total dried weight (Sajjadi et al., 2018).

Higher accumulation of lipids can be achieved with the manipulation of culture conditions, varying factors like irradiance, temperature, and nutrient availability (Bellou et al., 2014), which is an important advantage of microalgae for the production of lipids. However, as a higher lipid content is achieved when the microalgae cells are stressed, cell productivity rates decrease (Vasudevan & Briggs, 2008).

6.3 Methods for increased lipid production

6.3.1 Manipulation of cultivation conditions

Cultivation conditions optimization, which includes appropriate strain selection, type of bioreactor, nutrient concentration in the media, cultivation modes, and physical-chemical conditions, is essential for the production viability of microalgae biomass and/or bioproducts. Among these, increased lipid productivity has been a recurring demand in microalgae biotechnology.

The lipid increase in microalgae can be induced, mainly, by stress conditions applied to cultivations, including the limitation of nutrients, such as nitrogen and phosphorus (Chu et al., 2013; Li et al., 2014), increased temperature (Chokshi et al., 2015), salinity (Pancha et al., 2015a; Pandit, Fulekar, & Karuna, 2017; Xia et al., 2014), and light intensity (Cassuriaga, Freitas, Morais, & Costa, 2018), among others.

Nitrogen limitation in the cultivation of *P. tricornutum* provided increases in the relative content of saturated fatty acids and a decrease in PUFAs (Qiao et al., 2016). Likewise, Li et al. (2014) found that nitrogen deficiency associated with full phosphorus condition resulted in higher lipid productivity (224.14 mg L^{-1} d^{-1}) by *Chlorella protothecoides*. However, deprivation/limitation conditions of N and P would contribute to the decrease in PUFAs in microalgal biomass. San Pedro, González-López, Acién, and Molina-Grima (2014) evaluated lipid productivity and FA content in *Nannochloropsis gaditana* cultivations, produced in outdoor conditions. In this study, the use of a continuous cultivation strategy with nutrient restriction had no influence on lipid and FA productivities in *N. gaditana* biomass. On the other hand, it was observed that in nutrient-replete medium the highest lipid and FA productivities were observed at the dilution rate (DR) of 0.31 and 0.35 d^{-1}, with the highest content of PUFAs (34.3%) with a DR of 0.31 d^{-1}. Moreover, Camacho-Rodríguez, Cerón-García, Fernández-Sevilla, and Molina-Grima (2015) evaluated the effect of temperature (15°C–30°C), light intensity (250–1600 μmol s^{-1} m^{-2}), and nutrient DR (0.1–0.5 d^{-1}) in *N. gaditana* under outdoor cultivation conditions, in the production of biomass and EPA. EPA maximum productivity (17.2 mg L^{-1} d^{-1}) by *N. gaditana* was observed at 25°C, with a light intensity of 170 μmol s^{-1} m^{-2}, and 0.3 d^{-1} of specific growth rate.

An increase in salinity in culture media, using seawater, or addition of salt, can additionally promote the increase of lipids such as PUFAs in microalgae (Ho, Ye, Hasunuma, Chang, & Kondo, 2014). The addition of 50 g L^{-1} NaCl in *C. vulgaris* cultures increased the biomass lipid content (28% w w^{-1}) (Shen et al., 2015). Similar results were reported by Pandit et al. (2017), who observed that the increase in NaCl concentration from 60 to 400 mmol L^{-1} resulted in an increase in the lipid content from 31.5 to 49.5% w w^{-1} in *C. vulgaris* biomass. Lipids concentration in the *C. mexicana* biomass increased with the addition of 400 mmol L^{-1} NaCl to the medium; however, the highest lipid productivities were obtained in the assays with 100 and 200 mmol L^{-1} of NaCl (Salama et al., 2014).

Salinity, temperature, light intensity, and cultivation time can not only influence microalgal growth, but also lipid accumulation (Qiao et al., 2016). *P. tricornutum*, in batch cultivation, and medium f/2, showed a reduced percentage of PUFAs, while the concentration of MUFAs and the total fatty acid content (TFACs) increased with cultivation time. Reduced salinity (53% of seawater) caused a decline in TFACs. Furthermore, the DHA percentage increased with increasing light intensity, and EPA, DHA, and PUFAs content decreased with increasing temperature (Qiao et al., 2016).

Nannochloropsis sp. and *Tetraselmis* sp. have shown increased total lipid content and decreased PUFAs content when the cultivation time increases, using different cultivation medium. Among the microalgae evaluated, *Tetraselmis* sp. showed the highest lipid productivity (3.9–4.8 g m^{-2} d^{-1}) in any type of media during the exponential growth phase (Huerlimann et al., 2010).

pH and CO_2 flow rate also influence lipid productivity in microalgae. In this sense, pH adjustment (between 6.8 and 7.2) during *Chlorella zofingiensis* cultivation provided high biomass (66.9 mg L^{-1} d^{-1}) and lipid (37.5 mg L^{-1} d^{-1}) productivities (Zhu et al., 2014). The increase in CO_2 concentration (0.03% to 15% CO_2) enhanced lipid productivity in *Chlorella fusca* LEB 111 by 28% (Moraes, Santos, & Costa, 2019b). The increase in the CO_2 flow rate resulted in pH gradients, which caused stressful conditions, in *N. gaditana* outdoor cultivation, and increased the lipid content in the biomass. On the other hand, the reduction in CO_2 flow rate from 7.7 to 1.9 mL_{CO2} $L_{culture}$ min^{-1} promoted an increase in lipid productivity of *N. gaditana* by 44.3% (Moraes et al., 2019a) (Table 6.2).

CO_2 chemical absorbents have the potential to increase the concentration and productivity of lipids in *C. fusca* LEB 111. In this context, when batch cultivation was used, the addition of 100 mg L^{-1} monoethanolamine (MEA) caused the lipid productivity of this strain to increase by 10% compared to the control (Rosa, da Morais, & de Costa, 2018). However, the greatest increase in lipid productivity of *C. fusca* LEB 111 (51.0%) occurred with the use of fed-batch cultivation with 50 mg L^{-1} MEA and CO_2 (31.1 mg L^{-1} d^{-1}) (Rosa, Morais, & Costa, 2019).

In addition to the application of the above-mentioned conditions, other strategies are being investigated for the development of more efficient microalgal cultivation in terms of lipid productivity. According to Ho et al. (2014), the ideal process should include conditions that favor growth and also the accumulation of reserve biomolecules such as lipids. However, as mentioned previously, the accumulation of reserve polymers is generally accompanied by a reduction in biomass productivity, due to the change in metabolic routes in response to stress conditions.

Because of the demand for higher lipid productivity, a two-stage cultivation strategy has been proposed in order to increase the viability of biomass production with application to high added-value products, such as PUFAs (Pancha, Chokshi, Maurya, & Trivedi, 2015b; Ra, Kang, Na, Lee, & Kim, 2015; Wang et al., 2016). The double stage begins with the best cultivation conditions for greater biomass productivity (first stage); followed by stress conditions for reserve compounds accumulation (second stage). Among the applied strategies, nitrogen deprivation, sodium chloride addition, and ion concentration imbalance have been suggested as the most efficient

TABLE 6.2 Cultivation conditions applied to increase lipids average productivity (P_L, mg L^{-1} d^{-1}) in microalgae.

Microalgae	Applied condition	Effect	P_L (mg L^{-1} d^{-1})	References
Nannochloropsis sp.	Nitrogen deprivation + increased light intensity	Increased lipid productivity	76.6	Jiang, Luo, Fan, Yang, and Guo (2011)
Chlorella vulgaris AG10032	Nitrogen deprivation	Growth slows down slightly and the lipids increase	77.8	Mujtaba, Choi, Lee, and Lee (2012)
Neochloris oleoabundans	Nitrogen deprivation + increased light intensity + Fe	Growth rarely affected and increase in lipids	51.6	Sun et al. (2014)
Scenedesmus obtusus XJ-15	NaCl addition	Small effect on growth and increased lipid concentration	60.7	Xia, Ge, Zhou, Zhang, and Hu (2013)
Chlorella fusca LEB 111	Fed-batch with monoethanolamine and CO_2	Increased lipid productivity	46.9	Rosa et al. (2019)
Chlorella fusca LEB 111	Use of monoethanolamine to increase CO_2biofixation	Increased lipid productivity	55.0	Rosa et al. (2018)
Nannochloropsis gaditana	Different CO_2 flow rates for pH control	Increased lipid productivity	39.5	Moraes et al. (2019a)
Chlorella fusca LEB 111	Different concentrations of CO_2 as a carbon source	Increased lipid productivity	37.5	Moraes et al. (2019b)

(Ho et al., 2014). The two-stage strategy can be applied to different microalgal genera of commercial importance, mainly with Chlorophyta, such as *Scenedesmus* and *Chlorella* and Ochrophyta such as *Nannochloropsis*. Vidyashankar, Deviprasad, Chauhan, Ravishankar, and Sarada (2013) have observed that the use of double-stage cultivation with the addition of NaCl and nitrogen restriction increased the lipid concentration in *Scenedesmus dimorphus* biomass. Electrical stimulation has also been evaluated as a strategy for increasing the production of lipids, such as TAG and PUFAs. Choi et al. (2019) found increases of 36% and 57% in the contents of linolenic and LA acid, respectively, in *Chlorella* sp. KR-1 subjected to 4 V electrical treatment for 4 h, compared to untreated cells.

Spirulina species are often poor lipid producers, as the concentration found in the biomass is normally around 12.0% $w\,w^{-1}$ (Jesus et al., 2018; Rosa, Moraes, Souza, da, & Costa, 2016; Uebel, Costa, Olson, & Morais, 2019). However, as *Spirulina* is robust to changes in pH, temperature, and light, and shows high biomass productivity ($\sim 500\,mg\,L^{-1}\,d^{-1}$) (Soni, Sudhakar, & Rana, 2019), lipid productivity can also be relatively high ($\sim 60\,mg\,L^{-1}\,d^{-1}$).

6.3.2 Strain improvement for increased lipid production

Several strategies can be used for the development of improved microalgal strains for specific purposes: (1) enrichment in subpopulations or even in individual cells that underwent spontaneous mutations, or induced random mutagenesis, showing a given phenotype; or (2) via genetic transformation with specific genes. Either strategy can be used to trigger the development of strains with improved lipid contents—containing, for example, higher amounts of EPA and DHA. Nevertheless, the synthesis of these important fatty acids is dependent on the microalgae genetics as only some microalgal species can produce LC-PUFA. As highlighted in previous sections, environmental conditions can influence the synthesis and storage of fatty acids. The combination of the knowledge about their response to changing environments with genetic approaches will enable the full exploitation of strains with improved growth and/or the ability to accumulate higher amounts of these important fatty acids.

6.3.3 Random mutagenesis and strain selection

Strain improvement by random mutagenesis is a strategy that involves physical or chemical agents to induce changes in the DNA followed by a selection of mutants with the desired improved characteristics. Examples of physical agents are UV radiation, heavy-ion irradiation, and gamma- or X-rays, whereas chemical agents commonly used are ethyl methane sulfonate (EMS), methyl methane sulfate, nitrosoguanidine, ethyl-nitrosourea, and

N-methyl-*N*-nitrosourea (MNU). The latter kind of agents modify chemically the nucleotides by alkylation, inducing point mutations upon one round of replication. Ionizing radiation such as UV can also induce cytosine-to-thymine point mutations, but more often it causes larger DNA lesions affecting pairs of nucleotides in tandem (Ikehata & Ono, 2011). In random mutagenesis, as previous knowledge about the genetics of the microalgae is not required, techniques are often easy and fast to handle, and the new strains do not fall under the regulatory restrictions of genetically modified organisms (GMOs). However, this strategy is nonspecific, and the accumulation of target lipid classes or fatty acids may vary with the approach used and the species that was selected for improvement. For example, a EPA/DHA-rich strain of *P. lutheri* was obtained by UV mutagenesis, which produced over 30% more of these fatty acids than the wild type (WT) (Meireles, Guedes, & Malcata, 2002). In a different study on *Nannochloropsis* sp., the EMS-induced mutants showed increased lipid productivity, however, with the same fatty acid profile as the WT (Anandarajah, Mahendraperumal, Sommerfeld, & Hu, 2012). Therefore upon mutagenesis, it is crucial to develop efficient screening methods to isolate strains with an improved fatty acid profile or lipid synthesis, in particular enhanced levels of long-chain PUFAs such as EPA and/or DHA. For example, the herbicides cerulenin and quizalofop are inhibitors of fatty acid synthesis. The screening of strains resistant to these herbicides may contain mutations in key steps of this pathway leading to significant changes in fatty acid biosynthesis. Studies on *Nannochloropsis oculata* obtained after mutagenesis with either EMS, MNU, or UV, produced herbicide-resistant strains with improved EPA content (Chaturvedi & Fujita, 2006; Chaturvedi, Uppalapati, Alamsjah, & Fujita, 2004; Moha-León, Pérez-Legaspi, Ortega-Clemente, Rubio-Franchini, & Ríos-Leal, 2019). A different strategy is the selection of specific strains by staining lipids with lipophilic fluorescent dyes such Nile Red or BODIPY 505/515. This method can also be coupled with fluorescence-activated cell sorting (FACS) for a higher and faster throughput selection. Using this approach, an EMS-generated mutant of *N. salina* was obtained with increased contents of EPA and arachidonic acid (ARA) (Beacham, Macia, Rooks, White, & Ali, 2015).

6.3.4 Genetic transformation

Because of emerging technologies enabling faster, more reliable, and cheaper genome sequencing, more and more genomes have become available and key points of metabolic pathways can be targeted. One targeted pathway is the TAG assembly, which is not only beneficial for biofuel applications but also for nutritional purposes. Nevertheless, genetic engineering is still limited to a small number of species such as *Chlamydomonas reinhardtii*, diatoms (*P. tricornutum* and *Thalassiosira pseudonana*), and some species of *Nannochloropsis*. For example, in *P. tricornutum* the overexpression of type-2 diacylglycerolacyl transferase, a rate-limiting enzyme in the final step of TAG assembly, led not only to an increase in the number of oil bodies per cell but also to an over 70% increase in the content of EPA (Niu et al., 2013). The overexpression of a different enzyme of TAG assembly, 1-acyl-*sn*-glycerol-3-phosphate acyltransferase, led to similar results in *P. tricornutum*, namely a 1.8-fold increase of TAG accompanied by a 1.5-fold increase of EPA and DHA (Balamurugan et al., 2017). For the improvement of the production of PUFAs in particular, desaturases and elongases need to be targeted. Therefore coordinated expression of multiple enzymes such as malonyl CoA-acyl carrier protein transacylase and the desaturase 5b in *P. tricornutum* led to the highest content of EPA and DHA reported so far (Wang et al., 2017). In a different study on *Nannochloropsis oceanica*, the overexpression of the microsomal-like Δ12 desaturase, converting oleic acid to LA acid, yielded an increase in ARA-containing TAG under nitrogen starvation (Kaye et al., 2015). Another promising technology is transcriptional engineering, which is based on the overexpression of transcription factors resulting in the regulation of multiple components of a pathway (Bajhaiya, Moreira, & Pittman, Pittman, 2016). For example, the overexpression of the helix-loop-helix transcription factor (NsbHLH2) in *N. salina* which resulted in a higher growth rate and 33% increased EPA content compared with the WT (Kang et al., 2019).

6.3.5 Adaptive evolution

Adaptive evolution is a traditional and sustainable method of strain improvement, which has gained recent interest in the development of industrial microalgal strains. Strains are evolved by adapting to specified environmental conditions that favor a certain phenotype under iterative prolonged growth cycles (Dragosits & Mattanovich, 2013). This technique is based on spontaneous mutations that select for strains with the ability to grow under particular conditions. For the development of LC-PUFA-rich strains, conditions that favor PUFAs production (e.g., low light intensity or low temperatures) in the investigated species may be applied (Schüler et al., 2017). In a study on *P. tricornutum*, an evolved strain with high EPA content has been obtained by hyposalinity treatment maintained throughout 35 growth cycles (Wang et al., 2019). The advantage of adaptive evolution over genetic approaches is that strains can be grown in liquid media while clone selection usually takes place on solid media. Furthermore, genetic knowledge of the microalgae is not required, and the obtained strains are not considered GMOs. However, the selection process is laborious and time-consuming.

6.4 Microalgal lipid industrial production

The exploitation of microalgal lipids at an industrial scale relies on the production of microalgal biomass in large-scale production facilities. However, the production of microalgae biomass at higher scales displays some key particularities that are often neglected in lab-scale culture systems. With the increasing scale of production, problems arise (e.g., contamination), being an extraordinary challenge to maintain a monoalgal culture during a long production period. In addition, microalgae cells growing actively commonly display low lipid content, since the carbon metabolism is diverted to the production of sugars needed for cell division. In this context, the commonly used lipid induction strategies (e.g., nutrient starvation) normally focusing on two-stage growth approaches can be difficult to implement at an industrial scale.

There are two main systems widely used to effectively grow microalgae biomass in industrial-scale production facilities, namely, open and closed systems. Biomass production in open systems was first achieved in open raceway ponds, and later in thin-layer cascade systems (Borowitzka, 1999). Raceway ponds consist of horizontal low-depth recirculating ponds that commonly operate with a culture water column of 10–30 cm, mixed using a rotating paddlewheel (Chisti, 2012, 2016). The main advantage of using raceway ponds for microalgal production is the lower capital and operational costs over thin-layer cascades and closed production systems. Therefore open ponds are the production system chosen to produce microalgae for low-end applications, namely for bioremediation of effluents and CO_2 sequestration. However, since cultures are constantly exposed to the atmosphere and possible contaminants, only the microalgal strains that are able to grow in a challenging environment or display high growth rates are effectively cultivated in open systems. Typical examples of successful commercial-scale production in raceway ponds of strains that grow in extreme conditions, where common contaminants cannot proliferate, are *Dunaliella salina* (high salt concentration) and *Arthrospira* sp. (high pH).

Thin-layer cascades were developed in the Czech Republic in the 1960s by Šetlík, Sust, and Málek (1970). The production system consists of one or more sloped surface platforms, where a thin layer of microalgal suspension flows by gravity and cells perform the photosynthetic process. The term "cascade" comes from when the culture falls from the end of the slanted surface to a reservoir and the degassing process occurs; thereafter, a pump feeds the culture again to the upper section of the tilted surface. Because of the optimal light path and high efficiency of the degasification process of a thin-layer cascade, the biomass concentrations obtained are much higher than those observed for open raceway ponds and most closed production systems, reaching values up to 3.5 $g\,L^{-1}$ (Masojídek, Kopecky, Giannelli, & Torzillo, 2011). Moreover, due to the high production rates, more microalgae strains can be cultured, since the production cycles in batch conditions are carried out in a short period of time. However, higher capital and operational costs are observed in thin-layer cascades compared to raceway ponds. Overall, the main advantage of open production systems is the lower capital and operational costs compared to closed systems; however, since cultures in open systems are constantly exposed to the atmosphere, the culture of more sensitive microalgal strains and with lower growth rates can only be performed in closed systems.

Production of microalgae in closed production systems is performed in specialized growth systems, commonly named photobioreactors (PBRs). The cultures in this case are contained from the external environment and there is no direct contact with the atmosphere. There are several designs of PBRs, but three main configurations exist, namely, flat panel, vertical columns, and horizontal tubular PBRs. The main advantage of closed systems is that contamination of cultures is often prevented and greater control of most culture parameters, including thermal regulation, is effectively achieved. The effective control of process parameters enables high biomass concentrations and productivities of monoalgal cultures of most microalgal strains, including sensitive species for high-end applications (e.g., pharmaceutical grade). On the other hand, the main disadvantage of PBRs is the high capital and operational costs associated with these specialized culture systems, when compared to open production systems. Another constraining factor of PBRs is the high cell fragility of some microalgae that can undergo cell lysis due to the mechanical shear stress of standard pumps (e.g., *Pavlova* and *Isochrysis*) or air bubbling (e.g., *E. huxleyi*).

6.5 Harvesting of microalgae for lipid extraction

The choice of the appropriate harvesting method must take into account costs, cultivation conditions, and microalgal morphological characteristics (shape and size) (Brasil, Silva, & Siqueira, 2017). According to Vergini, Aravantinou, and Manariotis (2016), the most common harvesting methods are filtration, centrifugation, flocculation, sedimentation, and flotation.

Filtration is based on solid–liquid separation mechanisms, and uses a semipermeable barrier that retains microalgal cells (Gerardo, Van Den Hende, Vervaeren, Coward, & Skill, 2015). Biomass harvesting by filtration is a favorable process having the advantage of biomass recovery even when cell densities are low (Rickman, Pellegrino, & Davis, 2012). However, clogging or fouling in some systems can occur due to cell or biofilm deposition, which can prevent the liquid medium from being separated from microalgal biomass (Show & Lee, 2013).

Centrifugation is a harvesting method that uses centripetal acceleration to separate microalgae in regions of higher and lower density, which are separated from the water by drainage. Centrifugation can also be followed by sedimentation to separate the supernatant (Harun, Singh, Forde, & Danquah, 2010). Hydrocyclone, solid bowl decanter, spout-type centrifuge, and solid ejection discs are conventional centrifugation methods (Milledge & Heaven, 2013). Although centrifugation can collect efficiently most genera of microalgae, it is highly energy demanding, having high capital and operating costs (Yuan, Shen, Pei, Wu, & Mao, 2009). Centrifugation has an average capacity to harvest 12%–25% of microalgae biomass with energy demands higher than 1 kWh m^{-3} (Milledge & Heaven, 2013).

Flotation harvesting can be used due to the low density that microalgal cells normally present (Gerardo et al., 2015). This process consists of floating the cells to the surface, which can occur more quickly with the aid of air bubbles. However, higher efficiencies of this unitary operation are usually obtained with the use of a coagulating agent (Uduman, Qi, Danquah, Forde, & Hoadley, 2010). Microalgal biomass harvesting by flotation can achieve efficiencies ranging from 70% to 99% (Coward, Lee, & Caldwell, 2014; Zhang et al., 2014), as well as having low capital costs and requiring lower time compared to other methods (Gerardo et al., 2015). Different flotation methods are reported: dispersed air, microfiltration, foam flotation, dissolved air, vacuum gas, electro-flotation, and ozone flotation (Japar, Takriff, & Yasin, 2017).

Flocculation is a unit operation commonly used in conjunction with other recovery processes such as coagulation and flotation. Flocculation causes suspended microalgae cells to aggregate, increasing the sedimentation rate (Mata, Martins, & Caetano, 2010). There are four types of flocculation used: auto- (spontaneous), physical, physical-chemical, and biological flocculation. The autoflocculation process presents ~90% recovery and normally occurs when access to CO_2 by microalgal cells is stopped and the pH of the cultivation medium reaches values higher than 9 (Vandamme, Foubert, & Muylaert, 2013). This can occur with cyanobacteria like *Spirulina* due to the presence of gas vesicles. However, this process is highlighted as slow and requires the presence of calcium and magnesium ions (Gerardo et al., 2015).

Microalgae recovery from the liquid medium by sedimentation (solid–liquid) is one of the simplest forms of harvesting. This process is also known as gravitational sedimentation, and has as a prerequisite for separation that the density of the cells is higher than that of water (Peperzak, Colijn, Koeman, Gieskes, & Joordens, 2003). In comparison with other harvesting systems, sedimentation maintains microalgae morphological integrity (Zheng et al., 2012) and the energy demand is low (Pragya, Pandey, & Sahoo, 2013). The cells harvested in this operation can be used to obtain different final products free of external agents (Gao et al., 2010), but sedimentation without chemicals could be unenforceable depending on the strain due to the need for long periods for harvesting. Sedimentation may not be suitable for cells with high lipids concentration because cell density decreases (Japar et al., 2017). There are, however, notable exceptions; *Tetraselmis* sp. CTP4, for example, because of its large cell size (9–22 μm), settles down in 6 h with a medium removal efficiency of 80%, allowing a significant decrease in biomass dewatering costs even with high lipid content in biomass with a lipid productivity of 52.1 mg L^{-1} d^{-1} (Pereira et al., 2016).

6.6 Microalgal lipids extraction and refining

Lipid extraction is a key step in a biorefinery platform. Lipid extraction from oil-containing seeds (e.g., soybean, corn, rapeseed, and sunflower) is typically carried out by physical methods resorting to expellers, hydraulic pressing, or chemical methods like lipophilic solvent extractions (e.g., hexane, diethyl ether, benzene) or a combination of both. In the case of microalgae, the biomass needs to be dried prior to pressing (Lardon, Helias, Sialve, Steyer, & Bernard, 2009). The drying process is a considerably costly method because microalgae are cultivated in aquatic environments and, even after harvesting, microalgae paste still contains about 70%–80% water. Additionally, microalgae usually have strong cell coverings such as cell walls and thecae, which have a great influence on lipid extraction efficiency (Desai, Fernandez, Wijffels, & Eppink, 2019). Therefore a variety of cell disruption methods have been proposed for microalgae including mechanical (e.g., pressing, bead milling, and homogenization), physical (e.g., osmotic shock and ultrasound), thermal (e.g., microwave), autoclave, and electroflotation followed by solvent extractions (Silva et al., 2014). In order to avoid the use of toxic solvents such as those used in the Bligh and Dyer method (Bligh & Dyer, 1959), alternative green and food-grade water-miscible solvents such as ethanol and isopropanol have been successfully employed for the extraction of lipids from wet microalgal paste (Gangadhar et al., 2016a; Yang et al., 2014) as well as supercritical CO_2 (Lorenzen et al., 2017) and ionic liquids (Desai et al., 2019). Compared to a mechanical press, solvent, supercritical carbon dioxide, and solvent-free extractions are more appropriate for lipid extraction from wet biomass. However, selective extraction of lipids from microalgae is still in an early stage and under development. Hence, more efforts are required in the areas of solvent recycling, replacement of conventional organic solvents with green solvents, and the implementation of more sustainable processes.

Produced microalgal extracts are comprised of a complex mixture of lipid molecules, namely long-chain hydrocarbons, wax esters, esters of sterols, tri-(TAGs), di-, and monoacylglycerols, FFAs, pigments, vitamins, phospholipids, glycolipids, and carbohydrates among others, whereas vegetable oils are mostly composed of TAGs (90%) (Gangadhar et al., 2016b; López, Medina, Moreno, Cerdán, & Grima, 2016). Microalgae extracts, in particular, are rich in FFAs (Chen, Liu, Zhang, Chen, & Wang, 2012; Dong, Wang, Miao, Zheng, & Chen, 2013). Microalgae crude extracts generally have much lower percentages of TAG than vegetable oils, consisting of 53% TAGs, 29% phospholipids, and 16% glycolipids of the total lipids (Yang et al., 2014). On the other hand, microalgal lipids contain 10-fold higher unsaponifiable matter compared to vegetable-based lipids (Gangadhar et al., 2016b).

Microalgal crude extracts are often unsuitable as edible oils for human consumption, requiring a refining process to remove odor, particulate material, and chemical contaminants as well as to improve the color, clarity, and quality of the oils. A wide range of impurities can be found in algal extracts including chlorophyll, pigments and phospholipids, FFAs, wax esters, sterols, and trace metals among other contaminants. These unwanted impurities can be removed by: (1) neutralizing FFAs; (2) degumming to remove the phosphatides, sterols, etc.; (3) bleaching to remove color pigments, oxidation products, and trace metals; and (4) dewaxing (also known as winterization) to remove high melting waxes (Paisan, Chetpattananondh, & Chongkhong, 2017).

Biorefinery is a process which can produce multiple biochemicals from starting raw material by partitioning into multiple intermediates. However, biorefinery is technically more challenging due to the high proportions (i.e., qualitatively and quantitatively) of neutral lipids, FFAs, polar lipids, and water-soluble components in crude extracts of microalgae rather than in the refining of vegetable oils (Callejón et al., 2014). Hence, the refining of microalgal crude extracts into different raw materials is a less developed area in the context of a microalgae biorefinery ('t Lam, Vermuë, Eppink, Wijffels, & Van den Berg, 2018) and a great deal of effort is still needed to develop microalgae-based edible oils.

6.7 Marketed products

Consumers are constantly demanding new products, and the search for products rich in specific nutrients and from natural sources has been intensified through the search for a healthy diet. Among these, products made from microalgae are gaining markets and public attention. Most products of microalgal origin available on the market (around 70%) contain whole microalgal biomass, followed by products composed of oils extracted from biomass ($\approx$20%), or purified compounds such as DHA and EPA ($\approx$10%), pigments ($\approx$5%), and other fine chemicals (<1%). The greater the difficulty in extracting compounds from microalgal cells, the more expensive the product becomes due to the increase in production costs (Jacob-Lopes et al., 2019).

Food products supplemented with microalgae remain scarce, most being sold as nutritional supplements in the form of capsules and tablets of whole biomass. This market tends to change due to consumers' needs for easier sources of nutrients for their daily lives. The biggest challenge of products developed from microalgae is acceptability due to the deep green color and fish odor that often limit the inclusion of microalgal biomass to a certain concentration (Lafarga, 2019). Thus balance between nutritional supplementation and organoleptic factors is a major challenge for the food industry.

A few food products supplemented with microalgal oil are already on the market, among them are cooking oil (Thrive culinary oil) and a microalgae-derived liquid containing DHA and EPA that can be applied to food as a flavorant (Deva Vegan liquid DHA-EPA, lemon flavor). However, so far the largest markets selling products containing microalgal oil are cosmetics like body oil (The Ordinary "B" oil, REN Atlantic Kelp And Microalgae Anti Fatigue Bath Oil) and antiaging formulations (Algenist anti-aging repairing, Biodroga De-Stress Algae Sensitive Concentrate).

6.8 Challenges and future perspectives

The number of food products on the market using only microalgae oil remains low, probably because the use of whole biomass is cheaper. The costs of extracting microalgal oil, as well as the cultivation systems aimed at increasing lipid productivity are probably the main reason. An alternative would be the application of a biorefinery where more than one product is obtained from microalgal biomass, thus reducing the overall costs of lipids since other compounds with high added value are produced (Jacob-Lopes et al., 2019).

One of the great challenges remains the search for a balance between biomass productivity and an increase in lipid productivity with desirable characteristics. The increase in DHA and EPA content is an important point since they are highly nutritionally important compounds and must be consumed in the human diet (Kumar et al., 2019). The selection and isolation of strains that produce higher lipid content as well as the use of techniques such as FACS and random mutagenesis can be an interesting approach to increase the production of microalgal biomass and at the same time lipids (Pereira et al, 2018).

Another important factor is the methodologies for extracting lipidic compounds from biomass. Several techniques have already been reported and these will depend mainly on the characteristics of the microalgae cells used, as well as the amount of lipids present in the biomass. Most common methodologies usually involve the use of polar and nonpolar solvents which are not the best option since they can leave residues in the extracted oil that can be harmful to health in food applications (Bensalem et al, 2018). Mechanical extraction methodologies in dry or wet biomass, application of electric fields (Bensalem et al, 2018) to assist cell disruption, and supercritical extraction methods (Patel, Matsakas, Sartaj, & Chandra, 2020) have been studied with promising results.

Acknowledgments

The authors are indebted to the Foundation for Science and Technology (FCT), Portugal, for funding through UIDB/04326/2020 research program and a doctoral research grant (SFRH/BD/115325/2016) awarded to LS. Further funding was provided by the 0055

ALGARED + , ALGAVALOR (ALG-01-0247-FEDER-035234) 05 INTERREG V-A-España Portugal project and Greentreat (ALG-01-0145-FEDER-031567). KNG also is thankful to the FCT and University of Algarve for financial support under the transitional rule of Decree-Law no. 57/2016 as amended by Law No 57/2017.

References

Adarme-Vega, T. C., Lim, D. K. Y., Timmins, M., Vernen, F., Li, Y., & Schenk, P. M. (2012). Microalgal biofactories: A promising approach towards sustainable omega-3 fatty acid production. *Microbial Cell Factories*, *11*(96). Available from https://doi.org/10.1186/1475-2859-11-96.

Anandarajah, K., Mahendraperumal, G., Sommerfeld, M., & Hu, Q. (2012). Characterization of microalga *Nannochloropsis* sp. mutants for improved production of biofuels. *Applied Energy*, *96*, 371–377. Available from https://doi.org/10.1016/j.apenergy.2012.02.057.

Bajhaiya, A. K., Ziehe Moreira, J., & Pittman, J. K. (2016). Transcriptional engineering of microalgae: Prospects for high-value chemicals. *Trends in Biotechnology*, 1–4. Available from https://doi.org/10.1016/j.tibtech.2016.06.001.

Balamurugan, S., Wang, X., Wang, H. L., An, C. J., Li, H., Li, D. W., & Yang, W. D. (2017). Occurrence of plastidial triacylglycerol synthesis and the potential regulatory role of AGPAT in the model diatom *Phaeodactylum tricornutum*. *Biotechnology for Biofuels*, *10*, 1–14. Available from https://doi.org/10.1186/s13068-017-0786-0.

Beacham, T. A., Macia, V. M., Rooks, P., White, D. A., & Ali, S. T. (2015). Altered lipid accumulation in *Nannochloropsis salina* CCAP849/3 following EMS and UV induced mutagenesis. *Biotechnology Reports*, *7*, 87–94. Available from https://doi.org/10.1016/j.btre.2015.05.007.

Bellou, S., Baeshen, M. N., Elazzazy, A. M., Aggeli, D., Sayegh, F., & Aggelis, G. (2014). Microalgal lipids biochemistry and biotechnological perspectives. *Biotechnology Advances*, *32*, 1476–1493. Available from https://doi.org/10.1016/j.biotechadv.2014.10.003.

Bensalem, S., Lopes, F., Bodénèsa, P., Pareau, D., Français, O., & Pioufle, B. (2018). Understanding the mechanisms of lipid extraction from microalga *Chlamydomonas reinhardtii* after electrical field solicitations and mechanical stress within a microfluidic device. *Bioresource Technology*, *257*, 129–136. Available from https://doi.org/10.1016/j.biortech.2018.01.139.

Blažina, M., Haberle, I., Hrustić, E., Budiša, A., Petrić, I., Konjević, L., ... Geček, S. (2019). Growth aspects and biochemical composition of *Synechococcus* sp. MK568070 cultured in oil refinery wastewater. *Journal of Marine Science and Engineering*, 7. Available from https://doi.org/10.3390/jmse7060164.

Bligh, E. G., & Dyer, W. J. (1959). A rapid method of total lipid extraction and purification. *Canadian Journal of Biochemistry and Physiology*, *37*(8), 911–917. Available from https://doi.org/10.1139/y59-099.

Borowitzka, M. (1999). Commercial production of microalgae: ponds, tanks, tubes and fermenters. *Journal of Biotechnology*, *70*(1–3), 313–321. Available from https://doi.org/10.1016/S0168-1656(99)00083-8.

Brasil, B. S. A. F., Silva, F. C. P., & Siqueira, F. G. (2017). Microalgae biorefineries: The Brazilian scenario in perspective. *New Biotechnology*, *39*, 90–98. Available from https://doi.org/10.1016/j.nbt.2016.04.007.

Callejón, M. J. J., Medina, A. R., Sánchez, M. D. M., Peña, E. H., Cerdán, L. E., Moreno, P. A. G., & Grima, E. M. (2014). Extraction of saponifiable lipids from wet microalgal biomass for biodiesel production. *Bioresource Technology*, *169*, 198–205. Available from https://doi.org/10.1016/j.biortech.2014.06.106.

Camacho-Rodríguez, J., Cerón-García, M. C., Fernández-Sevilla, J. M., & Molina-Grima, E. (2015). The influence of culture conditions on biomass and high value product generation by *Nannochloropsis gaditana* in aquaculture. *Algal Research*, *11*, 63–73. Available from https://doi.org/10.1016/j.algal.2015.05.017.

Cardoso, C., Pereira, H., Franca, J., Matos, J., Monteiro, I., Pousão-Ferreira, P., ... Bandarra, N. M. (2020). Lipid composition and some bioactivities of 3 newly isolated microalgae (*Tetraselmis* sp. IMP3, *Tetraselmis* sp. CTP4, and *Skeletonema* sp.). *Aquaculture International*, *28*, 711–727. Available from https://doi.org/10.1007/s10499-019-00489-w.

Cassuriaga, A. P. A., Freitas, B. C. B., Morais, M. G., & Costa, J. A. V. (2018). Innovative polyhydroxybutyrate production by *Chlorella fusca* grown with pentoses. *Bioresource Technology*, *265*, 456–463. Available from https://doi.org/10.1016/j.biortech.2018.06.026.

Chaturvedi, R., & Fujita, Y. (2006). Isolation of enhanced eicosapentaenoic acid producing mutants of *Nannochloropsis oculata* ST-6 using ethyl methane sulfonate induced mutagenesis techniques and their characterization at mRNA transcript level. *Phycological Research*, *54*, 208–219. Available from https://doi.org/10.1111/j.1440-1835.2006.00428.x.

Chaturvedi, R., Uppalapati, S. R., Alamsjah, M. A., & Fujita, Y. (2004). Isolation of quizalofop-resistant mutants of *Nannochloropsis oculata* (Eustigmatophyceae) with high eicosapentaenoic acid following N-methyl-N-nitrosourea-induced random mutagenesis. *Journal of Applied Phycology*, *16*, 135–144. Available from https://doi.org/10.1023/B:JAPH.0000044826.70360.8e.

Chen, L., Liu, T., Zhang, W., Chen, X., & Wang, J. (2012). Biodiesel production from algae oil high in free fatty acids by two-step catalytic conversion. *Bioresource Technology*, *111*, 208–214. Available from https://doi.org/10.1016/j.biortech.2012.02.033.

Chen, Z., Wang, L., Qiu, S., & Ge, S. (2018). Determination of microalgal lipid content and fatty acid for biofuel production. *BioMed Research International*, *2018*, 1–17. Available from https://doi.org/10.1155/2018/1503126.

Chisti, Y. (2012). Raceways-based production of algal crude oil. In C. Posten, & C. Walter (Eds.), *Microalgal biotechnology: Potential and production* (pp. 113–146). Berlin: Gruyter.

Chisti, Y. (2016). Large-scale production of algal biomass: Raceway ponds. In F. Bux, & Y. Chisti (Eds.), *Algae biotechnology. Green energy and technology* (pp. 21–40). Cham: Springer.

Choi, S. A., Lee, S. Y., Lee, J., Cho, J. M., Lee, J. S., Kim, S. W., ... Oh, Y. K. (2019). Rapid induction of edible lipids in *Chlorella* by mild electric stimulation. *Bioresource Technology*, *292*, 121950. Available from https://doi.org/10.1016/j.biortech.2019.121950.

Chokshi, K., Pancha, I., Trivedi, K., George, B., Maurya, R., Ghosh, A., & Mishra, S. (2015). Biofuel potential of the newly isolated microalgae *Acutodesmus dimorphus* under temperature induced oxidative stress conditions. *Bioresource Technology*, *180*, 162–171. Available from https://doi.org/10.1016/j.biortech.2014.12.102.

Chu, F. F., Chu, P. N., Cai, P. J., Li, W. W., Lam, P. K. S., & Zeng, R. J. (2013). Phosphorus plays an important role in enhancing biodiesel productivity of *Chlorella vulgaris* under nitrogen deficiency. *Bioresource Technology*, *134*, 341–346. Available from https://doi.org/10.1016/j.biortech.2013.01.131.

Coward, T., Lee, J. G. M., & Caldwell, G. S. (2014). Harvesting microalgae by CTAB-aided foam flotation increases lipid recovery and improves fatty acid methyl ester characteristics. *Biomass Bioenergy*, *67*, 354–362. Available from https://doi.org/10.1016/j.biombioe.2014.05.019.

Custódio, L., Soares, F., Pereira, H., Barreira, L., Vizetto-Duarte, C., Rodrigues, M. J., ... Varela, J. (2014). Fatty acid composition and biological activities of *Isochrysis galbana* T-ISO, *Tetraselmis* sp. and *Scenedesmus* sp.: Possible application in the pharmaceutical and functional food industries. *Journal of Applied Phycology*, *26*, 151–161. Available from https://doi.org/10.1007/s10811-013-0037-0.

Desai, R. K., Fernandez, M. S., Wijffels, R. H., & Eppink, M. H. M. (2019). Mild fractionation of hydrophilic and hydrophobic components from neochloris oleoabundans using ionic liquids. *Frontiers in Bioengineering and Biotechnology*, *7*(284). Available from https://doi.org/10.3389/fbioe.2019.00284.

Dong, T., Wang, J., Miao, C., Zheng, Y., & Chen, S. (2013). Two-step in situ biodiesel production from microalgae with high free fatty acid content. *Bioresource Technology*, *136*, 8–15. Available from https://doi.org/10.1016/j.biortech.2013.02.105.

Dragosits, M., & Mattanovich, D. (2013). Adaptive laboratory evolution – Principles and applications for biotechnology. *Microbial Cell Factories*, *12*, 1–17. Available from https://doi.org/10.1186/1475-2859-12-64.

Durmaz, Y., Monteiro, M., Bandarra, N., Gökpinar, Ş., & Işik, O. (2007). The effect of low temperature on fatty acid composition and tocopherols of the red microalga, *Porphyridium cruentum*. *Journal of Applied Phycology*, *19*, 223–227. Available from https://doi.org/10.1007/s10811-006-9127-6.

El-Mowafy, A. M., Abdel-Dayem, M. A., Abdel-Aziz, A., El-Azab, M. F., & Said, S. A. (2011). Eicosapentaenoic acid ablates valproate-induced liver oxidative stress and cellular derangement without altering its clearance rate: Dynamic synergy and therapeutic utility. *Biochimica et Biophysica Acta - Molecular and Cell Biology of Lipids*, *1811*, 460–467. Available from https://doi.org/10.1016/j.bbalip.2011.04.014.

Fahy, E., Cotter, D., Sud, M., & Subramaniam, S. (2011). Lipid classification, structures and tools. *Biochimica et Biophysica Acta - Molecular and Cell Biology of Lipids*, *1811*, 637–647. Available from https://doi.org/10.1016/j.bbalip.2011.06.009.

Fakhry, E. M., & Maghraby, D. M. El (2013). Fatty acids composition and biodiesel Characterization of *Dunaliella salina*. *Journal of Water Resource and Protection*, *05*, 894–899. Available from https://doi.org/10.4236/jwarp.2013.59091.

Gangadhar, K. N., Pereira, H., Diogo, H. P., dos Santos, R. M. B., Prabhavathi Devi, B. L. A., Prasad, R. B. N., ... Barreira, L. (2016a). Assessment and comparison of the properties of biodiesel synthesized from three different types of wet microalgal biomass. *Journal of Applied Phycology*, *28*, 1571–1578. Available from https://doi.org/10.1007/s10811-015-0683-5.

Gangadhar, K. N., Pereira, H., Rodrigues, M. J., Custódio, L., Barreira, L., Malcata, F. X., & Varela, J. (2016b). Microalgae-based unsaponifiable matter as source of natural antioxidants and metal chelators to enhance the value of wet *Tetraselmis chuii* biomass. *Open Chemistry*, *14*, 299–307. Available from https://doi.org/10.1515/chem-2016-0029.

Gao, S., Yang, J., Tian, J., Ma, F., Tu, G., & Du, M. (2010). Electro-coagulation-flotation process for algae removal. *Journal of Hazardous Materials*, *177*, 336–343. Available from https://doi.org/10.1016/j.jhazmat.2009.12.037.

García, J. L., Vicente, M., & Galán, B. (2017). Microalgae, old sustainable food and fashion nutraceuticals. *Microbial Biotechnology*, *10*, 1017–1024. Available from https://doi.org/10.1111/1751-7915.12800.

Gerardo, M. L., Van Den Hende, S., Vervaeren, H., Coward, T., & Skill, S. C. (2015). Harvesting of microalgae within a biorefinery approach: A review of the developments and case studies from pilot-plants. *Algal Research*, *11*, 248–262. Available from https://doi.org/10.1016/j.algal.2015.06.019.

Gopalakrishnan, V., & Ramamurthy, D. (2014). Dyeing industry effluent system as lipid production medium of *Neochloris* sp. for biodiesel feedstock preparation. *BioMed Research International, 2014*. Available from https://doi.org/10.1155/2014/529560.

Guesnet, P., & Alessandri, J.-M. (2011). Docosahexaenoic acid (DHA) and the developing central nervous system (CNS) – Implications for dietary recommendations. *Biochimie, 93*, 7–12. Available from https://doi.org/10.1016/j.biochi.2010.05.005.

Guihéneuf, F., Mimouni, V., Ulmann, L., & Tremblin, G. (2009). Combined effects of irradiance level and carbon source on fatty acid and lipid class composition in the *microalga Pavlova lutheri* commonly used in mariculture. *Journal of Experimental Marine Biology and Ecology, 369*, 136–143. Available from https://doi.org/10.1016/j.jembe.2008.11.009.

Hamed, I. (2016). The evolution and versatility of microalgal biotechnology: A review. *Comprehensive Reviews in Food Science and Food Safety, 15*, 1104–1123. Available from https://doi.org/10.1111/1541-4337.12227.

Harun, R., Singh, M., Forde, G. M., & Danquah, M. K. (2010). Bioprocess engineering of microalgae to produce a variety of consumer products. *Renewable and Sustainable Energy Reviews, 14*, 1037–1047. Available from https://doi.org/10.1111/1541-4337.12227.

Ho, S. H., Ye, X., Hasunuma, T., Chang, J. S., & Kondo, A. (2014). Perspectives on engineering strategies for improving biofuel production from microalgae - A critical review. *Biotechnology Advances, 32*, 1448–1459. Available from https://doi.org/10.1016/j.biotechadv.2014.09.002.

Hooper, C., Barreto, P. S., Pahor, M., Weiner, M., & Vellas, B. (2018). The relationship of omega 3 polyunsaturated fatty acids in red blood cell membranes with cognitive function and brain structure: A review focussed on Alzheimer's disease. *The Journal of Prevention of Alzheimer's Disease, 5*, 78–84. Available from https://doi.org/10.14283/jpad.2017.19.

Huerlimann, R., de Nys, R., & Heimann, K. (2010). Growth, lipid content, productivity, and fatty acid composition of tropical microalgae for scale-up production. *Biotechnology and Bioengineering, 107*, 245–257. Available from https://doi.org/10.1002/bit.22809.

Ikehata, H., & Ono, T. (2011). The mechanisms of UV mutagenesis. *Journal of Radiation Research, 52*, 115–125. Available from https://doi.org/10.1269/jrr.10175.

Jacob-Lopes, E., Maroneze, M. M., Depra, M. C., Sartori, R. B., Dias, R. R., & Zepka, L. Q. (2019). Bioactive food compounds from microalgae: An innovative framework on industrial biorefineries. *Current Opinion in Food Science, 25*, 1–7. Available from https://doi.org/10.1016/j.cofs.2018.12.003.

Japar, A. S., Takriff, M. S., & Yasin, N. H. M. (2017). Harvesting microalgal biomass and lipid extraction for potential biofuel production: A review. *Journal of Environmental Chemical Engineering, 5*, 555–563. Available from https://doi.org/10.1016/j.jece.2016.12.016.

Jesus, C. S., Uebel, L., da, S., Costa, S. S., Miranda, A. L., Morais, E. G., ... Druzian, J. I. (2018). Outdoor pilot-scale cultivation of *Spirulina* sp. LEB-18 in different geographic locations for evaluating its growth and chemical composition. *Bioresource Technology, 256*, 86–94. Available from https://doi.org/10.1016/j.biortech.2018.01.149.

Jiang, L., Luo, S., Fan, X., Yang, Z., & Guo, R. (2011). Biomass and lipid production of marine microalgae using municipal wastewater and high concentration of CO2. *Applied Energy, 88*, 3336–3341. Available from https://doi.org/10.1016/j.apenergy.2011.03.043.

Jump, D. B., Lytle, K. A., Depner, C. M., & Tripathy, S. (2018). Omega-3 polyunsaturated fatty acids as a treatment strategy for nonalcoholic fatty liver disease. *Pharmacology & Therapeutics, 181*, 108–125. Available from https://doi.org/10.1016/j.pharmthera.2017.07.007.

Kang, N. K., Kim, E. K., Sung, M.-G., Kim, Y. U., Jeong, B., & Chang, Y. K. (2019). Increased biomass and lipid production by continuous cultivation of *Nannochloropsis salina* transformant overexpressing a bHLH transcription factor. *Biotechnology and Bioengineering, 116*, 555–568. Available from https://doi.org/10.1002/bit.26894.

Katiyar, R., & Arora, A. (2020). Health promoting functional lipids from microalgae pool: A review. *Algal Research, 46*. Available from https://doi.org/10.1016/j.algal.2020.101800.

Kaye, Y., Grundman, O., Leu, S., Zarka, A., Zorin, B., Didi-Cohen, S., … Boussiba, S. (2015). Metabolic engineering toward enhanced LC-PUFA biosynthesis in *Nannochloropsis oceanica*: Overexpression of endogenous Δ12 desaturase driven by stress-inducible promoter leads to enhanced deposition of polyunsaturated fatty acids in TAG. *Algal Research, 11*, 387–398. Available from https://doi.org/10.1016/j.algal.2015.05.003.

Kelly, O. J., Gilman, J. C., Kim, Y., & Ilich, J. Z. (2013). Long-chain polyunsaturated fatty acids may mutually benefit both obesity and osteoporosis. *Nutrition Research, 521-533*, 2013. Available from https://doi.org/10.1016/j.nutres.2013.04.012.

Khoeyi, Z. A., Seyfabadi, J., & Ramezanpour, Z. (2012). Effect of light intensity and photoperiod on biomass and fatty acid composition of the microalgae, Chlorella vulgaris. *Aquaculture International, 20*, 41–49. Available from https://doi.org/10.1007/s10499-011-9440-1.

Kim, K., Hoh, D., Ji, Y., Dŏ, H., Lee, B., & Holzapfel, W. (2013). Impact of light intensity, CO_2 concentration and bubble size on growth and fatty acid composition of *Arthrospira (Spirulina) platensis* KMMCC CY-007. *Biomass Bioenergy, 49*, 181–187. Available from https://doi.org/10.1016/j.biombioe.2012.12.021.

Kumar, B. R., Deviram, G., Mathimani, T., Duc, P. A., & Pugazhendhi, A. (2019). Microalgae as rich source of polyunsaturated fatty acids. *Biocatalysis and Agricultural Biotechnology, 17*, 583–588. Available from https://doi.org/10.1016/j.bcab.2019.01.017.

Lafarga, T. (2019). Effect of microalgal biomass incorporation into foods: Nutritional and sensorial attributes of the end products. *Algal Research, 41*, 101566. Available from https://doi.org/10.1016/j.algal.2019.101566.

Lang, I., Hodac, L., Friedl, T., & Feussner, I. (2011). Fatty acid profiles and their distribution patterns in microalgae: A comprehensive analysis of more than 2000 strains from the SAG culture collection. *BMC Plant Biology, 11*(124). Available from https://doi.org/10.1186/1471-2229-11-124.

Lardon, L., Helias, A., Sialve, B., Steyer, J. P., & Bernard, O. (2009). Life-cycle assessment of biodiesel production from microalgae. *Science and Technology, 17*, 6475–6481. Available from https://doi.org/10.1021/es900705j.

Li, Y., Han, F., Xu, H., Mu, J., Chen, D., Feng, B., & Zeng, H. (2014). Potential lipid accumulation and growth characteristic of the green alga *Chlorella* with combination cultivation mode of nitrogen (N) and phosphorus (P). *Bioresource Technology, 174*, 24–32. Available from https://doi.org/10.1016/j.biortech.2014.09.142.

López, E. N., Medina, A. R., Moreno, P. A. G., Cerdán, L. E., & Grima, E. M. (2016). Extraction of microalgal lipids and the influence of polar lipids on biodiesel production by lipase-catalyzed transesterification. *Bioresource Technology, 216*, 904–913. Available from https://doi.org/10.1016/j.biortech.2016.06.035.

Lorenzen, J., Igl, N., Tippelt, M., Stege, A., Qoura, F., Sohling, U., & Brück, T. (2017). Extraction of microalgae derived lipids with supercritical carbon dioxide in an industrial relevant pilot plant. *Bioprocess and Biosystems Engineering, 40*, 911–918. Available from https://doi.org/10.1007/s00449-017-1755-5.

Ma, Y., Wang, Z., Yu, C., Yin, Y., & Zhou, G. (2014). Evaluation of the potential of 9 *Nannochloropsis* strains for biodiesel production. *Bioresource Technology, 167*, 503–509. Available from https://doi.org/10.1016/j.biortech.2014.06.047.

Masojídek, J., Kopecky, J., Giannelli, L., & Torzillo, G. (2011). Productivity correlated to photobiochemical performance of *Chlorella* mass cultures grown outdoors in thin-layer cascades. *Journal of Industrial Microbiology & Biotechnology*, *38*(2), 307–317. Available from https://doi.org/10.1007/s10295-010-0774-x.

Mata, T. M., Martins, A. A., & Caetano, N. S. (2010). Microalgae for biodiesel production and other applications: A review. *Renewable and Sustainable Energy Reviews*, *14*(1), 217–232. Available from https://doi.org/10.1016/j.rser.2009.07.020.

Meireles, L. A., Guedes, A. C., & Malcata, F. X. (2002). Increase of the yields of eicosapentaenoic and docosahexaenoic acids by the microalga *Pavlova lutheri* following random mutagenesis. *Biotechnology and Bioengineering*, *81*, 50–55. Available from https://doi.org/10.1002/bit.10451.

Milledge, J. J., & Heaven, S. (2013). A review of the harvesting of micro-algae for biofuel production. *Reviews in Environmental Science and Biotechnology*, *12*, 165–178. Available from https://doi.org/10.1007/s11157-012-9301-z.

Mimouni, V., Ulmann, L., Pasquet, V., Mathieu, M., Picot, L., Bougaran, G., ... Schoefs, B. (2012). The potential of microalgae for the production of bioactive molecules of pharmaceutical interest. *Current Pharmaceutical Biotechnology*, *13*, 2733–2750. Available from https://doi.org/10.2174/138920112804724828.

Moha-León, J. D., Pérez-Legaspi, I. A., Ortega-Clemente, L. A., Rubio-Franchini, I., & Ríos-Leal, E. (2019). Improving the lipid content of *Nannochloropsis oculata* by a mutation-selection program using UV radiation and quizalofop. *Journal of Applied Phycology*, *31*, 191–199. Available from https://doi.org/10.1007/s10811-018-1568-1.

Moraes, L., Rosa, G. M., Morillas España, A., Santos, L. O., Morais, M. G., Molina Grima, E., ... Acién Fernández, F. G. (2019a). Engineering strategies for the enhancement of *Nannochloropsis gaditana* outdoor production: Influence of the CO_2 flow rate on the culture performance in tubular photobioreactors. *Process Biochemistry*, *76*, 171–177. Available from https://doi.org/10.1016/j.procbio.2018.10.010.

Moraes, L., Santos, L. O., & Costa, J. A. V. (2019b). Bioprocess strategies for enhancing biomolecules productivity in *Chlorella fusca* LEB 111 using CO_2 a carbon source. *Biotechnology Progress*. Available from https://doi.org/10.1002/btpr.2909.

Mujtaba, G., Choi, W., Lee, C. G., & Lee, K. (2012). Lipid production by *Chlorella vulgaris* after a shift from nutrient-rich to nitrogen starvation conditions. *Bioresource Technology*, *123*, 279–283. Available from https://doi.org/10.1016/j.biortech.2012.07.057.

Niu, Y.-F., Zhang, M.-H., Li, D.-W., Yang, W.-D., Liu, J.-S., Bai, W.-B., & Li, H.-Y. (2013). Improvement of neutral lipid and polyunsaturated fatty acid biosynthesis by overexpressing a type 2 diacylglycerol acyltransferase in marine diatom *Phaeodactylum tricornutum*. *Marine Drugs*, *11*, 4558–4569. Available from https://doi.org/10.3390/md11114558.

Olofsson, M., Lamela, T., Nilsson, E., Bergé, J. P., del Pino, V., Uronen, P., & Legrand, C. (2012). Seasonal variation of lipids and fatty acids of the microalgae *Nannochloropsis oculata* grown in outdoor large-scale photobioreactors. *Energies*, *5*, 1577–1592. Available from https://doi.org/10.3390/en5051577.

Paisan, S., Chetpattananondh, P., & Chongkhong, S. (2017). Assessment of water degumming and acid degumming of mixed algal oil. *Journal of Environmental Chemical Engineering*, *5*(5), 5115–5123. Available from https://doi.org/10.1016/j.jece.2017.09.045.

Pal, P., Chew, K. W., Yen, H. W., Lim, J. W., Lam, M. K., & Show, P. L. (2019). Cultivation of oily microalgae for the production of third-generation biofuels. *Sustainability*, *11*, 1–16. Available from https://doi.org/10.3390/su11195424.

Paliwal, C., Mitra, M., Bhayani, K., Bharadwaj, S. V. V., Ghosh, T., Dubey, S., & Mishra, S. (2017). Abiotic stresses as tools for metabolites in microalgae. *Bioresource Technology*, *244*, 1216–1226. Available from https://doi.org/10.1016/j.biortech.2017.05.058.

Pancha, I., Chokshi, K., Ghosh, T., Paliwal, C., Maurya, R., & Mishra, S. (2015a). Bicarbonate supplementation enhanced biofuel production potential as well as nutritional stress mitigation in the microalgae *Scenedesmus*. *Bioresource Technology*, *193*, 315–323. Available from https://doi.org/10.1016/j.biortech.2015.06.107.

Pancha, I., Chokshi, K., Maurya, R., & Trivedi, K. (2015b). Salinity induced oxidative stress enhanced biofuel production potential of microalgae *Scenedesmus* sp. CCNM 1077. *Bioresource Technology*, *189*, 341–348. Available from https://doi.org/10.1016/j.biortech.2015.04.017.

Pandit, P. R., Fulekar, M. H., & Karuna, M. S. L. (2017). Effect of salinity stress on growth, lipid productivity, fatty acid composition, and biodiesel properties in *Acutodesmus obliquus* and *Chlorella vulgaris*. *Environmental Science and Pollution Research*, *24*, 13437–13451. Available from https://doi.org/10.1007/s11356-017-8875-y.

Patel, A., Matsakas, L., Sartaj, K., & Chandra, R. (2020). Chapter 2 - Extraction of lipids from algae using supercritical carbon dioxide. In *Green sustainable process for chemical and environmental engineering and science*, 17–39. Available from https://doi.org/10.1016/B978-0-12-817388-6.00002-7.

Peperzak, L., Colijn, F., Koeman, R., Gieskes, W. W. C., & Joordens, J. C. A. (2003). Phytoplankton sinking rates in the rhine region of freshwater influence. *Journal of Plankton Research*, *25*, 365–383. Available from https://doi.org/10.1093/plankt/25.4.365.

Pereira, H., Barreira, L., Mozes, A., Florindo, C., Polo, C., Duarte, C. V., ... Varela, J. (2011). Microplate-based high throughput screening procedure for the isolation of lipid-rich marine microalgae. *Biotechnology for Biofuels*, *4*, 61. Available from https://doi.org/10.1186/1754-6834-4-61.

Pereira, H., Barreira, L., Figueiredo, F., Custódio, L., Vizetto-Duarte, C., Polo, C., ... Varela, J. (2012). Polyunsaturated fatty acids of marine macroalgae: Potential for nutritional and pharmaceutical applications. *Marine Drugs*, *10*, 1920–1935. Available from https://doi.org/10.3390/md10091920.

Pereira, H., Gangadhar, K. N., Schulze, P. S. C., Santos, T., Sousa, C. B., Schueler, L. M., ... Barreira, L. (2016). Isolation of a euryhaline microalgal strain, *Tetraselmis* sp. CTP4, as a robust feedstock for biodiesel production. *Science Report*, *6*, 35663. Available from https://doi.org/10.1038/srep35663.

Pereira, H., Schulze, P. S. C., Schüler, L. M., Santos, T., Barreira, L., & Varela, J. (2018). Fluorescence activated cell-sorting principles and applications in microalgal biotechnology. *Algae Research*, *30*, 113–120. Available from https://doi.org/10.1016/j.algal.2017.12.013.

Plaza, M., Herrero, M., Alejandro Cifuentes, A., & Ibáñez, E. (2009). Innovative natural functional ingredients from microalgae. *Journal of Agricultural and Food Chemistry*, *57*, 7159–7170. Available from https://doi.org/10.1021/jf901070g.

Pragya, N., Pandey, K. K., & Sahoo, P. K. (2013). A review on harvesting, oil extraction and biofuels production technologies from microalgae. *Renewable and Sustainable Energy Reviews*, *24*, 159–171. Available from https://doi.org/10.1016/j.rser.2013.03.034.

Qiao, H., Cong, C., Sun, C., Li, B., Wang, J., & Zhang, L. (2016). Effect of culture conditions on growth, fatty acid composition and DHA/EPA ratio of *Phaeodactylum tricornutum*. *Aquaculture (Amsterdam, Netherlands)*, *452*, 311–317. Available from https://doi.org/10.1016/j.aquaculture.2015.11.011.

Quelhas, P. M., Trovão, M., Silva, J. T., Machado, A., Santos, T., Pereira, H., ... Silva, J. L. (2019). Industrial production of *Phaeodactylum tricornutum* for CO_2 mitigation: Biomass productivity and photosynthetic efficiency using photobioreactors of different volumes. *Journal of Applied Phycology*, *31*, 2187–2196. Available from https://doi.org/10.1007/s10811-019-1750-0.

Ra, C. H., Kang, C., Na, K., Lee, C., & Kim, S. (2015). Cultivation of four microalgae for biomass and oil production using a two-stage culture strategy with salt stress. *Renewable Energy*, *80*, 117–122. Available from https://doi.org/10.1016/j.renene.2015.02.002.

Ranga Rao, A., Ravishankar, G. A., & Sarada, R. (2012). Cultivation of green alga *Botryococcus braunii* in raceway, circular ponds under outdoor conditions and its growth, hydrocarbon production. *Bioresource Technology*, *123*, 528–533. Available from https://doi.org/10.1016/j.biortech.2012.07.009.

Renaud, S. M., Thinh, L., Van Lambrinidis, G., & Parry, D. L. (2002). Effect of temperature on growth, chemical composition and fatty acid composition of tropical Australian microalgae grown in batch cultures. *Aquaculture (Amsterdam, Netherlands)*, *211*, 195–214. Available from https://doi.org/10.1016/S0044-8486(01)00875-4.

Rickman, M., Pellegrino, J., & Davis, R. (2012). Fouling phenomena during membrane filtration of microalgae. *Journal of Membrane Science*, *423–424*, 33–42. Available from https://doi.org/10.1016/j.memsci.2012.07.013.

Rimm, E. B., Appel, L. J., Chiuve, S. E., Djoussé, L., Engler, M. B., Kris-Etherton, P. M., … Lichtenstein, A. H. (2018). Seafood long-chain n-3 polyunsaturated fatty acids and cardiovascular disease: A science advisory from the American Heart Association. *Circulation*, *138*, e35–e47. Available from https://doi.org/10.1161/CIR.0000000000000574.

Rosa, G. M., Moraes, L., Souza, M., da, R. A. Z., & Costa, J. A. V. (2016). *Spirulina* cultivation with a CO_2 absorbent: Influence on growth parameters and macromolecule production. *Bioresource Technology*, *200*, 528–534. Available from https://doi.org/10.1016/j.biortech.2015.10.025.

Rosa, G. M., da Morais, M. G., & de Costa, J. A. V. (2018). Green alga cultivation with monoethanolamine: Evaluation of CO_2 fixation and macromolecule production. *Bioresource Technology*, *261*, 206–212. Available from https://doi.org/10.1016/j.biortech.2018.04.007.

Rosa, G. M., Morais, M. G., & Costa, J. A. V. (2019). Fed-batch cultivation with CO_2 and monoethanolamine: influence on *Chlorella fusca* LEB 111 cultivation, carbon biofixation and biomolecules production. *Bioresource Technology*, *273*, 627–633. Available from https://doi.org/10.1016/j.biortech.2018.11.010.

Sajjadi, B., Chen, W. Y., Raman, A. A. A., & Ibrahim, S. (2018). Microalgae lipid and biomass for biofuel production: A comprehensive review on lipid enhancement strategies and their effects on fatty acid composition. *Renewable and Sustainable Energy Reviews*, *97*, 200–232. Available from https://doi.org/10.1016/j.rser.2018.07.050.

Salama, E. S., Kim, H. C., Abou-Shanab, R. A. I., Ji, M. K., Oh, Y. K., Kim, S. H., & Jeon, B. H. (2013). Biomass, lipid content, and fatty acid composition of freshwater *Chlamydomonas mexicana* and *Scenedesmus obliquus* grown under salt stress. *Bioprocess and Biosystems Engineering*, *36*, 827–833. Available from https://doi.org/10.1007/s00449-013-0919-1.

Salama, E.-S., Abou-Shanab, R. A. I., Kim, J. R., Lee, S., Kim, S.-H., Oh, S.-E., … Jeon, B.-H. (2014). The effects of salinity on the growth and biochemical properties of *Chlamydomonas mexicana* GU732420 cultivated in municipal wastewater. *Environmental Technology*, *35*, 1491–1498. Available from https://doi.org/10.1080/09593330.2013.871350.

San Pedro, A., González-López, C. V., Acién, F. G., & Molina-Grima, E. (2014). Outdoor pilot-scale production of *Nannochloropsis gaditana*: Influence of culture parameters and lipid production rates in tubular photobioreactors. *Bioresource Technology*, *169*, 667–676. Available from https://doi.org/10.1016/j.biortech.2014.07.052.

Sayeda, M. A., Gamila, H. A., & Farouk, K. E. (2015). Potential production of omega fatty acids from microalgae. *International Journal of Applied Biology and Pharmaceutical*, *34*(2), 210–215.

Schüler, L. M., Schulze, P. S. C., Pereira, H., Barreira, L., León, R., & Varela, J. (2017). Trends and strategies to enhance triacylglycerols and high-value compounds in microalgae. *Algal Research*, *25*, 263–273. Available from https://doi.org/10.1016/j.algal.2017.05.025.

Šetlík, I., Sust, V., & Málek, I. (1970). Dual purpose open circulation units for large scale culture of algae in temperate zones. I. Basic design considerations and scheme of a pilot plant. *Algological Studies*, *1*, 111–164.

Sharma, K. K., Schuhmann, H., & Schenk, P. M. (2012). High lipid induction in microalgae for biodiesel production. *Energies*, *5*, 1532–1553. Available from https://doi.org/10.3390/en5051532.

Shen, Q. H., Gong, Y. P., Fang, W. Z., Bi, Z. C., Cheng, L. H., Xu, X. H., & Chen, H. L. (2015). Saline wastewater treatment by *Chlorella vulgaris* with simultaneous algal lipid accumulation triggered by nitrate deficiency. *Bioresource Technology*, *193*, 68–75. Available from https://doi.org/10.1016/j.biortech.2015.06.050.

Show, K. Y., & Lee, D. J. (2013). Algal biomass harvesting. In A. Pandey, D. J. Lee, Y. Chisti, & C. R. Soccol (Eds.), *Biofuels from algae* (pp. 85–110). Elsevier B.V.

Silva, A. P. F. S., Costa, M. C., Lopes, A. C., Neto, E. F. A., Leitão, R. C., Mota, C. R., & dos Santos, A. B. (2014). Comparison of pretreatment methods for total lipids extraction from mixed microalgae. *Renewable Energy*, *63*, 762–766. Available from https://doi.org/10.1016/j.renene.2013.10.038.

Soni, R. A., Sudhakar, K., & Rana, R. S. (2019). Comparative study on the growth performance of *Spirulina platensis* on modifying culture media. *Energy Reports*, *5*, 327–336. Available from https://doi.org/10.1016/j.egyr.2019.02.009.

Sun, X., Cao, Y., Xu, H., Liu, Y., Sun, J., Qiao, D., & Cao, Y. (2014). Effect of nitrogen-starvation, light intensity and iron on triacylglyceride/carbohydrate production and fatty acid profile of *Neochloris oleoabundans* HK-129 by a two-stage process. *Bioresource Technology*, *155*, 204–212. Available from https://doi.org/10.1016/j.biortech.2013.12.109.

't Lam, G. P., Vermuë, M. H., Eppink, M. H. M., Wijffels, R. H., & Van den Berg, C. (2018). Multi-product microalgae biorefineries: From concept towards reality. *Trends in Biotechnology*, *36*, 216–227. Available from https://doi.org/10.1016/j.tibtech.2017.10.011.

Uduman, N., Qi, Y., Danquah, M. K., Forde, G. M., & Hoadley, A. (2010). Dewatering of microalgal cultures: A major bottleneck to algae-based fuels. *Journal of Renewable and Sustainable Energy*, *2*, 012701. Available from https://doi.org/10.1063/1.3294480.

Uebel, L. S., Costa, J. A. V., Olson, A. C., & Morais, M. G. (2019). Industrial plant for production of *Spirulina* sp. LEB 18. *Brazilian Journal of Chemical Engineering*, *36*, 51–63. Available from https://doi.org/10.1590/0104-6632.20180361s20170284.

Vandamme, D., Foubert, I., & Muylaert, K. (2013). Flocculation as a low-cost method for harvesting microalgae for bulk biomass production. *Trends in Biotechnology*, *31*, 233–239. Available from https://doi.org/10.1016/j.tibtech.2012.12.005.

Vasudevan, P. T., & Briggs, M. (2008). Biodiesel production - Current state of the art and challenges. *Journal of Industrial Microbiology & Biotechnology*, *35*, 421–430. Available from https://doi.org/10.1007/s10295-008-0312-2.

Vergini, S., Aravantinou, A. F., & Manariotis, I. D. (2016). Harvesting of freshwater and marine microalgae by common flocculants and magnetic microparticles. *Journal of Applied Phycology*, *28*, 1041–1049. Available from https://doi.org/10.1007/s10811-015-0662-x.

Vidyashankar, S., Deviprasad, K., Chauhan, V. S., Ravishankar, G. A., & Sarada, R. (2013). Selection and evaluation of CO_2 tolerant indigenous microalga *Scenedesmus dimorphus* for unsaturated fatty acid rich lipid production under different culture conditions. *Bioresource Technology*, *144*, 28–37. Available from https://doi.org/10.1016/j.biortech.2013.06.054.

Wall, R., Ross, R. P., Fitzgerald, G. F., & Stanton, C. (2010). Fatty acids from fish: the anti-inflammatory potential of long-chain omega-3 fatty acids. *Nutrition Reviews*, *68*, 280–289. Available from https://doi.org/10.1111/j.1753-4887.2010.00287.x.

Wang, T., Tian, X., Liu, T., Wang, Z., Guan, W., Guo, M., ... Zhuang, Y. (2016). Enhancement of lipid productivity with a novel two-stage heterotrophic fed-batch culture of *Chlorella protothecoides* and a trial of CO_2 recycling by coupling with autotrophic process. *Biomass Bioenergy*, *95*, 235–243. Available from https://doi.org/10.1016/j.biombioe.2016.10.010.

Wang, X., Liu, Y., Wei, W., Zhou, X., Yuan, W., Balamurugan, S., ... Li, H. (2017). Enrichment of long-chain polyunsaturated fatty acids by coordinated expression of multiple metabolic nodes in the oleaginous microalga *Phaeodactylum tricornutum*. *Journal of Agricultural and Food Chemistry*, *65*, 7713–7720. Available from https://doi.org/10.1021/acs.jafc.7b02397.

Wang, X., Luo, S., Luo, W., Yang, W., Liu, J., & Li, H. (2019). Adaptive evolution of microalgal strains empowered by fulvic acid for enhanced polyunsaturated fatty acid production. *Bioresource Technology*, *277*, 204–210. Available from https://doi.org/10.1016/j.biortech.2018.12.116.

Woods, V. B., & Fearon, A. M. (2009). Dietary sources of unsaturated fatty acids for animals and their transfer into meat, milk and eggs: A review. *Livestock Science*, *126*, 1–3. Available from https://doi.org/10.1016/j.livsci.2009.07.002.

Xia, L., Ge, H., Zhou, X., Zhang, D., & Hu, C. (2013). Photoautotrophic outdoor two-stage cultivation for oleaginous microalgae *Scenedesmus obtusus* XJ-15. *Bioresource Technology*, *144*, 261–267. Available from https://doi.org/10.1016/j.biortech.2013.06.112.

Xia, L., Rong, J., Yang, H., He, Q., Zhang, D., & Hu, C. (2014). NaCl as an effective inducer for lipid accumulation in freshwater microalgae *Desmodesmus abundans*. *Bioresource Technology*, *161*, 402–409. Available from https://doi.org/10.1016/j.biortech.2014.03.063.

Yang, F., Xiang, W., Sun, X., Wu, H., Li, T., & Long, L. (2014). A novel lipid extraction method from wet microalga *Picochlorum* sp. at room temperature. *Marine Drugs*, *12*, 1258–1270. Available from https://doi.org/10.3390/md12031258.

Yuan, W., Shen, Y., Pei, Z. J., Wu, Q., & Mao, E. (2009). Microalgae mass production methods. *Transactions of the ASABE*, *52*(4), 1275–1287. Available from https://doi.org/10.13031/2013.27771.

Zhang, X., Hewson, J. C., Amendola, P., Reynoso, M., Sommerfeld, M., Chen, Y., & Hu, Q. (2014). Critical evaluation and modeling of algal harvesting using dissolved air flotation. *Biotechnology and Bioengineering*, *111*, 2477–2485. Available from https://doi.org/10.1002/bit.25300.

Zheng, H., Gao, Z., Yin, J., Tang, X., Ji, X., & Huang, H. (2012). Harvesting of microalgae by flocculation with poly (γ-glutamic acid). *Bioresource Technology*, *112*, 212–220. Available from https://doi.org/10.1016/j.biortech.2012.02.086.

Zhu, L., Hiltunen, E., Shu, Q., Zhou, W., Li, Z., & Wang, Z. (2014). Biodiesel production from algae cultivated in winter with artificial wastewater through pH regulation by acetic acid. *Applied Energy*, *128*, 103–110. Available from https://doi.org/10.1016/j.apenergy.2014.04.039.

Zhu, L. D., Li, Z. H., & Hiltunen, E. (2016). Strategies for lipid production improvement in microalgae as a biodiesel feedstock. *BioMed Research International*, *2016*, 7–9. Available from https://doi.org/10.1155/2016/8792548.

Chapter 7

Microalgae as a source of pigments for food applications

Marco García-Vaquero[1], Nigel Brunton[1] and Tomás Lafarga[2]
[1]*University College Dublin, School of Agriculture and Food Science, Dublin, Ireland,*
[2]*Department of Chemical Engineering, University of Almería, Almería, Spain*

7.1 Introduction

Food colors are defined by the European Food Safety Authority (EFSA) as additives used in food products with the purpose of: (1) restoring the original appearance of food that may be damaged or modified following processing, storage, packaging, and distribution; (2) making the food more visually appealing for consumers; or (3) giving color to otherwise colorless food products (EC, 2008). Food colors, together with sweeteners and texturants, are the most widely used additives in private households (Kozelová et al., 2013). These compounds play a major role in the food industry as they may directly influence the preference, selection, and eating desires of consumers (Gebhardt, Sperl, Carle, & Müller-Maatsch, 2020; Martins, Roriz, Morales, Barros, & Ferreira, 2016). Consumers' demands for nutritive and healthy foods, but also for attractive food products with high sensorial quality attributes, have increased in recent years, contributing to the specialization of the food coloring sector by developing new food aditives and ingredients (Garcia-Vaquero, Lopez-Alonso, & Hayes, 2017; Martins et al., 2016). Thereby, new colored molecules are currently used to correct the color variations of foodstuffs exposed to external conditions (i.e., light, air, temperature, moisture, and storage conditions), or to enhance or give color to foods targeting specific consumers by designing "fun foods" with increased atractiveness and global demand (Martins et al., 2016).

The initial colorants employed by the food industry were mainly extracted from natural sources including paprika, turmeric, indigo, saffron, and various flowers (Chen & Zhang, 2019; Sen, Barrow, & Deshmukh, 2019). These natural compounds were gradually replaced by synthetic molecules as the technological advances of the 1800s allowed to obtain high yields of chemically stable colored compounds, with larger ranges of hues and shades, and lower production costs

Cultured Microalgae for the Food Industry. DOI: https://doi.org/10.1016/B978-0-12-821080-2.00014-9

when compared to natural compounds (Chen & Zhang, 2019; Sen et al., 2019). However, recent reports on the allergenicity, toxicity, and carcinogenicity of several synthetic food colorants resulted in the banning of many synthetic compounds and an increased awareness and demand for natural colorants by consumers (Oplatowska-Stachowiak & Elliott, 2017; Sen et al., 2019).

The demand for natural colors is predicted to increase by 7% annually (Damant, 2011; Scotter, 2011, 2015), a fact that is also reflected in the global market value of these compounds which is expected to reach 1.77 billion USD by 2021 (Kannaujiya, Sundaram, & Sinha, 2017). Furthermore, natural colorants have additional nutritional and health benefits including antioxidant, antibacterial, and anticancer activities (Lafarga, Clemente, & García-Vaquero, 2019; Lila, Burton-Freeman, Grace, & Kalt, 2016). For example, carotenoids have powerful antioxidant and pro-vitamin A activities, playing a pivot role in the regulation of the immune system (Chew & Park, 2004). These promising health benefits have recently increased the interest in natural compounds in the food industry to develop nutraceuticals or compounds that display additional health benefits beyond those of basic nutrition when incorporated in food formulations (Garcia-Vaquero, Mora, & Hayes, 2019; Lafarga, Acién-Fernández, & Garcia-Vaquero, 2020).

Natural colors are primarily derived from plants, insects, mineral ores, and other alternative sources such as microalgae (Sen et al., 2019). The production and exploitation of natural compounds and pigments from microalgae offers certain advantages over other traditional natural sources. Microalgal cultures are able to perform more efficient bio-sequestration of atmospheric CO_2, require less arable land, and have lower post-harvest recovery periods compared to traditional crops (Matos, 2017). Moreover, the cultivation conditions of microalgae (i.e., light, temperature, CO_2, and other molecules added to the media) can be adapted and optimized to increase the production of pigments (Lafarga, 2019a; Liang, Wang, Wang, Zhu, & Jiang, 2019). In addition, consumers consider microalgae and microalgae-derived compounds as safe, sustainable, and healthy all key attributes for food products (Lafarga et al., 2021). As the demand for natural compounds increases, recent research has been also focusing on multiple strategies to reduce the downstream processing of natural compounds, increase the yields, and reduce the cost of production to be able to compete with chemically synthesized ingredients (Sen et al., 2019). Thereby, the application of ultrasounds, microwaves, or a combination of these technologies in a sequential or simultaneous manner has been succesfully explored to extract natural compounds from microalgae. Natural colorants produced by microalgae include chlorophylls, carotenoids, anthocyanins, and betanin that are responsible for green, yellow/orange/red, red/blue, and red colors when incorporated into food products, respectively. For example, β-carotene is commonly used as a food colorant at concentrations ranging from 2 to 50 ppm resulting in yellow and orange food products, respectively (Ribeiro, Barreto, & Coelho, 2011).

This chapter describes the current scenario and trends on the use of microalgal color agents in the food industry. The microalgal biomass and cultivation procedures currently utilized to produce the main chemical classes of food colorants (carotenoids, chlorophylls, and phycobiliproteins) at an industrial level are also discussed, together with the main compounds, promising health benefits, future uses, and current challenges faced by the industry of microalgal food colorants.

7.2 Carotenoids from microalgae

Carotenoids are one of the most widespread pigments in nature (Lafarga, Acién-Fernández, et al., 2019; Lafarga, Clemente, & García-Vaquero, 2019; Varela, Pereira, Vila, & León, 2015), being produced by all the photosynthetic organisms as well as by some nonphotosynthetic bacteria and fungi (Henríquez, Escobar, Galarza, & Gimpel, 2016). Carotenoids are lipophilic tetraterpenoids containing a C40 methyl branched hydrocarbon backbone and polyene chains with conjugated double bonds, responsible for the pigmentation and photosynthetic properties of these compounds (Henríquez et al., 2016). These molecules can be further classified based on their chemical nature into two main groups of hydrocarbons and oxygenated derivatives, namely carotenes and xanthophylls, respectively (Gong & Bassi, 2016; Henríquez et al., 2016). Due to the presence of hydroxyl and keto-groups at the end rings, xanthophylls are relatively more hydrophilic compounds compared to carotenes (Gong & Bassi, 2016).

More than 750 carotenoid structures have been isolated from different natural sources although no more than 30 of them have been described in eukaryotic microalgae (Varela et al., 2015). Xanthophylls produced by higher plants, such as violaxanthin, antheraxanthin, zeaxanthin, neoxanthin, and lutein, have also been described in microalgae together with other compounds not present in terrestrial plants such as loroxanthin, astaxanthin, and canthaxanthin (Henríquez et al., 2016). Based on the function of carotenoids in the microalgal cells, these compounds can also be classified as primary or secondary carotenoids. Primary carotenoids are those attached to membrane proteins or associated with light-harvesting complexes as they are necessary to maintain the functionality of phososynthesis in the cells (Gong & Bassi, 2016; Henríquez et al., 2016; Varela et al., 2015). Secondary carotenoids are mainly related to the protection of the cells against several stressors and can be localized in the membranes of the chloroplasts (i.e., β-carotene in *Dunaliella*) or in lipid globules (astaxanthin in *Haematococcus* species) (Varela et al., 2015).

Understanding the role and synthesis of carotenoids in microalgae is key for the future industrial exploitation of these compounds. In general, the synthesis of carotenoids in microalgae and terrestrial plants follows the metabolic pathway as described in Fig. 7.1, although the biosynthesis of

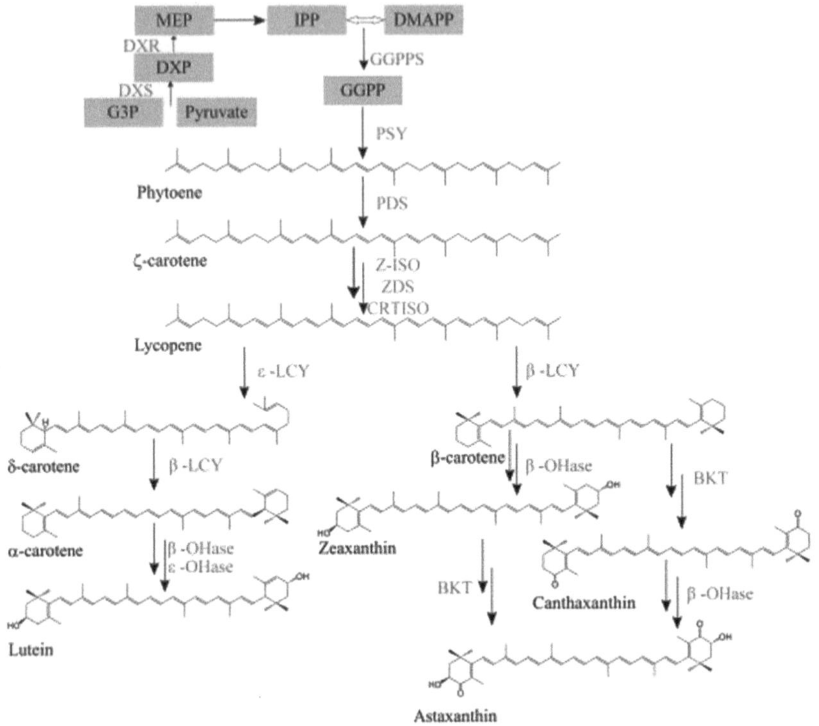

FIGURE 7.1 Biosynthetic pathways of carotenoids in microalgae. Abbreviations: *β-LCY*, β-cyclase; *β-OHase*, β-carotene hydroxylase; *CRTISO*, carotenoid isomerase; *DMAPP*, dimethylallyl diphosphate; *DXP*, deoxy-d-xylulose 5-phosphate; *DXS*, DXP synthase; *ε-LCY*, ε-cyclase; *ε-OHase*, ε-carotene hydroxylase; *G3P*, glyceraldehyde-3-phosphate; *GGPP*, geranylgeranyl diphosphate; GGPPS, GGPP synthase; *IPP*, isopentenyl pyrophosphate; *MEP*, methylerythritol 4-phosphate; *PDS*, phytoene desaturase; *PSY*, phytoene synthase; *ZDS*, ζ-carotene desaturase; *Z-ISO*, ζ-carotene isomerase. *Figure adapted from Nisar, Li, Lu, Khin, and Pogson (2015) Carotenoid metabolism in plants. Molecular Plant, 8(1), 68–82 by Gong and Bassi (2016) Carotenoids from microalgae: a review of recent developments. Biotechnology advances, 34(8), 1396–1412 and reproduced with permission from Elsevier.*

astaxanthin is only found in limited species of microalgae and slight differences in metabolic pathways can also be appreciated depending on the microalgal species.

The synthesis of carotenoids starts from the C5 building blocks isopentenyl pyrophosphate or dimethylallyl diphosphate (DMAPP) produced from acetyl-CoA via the mevalonic acid pathway in the cytosol or the methylerythritol 4-phosphate (MEP) pathway in plastids (Gong & Bassi, 2016; Varela et al., 2015). Successive enzymatic reactions achieve elongation of the chain and condensation of the molecules as seen in Fig. 7.1, leading to the production of lycopene which is the first colored carotenoid (Varela et al., 2015). Following the synthesis of α- or β-carotene, oxygenation of these molecules is required for the production of xanthophylls. Astaxanthin, not present in higher plants, is synthesized from canthaxanthin or zeaxanthin by microalgae at the end of the biosynthetic pathway as described in Fig. 7.1.

Carotenoids from microalgae have been traditionally used as a source of yellow, orange, and red colors for animal feed applications. For example, canthaxanthin, astaxanthin, and lutein from *Chlorella* have been included regularly in feed formulations in poultry and salmonid fish and trout to enhance the characteristic natural colors of egg yolks and fish meat, respectively (Guedes, Amaro, & Malcata, 2011). The increased evidence of the harmful effects of some synthetic colors has also contributed to an increased demand for natural pigments as the list of permitted synthetic colorants for the food industry is reduced (Singh, Khattar, Rajput, Chaudhary, & Singh, 2019). Moreover, carotenoids have shown important biological functions including antioxidant, pro-vitamin A, antiinflammatory, anticancer, and cardioprotective effects as demonstrated in multiple human studies (Table 7.1). The protective and health enhancement properties of carotenoids have increased the interest of the pharmaceutical and food industries in exploiting these compounds by developing multiple products such as food supplements and nutraceuticals and as ingredients in functional foods (Lafarga, 2019b).

The global market for carotenoids is estimated to reach values of over 1.5 billion USD by 2021 (Sathasivam & Ki, 2018), with β-carotene and asthaxanthine being the main drivers of this demand with prices ranging from 300–3000 and 2500–10,000 USD per kg of each compound, respectively (Taucher et al., 2016). Some species of *Dunaliella*, especially *D. salina* and *D. bardawil* have the ability to accumulate large amounts of carotenoids, mainly β-carotene and small amounts of α-carotene, lutein, zeaxanthin, cryptoxanthin, and neoxanthin (Ye, Jiang, & Wu, 2008). *Dunaliella salina* is the most common microalgae used to produce β-carotene as this species can accumulate up to 10% of this compound on a dry weight basis depending on the cultivation conditions (Lv et al., 2016). Lv et al. (2016) determined the influence of nutrient deprivation (nitrogen, sulfur, and phosphorus) or combinations of nutrients on the general appearance of microalgal cultures, cell morphology, and accumulation of carotenoids in the chloroplast of *Dunaliella salina* (Fig. 7.2).

Dunaliella salina is currently produced in open ponds or open raceway-type ponds mainly in Australia, Israel, India, and China (Borowitzka, 2013). The cultivation of this microalgae in closed systems or tubular photobioreactors has also been attempted, although none of the systems developed proved to be commercially viable (Borowitzka, 2013). Recent developments include hybrid helical-tubular photobioreactors, combining the airlift and tubular photobioreactors with an open pond system, while applying salt stressors to induce the production of β-carotene in *D. salina* (Hashemi, Moslemi, Pajoum Shariati, & Delavari Amrei, 2020). Under these conditions, Hashemi et al. (2020) achieved efficient cell growth and production of β-carotene (4.85 μg

TABLE 7.1 Health benefits of carotenoids confirmed by human studies.

Carotenoids	Health benefits	References
Astaxanthin	Antioxidant, antiinflammatory, anticancer, and cardioprotective effects	Chew, Park, Wong, and Wong (1999), Fasano et al. (2014), Li et al. (2011), Park, Chyun, Kim, Line, and Chew (2010), Pashkow, Watumull, and Campbell (2008).
Lutein	Prevention of cataract and age-related macular degeneration and cardiovascular diseases together with antioxidant and anticancer properties	Alves-Rodrigues and Shao (2004), Bone, Landrum, Guerra, and Ruiz (2003), Cha, Koo, and Lee (2008), Manayi et al. (2016), VijayaPadma, Ramyaa, Pavithra, and Krishnasamy (2014)
β-Carotene	Antioxidant properties and prevention against night blindness and liver fibrosis	Dufossé et al. (2005), Shaish et al. (2006), Virtamo et al. (2014)
Lycopene	Protection against radiation, prevention against cardiovascular diseases, and anticancer and antioxidant properties	Devasagayam et al. (2004), Srinivasan, Devipriya, Kalpana, and Menon (2009), Viuda-Martos et al. (2014)
Canthaxanthin	Antioxidant properties, immunomodulatory activity, and role in gap junction communication	Esatbeyoglu and Rimbach (2017), Zhang, Wang, Wang, and Liu (2014)
Fucoxanthin	Antiobesity and antioxidant properties	Abidov, Ramazanov, Seifulla, and Grachev (2010), Nanba et al. (2008), Petrushkina et al. (2017)

Source: Adapted from Gong and Bassi (2016) Carotenoids from microalgae: a review of recent developments. Biotechnology advances, 34(8), 1396–1412 and reproduced with permission from Elsevier.

per mg of dried microalgae) using 2.5 mol L^{-1} of salt after 21 days of cultivation. Chemically synthesized β-carotene is mainly dominated by all-*trans* stereoisomers, while the β-carotene produced naturally from *Dunaliella* contains mixtures of *trans* and *cis* stereoisomers that have better bioavailability and have more efficient antioxidant properties than all-*trans* stereoisomers (Gong & Bassi, 2016; Ye et al., 2008). However, due to the aforementioned technological limitations for the production of these compounds, natural compounds currently represent only 2% of the global market for carotenoids (Gong & Bassi, 2016).

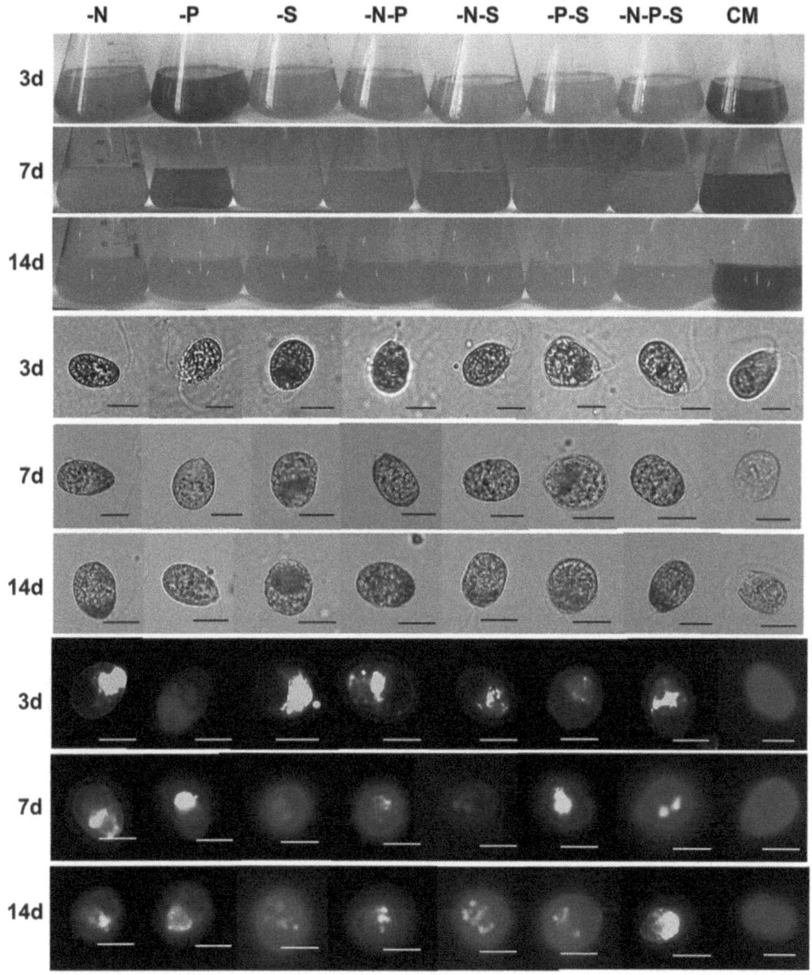

FIGURE 7.2 Changes in culture color and the morphology of *Dunaliella salina* cells under different nutrient deprivation contitions. Nitrogen deprivation (-N), phosphorous deprivation (-P), sulfur deprivation (-S), nitrogen and phosphorous deprivation (-N-P), nitrogen and sulfur deprivation (-N-S), phosphorous and sulfur deprivation (-P-S), nitrogen, phosphorous, and sulfur deprivation (-N-P-S), and complete medium (CM). *Reproduced from Lv et al. (2016), Analysis of the physiological and molecular responses of Dunaliella salina to macronutrient deprivation. PLoS ONE, 11(3).*

Haematococcus pluvialis is the main microalgae currently exploited for the production of asthaxanthin, as these cells are able to accumulate astaxanthin inside extraplastidial lipid bodies at a concentration of approximately 4% on a wet weight basis (Wayama et al., 2013). Astaxanthin derived from *Haematococcus pluvialis* was suggested as one of the top three main

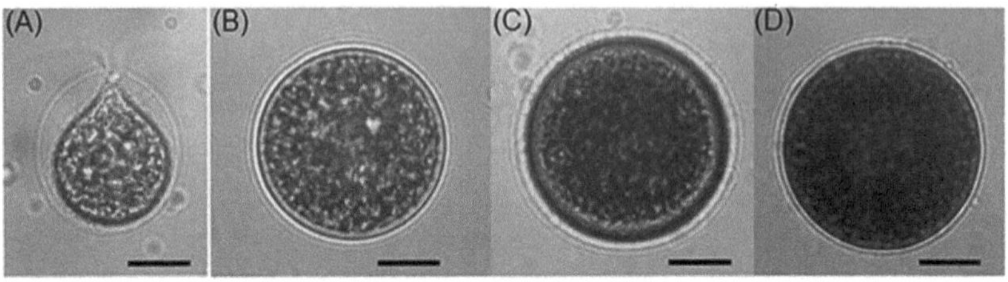

FIGURE 7.3 *Haematococcus pluvialis* images at "vegetative or green stage" (A, B) and "red or encysted phase" (C, D). Scale bars: 10 μm. *Reprinted from Shah et al. (2016), Astaxanthin-producing green microalga Haematococcus pluvialis: from single cell to high value commercial products. Frontiers in Plant Science, 7, 531.*

ingredients to watch during 2019 and a niche ingredient with huge potential (Lafarga, 2019b). The life cycle of *Haematococcus pluvialis* includes two clearly distinguishable phases namely "green or vegetative" and "red or encysted" with clear color differentiation due to differences in carotenoid comoposition as well as other morphological changes in the cells as described by Shah, Mahfuzur, Liang, Cheng, and Daroch (2016) (Fig. 7.3). *H. pluvialis* in the green phase contains mainly lutein (75%–80%) and β-carotene (10%–20%), while during the red phase the accumulation of asthaxanthin increases, representing between 80%–99% of the total amount of carotenoids of the cells (Shah et al., 2016).

Li, Zhu, Niu, Shen, and Wang (2011) estimated the cost of production of natural astaxanthin from *Haematococcus pluvialis* as 718 USD per kg, cheaper than the cost of production of synthetic astaxanthin that is currently manufactured from petrochemical sources, raising sustainability and safety concerns for consumers (Panis & Carreon, 2016). In fact, synthetic astaxanthin is only approved for its use as a feed additive in aquaculture and has not yet been approved for direct human consumption in food or supplements (Li et al., 2011; Panis & Carreon, 2016).

Overall, the application of abiotic stresses, omics-based technologies, and molecular approaches could contribute to lower the production cost of carotenoids from microalgae in the future (Liang et al., 2019). Significant research efforts have been made analyzing the influence of cultivation conditions (i.e., temperature, light irradiation, salinity, and nutrient concentration), the use of additives in the media (i.e., iron, acetate, and malonate) and the combination of stresses on the production of carotenoids in multiple microalgal species as reviewed in detail by Lafarga, Acién-Fernández, et al. (2019), Lafarga, Clemente, et al. (2019). Moreover, the production and final commercialization of carotenoids from microalgae will require further improvements in other downstream processing stages other than cultivation of the biomass (Table 7.2).

TABLE 7.2 Summary of the main downstream processing stages to obtain carotenoids from microalgae.

Microalgal cultivation
Open ponds, photobioreactors, or hybrid systems
Harvesting
Sedimentation, flocculation, centrifugation, or filtration
Pretreatments
Drying • Thermal drying • Freeze-drying Cell disruption • *Mechanical cell disruption techniques*: bead milling, high-pressure, pulse electric fields, ultrasounds or microwaves • *Nonmechanical cell disruption techniques*: osmotic shock, ionic liquids, acid/alkali hydrolysis, or enzymatic hydrolysis • *Other methods*: steam explosion, explosive decompression, atomic force microscopy, laser and microorganisms
Extraction
Conventional solvent extraction, pressurized solvent extraction, subcritical solvent extraction, or supercritical solvent extraction
Purification and storage
Willstatter method or chromatographic purification

Source: The content of this table was adapted from Lafarga, Acién-Fernández, et al. (2019), Effect of microalgae incorporation on the physicochemical, nutritional, and sensorial properties of an innovative broccoli soup. LWT, 111, 167-174, Lafarga, Clemente, et al. (2019), Carotenoids from microalgae. In C. Galanakis (Ed.), Carotenoids: Properties, Processing and Applications (pp. 149-188). India: Elsevier, and Lafarga, Mayre, et al. (2019), Potential of the microalgae Nannochloropsis and Tetraselmis for being used as innovative ingredients in baked goods. LWT, 115, 108439.

Further research on the application of efficient harvest or concentration techniques, innovative technologies to disrupt the cell walls of microalgae, and further purification and storage conditions will influence greatly the cost of production of microalgal carotenoids.

7.3 Chlorophylls

Approximately 1.2 billion tons of chlorophylls are produced naturally every year by photosynthetic organisms, both terrestrially and in the oceans (Humphrey, 2004). However, chlorophylls are probably the least studied of the naturally present food pigments. There are two main types of chlorophylls: chlorophyll *a* and chlorophyll *b*. Their structural differences, shown and described in Fig. 7.4, lead to chlorophyll *a* being blue/green with a maximum absorbance from 660 to 665 and chlorophyll *b* being green/yellow and presenting a maximum absorbance from 652 to 652 nm (Humphrey, 1980). Although there are only two main types of chlorophylls, there are more naturally present in nature—especially in microalgae. For example, the cyanobacteria *Acaryochloris marina* performs oxygenic photosynthesis but contains mostly chlorophyll *d* (and traces of chlorophyll *a*), providing a potential selective advantage because it allows this microalga to use infrared light (700–750 nm) which is not absorbed by chlorophyll *a* (Chen & Blankenship, 2011). Chlorophylls are present in nature because of their role in photosynthesis, the process by which plants and other photosynthetic organisms such as algae convert light energy into chemical energy. Chlorophylls absorb light more strongly in the blue portion of the electromagnetic spectrum followed by the red portion. They are poor absorbers of light in the green and near-green portion of the spectrum, which gets reflected leading to the green color of tissues containing chlorophyll.

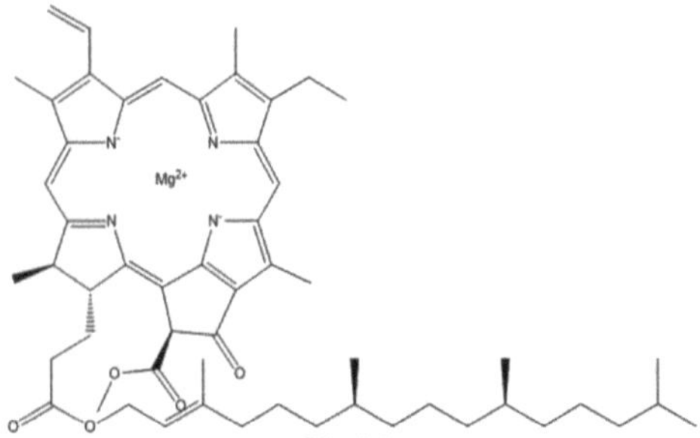

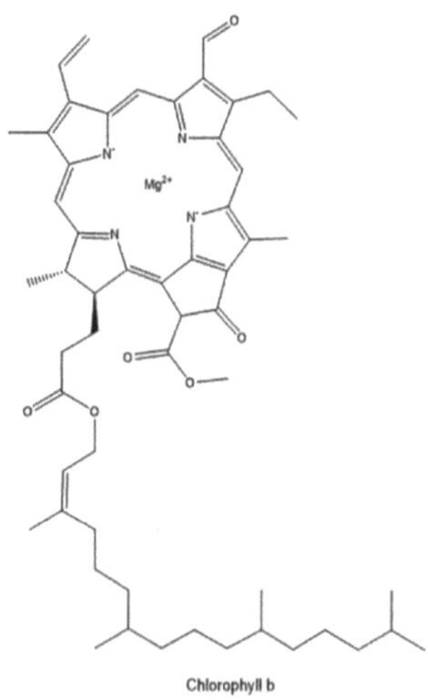

FIGURE 7.4 Chemical structures of chlorophyll *a* and chlorophyll *b*. There are over 100 different chlorophyll structures, but strictly speaking there are only five types of chlorophyll (*a*–*e*) (Queiroz Zepka, Jacob-Lopes, & Roca, 2019) although chlorophyll *f* has also been isolated from microalgae (Chen & Blankenship, 2011). Chlorophyll *a* and chlorophyll *b* are predominant in nature. The basic chemical structure of chlorophylls is a porphyrin macrocycle, which is a cyclic tetrapyrrole derivative (Hosikian et al., 2010). Besides the four pyrrole rings, these molecules also contain a characteristic iso-cyclic five-membered ring (Khan, Yusuf, Agarwal, & Prasad, 2018). In their center, chlorophylls contain four nitrogen facing inward creating an ideal place for binding metal ions (almost always magnesium). The main difference between chlorophyll *a* and chlorophyll *b* is the substituent on the second pyrrole ring at the C-7 position: chlorophyll *a* has a methyl group, while chlorophyll *b* has a formyl group (Khan et al., 2018).

Chlorophylls have been part of the human diet throughout history as an integrated part of plants and algae. Nowadays, chlorophyll is registered as a food additive as a colorant and its E number (code given to food additives for use within the European Union) is E140. Microalgae are mainly used in the food industry as green colorants and have been used to formulate a large number of green-colored foods (Lafarga, Acién-Fernández, et al., 2019; Lafarga, Mayre, et al., 2019). These molecules are especially interesting in the food industry not only because of their intense green color, but also for their high antioxidant properties and that of their derivatives (Lanfer-Marquez, Barros, & Sinnecker, 2005). Indeed, their high antioxidant capacity makes chlorophylls a protective factor against excessive UV radiation and protects algal tissues against oxidative stress (Yuan, 2007). Other potential health benefits of chlorophylls have been reviewed in a recent paper by Queiroz Zepka, Jacob-Lopes, & Roca, 2019 who summarized the known bioactive properties of chlorophylls and their catabolites which included antimutagenic, chemoprotective, antioxidant, antiinflammatory, and antimicrobial bioactivities among others. In that review paper the authors highlighted that probably the most promising health-promoting activity of chlorophylls is their potential role as drugs during photodynamic therapy acting as photosensitizers.

In order to quantify the chlorophyll content of microalgae it has to be first extracted. Because of the resistant cell wall of microalgae, it is also common to employ a disruption pretreatment before extraction. Simon and Helliwell (1998) reported that without a cell wall disruption step, only 25% of the potential chlorophyll *a* was able to be extracted. For this reason, methods including milling, sonication, pulsed electric fields, microwaves, or high-pressure homogenization are commonly used for cell disruption. For example, Luengo, Condón-Abanto, Álvarez, and Raso (2014) reported that pulsed electric fields processing (20 kV cm^{-1}, 75 μs) increased the yield of both chlorophyll *a* and chlorophyll *b* from *Chlorella vulgaris* by 0.7- and 0.8-fold, respectively. The observed increase was even higher when measured 1 h after processing, calculated as 1.6 and 2.1 times higher, respectively. In another study, Parniakov et al. (2015) reported higher chlorophyll recoveries after sonication (200 W, 24 kHz, 10 min, 30ºC) when compared to pulsed electric fields processing (20 kV/cm, 6 ms, 30ºC), which was ineffective at the conditions studied—the authors suggested that pulsed electric fields would be more effective for the extraction of water-soluble compounds. After disruption of the cell wall, chlorophylls are extracted using organic solvents. Different organic solvents have been used in the literature including methanol, ethanol, acetone, and dimethylformamide (Macías-Sánchez et al., 2007, 2009; Pasquet et al., 2011; Schumann, Häubner, Klausch, & Karsten, 2005). Kong et al. (2014) optimized the extraction of chlorophylls from the microalga *Chlorella* by conducting a response surface methodology and obtained the optimum extraction yields, measured as 31.1 mg g^{-1}, when operating at 61.4ºC, ultrasound potency of 200 W, and ethanol concentration of 79.4% (vol./vol.) during 78.7 min. Moreover, A. Catarina Guedes et al. (2013) evaluated the potential of using supercritical carbon dioxide to extract carotenoids and chlorophylls and observed that the highest extraction yields of chlorophylls were obtained when operating at 250 bar, 40ºC, and a flow rate of 4.3 g min^{-1}. Interestingly, in that same study the authors observed that the presence of ethanol at 7.7% (vol./vol.) led to increased chlorophyll extraction yields when compared to plain carbon dioxide. Other studies used water for extracting chlorophylls obtaining low extraction efficiencies as chlorophylls are insoluble in water (Parniakov et al., 2015). Overall, higher extraction efficiencies were observed when using dimethyl sulfoxide, although its toxic nature reduced its potential utilization as an efficient solvent (Hosikian, Lim, Halim, & Danquah, 2010).

One of the main disadvantages of natural colorants when compared to their synthetic counterparts is their low stability. The high susceptibility of chlorophyll to environmental stresses such as oxygen, light, temperature, and pH variations currently limits the commercial-scale utilization of chlorophylls as food colorants. Chlorophylls can show high resistance to temperature when suitable cooking strategies are used and food matrices selected. For example, Turkmen, Poyrazoglu, Sari, and Sedat Velioglu (2006) evaluated the thermal stability of chlorophylls after microwaving and conventional cooking methods of a number of green vegetables and observed that retention of chlorophylls varied between 19% and 100% depending on the food matrix and the cooking method and conditions selected. Much lower thermal stabilities were reported by Benlloch-Tinoco et al. (2015), who reported chlorophyll losses in the range 40%–100% after microwave and conventional thermal processing of kiwi. In the above-mentioned study by Turkmen et al. (2006), the authors reported that in 80% of the studied vegetables chlorophyll *a* was more resistant than chlorophyll *b*. Thermal stability of chlorophylls can be improved by, for example, microencapsulation. Kang, Lee, Kim, & Chang, 2019 recently encapsulated chlorophylls by spray-drying using different combinations of gum Arabic and maltodextrin and achieved a very high storage stability of the chlorophylls (94.7%–97.5%) during 10 days at either 4ºC, 20ºC, or even 40ºC. Similar results were reported by Hsiao et al. (2020), who achieved high stability of chlorophylls by using a polycaprolactone coating to encapsulate the chlorophyll and controlled the particle size through droplet microfluidics. Overall, although thermal degradation can be minimized by selecting suitable food matrices and cooking conditions, the use of chlorophylls in foods is still limited as a colorant because of their susceptibility to photodegradation and instability in acidic conditions (pH 3.5–5.0) (Rodriguez-Amaya, 2016). After chlorophyll extraction, its magnesium ion becomes unstable and can be easily displaced by a weak acid, and for this reason, magnesium ions are generally substituted with a copper ion forming a highly stable blue/green complex (Hosikian et al., 2010). There are still some controversial results in the literature because of the variable composition and purity of commercial preparations and further development of stable chlorophylls is needed and expected.

7.4 Phycobiliproteins

Phycobiliproteins are among the valuable compounds that can be obtained from microalgae—these are exclusive to the Cyanobacteria, red algae, and Cryptophyta and Glaucophyta (Borowitzka, 2013). Phycobiliproteins are water-soluble auxiliary pigments that consist of highly fluorescent proteins with linear prosthetic groups (bilins) that are linked to specific cysteine residues (Arashiro et al., 2020). These can be divided into three large classes depending on their composition and chromophore content: (1) phycocyanins (610–625 nm), (2) phycoerythrins (490–570 nm), and (3) allophycocyanins (650–660 nm) (Noreña-Caro & Benton, 2018). Phycobiliproteins are bright blue (phycocyanin and allophycoerythrin) and fuchsia (phycorerythrin) and have high commercial value as natural colorants in the nutraceutical, cosmetic, medical, and pharmaceutical industries as well as in the textile and food industries. Indeed, phycocyanin and allophycocyanin aqueous extracts obtained from the Cyanobacteria *Arthrospira platensis* (*Spirulina*) are approved by the EFSA and the United States FDA as a coloring foodstuff (Lauceri, Chini Zittelli, & Torzillo, 2019). The fluorescent properties of phycobiliproteins have also been used in the food industry to produce transparent candies and innovative alcoholic beverages that fluoresce at pH values in the range 5–6 (Tavanandi & Raghavarao, 2020).

As phycobiliproteins are water-soluble, they are generally extracted from the microalgal biomass using a saline buffer at different pH values in the range 6.5–8.0 (İlter et al., 2018; Tavanandi, Devi, & Raghavarao, 2018; Tavanandi & Raghavarao, 2020). Other studies used distilled water or even seawater as a solvent. For example, Sudhakar, Jagatheesan, Perumal, & Arunkumar, 2015 assessed the potential of seawater for being used as a solvent for phycobiliprotein extraction and obtained extracts with high purity after extraction and purification using a cellulose anion exchange column. Combinations of aqueous solutions and physical method such as pulsed electric fields or microwave processing have also been evaluated. Martínez, Luengo, Saldaña, Álvarez, & Raso, 2017 performed electroporation of *A. platensis* using varying electric field strengths and treatment durations and observed that: (1) the minimum electric field intensity for detecting phycocyanin in the extract was 15 kV cm^{-1} and (2) higher electric field strengths were required when shorter treatment times were applied. Moreover, Juin et al. (2015) evaluated the potential of microwave-assisted extraction of phycobiliproteins from *Porphyridium purpureum* and obtained a 180- to 1080-fold reduction of the extraction time when compared to a conventional soaking protocol. Extraction of phycobiliproteins from microalgae has been proved to exhibit a low carbon footprint (Papadaki, Kyriakopoulou, Tzovenis, & Krokida, 2017). The commercial value of phycobiliproteins depends largely on their purity grade, generally assessed by the ratio between the value of phycobiliprotein absorbance maximum, listed above, and the absorbance value at 280 nm, which is related to the total amount of proteins in the extract. Overall, it is accepted that purity values greater than 0.7 are food grade, greater than 1.5 are cosmetic grade, greater than 3.9 are reactive grade, and greater than 4.0 are analytical grade (Sala, Moraes, & Kalil, 2018). A large number of downstream processing strategies have been suggested to purify phycobiliproteins from microalgae. These have been recently reviewed by Lauceri, Chini Zittelli, & Torzillo, 2019 and include chromatographic methods (ion exchange chromatography, hydrophobic interaction chromatography, gel filtration chromatography, expanded bed adsorption chromatography, among others), precipitation using ammonium sulfate, chitosan affinity precipitation, isoelectric precipitation, microfiltration, and ultrafiltration.

As happens with other natural pigments, the stability of phycobiliproteins is low when compared to chemically synthesized colorants. Overall, phycobiliproteins are generally thermostable at temperatures ranging between 4°C and 40°C and show maximum stability at pH values in the range 6.0–7.0 (Rastogi, Sonani, & Madamwar, 2015). High temperatures (60°C–80°C) as well as extreme pH values significantly affect their fluorescence, although the functionality of phycoerythrin has been reported to be maintained even within the pH range 6.0–12.0 and under oxidative stress (Rastogi, Sonani, & Madamwar, 2015). Similar results were observed in previous studies that suggested that phycoerythrin obtained from *Porphyrium cruentum* has strong functional stability in the pH range 4.0–10.0, showing maximum stability at pH 7.0 (Gonzalez-Ramirez et al., 2014).

7.4.1 Phycobiliproteins: more than just a colorant for the food industry

As happens with chlorophylls and carotenoids, phycobiliproteins have the added advantage that besides having potential applications as food colorants, they have proved health benefits upon consumption. A large number of scientific publications have demonstrated several therapeutic activities of these molecules including anticarcinogenic, antiinflammatory, neuroprotective, immunomodulatory, and antioxidant properties. The antioxidant capacity of phycobiliproteins has been assessed both in vitro and in vivo using animal models (Sekar & Chandramohan, 2008; Sonani, Rastogi, Patel, & Madamwar, 2016; Yu, Wu, Wang, Jia, & Zhang, 2017). For example, Rodríguez-Sánchez, Ortiz-Butrón, Blas-Valdivia, Hernández-García, and Cano-Europa (2012) evaluated whether phycobiliproteins obtained from

Spirulina can protect renal cells against mercury-caused oxidative stress and cellular damage in the kidney using a rat model. In that study, the authors observed that all doses of phycobiliproteins (or phycocyanin) prevented enhancement of oxidative markers and protected the cells against $HgCl_2$-derived cellular damage. Studied phycocyanin of phycobiliprotein doses were either 50 or 100 mg kg^{-1} and were orally administered. Similar results were observed by Cano-Europa et al. (2010) who produced an extract containing 2% phycocyanin, 9% allophycocyanin, and 89% phycoerythrin from *Pseudanabaena tenuis* and demonstrated that all the studied doses (10–100 mg kg^{-1} of body weight, orally administered) prevented the increase of oxidative markers and partially protected cells of mice against $HgCl_2$-induced oxidative stress and cellular damage. Sonani, Singh, Kumar, Thakar, & Madamwar, (2014) also demonstrated in vitro antioxidant effects of phycobiliproteins from *Lyngbya* sp. A09DM as well as the antiaging potential of phycoerythrin in *Caenorhabditis elegans*, attributed to their antioxidant capacity.

Other health-promoting properties attributed to phycobiliproteins include antiulcerogenic and anticarcinogenic activities. Indeed, Guzmán-Gómez et al. (2018) reported that *Spirulina*-derived phycobiliproteins could show antiulcerogenic properties against ethanol-induced gastric ulcers in a rat model. Studied phycobiliprotein doses ranged between 100 and 400 mg kg^{-1} of body weight (intragastric administration). The greatest protection was observed at 400 mg kg^{-1} and the authors attributed the observed effects to the antioxidant and antiinflammatory properties of phycobiliproteins. Moreover, phycocyanin also showed potential for being used for the prevention of fibrotic diseases. Pattarayan, Rajarajan, Sivanantham, Palanichamy, and Rajasekaran (2017) concluded that phycocyanin can suppress transforming growth factor-β1-induced epithelial–mesenchymal transition (EMT) and EMT-associated proliferation in human epithelial cell lines, which has been shown to contribute to the development of fibrotic disease. Results obtained by An, Park, and Lee (2016) support this hypothesis as the authors of that study demonstrated that phycocyanin could reduce the overexpression of connective tissue growth factor and α-smooth muscle actins (α-SMA), alleviating fibrotic contracture. Phycocyanin (6 nM) inhibited α-SMA expression by 70% and reduced collagen contraction by 29%. Other studies also suggested that phycobiliproteins (phycocyanobilin) can reduce brain injury after endothelial-1-induced focal cerebral ischemia (Pavón-Fuentes et al., 2020).

7.5 Conclusions

Microalgae are rich sources of natural colored molecules that can be used as food additives. They include mainly chlorophylls, carotenoids, and phycobiliproteins. Despite not being as stable as their synthetic counterparts, microalgal-derived pigments show potential applications in the food industry not only as colorants but also as ingredients with health-promoting properties. Microalgae-derived pigments have potential for being used as novel ingredients that could satisfy consumer demands, which include foods that are natural, healthy, convenient, and free from synthetic additives. Microalgal biomass and microalgae-derived pigments are positioning firmly in the market and several of these pigments have been authorized for being used as food additives in the United States and EU. Future studies are needed to completely understand their potential health-promoting properties and to assess their safety, although results reported so far are promising and forecast a bright future for microalgae biomass and microalgae-derived pigments in the food industry.

Acknowledgments

This work forms part of the SABANA Project, funded by the European Union's Horizon 2020 Research and Innovation Programme (Grant # 727874) and the ALGALPROT Project funded by CEIMAR (CEIJ-002). T. Lafarga is in receipt of a *Juan de la Cierva Incorporación* postdoctoral grant awarded by the Spanish Ministry of Science, Innovation, and Universities (IJC2018-035287-I) and a Leonardo 2020 Grant for Researchers and Cultural Creators.

References

Abidov, M., Ramazanov, Z., Seifulla, R., & Grachev, S. (2010). The effects of Xanthigen™ in the weight management of obese premenopausal women with non-alcoholic fatty liver disease and normal liver fat. *Diabetes, Obesity and Metabolism, 12*(1), 72–81.

Alves-Rodrigues, A., & Shao, A. (2004). The science behind lutein. *Toxicology Letters, 150*(1), 57–83.

An, E., Park, H., & Lee, A.-R. C. (2016). Inhibition of fibrotic contraction by C-phycocyanin through modulation of connective tissue growth factor and α-smooth muscle actin expression. *Tissue Engineering and Regenerative Medicine, 13*(4), 388–395.

Arashiro, L. T., Boto-Ordóñez, M., Van Hulle, S. W. H., Ferrer, I., Garfí, M., & Rousseau, D. P. L. (2020). Natural pigments from microalgae grown in industrial wastewater. *Bioresource Technology, 303*, 122894.

Benlloch-Tinoco, M., Kaulmann, A., Corte-Real, J., Rodrigo, D., Martínez-Navarrete, N., & Bohn, T. (2015). Chlorophylls and carotenoids of kiwifruit puree are affected similarly or less by microwave than by conventional heat processing and storage. *Food Chemistry, 187*, 254–262.

Bone, R. A., Landrum, J. T., Guerra, L. H., & Ruiz, C. A. (2003). Lutein and zeaxanthin dietary supplements raise macular pigment density and serum concentrations of these carotenoids in humans. *The Journal of Nutrition, 133*(4), 992–998.

Borowitzka, M. A. (2013). High-value products from microalgae—their development and commercialisation. *Journal of Applied Phycology, 25*(3), 743–756.

Cano-Europa, E., Ortiz-Butrón, R., Gallardo-Casas, C. A., Blas-Valdivia, V., Pineda-Reynoso, M., Olvera-Ramírez, R., & Franco-Colin, M. (2010). Phycobiliproteins from Pseudanabaena tenuis rich in c-phycoerythrin protect against HgCl2-caused oxidative stress and cellular damage in the kidney. *Journal of Applied Phycology, 22*(4), 495–501.

Cha, K. H., Koo, S. Y., & Lee, D.-U. (2008). Antiproliferative effects of carotenoids extracted from Chlorella ellipsoidea and Chlorella vulgaris on human colon cancer cells. *Journal of Agricultural and Food Chemistry*, *56*(22), 10521–10526.

Chen, L., & Zhang, B. (2019). *Microbial production of natural food colorants. Systems and Synthetic Biotechnology for Production of Nutraceuticals* (pp. 129–158). Springer.

Chen, M., & Blankenship, R. E. (2011). Expanding the solar spectrum used by photosynthesis. *Trends in Plant Science*, *16*(8), 427–431.

Chew, B., Park, J., Wong, M., & Wong, T. (1999). A comparison of the anticancer activities of dietary beta-carotene, canthaxanthin and astaxanthin in mice in vivo. *Anticancer Research*, *19*(3A), 1849–1853.

Chew, B. P., & Park, J. S. (2004). Carotenoid action on the immune response. *The Journal of Nutrition*, *134*(1), 257S–261S.

Damant, A. (2011). *Food colourants. Handbook of Textile and Industrial Dyeing* (pp. 252–305). Elsevier.

Devasagayam, T., Tilak, J., Boloor, K., Sane, K. S., Ghaskadbi, S. S., & Lele, R. (2004). Free radicals and antioxidants in human health: current status and future prospects. *JAPI*, *52* (794804), 4.

Dufossé, L., Galaup, P., Yaron, A., Arad, S. M., Blanc, P., Murthy, K. N. C., & Ravishankar, G. (2005). Microorganisms and microalgae as sources of pigments for food use: a scientific oddity or an industrial reality? *Trends in Food Science and Technology*, *16*(9), 389–406.

EC. (2008). *Regulation (EC) No 1333/2008 of the European Parliament and of the Council of 16 December 2008 on food additives*.

Esatbeyoglu, T., & Rimbach, G. (2017). Canthaxanthin: from molecule to function. *Molecular Nutrition and Food Research*, *61*(6), 1600469.

Fasano, E., Serini, S., Mondella, N., Trombino, S., Celleno, L., Lanza, P., & Calviello, G. J. B. R. I. (2014). Antioxidant and anti-inflammatory effects of selected two human immortalized keratinocyte lines. *BioMed Research International*, *2014*(2014), 327452.

Garcia-Vaquero, M., Lopez-Alonso, M., & Hayes, M. (2017). Assessment of the functional properties of protein extracted from the brown seaweed Himanthalia elongata (Linnaeus) SF Gray. *Food Research International*, *99*, 971–978.

Garcia-Vaquero, M., Mora, L., & Hayes, M. (2019). In vitro and in silico approaches to generating and identifying angiotensin-converting enzyme I inhibitory peptides from green macroalga Ulva lactuca. *Marine Drugs*, *17*(4), 204.

Gebhardt, B., Sperl, R., Carle, R., & Müller-Maatsch, J. (2020). Assessing the sustainability of natural and artificial food colorants. *Journal of Cleaner Production*, 120884.

Gong, M., & Bassi, A. (2016). Carotenoids from microalgae: a review of recent developments. *Biotechnology advances*, *34*(8), 1396–1412.

Gonzalez-Ramirez, E., Andujar-Sanchez, M., Ortiz-Salmeron, E., Bacarizo, J., Cuadri, C., Mazzuca-Sobczuk, T., & Martinez-Rodriguez, S. (2014). Thermal and pH stability of the B-phycoerythrin from the red algae Porphyridium cruentum. *Food Biophysics*, *9*(2), 184–192.

Guedes, A. C., Amaro, H. M., & Malcata, F. X. (2011). Microalgae as sources of carotenoids. *Marine Drugs*, *9*(4), 625–644.

Guedes, A. C., Gião, M. S., Matias, A. A., Nunes, A. V. M., Pintado, M. E., Duarte, C. M. M., & Malcata, F. X. (2013). Supercritical fluid extraction of carotenoids and chlorophylls a, b and c, from a wild strain of Scenedesmus obliquus for use in food processing. *Journal of Food Engineering*, *116*(2), 478–482.

Guzmán-Gómez, O., García-Rodríguez, R. V., Quevedo-Corona, L., Pérez-Pastén-Borja, R., Rivero-Ramírez, N. L., Ríos-Castro, E., & Chamorro-Cevallos, G. A. (2018). Amelioration of ethanol-induced gastric ulcers in rats pretreated with phycobiliproteins of Arthrospira (Spirulina) Maxima. *Nutrients*, *10*(6), 763.

Hashemi, A., Moslemi, M., Pajoum Shariati, F., & Delavari Amrei, H. (2020). Beta-carotene production within Dunaliella salina cells under salt stress condition in an indoor hybrid helical-tubular photobioreactor. *The Canadian Journal of Chemical Engineering*, *98*(1), 69–74.

Henríquez, V., Escobar, C., Galarza, J., & Gimpel, J. (2016). *Carotenoids in microalgae. Carotenoids in Nature* (pp. 219–237). Springer.

Hosikian, A., Lim, S., Halim, R., & Danquah, M. K. (2010). Chlorophyll extraction from microalgae: a review on the process engineering aspects. *International Journal of Chemical Engineering*, *2010*.

Hsiao, C.-J., Lin, J.-F., Wen, H.-Y., Lin, Y.-M., Yang, C.-H., Huang, K.-S., & Shaw, J.-F. (2020). Enhancement of the stability of chlorophyll using chlorophyll-encapsulated polycaprolactone microparticles based on droplet microfluidics. *Food Chemistry*, *306*, 125300.

Humphrey, A. (1980). Chlorophyll. *Food Chemistry*, *5*(1), 57–67.

Humphrey, A. (2004). Chlorophyll as a color and functional ingredient. *Journal of Food Science*, *69*(5), C422–C425.

İlter, I., Akyıl, S., Demirel, Z., Koç, M., Conk-Dalay, M., & Kaymak-Ertekin, F. (2018). Optimization of phycocyanin extraction from Spirulina platensis using different techniques. *Journal of Food Composition and Analysis*, *70*, 78–88.

Juin, C., Chérouvrier, J.-R., Thiéry, V., Gagez, A.-L., Bérard, J.-B., Joguet, N., & Picot, L. (2015). Microwave-assisted extraction of phycobiliproteins from Porphyridium purpureum. *Applied biochemistry biotechnology*, *175*(1), 1–15.

Kang, Y.-R., Lee, Y.-K., Kim, Y. J., & Chang, Y. H. (2019). Characterization and storage stability of chlorophylls microencapsulated in different combination of gum Arabic and maltodextrin. *Food Chemistry*, *272*, 337–346.

Kannaujiya, V. K., Sundaram, S., & Sinha, R. P. (2017). *Phycobiliproteins: Recent Developments and Future Applications*. Springer.

Khan, S. A., Yusuf, M., Agarwal, P., & Prasad, L. (2018). Chlorophylls as pigment: a contemporary approach. In M. Yusuf (Ed.), *Handbook of Renewable Materials for Coloration and Finishing* (pp. 115–124). NJ, USA: Wiley.

Kong, W., Liu, N., Zhang, J., Yang, Q., Hua, S., Song, H., & Xia, C. (2014). Optimization of ultrasound-assisted extraction parameters of chlorophyll from Chlorella vulgaris residue after lipid separation using response surface methodology. *Journal of Food Science and Technology*, *51*(9), 2006–2013.

Kozelová, D., Fikselová, M., Dodoková, S., Mura, L., Mendelová, A., & Vietoris, V. (2013). Analysis of consumer preferences focused on food additives. *Acta universitatis Agriculturae et Silviculturae Mendelianae Brunensis*, *60*(6), 197–204.

Lafarga, T. (2019a). Cultured microalgae and compounds derived thereof for food applications: strain selection and cultivation, drying, and processing strategies. *Food Reviews International*, 1–26.

Lafarga, T. (2019b). Effect of microalgal biomass incorporation into foods: nutritional and sensorial attributes of the end products. *Algal Research*, *41*, 101566. Available from https://doi.org/10.1016/j.algal.2019.101566.

Lafarga, T., Acién-Fernández, F. G., Castellari, M., Villaró, S., Bobo, G., & Aguiló-Aguayo, I. (2019). Effect of microalgae incorporation on the physicochemical, nutritional, and sensorial properties of an innovative broccoli soup. *LWT*, *111*, 167–174.

Lafarga, T., Acién-Fernández, F. G., & Garcia-Vaquero, M. (2020). Bioactive peptides and carbohydrates from seaweed for food applications: natural occurrence, isolation, purification, and identification. *Algal Research, 48*, 101909.

Lafarga, T., Clemente, I., & García-Vaquero, M. (2019). Carotenoids from microalgae. In C. Galanakis (Ed.), *Carotenoids: Properties, Processing and Applications* (pp. 149–188). India: Elsevier.

Lafarga, T., Mayre, E., Echeverria, G., Viñas, I., Villaró, S., Acién-Fernández, F. G., & Aguiló-Aguayo, I. (2019). Potential of the microalgae Nannochloropsis and Tetraselmis for being used as innovative ingredients in baked goods. *LWT, 115*, 108439.

Lafarga, T., Rodríguez-Bermúdez, R., Morillas-España, A., Villaró, S., García-Vaquero, M., Morán, L., ... Acién-Fernández, F. G. (2021). Consumer knowledge and attitudes towards microalgae as food: The case of Spain. *Algal Research, 54*, 102174.

Lanfer-Marquez, U. M., Barros, R. M. C., & Sinnecker, P. (2005). Antioxidant activity of chlorophylls and their derivatives. *Food Research International, 38*(8), 885–891.

Lauceri, R., Chini Zittelli, G., & Torzillo, G. (2019). A simple method for rapid purification of phycobiliproteins from Arthrospira platensis and Porphyridium cruentum biomass. *Algal Research, 44*, 101685.

Li, J., Zhu, D., Niu, J., Shen, S., & Wang, G. (2011). An economic assessment of astaxanthin production by large scale cultivation of Haematococcus pluvialis. *Biotechnology Advances, 29*(6), 568–574.

Liang, M.-H., Wang, L., Wang, Q., Zhu, J., & Jiang, J.-G. (2019). High-value bioproducts from microalgae: strategies and progress. *Critical Reviews in Food Science and Nutrition, 59*(15), 2423–2441.

Lila, M. A., Burton-Freeman, B., Grace, M., & Kalt, W. (2016). Unraveling anthocyanin bioavailability for human health. *Annual Review of Food Science and Technology, 7*, 375–393.

Luengo, E., Condón-Abanto, S., Álvarez, I., & Raso, J. (2014). Effect of pulsed electric field treatments on permeabilization and extraction of pigments from Chlorella vulgaris. *The Journal of Membrane Biology, 247*(12), 1269–1277.

Lv, H., Cui, X., Wahid, F., Xia, F., Zhong, C., & Jia, S. (2016). Analysis of the physiological and molecular responses of Dunaliella salina to macronutrient deprivation. *PLoS ONE, 11*(3).

Macías-Sánchez, M., Mantell, C., Rodriguez, M., de la Ossa, E. M., Lubián, L., & Montero, O. (2009). Comparison of supercritical fluid and ultrasound-assisted extraction of carotenoids and chlorophyll a from Dunaliella salina. *Talanta, 77*(3), 948–952.

Macías-Sánchez, M. D., Mantell, C., Rodríguez, M., Martínez de la Ossa, E., Lubián, L. M., & Montero, O. (2007). Supercritical fluid extraction of carotenoids and chlorophyll a from Synechococcus sp. *The Journal of Supercritical Fluids, 39*(3), 323–329.

Manayi, A., Abdollahi, M., Raman, T., Nabavi, S. F., Habtemariam, S., Daglia, M., & Nabavi, S. M. (2016). Lutein and cataract: from bench to bedside. *Critical Reviews in Biotechnology, 36*(5), 829–839.

Martínez, J. M., Luengo, E., Saldaña, G., Álvarez, I., & Raso, J. (2017). C-phycocyanin extraction assisted by pulsed electric field from Artrosphira platensis. *Food Research International, 99*, 1042–1047.

Martins, N., Roriz, C. L., Morales, P., Barros, L., & Ferreira, I. C. (2016). Food colorants: Challenges, opportunities and current desires of agro-industries to ensure consumer expectations and regulatory practices. *Trends in Food Science and Technology, 52*, 1–15.

Matos, Â. P. (2017). The impact of microalgae in food science and technology. *Journal of the American Oil Chemists' Society, 94*(11), 1333–1350.

Nanba, K., Toyooka, S., Soh, J., Tsukuda, K., Yamamoto, H., Sakai, A., & Koide, N. (2008). The allelic distribution of a single nucleotide polymorphism in the PDCD5 gene locus of Japanese non-small cell lung cancer patients. *Molecular Medicine Reports*, *1*(5), 667–671.

Nisar, N., Li, L., Lu, S., Khin, N. C., & Pogson, B. J. (2015). Carotenoid metabolism in plants. *Molecular Plant*, *8*(1), 68–82.

Noreña-Caro, D., & Benton, M. G. (2018). Cyanobacteria as photoautotrophic biofactories of high-value chemicals. *Journal of CO2 Utilization*, *28*, 335–366.

Oplatowska-Stachowiak, M., & Elliott, C. T. (2017). Food colors: existing and emerging food safety concerns. *Critical Reviews in Food Science and Nutrition*, *57*(3), 524–548.

Panis, G., & Carreon, J. R. (2016). Commercial astaxanthin production derived by green alga Haematococcus pluvialis: a microalgae process model and a techno-economic assessment all through production line. *Algal Research*, *18*, 175–190.

Papadaki, S., Kyriakopoulou, K., Tzovenis, I., & Krokida, M. (2017). Environmental impact of phycocyanin recovery from Spirulina platensis cyanobacterium. *Innovative Food Science & Emerging Technologies*, *44*, 217–223.

Park, J. S., Chyun, J. H., Kim, Y. K., Line, L. L., & Chew, B. P. (2010). Astaxanthin decreased oxidative stress and inflammation and enhanced immune response in humans. *Nutrition Metabolism*, *7*(1), 18.

Parniakov, O., Barba, F. J., Grimi, N., Marchal, L., Jubeau, S., Lebovka, N., & Vorobiev, E. (2015). Pulsed electric field and pH assisted selective extraction of intracellular components from microalgae Nannochloropsis. *Algal Research*, *8*, 128–134.

Pashkow, F. J., Watumull, D. G., & Campbell, C. L. (2008). Astaxanthin: a novel potential treatment for oxidative stress and inflammation in cardiovascular disease. *The American Journal of Cardiology*, *101*(10), S58–S68.

Pasquet, V., Chérouvrier, J.-R., Farhat, F., Thiéry, V., Piot, J.-M., Bérard, J.-B., & Picot, L. (2011). Study on the microalgal pigments extraction process: performance of microwave assisted extraction. *Process Biochemistry*, *46*(1), 59–67.

Pattarayan, D., Rajarajan, D., Sivanantham, A., Palanichamy, R., & Rajasekaran, S. (2017). C-phycocyanin suppresses transforming growth factor-β1-induced epithelial mesenchymal transition in human epithelial cells. *Pharmacological Reports*, *69*(3), 426–431.

Pavón-Fuentes, N., Marín-Prida, J., Llópiz-Arzuaga, A., Falcón-Cama, V., Campos-Mojena, R., Cervantes-Llanos, M., & Pentón-Rol, G. (2020). Phycocyanobilin reduces brain injury after endothelin-1-induced focal cerebral ischaemia. *Clinical Experimental Pharmacology Physiology*, *47*(3), 383–392.

Petrushkina, M., Gusev, E., Sorokin, B., Zotko, N., Mamaeva, A., Filimonova, A., & Guglya, E. (2017). Fucoxanthin production by heterokont microalgae. *Algal Research*, *24*, 387–393.

Queiroz Zepka, L., Jacob-Lopes, E., & Roca, M. (2019). Catabolism and bioactive properties of chlorophylls. *Current Opinion in Food Science*, *26*, 94–100.

Rastogi, R. P., Sonani, R. R., & Madamwar, D. (2015). Physico-chemical factors affecting the in vitro stability of phycobiliproteins from Phormidium rubidum A09DM. *Bioresource Technology*, *190*, 219–226.

Ribeiro, B. D., Barreto, D. W., & Coelho, M. A. Z. (2011). Technological aspects of β-carotene production. *Food and Bioprocess Technology*, *4*(5), 693–701.

Rodriguez-Amaya, D. B. (2016). Natural food pigments and colorants. *Current Opinion in Food Science*, *7*, 20–26.

Rodríguez-Sánchez, R., Ortiz-Butrón, R., Blas-Valdivia, V., Hernández-García, A., & Cano-Europa, E. (2012). Phycobiliproteins or C-phycocyanin of Arthrospira (Spirulina) maxima protect against HgCl2-caused oxidative stress and renal damage. *Food Chemistry*, *135*(4), 2359–2365.

Sala, L., Moraes, C. C., & Kalil, S. J. (2018). Cell pretreatment with ethylenediaminetetraacetic acid for selective extraction of C-phycocyanin with food grade purity. *Biotechnology Progress*, *34*(5), 1261–1268.

Sathasivam, R., & Ki, J.-S. (2018). A review of the biological activities of microalgal carotenoids and their potential use in healthcare and cosmetic industries. *Marine Drugs*, *16*(1), 26.

Schumann, R., Häubner, N., Klausch, S., & Karsten, U. (2005). Chlorophyll extraction methods for the quantification of green microalgae colonizing building facades. *International Biodeterioration Biodegradation*, *55*(3), 213–222.

Scotter, M. (2015). *Overview of EU regulations and safety assessment for food colours. Colour Additives for Foods and Beverages* (pp. 61–74). Elsevier.

Scotter, M. J. (2011). Methods for the determination of European Union-permitted added natural colours in foods: a review. *Food Additives and Contaminants*, *28*(5), 527–596.

Sekar, S., & Chandramohan, M. (2008). Phycobiliproteins as a commodity: trends in applied research, patents and commercialization. *Journal of Applied Phycology*, *20*(2), 113–136.

Sen, T., Barrow, C. J., & Deshmukh, S. K. (2019). Microbial pigments in the food industry—challenges and the way forward. *Frontiers in Nutrition*, *6*.

Shah, M., Mahfuzur, R., Liang, Y., Cheng, J. J., & Daroch, M. (2016). Astaxanthin-producing green microalga Haematococcus pluvialis: from single cell to high value commercial products. *Frontiers in Plant Science*, *7*, 531.

Shaish, A., Harari, A., Hananshvili, L., Cohen, H., Bitzur, R., Luvish, T., & Gavish, D. (2006). 9-cis β-carotene-rich powder of the alga Dunaliella bardawil increases plasma HDL-cholesterol in fibrate-treated patients. *Atherosclerosis*, *189*(1), 215–221.

Simon, D., & Helliwell, S. (1998). Extraction and quantification of chlorophyll a from freshwater green algae. *Water Research*, *32*(7), 2220–2223.

Singh, D. P., Khattar, J. S., Rajput, A., Chaudhary, R., & Singh, R. (2019). High production of carotenoids by the green microalga Asterarcys quadricellulare PUMCC 5.1. 1 under optimized culture conditions. *PLoS ONE*, *14*(9).

Sonani, R. R., Rastogi, R. P., Patel, R., & Madamwar, D. (2016). Recent advances in production, purification and applications of phycobiliproteins. *World Journal of Biological Chemistry*, *7* (1), 100.

Sonani, R. R., Singh, N. K., Kumar, J., Thakar, D., & Madamwar, D. (2014). Concurrent purification and antioxidant activity of phycobiliproteins from Lyngbya sp. A09DM: an antioxidant and anti-aging potential of phycoerythrin in Caenorhabditis elegans. *Process Biochemistry*, *49*(10), 1757–1766.

Srinivasan, M., Devipriya, N., Kalpana, K., & Menon, V. P. (2009). Lycopene: an antioxidant and radioprotector against γ-radiation-induced cellular damages in cultured human lymphocytes. *Toxicology*, *262*(1), 43–49.

Sudhakar, M. P., Jagatheesan, A., Perumal, K., & Arunkumar, K. (2015). Methods of phycobiliprotein extraction from Gracilaria crassa and its applications in food colourants. *Algal Research*, *8*, 115–120.

Taucher, J., Baer, S., Schwerna, P., Hofmann, D., Hümmer, M., Buchholz, R., & Becker, A. (2016). Cell disruption and pressurized liquid extraction of carotenoids from microalgae. *Journal of Thermodynamics and Catalysis*, *7*, 158–165.

Tavanandi, H. A., Devi, A. C., & Raghavarao, K. (2018). A newer approach for the primary extraction of allophycocyanin with high purity and yield from dry biomass of Arthrospira platensis. *Separation and Purification Technology*, *204*, 162–174.

Tavanandi, H. A., & Raghavarao, K. S. M. S. (2020). Ultrasound-assisted enzymatic extraction of natural food colorant C-Phycocyanin from dry biomass of Arthrospira platensis. *LWT*, *118*, 108802.

Turkmen, N., Poyrazoglu, E. S., Sari, F., & Sedat Velioglu, Y. (2006). Effects of cooking methods on chlorophylls, pheophytins and colour of selected green vegetables. *International Journal of Food Science and Technology*, *41*(3), 281–288.

Varela, J. C., Pereira, H., Vila, M., & León, R. (2015). Production of carotenoids by microalgae: achievements and challenges. *Photosynthesis Research*, *125*(3), 423–436.

VijayaPadma, V., Ramyaa, P., Pavithra, D., & Krishnasamy, R. (2014). Protective effect of lutein against benzo (a) pyrene-induced oxidative stress in human erythrocytes. *Toxicology Industrial Health*, *30*(3), 284–293.

Virtamo, J., Taylor, P. R., Kontto, J., Männistö, S., Utriainen, M., Weinstein, S. J., & Albanes, D. (2014). Effects of α-tocopherol and β-carotene supplementation on cancer incidence and mortality: 18-year postintervention follow-up of the alpha-tocopherol, beta-carotene cancer prevention study. *International Journal of Cancer*, *135*(1), 178–185.

Viuda-Martos, M., Sanchez-Zapata, E., Sayas-Barberá, E., Sendra, E., Pérez-Álvarez, J., & Fernández-López, J. (2014). Tomato and tomato byproducts. Human health benefits of lycopene and its application to meat products: a review. *Critical Reviews in Food Science and Nutrition*, *54*(8), 1032–1049.

Wayama, M., Ota, S., Matsuura, H., Nango, N., Hirata, A., & Kawano, S. (2013). Three-dimensional ultrastructural study of oil and astaxanthin accumulation during encystment in the green alga *Haematococcus pluvialis*. *PloS one*, *8*(1), e53618.

Ye, Z.-W., Jiang, J.-G., & Wu, G.-H. (2008). Biosynthesis and regulation of carotenoids in Dunaliella: progresses and prospects. *Biotechnology Advances*, *26*(4), 352–360.

Yu, P., Wu, Y., Wang, G., Jia, T., & Zhang, Y. (2017). Purification and bioactivities of phycocyanin. *Critical Reviews in Food Science and Nutrition*, *57*(18), 3840–3849.

Yuan, Y. V. (2007). Antioxidants from edible seaweeds. In F. Shahidi, & C. T. Ho (Eds.), *Antioxidant Measurement and Applications* (pp. 268–301). NJ, USA: American Chemical Society.

Zhang, W., Wang, J., Wang, J., & Liu, T. (2014). Attached cultivation of Haematococcus pluvialis for astaxanthin production. *Bioresource Technology*, *158*, 329–335.

Chapter 8

Current utilization of microalgae in the food industry beyond direct human consumption

Ana F. Esteves, José C.M. Pires and Ana L. Gonçalves

LEPABE—Laboratory for Process Engineering, Environment, Biotechnology and Energy, Faculty of Engineering, University of Porto, Porto, Portugal

8.1 Introduction

The global population has grown rapidly in recent decades (from six billion in 1999 to seven billion in 2011), and it is estimated that it will increase to nine billion by 2050 (Pathak et al., 2018; Singh et al., 2017). This increase is associated with a higher demand for several resources, especially in terms of food supplies (Chiaiese, Corrado, Colla, Kyriacou, & Rouphael, 2018; Pathak et al., 2018; Ronga et al., 2019; Singh et al., 2017), with impacts on different fields of the food industry (Lum, Kim, & Lei, 2013; Pathak et al., 2018; Porwal, Mane, & Velhal, 2015; Ronga et al., 2019), including: (1) an increase in the volumes of water processed by food industries and agro-industrial practices, and hence, an increase in the volumes of produced wastewater; (2) the necessity of improving agricultural productivities and developing sustainable agricultural practices; and (3) the need to enhance the production of livestock for human food. In the case of higher volumes of produced wastewater, effective and sustainable methods to treat these effluents are required. On the other hand, to improve agricultural practices, it is important to find biologically based substances that can improve soil quality and fertility and also crop productivities, to replace the commonly used synthetic fertilizers. Finally, to increase livestock productivities, it is essential to find alternative feed sources, to avoid the competition between crops for food and feed production.

Although these challenges seem to be quite different, they can be addressed using a similar statergy: the use of microalgal cultures. Microalgae are photosynthetic microorganisms that, due to their simple cellular structure, are easy to cultivate and are capable of being produced quickly, even under unfavorable conditions (Buono et al., 2016). The use of these microorganisms have numerous advantages (Goh et al., 2019; Mobin & Alam, 2017): (1) being

Cultured Microalgae for the Food Industry. DOI: https://doi.org/10.1016/B978-0-12-821080-2.00005-8

photosynthetic organisms, microalgae assimilate CO_2 as a carbon source, contributing to a reduction in the concentration of this greenhouse gas from the atmosphere or industrial gaseous emissions; (2) microalgae require nitrogen and phosphorus (typically found in wastewaters from different sources) as nutrients, indicating their ability to promote nutrient removal from these effluents; (3) microalgae can grow on non-arable land, thus not competing with the area available for food crop production; and (4) microalgal biochemical composition is very rich in diversified metabolites (e.g., carbohydrates, lipids, proteins, pigments, antioxidants, vitamins, and essential minerals), which make microalgae a valuable resource for application in agriculture or in the food, cosmetics, pharmaceutical, and biofuel industries. An overview of the main applications described for microalgae and microalgal biomass is presented in Fig. 8.1.

Considering the above-mentioned advantages, the use of microalgal cultures to respond to the three challenges highlighted as a result of the worldwide population increase becomes clear. Through their ability to assimilate nitrogen and phosphorus from wastewaters, microalgae can be applied in the tertiary treatment step of food industry and agro-industrial effluents, with the additional possibility to recover the produced biomass for further applications (Shahid et al., 2020). The unique properties of microalgae (ability to fix nitrogen, and high nitrogen, phosphorus, and mineral contents) and microalgal metabolites (antioxidant, antimicrobial, and structural activity) confer microalgae the ability to act as biofertilizers and/or biostimulants in food crops, thus improving agricultural productivities (Pan, Jeevanandam, & Danquah, 2019; Renuka, Guldhe, Prasanna, Singh, & Bux, 2018). Finally, several microalgal strains present high protein content (above 50% w/w) (Jacob-Lopes et al., 2019) and are an important source of vitamins B, C, and E (Khan, Shin, & Kim, 2018). They can be

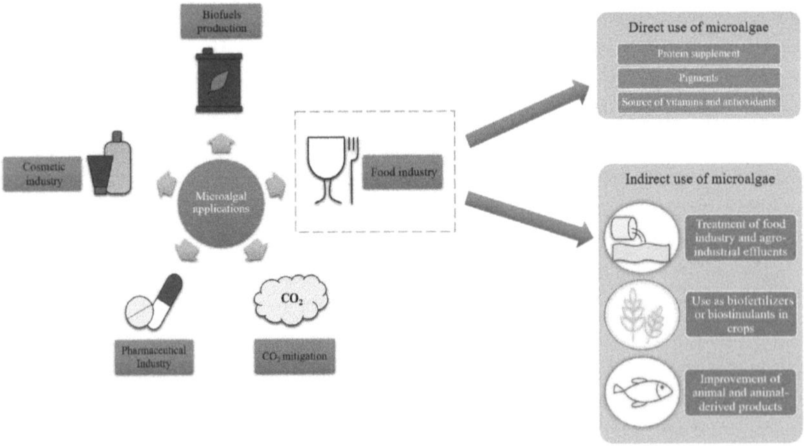

FIGURE 8.1 Main applications described for microalgae, with emphasis on those beyond direct human consumption.

used to enhance the nutritional values of feed products, thus improving animal and animal-derived products and contributing to an increase in livestock production for human food. This chapter presents in detail how microalgal cultures can address these challenges, with a focus on the most recent studies carried out in this field.

8.2 Microalgae in the treatment of food industry and agro-industrial effluents

Food industry and agro-industrial effluents have a high load of nutrients (nitrogen and phosphorus) and, when not properly treated, their discharge increases the concentrations of those nutrients in watercourses. As a result, the eutrophication phenomenon may occur, leading to the loss of certain aquatic species and total degradation of these ecosystems (Ruiz et al., 2013). To avoid this problem, wastewaters containing high levels of nitrogen and phosphorus should be treated before their discharge. Wastewater treatment processes have four main phases: preliminary, primary, secondary, and tertiary treatment. In the initial step, solids, oils, and fats are removed. Then, primary treatment aims at removing suspended solids through physical processes. In the secondary treatment, the dissolved organic matter is removed through biological processes. Finally, tertiary treatment is used to remove nutrients that are in excess, which is the case of nitrogen and phosphorus, and pathogens (Gonçalves, Pires, & Simões, 2017; Umamaheswari & Shanthakumar, 2016). Chemical or physical methods are commonly used to remove the nutrients in food industry wastewaters. However, these methodologies have some disadvantages like high operating costs and the use of chemicals during the process (S. Li et al., 2019). On the other hand, nutrients removal using microalgae provides significant benefits (Elshobary et al., 2019; K. Li et al., 2019; S. Li et al., 2019; Molinuevo-Salces, Riaño, Hernández, & García-González, 2019; Nur & Buma, 2018; Umamaheswari & Shanthakumar, 2016; Riaño, Molinuevo, & García-González, 2011): (1) microalgae can effectively remove these nutrients from the liquid effluent; (2) the biomass produced can be further exploited for other applications, thus not being considered sludge from the treatment process; and (3) due to their ability to fix CO_2, coupling microalgal cultures in the tertiary step may result in a decrease in the overall CO_2 emissions of conventional wastewater treatment plants.

8.2.1 Characterization of food industry and agro-industrial effluents

The food industry effluents are commonly characterized by: (1) high concentration of suspended solids, biological oxygen demand, and chemical oxygen demand (COD); (2) the presence of fats, oils, and nutrients, such as nitrogen and phosphorus; and (3) low toxicity (Ramjeawon, 2000). However, the effluent characteristics and volumes produced per day are industry-specific and can

TABLE 8.1 Typical compositions of different food industry and agro-industrial effluents in terms of nitrogen, phosphorus, and organic carbon.

Wastewater type	Nitrogen (mg L^{-1})	Phosphorus (mg L^{-1})	Organic carbon (mg L^{-1})	References
Potato processing	69 TN 12–190 NH_4-N	4–6 TP 103 PO_4-P	327 TOC 745–37,000 COD	Hernández, Riaño, Coca, and García-González (2013), Muniraj, Xiao, Hu, Zhan, and Shi (2013), Posadas et al. (2014)
Starch processing	204–1042 TN 49–115 NH_4-N	50–385 TP	2470–15440 COD	Sklyar, Epov, Gladchenko, Danilovich, and Kalyuzhnyi (2003)
Fish processing	82 TN 9–17 NH_4-N	3–11 TP	381 TOC 286–541 COD	Posadas et al. (2014), Riaño et al. (2011)
Aquaculture	1–1024 TN 1–102 NH_4-N	1–459 TP	18–2494 COD	Boopathy, Bonvillain, Fontenot, and Kilgen (2007), Gao et al. (2016), Han, Lu, Fan, and Zhou (2019), Schulz, Gelbrecht, and Rennert (2004)
Rice culture	25–95 TN	12–94 PO_4-P	2578–6480 COD	Queiroz, Lopes, Zepka, Bastos, and Goldbeck (2007)
Piggery	1388 NH_4-N	158 TP	11,000 COD	Wang et al. (2012)
Animal feed production	197 TN 189 NH_4-N	27 TP	959 TOC	Posadas et al. (2014)
Soybean processing	108–2550 TN 52–270 NH_4-N	56–183 TP	182–22,700 COD	Hongyang, Yalei, Chunmin, Xuefei, and Jinpeng (2011), Li et al. (2014), Qiu et al. (2019), Zhu, Li, Wu, Jin, and Wang (2008)

Processed food products	146 TN 112 NH_4-N 42 NO_3-N	51 TP 24 PO_4	5040–6720 COD	Gupta and Pawar (2018), Gupta, Pandey, and Pawar (2016)
Processed food products (anaerobically digested)	178 TN 168 NH_4-N	23 TP	139 COD	Gupta and Pawar (2018)
Dairy	176–283 TN 121–277 NH_4-N	10–116 TP 6 PO_4	2000–6000 COD	Chokshi, Pancha, Ghosh, and Mishra (2016), Karadag, Köroğlu, Ozkaya, and Cakmakci (2015), W. Lu, Wang, Wang, and Yuan (2015), Qin et al. (2016)
Coffee manufacturing	766 TN 101 NH_4-N	59 TP	8532 TOC	Posadas et al. (2014)
Brewery	25–450 TN 3–103 NH_4-N 2–11 NO_3-N	1–216 TP 2–35 PO_4-P 57–326 PO_4	565–32,500 COD	Arantes, Alves, Sequinel, and Da Silva (2017), Raposo, Oliveira, Castro, Bandarra, and Morais (2010)
Meat processing	49–328 TN 2–193 NO_3-N	6–70 TP	734–6720 COD	Q. Lu et al. (2015), Sroka, Kamiński, and Bohdziewicz (2004)
Yeast production (anaerobically digested)	703 TN 656 NH_4-N	7 TP	1186 TOC	Posadas et al. (2014)
Oil mill	0.2 TN	n.s.	51,000 COD	Sampaio, Gonçalves, and Marques (2011)

TN, total nitrogen (mg N L^{-1}); NH_4-N, ammonium-nitrogen (mg N L^{-1}); NO_3-N, nitrate-nitrogen (mg N L^{-1}); TP, total phosphorus (mg P L^{-1}); PO_4-P, phosphate-phosphorus (mg P L^{-1}); PO_4, phosphate (mg PO_4 L^{-1}); TOC, total organic carbon (mg C L^{-1}); COD, chemical oxygen demand (mg O_2 L^{-1}); n.s., not specified.

fluctuate throughout the year (Posadas et al., 2014). Table 8.1 shows the typical compositions of different food industry and agro-industrial effluents. In general, wastewaters resulting from the piggery, starch, animal feed, soybean, coffee, and brewery industries have high concentrations of nitrogen and phosphorus. The industries related to oil mills, starch, rice, piggeries, soybean, dairy, coffee, and meat processing present high COD levels. Microalgae can be used to treat raw and anaerobically digested wastewater (Gupta, Pawar, & Pandey, 2019). Food industry effluents are commonly treated through anaerobic digestion to stabilize the organic matter present in the effluent, with most of the organic carbon being transformed into methane. However, even after this process, the wastewater still needs additional treatments, due to the high levels of nutrients that remain in the water (Gupta & Pawar, 2018). Furthermore, some effluents present high levels of ammonia, which can be prejudicial to microalgal growth. In these cases, the effluent cannot be directly used for microalgal growth, and so it is common to dilute it to obtain non-toxic ammonia levels (Wang, Xiong, Hui, & Zeng, 2012).

8.2.2 Role of microalgae and microalgal consortia in the treatment of food industry and agro-industrial effluents

Microalgae need nutrients for their growth, mainly nitrogen, phosphorus, and carbon (as well as other micronutrients). These nutrients are present in high levels in food industry and agro-industrial wastewaters, so it is possible to supply the main nutrients required by microalgae through cultivation in these wastewaters (Shahid et al., 2020). Moreover, these effluents are adequate for microalgal growth, due to their low toxicity (S. Li et al., 2019).

Microalgae can treat raw and anaerobically digested food industry wastewaters, but in some cases, these effluents cannot be directly used in microalgal cultivation due to their composition, requiring a pretreatment step. The most common pretreatment methods include filtration, dilution, centrifugation, autoclaving, or anaerobic digestion (Gupta et al., 2019). Microalgae can be used in wastewater treatment as monocultures or microalgae and microalgae–bacteria consortia. There are some advantages in using microalgal consortia instead of monocultures, such as the cooperative interactions established within the consortia, that typically result in higher productivities and contamination resistance (Mendes & Vermelho, 2013). These cooperative interactions are more pronounced in microalgae–bacteria consortia, positively affecting wastewater treatment processes (Fito & Alemu, 2019). When growing autotrophically, microalgae need CO_2 and nutrients to produce O_2 through photosynthesis. At the same time, bacteria use the O_2 provided by microalgae for their metabolism, degrade the organic matter present in wastewater and produce the CO_2 and nutrients by microalgae. In these systems, it is possible to remove simultaneously organic carbon, nitrogen, and phosphorus from the wastewater (Gonçalves et al., 2017; Mujtaba & Lee, 2017; Zhu et al., 2019). These systems can also enhance microalgal bioflocculation, which makes the harvesting and further biomass exploitation easier and cheaper (K. Li et al., 2019).

Several studies have focused on the use of microalgae-bacteria and microalgal consortia for nitrogen and phosphorus removal from food industry and agro-industrial wastewaters (Table 8.2). For example, Daneshvar et al. (2019) determined the efficiency of *Scenedesmus quadricauda*, a freshwater microalga, and *Tetraselmis suecica*, a marine microalga, on nutrients removal from a dairy wastewater. In these experiments, two cultivation modes were tested: heterotrophic, in the absence of light, and mixotrophic, in the presence of light. Microalgae were cultivated during 12 d with a light intensity of 90 μmol m^{-2} s^{-1} and a photoperiod of 12:12. The working volume used was 1 L and, during the experiment, the pH was maintained at 6.30 and temperature at 22°C. For *T. suecica* cultivation, sea salt was added to the dairy wastewater. The authors reported a total nitrogen removal of 86.7% (for *S. quadricauda*) and 66.8% (for *T. suecica*), in mixotrophic conditions, and 46.7% (for *S. quadricauda*) and 44.9% (for *T. suecica*), in heterotrophic conditions. Phosphate removal reported in mixotrophic cultivation was 71.2% and 41.6% for *S. quadricauda* and *T. suecica*, respectively, and in heterotrophic cultivation, there was no phosphate removal. The total organic carbon (TOC) removal efficiency was also evaluated, and the highest values were obtained by *S. quadricauda* in mixotrophic and heterotrophic modes (69.1% and 76.4%, respectively). Gao et al. (2018) evaluated the growth and nutrient removal of *Chlorella* sp. in seafood processing wastewater with different pretreatments (raw, diluted, and aerated effluents were used). Dilution of the wastewater was applied to lower the ammonia concentration, and the goal of aeration was to convert part of the ammonia into nitrite or nitrate through nitrification, also aiming at reducing the levels of ammonia in the wastewater. The experiments were performed in batch mode for 45 d, with a working volume of 0.5 L. Temperature was set at 25°C and light was continuously supplied at an intensity of 135 μmol m^{-2} s^{-1}. The authors reported that *Chlorella* sp. had the highest specific growth rate and biomass productivity in the aerated wastewater (0.156 d^{-1} and 77.7 mg L^{-1} d^{-1}, respectively) and the lowest in the raw effluent (0.040 d^{-1} and 20.4 mg L^{-1} d^{-1}, respectively). The nitrogen removal rate was higher (4.98 mg L^{-1} d^{-1}) in the aerated effluent and lower (2.26 mg L^{-1} d^{-1}) in the diluted one. *Chlorella* sp. achieved the highest phosphorus removal rate (1.91 mg L^{-1} d^{-1}) in the aerated effluent and the lower rate (0.89 mg L^{-1} d^{-1}) in the raw wastewater. Additionally, the authors evaluated lipid production, concluding that in the aerated effluent, the lipid productivity was the highest, reaching 20.4 mg L^{-1} d^{-1}, which is 3.5-fold higher than the value obtained in the raw effluent.

TABLE 8.2 Microalgal efficiency in nutrients removal (nitrogen, phosphorus, and organic carbon) from food industry and agro-industrial wastewaters.

Effluent	Microorganisms	Operational conditions	Nitrogen		Phosphorus		Organic carbon		References
			C_i (mg L^{-1})	RE (%)	C_i (mg L^{-1})	RE (%)	C_i (mg L^{-1})	RE (%)	
Piggery	*Chlorella sorokiniana* and activated sludge native bacteria	OM: continuous HRT: 90 d LI: 10,000 lux L:D: 24:0 V: 7.5 L	656 NH_4-N	94–100	117 PO_4-P	70–90	1247 TOC	44	De Godos, González, Becares, García-Encina, and Muñoz (2009)
	Desmodesmus sp.	OM: batch HRT: 7 d LI: 100–200 μmol m^{-2} s^{-1} T: 15°C–35°C V: 0.6 L	112.1 NH_4-N	78.46	3.15 TP	91.66	n.s.	n.s.	Luo, Shao, Luo, Zeng, and Tian (2019)
Piggery (anaerobically digested)	*C. sorokiniana* and activated sludge native bacteria	OM: semicontinuous HRT: 10 d LI: 6000 lux L:D: 12:12 T: 24.1°C V: 5 L	12.3 NH_4-N	82.7	47.5 SP	58.0	465 COD	58.1	Hernández et al. (2013)
	Chlamydomonas mexicana	OM: batch HRT: 20 d pH: 8.4 LI: 40 μmol m^{-2} s^{-1} L:D: 24:0 T: 27°C V: 0.25 L	56 TN	60–65	13.5 TP	60–65	n.s.	n.s.	Abou-Shanab, Ji, Kim, Paeng, and Jeon (2013)

	Chlorella vulgaris	OM: batch HRT: 20 d pH: 8.4 LI: 40 μmol m^{-2} s^{-1} L:D: 24:0 T: 27°C V: 0.25 L	56 TN	50	13.5 TP	50–60	n.s.	n.s.	Abou-Shanab et al. (2013)
Processed food products	*C. sorokiniana*	OM: batch HRT: 8 d pH: 7 LI: 80 μmol m^{-2} s^{-1} L:D: 10:14 T: 29°C V: 0.5 L	30 TN	100	n.s.	n.s.	1500 COD	92	Gupta and Pawar (2018)
	Scenedesmus obliquus	OM: batch HRT: 8 d pH: 7 LI: 80 μmol m^{-2} s^{-1} L:D: 10:14 T: 29°C V: 0.5 L	30 TN	100	n.s.	n.s.	1500 COD	88.4	Gupta and Pawar (2018)
Processed food products (anaerobically digested)	*C. sorokiniana*	OM: batch HRT: 8 d pH: 7 LI: 80 μmol m^{-2} s^{-1} L:D: 10:14 T: 29°C V: 0.5 L	172 TN	49.1	n.s.	n.s.	128 COD	32.8	Gupta and Pawar (2018)
	S. obliquus	OM: batch HRT: 8 d pH: 7 LI: 80 μmol m^{-2} s^{-1}	172 TN	44.3	n.s.	n.s.	128 COD	39.8	Gupta and Pawar (2018)

(Continued)

TABLE 8.2 (Continued)

Effluent	Microorganisms	Operational conditions	Nitrogen		Phosphorus		Organic carbon		References
			C_i (mg L^{-1})	RE (%)	C_i (mg L^{-1})	RE (%)	C_i (mg L^{-1})	RE (%)	
		L:D: 10:14 T: 29°C V: 0.5 L							
Soybean processing	*Chlorella* sp.	OM: batch HRT: 12 d LI: 6000 lux T: 22°C V: 0.25 L	95 TN	60–80	12 TP	90–95	2270 COD	55–70	Qiu et al. (2019)
Potato processing	*C. sorokiniana* and activated sludge native bacteria	OM: semicontinuous HRT: 10 d LI: 6000 lux L:D: 12:12 T: 24.1°C V: 5 L	12.1 NH_4-N	≥95	3.4 SP	80.7	745 COD	86.1	Hernández et al. (2013)
Fish processing	*Oocystis* sp.	OM: batch HRT: 5 d pH: 8.2–9.2 LI: 12,000 lux L:D: 24:0 T: 23°C–31°C V: 3 L	17.3 NH_4-N	>95	6.1 SP	71–74	541 COD	69–71	Riaño et al. (2011)

Brewery	*S. obliquus*	OM: batch HRT: 14 d pH: 6.5 LI: 4500 lux L:D: 12:12 T: 30°C V: 0.25 L	54 TN	15.2	n.s.	n.s.	3635 COD	33.9	Mata, Melo, Simões, and Caetano (2012)
		OM: batch HRT: 14 d pH: 6.5 LI: 4500 lux L:D: 24:0 T: 30°C V: 0.25 L	54 TN	11.0	n.s.	n.s.	3635 COD	66.8	Mata et al. (2012)
		OM: batch HRT: 14 d pH: 6.5 LI: 12,000 lux L:D: 12:12 T: 30°C V: 0.25 L	54 TN	20.8	n.s.	n.s.	3635 COD	57.5	Mata et al. (2012)
		OM: batch HRT: 14 d pH: 6.5 LI: 12,000 lux	54 TN	11.6	n.s.	n.s.	3635 COD	40.0	Mata et al. (2012)

(*Continued*)

TABLE 8.2 (Continued)

Effluent	Microorganisms	Operational conditions	Nitrogen		Phosphorus		Organic carbon		References
			C_i (mg L^{-1})	RE (%)	C_i (mg L^{-1})	RE (%)	C_i (mg L^{-1})	RE (%)	
		L:D: 24:0 T: 30°C V: 0.25 L							
Brewery (anaerobically digested)	*S. obliquus*	OM: batch LI: 43.2 μmol m^{-2} s^{-1} L:D: 24:0 T: 23°C–25°C V: 4 L	29.4 NH_4-N	81.0	14.2 PO_4-P	29.7	295.7 COD	70.8	Ferreira et al. (2019)
Dairy	Microalgae consortia (*Chlorella* sp., *Scenedesmus* spp. and *Chlorella zofingiensis*)	OM: batch HRT: 4 d LI: 150 μmol m^{-2} s^{-1} L:D: 24:0 T: 30°C V: 0.4 L	176.0 TN	80–90	39.6 TP	70–80	2128 COD	40–50	Qin et al. (2016)
	Acutodesmus dimorphus	OM: batch HRT: 8 d pH: 7.8 LI: 60 μmol m^{-2} s^{-1} L:D: 12:12 T: 35°C V: 1 L	277 NH_4-N	100	6 PO_4	100	2593 COD	90.1	Chokshi et al. (2016)

	Scenedesmus quadricauda	OM: batch HRT: 12 d pH: 6.30 LI: 90 μmol m^{-2} s^{-1} L:D: 12:12 T: 22°C V: 1 L	86.0 TN	86.7	8.75 PO_4	71.2	170.1 TOC	69.1	Daneshvar et al. (2019)
	Tetraselmis suecica	OM: batch HRT: 12 d pH: 6.30 LI: 90 μmol m^{-2} s^{-1} L:D: 12:12 T: 22°C V: 1 L	86.7 TN	66.8	9.50 PO_4	41.6	179.4 TOC	38.2	
Agro-industrial mixture (anaerobically digested)	*Parachlorella kessleri*	OM: batch HRT: 25 d LI: 200 μmol m^{-2} s^{-1} L:D: 24:0 T: 25°C V: 0.4 L	330.6 NH_4-N	>98	13.5 TP	80–90	1134 COD	39.1	Koutra, Grammatikopoulos, and Kornaros (2018)
Meat processing	*Chlorella* sp.	OM: batch HRT: 6 d LI: 120 μmol m^{-2} s^{-1} L:D: 24:0 T: 25°C V: 1 L	327.6 TN 8.2 NH_4-N	33.3 45.6	45.9 TP	65.4	1019 COD	52.3	Q. Lu et al. (2015)
	Chlorella sp.	OM: batch HRT: 6 d LI: 120 μmol m^{-2} s^{-1} L:D: 24:0 T: 25°C V: 1 L	117.5 TN 101.7 NH_4-N	33.7 60.5	5.6 TP	100	1016 COD	39.6	Q. Lu et al. (2015)

(*Continued*)

TABLE 8.2 (Continued)

Effluent	Microorganisms	Operational conditions	Nitrogen		Phosphorus		Organic carbon		References
			C_i (mg L^{-1})	RE (%)	C_i (mg L^{-1})	RE (%)	C_i (mg L^{-1})	RE (%)	
Aquaculture	*C. vulgaris*	OM: continuous HRT: 1 d pH: 6.8–7.2 LI: 9000 lux T: 25°C V: 4 L	6.81 TN	86.1	0.42 TP	82.7	n.s.	n.s.	Gao et al. (2016)
	C. sorokiniana	OM: batch HRT: 14 d pH: 7.3 LI: 120 μmol m^{-2} s^{-1} T: 25°C V: 0.5 L	5.32 NH_4-N 40.7 NO_3-N	77.7	8.82 PO_4-P	100	96 COD	42	Ansari, Singh, Guldhe, and Bux (2017)
	S. obliquus	OM: batch HRT: 14 d pH: 7.3 LI: 120 μmol m^{-2} s^{-1} T: 25°C V: 0.5 L	5.32 NH_4-N 40.7 NO_3-N	75.8	8.82 PO_4-P	100	96 COD	69	Ansari et al. (2017)

Swine manure (anaerobically digested)	*C. vulgaris*	OM: batch HRT: 7 d LI: 50 μmol: m^{-2} s^{-1} L:D: 24:0 T: 25°C V: 0.5 L	102.5 TN 83.7 NH_4-N	49.2 83.7	57.0 TP	22.4	9500 COD	34.0	Deng et al. (2017)
Seafood processing	*Chlorella* sp.	OM: batch HRT: 45 d pH: 7.92 LI: 135 μmol m^{-2} s^{-1} L:D: 24:0 T: 25°C V: 0.5 L	121.1 TN	96.6	57.3 TP	68.4	n.s.	n.s.	Gao et al. (2018)

OM, operation mode; HRT, hydraulic residence time; LI, light intensity; L:D, light:dark cycle; T, temperature; V, volume processed; RE, recovery efficiency; Ci, initial concentration; TN, total nitrogen (mg N L^{-1}); NH_4-N, ammonium-nitrogen (mg N L^{-1}); NO_3-N, nitrate-nitrogen (mg N L^{-1}); TP, total phosphorus (mg P L^{-1}); SP, soluble phosphorus (mg P L^{-1}); PO_4-P, phosphate-phosphorus (mg P L^{-1}); PO_4: phosphate (mg PO_4 L^{-1}); TOC, total organic carbon (mg C L^{-1}); COD, chemical oxygen demand (mg O_2 L^{-1}); n.s., not specified.

8.3 Microalgae in the improvement of agricultural practices

The increase in worldwide population has led to a higher demand for food. Social and economic activities have contributed to a reduction in the area available for cultivation, water limitation, accumulation of xenobiotic compounds in the soil, and soil corrosion. At the same time, climate changes have resulted in the degradation of soil quality and fertility (Pathak et al., 2018). To meet the increased demand for food and land management challenges, an improvement in agricultural productivity and sustainability is a crucial step. Enhanced agricultural productivities may be achieved either by an increase in crop yields or by the prevention of culture loss due to biotic and abiotic stress conditions (Dmytryk & Chojnacka, 2018; Pathak et al., 2018). Chemically-based products have been extensively used as plant growth promoters and/or plant protection agents (Baweja, Kumar, & Kumar, 2019; Dmytryk & Chojnacka, 2018; Pan et al., 2019; Win, Barone, Secundo, & Fu, 2018). However, current regulations tend to limit the use of mineral fertilizers and chemical products, with impacts on the fulfillment of productivity targets (Dmytryk & Chojnacka, 2018; Pan et al., 2019). Considering this and the adverse effects associated with the use of chemicals in plants, increased attention is being diverted into the use of biological products for plant growth promotion and protection (Chiaiese et al., 2018; Dmytryk & Chojnacka, 2018; Pan et al., 2019; Win et al., 2018). For this purpose, microalgae and cyanobacteria have emerged as potential solutions for the deployment of sustainable agriculture, as they can act as both biofertilizers and biostimulants (Chiaiese et al., 2018; Dmytryk & Chojnacka, 2018; Górka, Korzeniowska, Lipok, & Wieczorek, 2018; Pan et al., 2019; Pathak et al., 2018; Singh et al., 2017). Algae and cyanobacteria are known for their ability to produce a huge variety of biologically-active compounds, which can improve agriculture by acting either at the soil or plant level (Pan et al., 2019; Singh et al., 2017). Taking into account their benefits for the development of sustainable agriculture, both biomass and extracts from microalgae and cyanobacteria are commercially available (Górka et al., 2018). The next sections present in detail how algal biomass and extracts can improve crop production to meet global food demand.

8.3.1 Biofertilizers versus biostimulants

Biofertilizers and biostimulants comprise a wide variety of biologically-derived organic and inorganic compounds, such as phenolic compounds, polysaccharides, hormone-like substances, and proteins, as well as living organisms, both prokaryotes (e.g., plant growth-promoting bacteria and nitrogen-fixing cyanobacteria) and eukaryotes (microalgae and macroalgae or seaweeds) (Chiaiese et al., 2018). Biofertilizers and biostimulants have huge potential in the development of sustainable agriculture, in opposition to mineral and chemical substances (Ronga et al., 2019). However, these compounds act on plant growth promotion and development at different levels.

Biofertilizers are mainly recognized by their activity at the soil level: compounds and/or microorganisms used as biofertilizers promote plant growth by supplying the cultures with essential nutrients, such as nitrogen, phosphorus, and potassium, which improve the biological and chemical properties of the soil and restore soil fertility (Kusvuran & Kusvuran, 2019; Reddy & Saravanan, 2013; Ronga et al., 2019; Win et al., 2018). Considering the benefits of biofertilizers in crop nutrition, the main advantages of using these substances/formulations include (Carvajal-Muñoz & Carmona-Garcia, 2012; Vessey, 2003): (1) increased crop productivities per unit of area and time; (2) reduced energetic requirements; (3) control and maintenance of adequate soil properties and fertility; (4) lower risks of soil and water contamination; and (5) crop protection against pathogenic organisms.

Biostimulants have a positive effect on plant metabolism, enhancing respiration, photosynthetic activity, nucleic acid synthesis, and ion uptake and, hence, plant growth and development (Górka et al., 2018). This positive effect can be observed under both optimal and adverse conditions, which attribute to biostimulants an undeniable potential in promoting plant resistance and tolerance against stress conditions (Chiaiese et al., 2018; Khan et al., 2009; Ronga et al., 2019). Other advantages of using biostimulants include (Chiaiese et al., 2018; Ronga et al., 2019): (1) increased crop productivities; (2) increased nutrient utilization efficiencies; and (3) enhanced crop quality.

Although biofertilizers and biostimulants can be obtained from different sources, microalgal-based products, which include biomass and extracts from the eukaryotic microalgae and the prokaryotic cyanobacteria, are considered the most promising, due to their benefits in plant growth promotion and the advantages of microalgal biomass production over other biological resources (Chiaiese et al., 2018; Ronga et al., 2019). Commercial agricultural products based on microalgae are already available. Besides the use as biofertilizers and biostimulants, other biologically active compounds from microalgal biomass, with demonstrated antimicrobial-activity, can also be applied as biopesticides, enhancing crop production through the inactivation of pathogenic organisms.

8.3.2 Microalgal metabolites and phytohormones with potential interest for agriculture

Microalgae are known for their diversified biochemical composition, which enables the achievement of a wide range of metabolites that differ in their chemical structure and biological activity. Among these compounds, some of them are of extreme relevance to agricultural practices, as they play an important role in (Górka et al., 2018; Ronga et al., 2019): (1) soil decontamination and fertilization; (2) plant protection against biotic and abiotic stress factors; and (3) plant development. These include phenolic compounds, terpenoids, free fatty acids, polysaccharides, and carotenoids. Microalgae also present phytohormones, which are known for their activity as plant growth promoters (Han, Zeng, Bartocci, Fantozzi, & Yan, 2018; Lu & Xu, 2015; Pan et al., 2019). Table 8.3

TABLE 8.3 Main algal metabolites with potential interest for agriculture.

Metabolites	Examples	Algal sources	Biological activity	Role in agriculture	References
Phenolic compounds	Polyphenols phenolic acids flavonoids phenylpropanoids	*Botryococcus braunii* *Chaetoceros calcitrans* *Chlorella vulgaris* *Isochrysis galbana* *Isochrysis* sp. *Neochloris oleoabundans* *Odontella sinensis* *Phaeodactylum tricornutum* *Saccharina japonica* *Skeletonema costatum* *Tetraselmis suecica*	Antimicrobial antioxidant fungicide	Crop protection against pathogens or other biotic and abiotic stress conditions	Esquivel-Hernández et al. (2017), Foo et al. (2017), Goiris et al. (2012), Khoddami, Wilkes, and Roberts (2013), Michalak, Chojnacka, and Saeid (2017), Pan et al. (2019), Singh et al. (2017), Sytar, Brestic, Rai, and Shao (2012)
Terpenoids	Hemiterpenes monoterpenes sesquiterpenes diterpenes triterpenes polyterpenes	*Chondrococcus hornemanni* *Hypnea pannosa* *Oscillatoria peronata* *Planktothricoids raciborskii* *Plocamium cornutum* *Plocamium leptophyllum* *Portieria hornemann* *Pseudanabaena articulate*	Antibacterial anticancer antioxidant	Crop protection against bacteria, insects, and other organisms Stimulation of preliminary growth and development of plants Attraction of pollinators	Awasthi, Upadhyay, Singh, Pandey, and Dwivedi (2018), Betterle and Melis (2019), Gershenzon and Dudareva (2007), Pan et al. (2019), Pattanaik and Lindberg (2015), Pavela and Benelli (2016), Rodriguez-Garcia, Hosseini, Martinez-Chapa, and Cordell

		Pseudanabaena sp. *Sphaerococcus coronopifolius* *Synechocystis* sp. *Thermosynechococcus elongates*			(2017), Singh et al. (2017), Wei et al. (2019)
Free fatty acids	Saturated fatty acids (e.g., *cis*- and *trans*-fatty acids) unsaturated fatty acids (e.g., lignoceric, palmitic, capric, behenic, cerotic, caprylic, lauric, stearic, myristic, arachidic acids)	Microalgae from the genera: *Anabaena* *Arthrospira* *Chlorella* *Dunaliella* *Nannochloropsis* *Porphyridium* *Scenedesmus*	Antibiotic anticarcinogenic antifungal antioxidant antiviral	Crop protection against pathogens or other biotic and abiotic stress conditions	Demirbas and Demirbas (2011), Desbois and Smith (2010), El-Baz, El-Senousy, El-Sayed, and Kamel (2013), Feller et al. (2018), Lam and Lee (2012), Pan et al. (2019), Singh et al. (2017)
Polysaccharides	Extracellular polysaccharides structural polysaccharides (e.g., arabinoxylans, cellulose, chitin and pectins) energy-storage polysaccharides (e.g., starch and glycogen)	Microalgae from the genera: *Aphanothece* *Arthrospira* *Chlamydomonas* *Chlorella* *Cylindrotheca* *Dunaliella* *Navicula* *Nostoc* *Phaeodactylum* *Porphyridium*	Antibacterial anticancer anticoagulant antiinflammatory antioxidant	Improvement of soil quality (soil conditioners, nutrient carriers and safe release systems for agrochemicals) Plant growth stimulation Crop protection against biotic and	Arroussi et al. (2018), Campos, De Oliveira, Fraceto, and Singh (2015), Chanda et al. (2019), Delattre, Pierre, Laroche, and Michaud (2016), Dvir, Stark, Chayoth, Madar, and Arad (2009), Elarroussia et al. (2016), Farid et al. (2019), González, Castro,

(*Continued*)

TABLE 8.3 (Continued)

Metabolites	Examples	Algal sources	Biological activity	Role in agriculture	References
		Rhodella *Scytonema*		abiotic stress conditions	Vera, and Moenne (2013), Guilherme et al. (2015), Guzmán, Gato, Lamela, Freire-Garabal, and Calleja (2003), Pan et al. (2019), Rechter et al. (2006), Singh et al. (2017), Usman, Khalid, Usman, Hussain, and Wang (2017), Vera, Castro, Gonzalez, and Moenne (2011)
Carotenoids	Alpha-carotene beta-carotene lutein lycopene astaxanthin zeaxanthin	*Chlorella protothecoides* *Chlorella pyrenoidosa* *Chlorella zofingiensis* *Dunaliella salina* *Haematococcus pluvialis* *Muriellopsis* sp. *P. tricornutum* *Spiruina* sp.	Anticancer antiinflammatory antioxidant	Soil bioremediation and fertilization Crop protection against bacteria, insects, and other biotic and abiotic stress conditions Crop fortification (increase in the availability of provitamin A)	Cezare-Gomes et al. (2019), Galasso, Corinaldesi, and Sansone (2017), Guedes, Amaro, and Malcata (2011), Han, Zhao, and Wang (2016), Pan et al. (2019), Rajesh, Rohit, and Venkata Mohan (2017), Raposo, De Morais, and De Morais (2015), Sakamoto et al. (2017)

Phytohormones	Auxins abscisic acid citokinins ethylene gibberellins	Microalgae from the genera: *Arthrospira* *Chlamydomonas* *Chlorella* *Phormidium* *Protococcus* *Scenedesmus*	Chemical messengers	Plant growth stimulation (root growth and development, stem elongation, seed germination and leaf and flower development) Regulation of cellular activities in crops (cell division, photosynthesis, nucleic acid synthesis, respiration, and nutrient uptake) Crop response to stress conditions (phototropism, gravitropism and induction or reduction of ethylene synthesis)	Han et al. (2018), Lu and Xu (2015), Ördög, Stirk, Van Staden, Novák, and Strnad (2004), Pan et al. (2019), Ronga et al. (2019), Singh et al. (2017)

presents the main biologically-active compounds that can be extracted from algae for their benefits in agriculture. A more detailed description of these compounds and their role in agriculture is presented in Singh et al. (2017) and Pan et al. (2019).

8.3.3 Role of algae and algal extracts in the improvement of agricultural practices

Biologically-active compounds produced by algae have been proposed as effective substances for plant protection, production, and development, contributing to the improvement of agricultural productivities. Their effect on crop production can be observed either by their use as pure extracts or as a crude algal compost and their mode of action impacts productivities at three main levels: (1) soil improvement; (2) crop protection; and (3) direct growth stimulation. The next sections highlight how algae and algal extracts improve agricultural practices at these three levels and Table 8.4 presents some application examples.

8.3.3.1 Soil improvement

Cyanobacteria have the ability to fix atmospheric inorganic nitrogen (N_2) into organic nitrogen, which can be easily assimilated by higher plants. Although cyanobacteria also require nitrogen for their growth, when applied in cultivable soils, these microorganisms act in symbiosis with higher plants, by increasing the availability of nitrogen in soils (Baweja et al., 2019; Renuka et al., 2018; Sharma, Khokhar, Jat, & Khandelwal, 2012; Win et al., 2018). Several studies have demonstrated that the presence of cyanobacteria in agricultural fields contributed to an increase in nitrogen levels in soils, with a direct impact on crop productivities (Renuka et al., 2016). Jha and Prasad (2006) applied a cyanobacterial inoculum in a rice paddy and evaluated its effect on grain and straw yield and on soil composition. The cyanobacterial inoculum (both alone and combined with chemical fertilizer) improved grain and straw yields and increased the availability of nitrogen in the soil. Singh and Datta (2007) applied wild-type and herbicide-resistant strains of *Anabaena variabilis* to a rice field, observing a positive effect in terms of plant height, leaf length, and seed, grain, and straw productivities. Although the majority of studies report considerable benefits in rice fields, more recent studies have also shown the positive effect of cyanobacteria in nitrogen availability to other cultures, such as pea plant, wheat, soybean, and mungbean (Osman, El-Sheekh, El-Naggar, & Gheda, 2010; Prasanna et al., 2014; Swarnalakshmi et al., 2013). Taking into account these benefits, it is estimated that the use of cyanobacteria can replace chemical nitrogen fertilizers, contributing to savings of between 25% and 40%. Furthermore, cyanobacteria excrete polysaccharides and form soil crusts, which prevent nitrogen

TABLE 8.4 Impact of algae and algae-based products on food crop productivity.

Mode of action	Algae/algal extracts	Target crop/soil	Observed effect	References
Nitrogen fixation	Cyanobacterial inoculum composed by *Aulosira fertilissima, Anabaena sphaerica, Nostoc hatei, Cylindrospermum majus* and *Westiellopsis prolifica*	Rice	Increase in nitrogen availability in the soil Increase in grain and straw yields	Jha and Prasad (2006)
	Nostoc entophytum and *Oscillatoria angustissima*	Pea	Increase in nitrogen fixation Increase in growth parameters, germination percentage and photosynthetic pigments Increase in nutritional value of pea seeds	Osman et al. (2010)
	Anabaena torulosa biofilm	Wheat	Increase in nitrogen availability in the soil	Swarnalakshmi et al. (2013)
Nutrients availability in soils	*Calothrix ghosei, Hapalosiphon intricatus* and *Nostoc* sp.	Wheat	Increase in organic carbon content in the soil Increase in grain yield	Karthikeyan et al. (2007)
	Microalgal–bacterial flocs and *Nannochloropsis oculata*	Tomato	Increase in ammonium, phosphorus, and potassium availability in the soil Improvement of fruit quality (increase in sugar and carotenoids content)	Coppens et al. (2016)
	Consortia and biofilms including the species: *Azotobacter* sp., *Anabaena* sp., *Providencia* sp., and *Calothrix* sp.	Okra	Increase in zinc and iron availability in the soil Beneficial changes in the microbiome Increase in root yield and weight	Manjunath et al. (2016)

(*Continued*)

TABLE 8.4 (Continued)

Mode of action	Algae/algal extracts	Target crop/soil	Observed effect	References
Soil physical and chemical amendments	*Microcoleus vaginatus, Phormidium tenue, Scytonema javanicum, Nostoc* sp. and *Desmococcus olivaceus*	Unconsolidated sand	Increase in crust cohesion Increase in the resistance to wind erosion	Hu et al. (2002)
	Nostoc strains	Poorly aggregated tropical soils	Increase in aggregate stability	Issa et al. (2007)
	Phycocyanin extract from *Arthrospira platensis* and inactive biomass of *A. platensis*	Soil contaminated with diesel and biodiesel	Reduction in diesel and biodiesel concentration using *A. platensis* and phycocyanin, respectively	Decesaro et al. (2017)
Protection against biotic factors	*Oscillatoria, Anabaena, Nostoc, Nodularia* and *Calothrix* species	Pepper	Fungicidal activity against *Alternaria alternate* and *Botrytis cinerea*	Kim (2006)
	Oscillatoria chlorina	Tomato	Fungicidal activity against *Meloidogyne arenaria* Improved plant growth	Khan, Kim, Kim, and Kim (2007)
	A. torulosa, Anabaena laxa, Anabaena azollae, Anabaena oscillarioides and *Calothrix* sp.	Wheat	Increase in nitrogen fixation Increase in the activity of hydrolytic and defense enzymes (endoglucanase, peroxidase, polyphenol oxidase, and phenylalanine ammonia lyase) Increase in fresh and dry weight of plant	Babu, Prasanna, Bidyarani, and Singh (2015)

Protection against abiotic factors	Seaweed extracts	Cucumber	Low-temperature stress was not observed in treated plants Improvement of plant productivity (increase in number of fruit per plant, fruit weight and total yield)	Sarhan and Ismael (2014)
	Polysaccharides extracted from *Dunaliella salina*	Tomato	Alleviation of salt stress Activation of metabolic pathways involved in plant tolerance to stress Mitigation of the decrease in length and dry weight of shoots and roots	Arroussi et al. (2018)
	Oscillatoria agardhii	Wheat	Alleviation of drought stress Increase in the activity of hydrolytic and defense enzymes (catalase, superoxide dismutase, and peroxidase) Increase in grain yield	Haggag et al. (2018)
Direct growth stimulation	Exopolysaccharydes extracts from *D. salina*	Wheat	Increase in germination rate and seedling growth Increase in root and coleoptiles height Increase in tolerance to salt stress	El Arroussi et al. (2015)
	A. platensis	Lettuce	Increase in seedling growth Increase in spermine content in leaves	Mógor, Ördög, Lima, Molnár, and Mógor (2018)
	Chlorella vulgaris and *A. platensis*	Maize	Increase in germination rate and plant yield Increase in fresh and dry weights of shoot, root, and whole plant Increase in shoot length and in the number of leaves	Dineshkumar et al. (2019)

leaching/runoff and, thus, reduce the risks of water contamination (Renuka et al., 2018).

After nitrogen, phosphorus is the second essential nutrient for plants and microorganisms. Photosynthetic microorganisms can uptake both nitrogen and phosphorus from different environments, storing high amounts of these compounds in their biomass. Therefore the use of algal biomass in soils may act as a source for essential nutrients like nitrogen and phosphorus, which constitutes a promising replacement for chemical fertilizers (Ronga et al., 2019; Sharma et al., 2012; Win et al., 2018). Algal biomass is also a source of other microelements, such as potassium, magnesium, sulfur, and iron. These nutrients are essential for crop development, as they play an important role as coenzymes for redox reactions, which are essential for plant metabolism. Accordingly, the use of microalgae guarantees the availability of both macro- and micronutrients to higher plants, as has been reported in the literature (Coppens et al., 2016; Manjunath et al., 2016; Marks, Montero, & Rad, 2019; Renuka et al., 2017).

Besides the availability of essential macro- and micro-nutrients for the development of higher plants, the presence of organic matter in adequate concentrations is also very important for agriculture sustainability (Renuka et al., 2018; Sharma et al., 2012). Microalgae and cyanobacteria are considered important sources of organic matter in soils: (1) as photosynthetic organisms, they can fix atmospheric CO_2 and convert it into organic biomass; (2) they excrete extracellular polymeric substances, which increase organic carbon levels and the growth of other microorganisms; and (3) they are subjected to degradation and grazing, which also increases organic matter contents. Several studies have reported an increase in organic carbon contents in soils after the application of algal biomass (Karthikeyan, Prasanna, Nain, & Kaushik, 2007; Nisha, Kaushik, & Kaushik, 2007; Renuka et al., 2017). For example, Renuka et al. (2017) evaluated the effect of unicellular and filamentous microalgae–cyanobacteria consortia on soil composition and grain quality of a wheat crop. Besides the increase in micronutrient availability in the soil (mainly zinc, iron, copper, and manganese), the studied consortia enhanced the organic carbon content of the soil. This improvement in soil quality resulted in increased grain yield and higher quality product, as obtained grains presented an enhanced nutritional composition.

Microalgal biomass also plays an important role in soil reclamation, either by controlling its pH and salinity or by sequestering heavy metals and other contaminants present in soils (Kumar, Dahms, Won, Lee, & Shin, 2015; Pan et al., 2019; Priya, Gurung, Mukherjee, & Bose, 2014; Sharma et al., 2012). Although some reports have demonstrated opposite effects of algal biomass in soils, the role of these organisms in modulating soils' acidic and basic state is noteworthy (Chatterjee et al., 2017; Pan et al., 2019). Regarding salinity, several studies have referred to the potential of microalgae in reducing salinity levels or, at least, promoting plant growth and development under these harsh conditions. Salinity levels are typically reduced by improving the water retention capacity of soils. On the other hand, gibberellic acid, a phytohormone, is responsible for an increase in salt tolerance by plants, which is of extreme importance, especially in large soil areas (Sharma et al., 2012). Finally, the ability of microalgae to sequester trace elements (e.g., iron, zinc, copper, and manganese), heavy metals (e.g., cadmium, lead, and chromium), hydrocarbons, and other compounds, contribute to soils' remediation, reducing the risks of erosion, cation adsorption, and restoring soil quality and fertility (Kumar et al., 2015; Pan et al., 2019; Priya et al., 2014; Win et al., 2018).

Finally, microalgae can be successfully applied in the improvement of some soil characteristics, such as aggregation, porosity, permeability, ventilation, and humidity level (Baweja et al., 2019; Chatzissavvidis & Therios, 2014; Pan et al., 2019; Sharma et al., 2012). The majority of these properties has been associated with the polysaccharides excreted by several algal species (Baweja et al., 2019; Sharma et al., 2012). These polysaccharides, when bonded to the soil, form an adhesive sheath, which improves soil aggregation and prevents soil erosion (Baweja et al., 2019). Also, the presence of these polysaccharides promotes soil aeration and the ability of soils to retain water/moisture content, which is crucial for the maintenance of soil temperature and pH (Baweja et al., 2019; Chatzissavvidis & Therios, 2014; Pan et al., 2019; Sharma et al., 2012). Improving these properties has a direct impact on crop production, as they stimulate root growth and boost soil microbial activity. Accordingly, microalgae have been widely used in soil amendment for promoting soil fertility and recovering damaged soils (Hu, Liu, Song, & Zhang, 2002; Issa et al., 2007).

8.3.3.2 Crop protection

The presence of organisms, such as insects, nematodes, bacteria, and fungi, can negatively impact crop productivities, as these organisms can exert pathogenicity on higher plants (Chanda, Merghoub, & Arroussi, 2019; Hamed, Abd El-Rhman, Abdel-Raouf, & Ibraheem, 2018; Pan et al., 2019). In plants, defense mechanisms against these organisms are promoted by the detection of signaling molecules present in the pathogenic organism cell wall (Chanda et al., 2019). Typically, polysaccharides act as recognizing centers for these molecules, inducing a wide range of responses, such as the regulation of signaling pathways, gene expression, and specific biosynthetic pathways, which result in the production of phenolic compounds, terpenoids, and other compounds that are known for their activity as antioxidants, antimicrobials, fungicides, and pesticides (Chanda et al., 2019; Farid et al., 2019; Mercier et al., 2001; Pan et al., 2019; Renuka et al., 2018). Taking into account that algal biomass represents a great source of polysaccharides, its application in crops induces these responses and, hence, plant defense mechanisms (Chanda et al., 2019; Pardee, Ellis, Bouthillier, Towers, & French, 2004). When inoculating seeds of spice crops with *Anabaena laxa* and *Calothrix elenkinii*, Kumar et al. (2013) demonstrated a significant increase in the activity of β-1,3-endoglucanase in shoots and roots, a protein responsible for the breakdown of the cell wall components of pathogens. As a consequence, an increase in fungicidal activity, plant dry weight, and shoot and root length was observed.

Besides the ability of algal polysaccharides to trigger these mechanisms, algal biomass already contains a wide diversity of bioactive metabolites that can also act on crop protection (Hamed et al., 2018). Several authors have indicated that antimicrobial properties demonstrated by various algal extracts may be associated with their fatty acids contents, namely the type and amount of myristic, palmitic, oleic, and eicosapentaenoic acids (Benkendorff, Davis, Rogers, & Bremner, 2005; Gerasimenko, Martyyas, Logvinov, & Busarova, 2014). On the other hand, phenolic compounds exert their activity by degrading the cell walls of pathogenic organisms (Hamed et al., 2018). Finally, carotenoids may exhibit antibacterial properties, due to their antioxidant nature (Esserti et al., 2017). Accordingly, direct application of algal biomass and extracts may also be involved in plant protection against biotic factors, due to the activity of the referred biologically-active compounds.

Apart from biotic factors, plants are exposed to a wide variety of environmental conditions. Among these, the most extreme ones (very cold and hot temperatures, high salinity conditions, and drought) may have a negative impact on crop production and development (Bulgari, Franzoni, & Ferrante, 2019; Górka et al., 2018). For example, exposure to low temperatures reduces plant metabolism and may cause damage to cell membranes due to phospholipid layer instability. On the other hand, high temperatures influence the synthesis and activity of proteins and enzymes and are also responsible for membrane damage (Bulgari et al., 2019). Salinity promotes osmotic stress in plants, affecting their metabolism and growth, as well as plant nutritional composition (typically, salt-stressed plants present significantly lower chlorophyll contents than nonstressed ones) (Bulgari et al., 2019; Górka et al., 2018; Kaya, Higgs, & Sakar, 2002). Although drought can be associated with salinity (drought can be responsible for the osmotic stress), it also influences gas exchange in plants, with impacts on photosynthetic and transpiration rates and, hence, on plant yield (Bulgari et al., 2019; Górka et al., 2018). Accordingly, it is important to improve plants' tolerance to these conditions, so that agricultural productivities are not compromised. Considering the potential of biostimulants in modulating several plant responses, their use for plant protection purposes has been widely explored, with numerous studies concluding that biostimulants can counteract the above-discussed abiotic stress factors in several ways (Bulgari et al., 2019; Du Jardin, 2015), including: (1) interaction with processes involved in plant responses to stress; (2) increase in the accumulation of antioxidant compounds, which reduce plant sensitivity to stress; and (3) improvement of plant performance and growth. Among biostimulants, increased attention is being diverted into algal-based ones, especially due to their content in biologically-active compounds and their ability to trigger several metabolic pathways in plants. Application of algal-based products has been successfully applied in food crops under temperature, salt, and drought stress conditions (Arroussi et al., 2018; Haggag, Hoballah, & Ali, 2018; Latef, Srivastava, Saber, Alwaleed, & Tran, 2017; Mansori et al., 2015; Rodríguez, Stella, Storni, Zulpa, & Zaccaro, 2006; Sarhan & Ismael, 2014; Stamatiadis et al., 2015). For example, Arroussi et al. (2018) demonstrated that polysaccharides excreted by *Dunaliella salina* were able to reduce salt stress damage in tomato cultures, by inducing several metabolic pathways involved in plants' responses to stress, such as the jasmonic acid pathway. With this algal-based treatment, an improvement in root and shoot length was also observed. Treatment of bean plants under drought stress with seaweed extracts from *Ulva rigida* and *Fucus spiralis* allowed plants to overcome these stress conditions and promoted an improved growth of bean plants (Mansori et al., 2015).

8.3.3.3 Direct growth stimulation

Algae and algal products can have a positive impact on plant growth and development, as they trigger several metabolic processes, such as respiration, photosynthesis, nucleic acid synthesis, chlorophyll production, and ion uptake (Górka et al., 2018). These processes may result in increased germination rates and improved plant characteristics (e.g., increased shoot and root length, increased leaf area, and higher nutritional contents), being mainly associated with biologically-active substances that can be obtained from algae. These include plant growth-promoting factors, such as phytohormones, but also some amino acids, vitamins (vitamin B12 and biotin), polysaccharides, and polypeptides (Sharma et al., 2012). Elarroussia et al. (2016) evaluated the effect of applying a solution of total polysaccharide extracts obtained from *Arthrospira platensis* in tomato and pepper cultures. With this study, the authors demonstrated that application of the extracts' solution resulted in an increase in the size of both tomato and pepper plants. In addition, increased root weight, size, and number of nodes was registered for the treated plants.

8.4 Microalgae in the improvement of animal and animal-derived products

The increase in livestock for human food as a response to the population increase (world meat production is expected to double by 2050) leads to competition for agricultural resources between food and feed industries

(Lum et al., 2013). Therefore it is important to find sustainable solutions to increase the resources for the cited industries.

As was previously referred to, microalgae can also be applied as a source for animal and aquatic feed production (Aratboni, Rafiei, Garcia-Granados, Alemzadeh, & Morones-Ramirez, 2019; Khan et al., 2018; Pires, Alvim-Ferraz, Martins, & Simoes, 2012), promoting animal growth and improving the composition of animal-derived products. Also, the study of microalgae for feed production may achieve important sustainability targets in this field (Spruijt, Van Der Weide, & Van Krimpen, 2016): (1) to give sustainable resources for feed production; (2) to close mineral cycles of nitrogen and phosphorus; and (3) to provide new protein sources through the bio-based economy.

8.4.1 Microalgal composition and beneficial effects

Microalgae present high protein composition, carotenoids (e.g., astaxanthin, lutein, β-carotene, and chlorophyll), lipids [polyunsaturated fatty acids (PUFAs) like eicosapentaenoic acid (EPA), 20:5, ω-3- and docosahexaenoic acid (DHA), 22:6, ω-6], essential vitamins, among other biologically-active compounds (Gouveia, Batista, Sousa, Raymundo, & Bandarra, 2008; Yaakob, Ali, Zainal, Mohamad, & Takriff, 2014). Their composition depends on species, cultivation conditions, treatment after harvesting, and drying methods. Therefore, their biomass can enhance the nutritional content of feed formulations, helping to improve animal health. Besides the variability of chemical composition, the nutritive value of microalgae for animals is highly dependent on the adaptation of animals to the ingredient. The main concern of introducing microalgae in new feedstuffs is acceptability by the animal (taste and poor oxidative stability have limited their applicability). Van Emon, Loy, and Hansen (2015) evaluated the digestibility and preference of steers when fed with increasing concentrations of microalgae in diet formulations. They observed that the animals consumed diets containing microalgae at concentrations up to 30%. However, for higher concentrations (45%), the given diet was consumed more slowly.

The addition of microalgae to livestock feeds may have important beneficial effects. Several microalgal species have shown antibacterial characteristics (Spruijt et al., 2016). Thus, the use of antibiotics in livestock may be decreased, as well as the resistance of bacteria against these antibiotics, both in humans and livestock. Moreover, microalgal biomass is composed of several types of antioxidants (e.g., carotenoids) that have antiinflammatory characteristics and could prevent degenerative diseases. Microalgae can also improve the immune responses of animals, reproduction capacity, and external characteristics (e.g., healthy skin, shiny coat). Research studies showed that the addition of microalgae as feed supplement increased the levels of PUFA in eggs of laying hens, the meat of broilers, and milk of dairy cows

(Moran, Currie, Keegan, & Knox, 2018; Park, Upadhaya, & Kim, 2015; Spruijt et al., 2016). Therefore microalgae may be an essential resource in the future, as they constitute an alternative to conventional feed sources (corn and soybean), presenting a sustainable solution to the competition between the food, feed, and biofuel industries.

8.4.2 Opportunities for microalgae as an ingredient in animal diets

The microalgae from the genera *Chlorella*, *Schizochytrium*, *Arthrospira*, *Nannochloropsis*, *Scenedesmus*, *Euglena*, and *Haematococcus* can be applied in terrestrial and aquatic animal feed (Camacho, Macedo, & Malcata, 2019; Gouveia et al., 2008). Considering this application, several research studies have already been performed, aiming at analyzing the effect of microalgae on animal feed. Table 8.5 presents research findings from the effect of introducing microalgae in animal diets on their growth and on animal-derived products' quality.

In poultry farming, microalgae have been studied for partial protein replacement, improvement of meat and egg yolk quality, and immune response (Gatrell, Kim, Derksen, O'Neil, & Lei, 2015; Ginzberg et al., 2000; Toyomizu, Sato, Taroda, Kato, & Akiba, 2001). The replacement of protein in the diets has been proposed with different microalgal species with percentages usually lower than 10%. Also, the fatty acid profiles of chicken meat are highly dependent on the fatty acid content of the animal diet. Marine products (fish oil and fish meal) have been used to increase the composition in n-3 (ω-3) fatty acids. However, the high cost and demand of these products led to the search for alternative and more sustainable n-3 fatty acid sources for animal feed. Microalgae contain higher composition in EPA and DHA than traditional feed protein supplements. Gatrell et al. (2015) evaluated the effect of oil extracted from microalgal biomass (*Nannochloropsis oceanica*) as a feed supplement for chicken, trying to valorize a coproduct of biofuel production (to increase the commercial value of microalgal biomass). The inclusion of microalgae in the diet did not change total fat or PUFAs but increased the total n-3 fatty acids (lower n-6/n-3 fatty acid ratio), EPA, and DHA in plasma, liver, breast, and thigh. The inclusion of 2% of microalgae in the diet was enough to enhance the composition of chicken meat. Ginzberg et al. (2000) studied the effect of the addition of the microalga *Porphyridium* sp. (5% and 10%) in chickens' diet, evaluating the blood and cholesterol levels. Comparing body weight, egg number, and weight, no significant differences were found between diets. In the diet of 10% microalgae, the following characteristics were observed: (1) high feed conversion; (2) low cholesterol in egg yolk (10% less); (3) high linoleic acid level; (4) high arachidonic acid level; and (5) darker egg color due to high production of carotenoids.

TABLE 8.5 Research findings from the effect of microalgae addition as a feed supplement.

Animal	Experimental setup	Microalgae effect	References
Chicken	N: 37 AA: 210 d ED: 10 d. Diets (3): no supplementation 5% and 10% *Porphyridium* sp.	No differences in body weight, egg number, and egg weight Low serum cholesterol levels Reduced cholesterol levels in egg yolk of chicken Darker egg yolk due to higher carotenoid levels	Ginzberg et al. (2000)
Broiler	N: 24 AA: 21 d ED: 16 d. Diets (3): no supplementation 4% and 8% *Arthrospira*	No significant differences in body weights Influence on both the yellowness and redness of broiler flesh	Toyomizu et al. (2001)
	N: 300 AIW: 42.6 ± 0.1 g AA: 2 d ED: 35 d. Diets (3): no supplementation 0.1% and 8% *Schizochytrium* JB5	No effect on growth performance Reduced ω-6/ω-3 fatty acid ratio, saturated fatty acid contents Potential source of ω-3 fatty acid in the broiler's diet	Yan and Kim (2013)
	N: 120 AA: 1 d ED: 14 d. Diets (4): 21% crude protein 17% crude protein the mentioned formulations adding 7.4% extract of *Schizochytrium*	Positive impact on broiler body weight, carcass yield, and meat fatty acid profile Higher carcass yield Low acceptability of chicken meat by consumers (decrease in meat oxidative stability)	Ribeiro et al. (2014)
Hens	N: 12 AA: 175 d ED: 56 d. Diets (3): no supplementation 5% and 10% *Nannochloropsis gaditana*	The color of egg yolk shifted from yellow to more orange-red (transfer of microalgal carotenoids to egg yolk) Alternative to current sources for the production of DHA-enriched eggs	Bruneel et al. (2013)

Pekin duck	N: 300 AA: 1 d ED: 42 d. Diets (3): no supplementation 0.1% and 0.2% fermented *Chlorella vulgaris*	No effect on feed efficiency Influence on the total microbes, lactic acid bacteria, and coliforms in fecal contents Enhanced meat qualities in Pekin ducks	Oh et al. (2015)
Piglet	N: 72 AIW: 9.1 ± 1.1 kg (trial 1) N: 24 AIW: 9.1 ± 0.9 kg (trial 2) AA: 28 d ED: 14 d (after weaning). Diets (4): no supplementation 1% *Arthrospira* 1% *Chlorella* 0.2% colistin (antibiotic, positive control)	No beneficial effect of colistin on diarrhea incidence Reduced diarrhea incidence by *Chlorella* supplementation Improved mucosal architecture from the intestine and nutrient digestibility	Furbeyre et al. (2017)
Pig	N: 96 AIW: 26.58 ± 1.41 kg ED: 6 weeks. Diets (4): basal diet 0.05% tylosin 0.1% and 0.2% fermented *C. vulgaris*	Higher ($p < .05$) *Lactobacillus* concentration and lower *Escherichia coli* concentration Reduced fecal NH_3 and H_2S content Improved growth performance, nutrient digestibility, fecal microbial shedding (lower *E. coli* and higher *Lactobacillus*), and decreased the fecal noxious gas emission	Yan et al. (2012)
	N: 50 AIW: 75 kg ED: 45 d. Diets (5): 0.6% linseed oil 0.3%, 0.6% and 1.2% *Schizochytrium* sp. 0.4% soybean oil	Increased DHA concentration in the loin and dry-cured ham Increased nutritional value of these meat products No effect on carcass quality and oxidative stability of the fresh meat	Vossen, Raes, Van Mullem, and De Smet (2017)
Lamb	N: 33 AIW: 16.3 ± 0.3 kg	Lower average daily gain and greater slaughter age Increased EPA and DHA content in the studied tissues	Urrutia et al. (2016)

(*Continued*)

TABLE 8.5 (Continued)

Animal	Experimental setup	Microalgae effect	References
	AA: 55.1 ± 1.5 d. Diets (3): control (barley and soybean) L diet (barley, soybean, 10% linseed) and L-A diet (barley, soybean, 5% linseed, 3.89% *Schizochytrium* spp.)	Adverse effects on meat quality (greater lipid oxidation and reduced ratings for odor and flavor)	
	N: 44 AIW: 14.7 ± 0.5 kg ED: 10 d. Diets (4): control 8.75% extruded linseed 10.7% extruded linseed + 4% *Isochrysis* sp. 3.3% fish oil	C18:3n-3 meat enrichment	de la Fuente-Vazquez et al. (2014)
	N: 44 AIW: 22.7 ± 3.9 kg ED: 14 d. Diets (4): no supplementation 1%, 2% and 3% *Schizochytrium* spp.	Decreased n-6/n-3 ratio in all tissues	Meale, Chaves, He, and Mcallister (2014)
	N: 24 AIW: 15.3 ± 0.4 kg AA: 7–8 weeks ED: 8 d. Diets (2): no supplementation 2% microalgae	Enhanced the nutritional quality of lamb muscle, increasing DHA levels and total n-3 fatty acids in intramuscular fat	Díaz et al. (2017)
Rabbit	N: 40 AIW: 2034 ± 174 g AA: 9 weeks ED: 24 d. Diets (4): no supplementation 5%, 10%, and 15% *Arthrospira platensis*	No effect on final weight, weight gain and feed efficiency Lower digestibility Increased feed consumption	Peiretti and Meineri (2008)

	N: 294 AA: 35 d ED: 42 d. Diets (3): no supplementation 5% *Arthrospira* 3% *Thymus vulgaris*	Increased γ-linolenic acid content in meat Improved oxidative stability of raw and freeze-dried meat	Zotte et al. (2014)
Rainbow trout	N: 144 AIW: 27.3 ± 1.9 g ED: 30 d. Diets (3): 0.1%, 0.3%, and 1% *Haematococcus pluvialis*	Triglyceride and cholesterol levels showed a significant decrease with 0.3%	Sheikhzadeh, Tayefi-Nasrabadi, Oushani, and Enferadi (2012)
Coral trout	N: 540 AIW: 18.0 ± 2.0 g ED: 56 d. Diets (6): no supplementation 2%, 4%, 6%, 8%, and 10% *A. platensis*	Positive effects on growth, hematological parameters, and immune responses Activation and suppression of different hepatic antioxidant parameters	Yu et al. (2018)
Seabass	N: 48 AIW: 21.76 ± 4.03 g ED: 93 d. Diets (4): control 5%, 10% and 15% of lipids extracted from *Nannochloropsis* sp.	High feed conversion ratio No effect on composition and growth performance	Valente, Custodio, Batista, Fernandes, and Kiron (2019)
Nile tilapia	N: 120 AIW: 30 g ED: 84 d. Diets (4): control 3%, 5.5% and 8% of lipids extracted from *Nannochloropsis* sp.	No effect on fish growth, feed conversion, and fish survival Lower digestibility values for crude protein than the whole microalgal cells	Sarker et al. (2018)
Tilapia	N: 350 AIW: 158 ± 2 g. Diets (7): control 1%, 3% and 5% fish oil 1.75%, 5.26%, and 8.77% *Schizochytrium* sp.	Improved fatty acid profiles (higher n-3, lower n-6, n-6/n-3) In the fillet, total n-3 was increased	Stoneham et al. (2018)

N, number of animals; AIW, animal initial weight; AA, animal age; ED, experimental duration; EPA, eicosapentaenoic acid; DHA, docosahexaenoic acid.

In pig farming, microalgae were also studied for protein replacement, improvement of meat quality, and health enhancement (Furbeyre, Van Milgen, Mener, Gloaguen, & Labussiere, 2017; Yan, Lim, & Kim, 2012). Digestive disorders associated with weaning are usual for these animals, requiring the use of antibiotics to maintain intestinal health. Furbeyre et al. (2017) studied the use of microalgae as an alternative to antibiotics by testing four diets: (1) control without microalgae supplementation (D1); (2) standard diet supplemented with 1% *Spirulina* (D2); (3) standard diet with 1% *Chlorella* (D3); and (4) standard diet with 0.2% of colistin (D4). In D1, 94% of pigs had diarrhea after weaning, and the addition of colistin did not show any beneficial effect in D4. In D2 and D3, the addition of microalgae did not show a significant effect on average daily feed intake, average daily gain, and gain:feed ratio. With D3, the diarrhea incidence was significantly lower, showing that microalgae could also contribute to healthy intestinal development and a reduction of digestive disorders.

In dairy farming, the studies of microalgae supplements were mainly performed to evaluate the effect on milk production and quality improvement. Boeckaert et al. (2008) evaluated the effect of DHA-enriched microalgae (*Schizochytrium* sp. with 4.3%) on milk acid composition. The diet with microalgae contributed to the decrease of milk fat composition and the increase of linoleic acid and DHA levels. Till et al. (2019) analyzed the effect of microalgal feed supplementation on the sensory characteristics and fatty acid composition of cheese. Both milk and cheese DHA and PUFA contents increased linearly with microalgae dose in the diet. Other characteristics of these products were also observed: (1) reduction of saturated fatty acids and n-6/n-3 fatty acids ratio; (2) change in 20 of the 32 sensory attributes (increase in cheese air holes, nutty flavor, and dry mouth aftertaste); and (3) no effect on dry matter intake or milk yield.

In the last few years, aquaculture has become an important activity for food production (accounting for half of global food fish production) due to the environmental concerns associated with open ocean fishing (Anemaet, Bekker, & Hellingwerf, 2010). Aquaculture is the activity where microalgae have been most extensively studied for feed supplementation. Microalgae can contribute to (Yaakob et al., 2014; Yu et al., 2018): (1) the increase of reproductive performance and egg quality; (2) growth performance and survival; (3) stress tolerance and disease resistance; and (4) skin and flesh pigmentation. Astaxanthin is one of the most commercial interesting compounds of microalgae. In aquaculture, it is used as a feed additive to enhance the muscle tissue color of salmon, trout, and shrimp (Hu, Nagarajan, Zhang, Chang, & Lee, 2018; Yaakob et al., 2014). Yu et al. (2018) evaluated the effect of *A. platensis* diets (with percentages between 0% and 10%) on the growth parameters of coral trout. The addition of microalgae had positive effects on growth, hematological parameters, and immune responses, including: (1) red cell count; (2) white cell count; (3) hemoglobin; (4) total antioxidant capacity; (5) lysozyme and respiratory burst activities; (6) immunoglobulin content; and (7) resistance to *Vibrio harveyi* infection.

8.5 Challenges to the application of microalgae in these applications

The application of microalgal cultures presents some drawbacks, which are mainly related to the high costs associated with the microalgal production process.

Besides this common drawback, in wastewater treatment processes, other parameters related to microalgal growth should be taken into account, including the light supply, the nutrient concentration (mainly carbon, nitrogen, and phosphorus), and the operational conditions (such as pH, temperature, and mixing). Furthermore, in some cases, it is necessary to modify the wastewater composition to meet the growth conditions required by microalgae. For example, when nutrient concentration in the wastewater is lower than optimal, the least available nutrients should be added to the effluent to be treated (K. Li et al., 2019). On the other hand, when the nutrient concentration is too high for microalgae, the effluent can be diluted (Wang et al., 2012). Also, some compounds present in the wastewater can negatively affect microalgal growth, with high concentrations of ammonia being the most common toxic compound present in food industry effluents (Molinuevo-Salces et al., 2019). In this case, it becomes necessary to pretreat or dilute the effluent. Also, this type of effluent usually presents high levels of suspended solids, which result in high turbidity, and, in some cases, the effluent also presents its own color. These two factors can cause the amount of light that penetrates the culture to decrease, reducing the photosynthetic activity of microalgae and, hence, nutrients removal efficiency (Amenorfenyo et al., 2019; Molinuevo-Salces et al., 2019). To lower the concentration of suspended solids and remove color, it is also possible to dilute or pretreat the effluent (through physical and/or chemical methods). Another drawback in using microalgae for the treatment of food industry wastewaters is the presence of biological contaminants, such as protozoa, some strains of bacteria, and zooplankton that feed on microalgae (the protozoa) and compete for nutrients, decreasing the efficiency of the remediation process (Gupta et al., 2019). In these cases, it is important to plan wastewater pretreatment approaches prior to microalgal cultivation. To maintain monoculture cultivation, sterilization and production in closed photobioreactors are preferred (Mendes & Vermelho, 2013).

Regarding the use of microalgae and microalgal extracts to improve agricultural practices, the main limitation is associated with the economics of the process. On one hand, biomass recovery after the production process represents a large fraction of microalgal biomass production costs. On the other hand, if the extraction of the biologically-active compounds is required, the commonly used methods tend to be quite expensive (Pan et al., 2019; Renuka et al., 2018).

These high costs, combined with microalgal biomass production costs, result in an overall increase in biofertilizer/biostimulant price to values that do not compete with the price of chemical fertilizers. Finally, the diversity of microalgal species already isolated and identified, together with the huge number and complexity of metabolites that can be extracted from microalgal biomass, makes it difficult for the complete understanding of the interactions established between microalgae/microalgal extracts, plants, and the environment (Pan et al., 2019). Understanding these interactions will be essential for a clear elucidation on the effect of each metabolite/biomass on crops, allowing the definition of adequate substances for agricultural activities.

The cost of microalgal-based feed products is also an issue in the obtention of feedstock for human food from microalgal biomass. Currently, the cost of microalgal protein is higher when compared with conventional sources (e.g., soybean). To be economically competitive, besides the increase in microalgal production scale, the biomass produced should be exploited for several applications at the same time (Spruijt et al., 2016). Considering the biorefinery concept, animal feed can be considered as one of several products that can be obtained from biomass. Accordingly, taking as an example the conventional process of sunflower and soybean oil production, microalgae lipids may be first extracted for biodiesel production and the remaining (protein-rich) biomass should be used as an animal feed supplement. The use of microalgae in animal diets presents another challenge: the digestibility of microalgal biomass, which depends on the biomass chemical composition and digestive physiology of the species (Van Emon et al., 2015). Since disruption of the cell wall significantly increases the nutrient availability of microalgal biomass, enzymatic or physical treatments to enhance digestibility can be applied.

8.6 Conclusions

This chapter has presented an overview of how microalgae can improve food industry processes beyond direct human consumption. The main issues addressed included the use of microalgae for: (1) the remediation of wastewaters resulting from food-processing industries and agro-industrial activities; (2) the improvement of crop yields and soil fertilization, toward an increase in agricultural productivities; and (3) the improvement of animal and animal-derived products, when introduced as part of animal diets. Considering the high nutrient concentration present in food industry wastewaters and the nontoxic nature of these effluents, microalgal cultures can be applied in their remediation. Regarding the second application, microalgal biomass and extracts can be an important alternative to the chemically-based biofertilizers, acting on soil quality improvement, crop protection against biotic and abiotic factors, and plant growth stimulation. Finally, the use of microalgae as part of animal diets promotes an improvement of animal and animal-derived products, reducing the requirements for crop-based products for animal nutrition. The successful application of microalgae in these fields, together with their use in direct human consumption, constitutes a new opportunity for integrating microalgal cultures in all steps of the food industry value-chain, which will be crucial to meeting the increasing demands for food supplies.

Acknowledgments

This work was financially supported by: (1) Base Funding—UIDB/00511/2020 of the Laboratory for Process Engineering, Environment, Biotechnology and Energy—LEPABE—funded by national funds through the FCT/MCTES (PIDDAC); and (2) Project PTDC/BTA-BTA/31736/2017-POCI-01-0145-FEDER-031736—funded by FEDER funds through COMPETE2020—Programa Operacional Competitividade e Internacionalização (POCI), and with financial support of FCT/MCTES through national funds (PIDDAC). A.F. Esteves acknowledges the FCT Scholarship 2020.05477.BD and J.C.M. Pires acknowledges the FCT Investigator 2015 Programme (IF/01341/2015).

References

Abou-Shanab, R. A., Ji, M.-K., Kim, H.-C., Paeng, K.-J., & Jeon, B.-H. (2013). Microalgal species growing on piggery wastewater as a valuable candidate for nutrient removal and biodiesel production. *Journal of Environmental Management*, *115*, 257–264.

Amenorfenyo, D. K., Huang, X., Zhang, Y., Zeng, Q., Zhang, N., Ren, J., & Huang, Q. (2019). Microalgae brewery wastewater treatment: Potentials, benefits and the challenges. *International Journal of Environmental Research and Public Health*, *16*, 1910–1928.

Anemaet, I. G., Bekker, M., & Hellingwerf, K. J. (2010). Algal photosynthesis as the primary driver for a sustainable development in energy, feed, and food production. *Marine Biotechnology*, *12*, 619–629.

Ansari, F. A., Singh, P., Guldhe, A., & Bux, F. (2017). Microalgal cultivation using aquaculture wastewater: Integrated biomass generation and nutrient remediation. *Algal Research*, *21*, 169–177.

Arantes, M. K., Alves, H. J., Sequinel, R., & Da Silva, E. A. (2017). Treatment of brewery wastewater and its use for biological production of methane and hydrogen. *International Journal of Hydrogen Energy*, *42*, 26243–26256.

Aratboni, H. A., Rafiei, N., Garcia-Granados, R., Alemzadeh, A., & Morones-Ramirez, J. R. (2019). Biomass and lipid induction strategies in microalgae for biofuel production and other applications. *Microbial Cell Factories*, *18*, 178–194.

Arroussi, H. E., Benhima, R., Elbaouchi, A., Sijilmassi, B., Mernissi, N. E., Aafsar, A., ... Smouni, A. (2018). *Dunaliella salina* exopolysaccharides: A promising biostimulant for salt stress tolerance in tomato (*Solanum lycopersicum*). *Journal of Applied Phycology*, *30*, 2929–2941.

Awasthi, M., Upadhyay, A. K., Singh, S., Pandey, V. P., & Dwivedi, U. N. (2018). Terpenoids as promising therapeutic molecules against Alzheimer's disease: Amyloid beta-and acetylcholinesterase-directed pharmacokinetic and molecular docking analyses. *Molecular Simulation*, *44*, 1–11.

Babu, S., Prasanna, R., Bidyarani, N., & Singh, R. (2015). Analysing the colonisation of inoculated cyanobacteria in wheat plants using biochemical and molecular tools. *Journal of Applied Phycology*, *27*, 327–338.

Baweja, P., Kumar, S., & Kumar, G. (2019). Organic fertilizer from algae: A novel approach towards sustainable agriculture. In B. Giri, R. Prasad, Q.-S. Wu, & A. Varma (Eds.), *Biofertilizers for sustainable agriculture and environment*. Switzerland: Springer.

Benkendorff, K., Davis, A. R., Rogers, C. N., & Bremner, J. B. (2005). Free fatty acids and sterols in the benthic spawn of aquatic molluscs, and their associated antimicrobial properties. *Journal of Experimental Marine Biology and Ecology*, *316*, 29–44.

Betterle, N., & Melis, A. (2019). Photosynthetic generation of heterologous terpenoids in cyanobacteria. *Biotechnology and Bioengineering*, *116*, 2041–2051.

Boeckaert, C., Vlaeminck, B., Dijkstra, J., Issa-Zacharia, A., Van Nespen, T., Van Straalen, W., & Fievez, V. (2008). Effect of dietary starch or micro algae supplementation on rumen fermentation and milk fatty acid composition of dairy cows. *Journal of Dairy Science*, *91*, 4714–4727.

Boopathy, R., Bonvillain, C., Fontenot, Q., & Kilgen, M. (2007). Biological treatment of low-salinity shrimp aquaculture wastewater using sequencing batch reactor. *International Biodeterioration & Biodegradation*, *59*, 16–19.

Bruneel, C., Lemahieu, C., Fraeye, I., Ryckebosch, E., Muylaert, K., Buyse, J., & Foubert, I. (2013). Impact of microalgal feed supplementation on omega-3 fatty acid enrichment of hen eggs. *Journal of Functional Foods*, *5*, 897–904.

Bulgari, R., Franzoni, G., & Ferrante, A. (2019). Biostimulants application in horticultural crops under abiotic stress conditions. *Agronomy*, *9*, 306–335.

Buono, S., Colucci, A., Angelini, A., Langellotti, A. L., Massa, M., Martello, A., ... Dibenedetto, A. (2016). Productivity and biochemical composition of *Tetradesmus obliquus* and *Phaeodactylum tricornutum*: Effects of different cultivation approaches. *Journal of Applied Phycology*, *28*, 3179–3192.

Camacho, F., Macedo, A., & Malcata, F. (2019). Potential industrial applications and commercialization of microalgae in the functional food and feed industries: A short review. *Marine Drugs*, *17*, 312–336.

Campos, E. V. R., De Oliveira, J. L., Fraceto, L. F., & Singh, B. (2015). Polysaccharides as safer release systems for agrochemicals. *Agronomy for Sustainable Development*, *35*, 47–66.

Carvajal-Muñoz, J., & Carmona-Garcia, C. (2012). Benefits and limitations of biofertilization in agricultural practices. *Livestock Research for Rural Development*, *24*, 1–8.

Cezare-Gomes, E. A., Mejia-Da-Silva, L. D. C., Pérez-Mora, L. S., Matsudo, M. C., Ferreira-Camargo, L. S., Singh, A. K., & De Carvalho, J. C. M. (2019). Potential of microalgae carotenoids for industrial application. *Applied Biochemistry and Biotechnology*, *188*, 602–634.

Chanda, M.-J., Merghoub, N., & Arroussi, H. E. (2019). Microalgae polysaccharides: The new sustainable bioactive products for the development of plant bio-stimulants? *World Journal of Microbiology and Biotechnology*, *35*, 177–186.

Chatterjee, A., Singh, S., Agrawal, C., Yadav, S., Rai, R., & Rai, L. C. (2017). Role of algae as a biofertilizer. In R. P. Rastogi, D. Madamwar, & A. Pandey (Eds.), *Algal green chemistry*. Netherlands: Elsevier.

Chatzissavvidis, C., & Therios, I. (2014). Role of algae in agriculture. In V. H. Pomin (Ed.), *Seaweeds: Agricultural uses, biological and antioxidant agents*. USA: Nova Science Publishers, Inc.

Chiaiese, P., Corrado, G., Colla, G., Kyriacou, M. C., & Rouphael, Y. (2018). Renewable sources of plant biostimulation: Microalgae as a sustainable means to improve crop performance. *Frontiers in Plant Science*, *9*, 1782–1787.

Chokshi, K., Pancha, I., Ghosh, A., & Mishra, S. (2016). Microalgal biomass generation by phycoremediation of dairy industry wastewater: An integrated approach towards sustainable biofuel production. *Bioresource Technology*, *221*, 455–460.

Coppens, J., Grunert, O., Van Den Hende, S., Vanhoutte, I., Boon, N., Haesaert, G., & De Gelder, L. (2016). The use of microalgae as a high-value organic slow-release fertilizer results in tomatoes with increased carotenoid and sugar levels. *Journal of Applied Phycology*, *28*, 2367–2377.

Daneshvar, E., Zarrinmehr, M. J., Koutra, E., Kornaros, M., Farhadian, O., & Bhatnagar, A. (2019). Sequential cultivation of microalgae in raw and recycled dairy wastewater: Microalgal growth, wastewater treatment and biochemical composition. *Bioresource Technology*, *273*, 556–564.

Decesaro, A., Rampel, A., Machado, T. S., Thomé, A., Reddy, K., Margarites, A. C., & Colla, L. M. (2017). Bioremediation of soil contaminated with diesel and biodiesel fuel using biostimulation with microalgae biomass. *Journal of Environmental Engineering*, *143*, 1–8.

De Godos, I., González, C., Becares, E., García-Encina, P. A., & Muñoz, R. (2009). Simultaneous nutrients and carbon removal during pretreated swine slurry degradation in a tubular biofilm photobioreactor. *Applied Microbiology and Biotechnology*, *82*, 187–194.

de la Fuente-Vazquez, J., Diaz-Diaz-Chiron, M. T., Perez-Marcos, C., Caneque-Martinez, V., Sanchez-Gonzalez, C. I., Alvarez-Acero, I., ... Lauzurica-Gomez, S. (2014). Linseed, microalgae or fish oil dietary supplementation affects performance and quality characteristics of light lambs. *Spanish Journal of Agricultural Research*, *12*, 436–447.

Delattre, C., Pierre, G., Laroche, C., & Michaud, P. (2016). Production, extraction and characterization of microalgal and cyanobacterial exopolysaccharides. *Biotechnology Advances*, *34*, 1159–1179.

Demirbas, A., & Demirbas, M. F. (2011). Importance of algae oil as a source of biodiesel. *Energy Conversion and Management*, *52*, 163–170.

Deng, X.-Y., Gao, K., Zhang, R.-C., Addy, M., Lu, Q., Ren, H.-Y., ... Ruan, R. (2017). Growing *Chlorella vulgaris* on thermophilic anaerobic digestion swine manure for nutrient removal and biomass production. *Bioresource Technology*, *243*, 417–425.

Desbois, A. P., & Smith, V. J. (2010). Antibacterial free fatty acids: Activities, mechanisms of action and biotechnological potential. *Applied Microbiology and Biotechnology*, *85*, 1629–1642.

Díaz, M. T., Pérez, C., Sánchez, C. I., Lauzurica, S., Cañeque, V., González, C., & de la Fuente, J. (2017). Feeding microalgae increases omega-3 fatty acids of fat deposits and muscles in light lambs. *Journal of Food Composition and Analysis*, *56*, 115–123.

Dineshkumar, R., Subramanian, J., Gopalsamy, J., Jayasingam, P., Arumugam, A., Kannadasan, S., & Sampathkumar, P. (2019). The impact of using microalgae as biofertilizer in maize (*Zea mays* L.). *Waste and Biomass Valorization*, *10*, 1101–1110.

Dmytryk, A., & Chojnacka, K. (2018). Algae as fertilizers, biostimulants, and regulators of plant growth. In K. Chojnacka, P. P. Wieczorek, G. Schroeder, & I. Michalak (Eds.), *Algae biomass: Characteristics and applications*. Switzerland: Springer.

Du Jardin, P. (2015). Plant biostimulants: Definition, concept, main categories and regulation. *Scientia Horticulturae*, *196*, 3–14.

Dvir, I., Stark, A. H., Chayoth, R., Madar, Z., & Arad, S. M. (2009). Hypocholesterolemic effects of nutraceuticals produced from the red microalga *Porphyridium* sp. in rats. *Nutrients*, *1*, 156–167.

El Arroussi, H., Elbaouchi, A., Benhima, R., Bendaou, N., Smouni, A., & Wahby, I. (2015). Halophilic microalgae *Dunaliella salina* extracts improve seed germination and seedling growth of *Triticum aestivum* L. under salt stress. *Acta Horticulturae*, *1148*, 13–26.

Elarroussia, H., Elmernissia, N., Benhimaa, R., El Kadmiria, I. M., Bendaou, N., Smouni, A., & Wahbya, I. (2016). Microalgae polysaccharides a promising plant growth biostimulant. *Journal of Algal Biomass Utilization*, *7*, 55–63.

El-Baz, F., El-Senousy, W., El-Sayed, A., & Kamel, M. (2013). *In vitro* antiviral and antimicrobial activities of *Spirulina platensis* extract. *Journal of Applied Pharmaceutical Science*, *3*, 52–56.

Elshobary, M. E., Abo-Shady, A. M., Khairy, H. M., Essa, D., Zabed, H. M., Qi, X., & Abomohra, A. E.-F. (2019). Influence of nutrient supplementation and starvation conditions on the biomass and lipid productivities of *Micractinium reisseri* grown in wastewater for biodiesel production. *Journal of Environmental Management, 250*, 109529–109537.

Esquivel-Hernández, D. A., Ibarra-Garza, I. P., Rodríguez-Rodríguez, J., Cuéllar-Bermúdez, S. P., Rostro-Alanis, M. D. J., Alemán-Nava, G. S., … Parra-Saldívar, R. (2017). Green extraction technologies for high-value metabolites from algae: A review. *Biofuels, Bioproducts and Biorefining, 11*, 215–231.

Esserti, S., Smaili, A., Rifai, L. A., Koussa, T., Makroum, K., Belfaiza, M., … Faize, M. (2017). Protective effect of three brown seaweed extracts against fungal and bacterial diseases of tomato. *Journal of Applied Phycology, 29*, 1081–1093.

Farid, R., Mutale-Joan, C., Redouane, B., Najib, E. M., Abderahime, A., Laila, S., & Hicham, E. A. (2019). Effect of microalgae polysaccharides on biochemical and metabolomics pathways related to plant defense in *Solanum lycopersicum*. *Applied Biochemistry and Biotechnology, 188*, 225–240.

Feller, R., Matos, Â. P., Mazzutti, S., Moecke, E. H. S., Tres, M. V., Derner, R. B., … Junior, A. F. (2018). Polyunsaturated ω-3 and ω-6 fatty acids, total carotenoids and antioxidant activity of three marine microalgae extracts obtained by supercritical CO_2 and subcritical n-butane. *The Journal of Supercritical Fluids, 133*, 437–443.

Ferreira, A., Ribeiro, B., Ferreira, A. F., Tavares, M. L., Vladic, J., Vidović, S., … Goginyan, V. (2019). *Scenedesmus obliquus* microalga-based biorefinery–from brewery effluent to bioactive compounds, biofuels and biofertilizers–aiming at a circular bioeconomy. *Biofuels, Bioproducts and Biorefining, 13*, 1169–1186.

Fito, J., & Alemu, K. (2019). Microalgae-bacteria consortium treatment technology for municipal wastewater management. *Nanotechnology for Environmental Engineering, 4*, 4–12.

Foo, S. C., Yusoff, F. M., Ismail, M., Basri, M., Yau, S. K., Khong, N. M., … Ebrahimi, M. (2017). Antioxidant capacities of fucoxanthin-producing algae as influenced by their carotenoid and phenolic contents. *Journal of Biotechnology, 241*, 175–183.

Furbeyre, H., Van Milgen, J., Mener, T., Gloaguen, M., & Labussiere, E. (2017). Effects of dietary supplementation with freshwater microalgae on growth performance, nutrient digestibility and gut health in weaned piglets. *Animal, 11*, 183–192.

Galasso, C., Corinaldesi, C., & Sansone, C. (2017). Carotenoids from marine organisms: Biological functions and industrial applications. *Antioxidants, 6*, 96–128.

Gao, F., Li, C., Yang, Z.-H., Zeng, G.-M., Feng, L.-J., Liu, J.-Z., … Cai, H.-W. (2016). Continuous microalgae cultivation in aquaculture wastewater by a membrane photobioreactor for biomass production and nutrients removal. *Ecological Engineering, 92*, 55–61.

Gao, F., Peng, Y.-Y., Li, C., Yang, G.-J., Deng, Y.-B., Xue, B., & Guo, Y.-M. (2018). Simultaneous nutrient removal and biomass/lipid production by *Chlorella* sp. in seafood processing wastewater. *Science of the Total Environment, 640*, 943–953.

Gatrell, S. K., Kim, J., Derksen, T. J., O'Neil, E. V., & Lei, X. G. (2015). Creating omega-3 fatty-acid-enriched chicken using defatted green microalgal biomass. *Journal of Agricultural and Food Chemistry, 63*, 9315–9322.

Gerasimenko, N., Martyyas, E., Logvinov, S., & Busarova, N. (2014). Biological activity of lipids and photosynthetic pigments of *Sargassum pallidum* C. Agardh. *Applied Biochemistry and Microbiology, 50*, 73–81.

Gershenzon, J., & Dudareva, N. (2007). The function of terpene natural products in the natural world. *Nature Chemical Biology, 3*, 408–414.

Ginzberg, A., Cohen, M., Sod-Moriah, U. A., Shany, S., Rosenshtrauch, A., & Arad, S. (2000). Chickens fed with biomass of the red microalga *Porphyridium* sp. have reduced blood cholesterol level and modified fatty acid composition in egg yolk. *Journal of Applied Phycology*, *12*, 325–330.

Goh, B. H. H., Ong, H. C., Cheah, M. Y., Chen, W.-H., Yu, K. L., & Mahlia, T. M. I. (2019). Sustainability of direct biodiesel synthesis from microalgae biomass: A critical review. *Renewable & Sustainable Energy Reviews*, *107*, 59–74.

Goiris, K., Muylaert, K., Fraeye, I., Foubert, I., De Brabanter, J., & De Cooman, L. (2012). Antioxidant potential of microalgae in relation to their phenolic and carotenoid content. *Journal of Applied Phycology*, *24*, 1477–1486.

Gonçalves, A. L., Pires, J. C., & Simões, M. (2017). A review on the use of microalgal consortia for wastewater treatment. *Algal Research*, *24*, 403–415.

González, A., Castro, J., Vera, J., & Moenne, A. (2013). Seaweed oligosaccharides stimulate plant growth by enhancing carbon and nitrogen assimilation, basal metabolism, and cell division. *Journal of Plant Growth Regulation*, *32*, 443–448.

Górka, B., Korzeniowska, K., Lipok, J., & Wieczorek, P. P. (2018). The biomass of algae and algal extracts in agricultural production. In K. Chojnacka, P. P. Wieczorek, G. Schroeder, & I. Michalak (Eds.), *Algae biomass: Characteristics and applications*. Switzerland: Springer.

Gouveia, L., Batista, A. P., Sousa, I., Raymundo, A., & Bandarra, N. M. (2008). Microalgae in novel food products. In K. N. Papadopoulos (Ed.), *Food chemistry research developments*. USA: Nova Science Publishers, Inc.

Guedes, A. C., Amaro, H. M., & Malcata, F. X. (2011). Microalgae as sources of carotenoids. *Marine Drugs*, *9*, 625–644.

Guilherme, M. R., Aouada, F. A., Fajardo, A. R., Martins, A. F., Paulino, A. T., Davi, M. F., … Muniz, E. C. (2015). Superabsorbent hydrogels based on polysaccharides for application in agriculture as soil conditioner and nutrient carrier: A review. *European Polymer Journal*, *72*, 365–385.

Gupta, S., Pandey, R., & Pawar, S. B. (2016). Microalgal bioremediation of food-processing industrial wastewater under mixotrophic conditions: Kinetics and scale-up approach. *Frontiers of Chemical Science and Engineering*, *10*, 499–508.

Gupta, S., & Pawar, S. B. (2018). An integrated approach for microalgae cultivation using raw and anaerobic digested wastewaters from food processing industry. *Bioresource Technology*, *269*, 571–576.

Gupta, S., Pawar, S. B., & Pandey, R. (2019). Current practices and challenges in using microalgae for treatment of nutrient rich wastewater from agro-based industries. *The Science of the Total Environment*, *687*, 1107–1126.

Guzmán, S., Gato, A., Lamela, M., Freire-Garabal, M., & Calleja, J. M. (2003). Anti-inflammatory and immunomodulatory activities of polysaccharide from *Chlorella stigmatophora* and *Phaeodactylum tricornutum*. *Phytotherapy Research*, *17*, 665–670.

Haggag, W., Hoballah, M., & Ali, R. (2018). Applications of nano biotechnological microalgae product for improve wheat productivity in semai aird areas. *International Journal of Agricultural Technology*, *14*, 675–692.

Hamed, S. M., Abd El-Rhman, A. A., Abdel-Raouf, N., & Ibraheem, I. B. M. (2018). Role of marine macroalgae in plant protection & improvement for sustainable agriculture technology. *Beni-Suef University Journal of Basic and Applied Sciences*, *7*, 104–110.

Han, P., Lu, Q., Fan, L., & Zhou, W. (2019). A review on the use of microalgae for sustainable aquaculture. *Applied Sciences*, *9*, 2377–2396.

Han, T., Zhao, Z., & Wang, Y. (2016). The effect of ryegrass and fertilizer on the petroleum contaminated soil remediation. *Fresenius Environmental Bulletin*, *25*, 2243–2250.

Han, X., Zeng, H., Bartocci, P., Fantozzi, F., & Yan, Y. (2018). Phytohormones and effects on growth and metabolites of microalgae: A review. *Fermentation*, *4*, 25–39.

Hernández, D., Riaño, B., Coca, M., & García-González, M. (2013). Treatment of agro-industrial wastewater using microalgae–bacteria consortium combined with anaerobic digestion of the produced biomass. *Bioresource Technology*, *135*, 598–603.

Hongyang, S., Yalei, Z., Chunmin, Z., Xuefei, Z., & Jinpeng, L. (2011). Cultivation of *Chlorella pyrenoidosa* in soybean processing wastewater. *Bioresource Technology*, *102*, 9884–9890.

Hu, C., Liu, Y., Song, L., & Zhang, D. (2002). Effect of desert soil algae on the stabilization of fine sands. *Journal of Applied Phycology*, *14*, 281–292.

Hu, J. J., Nagarajan, D., Zhang, Q. G., Chang, J. S., & Lee, D. J. (2018). Heterotrophic cultivation of microalgae for pigment production: A review. *Biotechnology Advances*, *36*, 54–67.

Issa, O. M., Défarge, C., Le Bissonnais, Y., Marin, B., Duval, O., Bruand, A., ... Annerman, M. (2007). Effects of the inoculation of cyanobacteria on the microstructure and the structural stability of a tropical soil. *Plant and Soil*, *290*, 209–219.

Jacob-Lopes, E., Maroneze, M. M., Deprá, M. C., Sartori, R. B., Dias, R. R., & Zepka, L. Q. (2019). Bioactive food compounds from microalgae: An innovative framework on industrial biorefineries. *Current Opinion in Food Science*, *25*, 1–7.

Jha, M. N., & Prasad, A. N. (2006). Efficacy of new inexpensive cyanobacterial biofertilizer including its shelf-life. *World Journal of Microbiology and Biotechnology*, *22*, 73–79.

Karadag, D., Köroğlu, O. E., Ozkaya, B., & Cakmakci, M. (2015). A review on anaerobic biofilm reactors for the treatment of dairy industry wastewater. *Process Biochemistry*, *50*, 262–271.

Karthikeyan, N., Prasanna, R., Nain, L., & Kaushik, B. D. (2007). Evaluating the potential of plant growth promoting cyanobacteria as inoculants for wheat. *European Journal of Soil Biology*, *43*, 23–30.

Kaya, C., Higgs, D., & Sakar, E. (2002). Response of two leafy vegetables grown at high salinity to supplementary potassium and phosphorus during different growth stages. *Journal of Plant Nutrition*, *25*, 2663–2676.

Khan, M. I., Shin, J. H., & Kim, J. D. (2018). The promising future of microalgae: Current status, challenges, and optimization of a sustainable and renewable industry for biofuels, feed, and other products. *Microbial Cell Factories*, *17*, 36–56.

Khan, W., Rayirath, U. P., Subramanian, S., Jithesh, M. N., Rayorath, P., Hodges, D. M., ... Prithiviraj, B. (2009). Seaweed extracts as biostimulants of plant growth and development. *Journal of Plant Growth Regulation*, *28*, 386–399.

Khan, Z., Kim, Y. H., Kim, S. G., & Kim, H. W. (2007). Observations on the suppression of root-knot nematode (*Meloidogyne arenaria*) on tomato by incorporation of cyanobacterial powder (*Oscillatoria chlorina*) into potting field soil. *Bioresource Technology*, *98*, 69–73.

Khoddami, A., Wilkes, M. A., & Roberts, T. H. (2013). Techniques for analysis of plant phenolic compounds. *Molecules*, *18*, 2328–2375.

Kim, J.-D. (2006). Screening of cyanobacteria (blue-green algae) from rice paddy soil for antifungal activity against plant pathogenic fungi. *Mycobiology*, *34*, 138–142.

Koutra, E., Grammatikopoulos, G., & Kornaros, M. (2018). Selection of microalgae intended for valorization of digestate from agro-waste mixtures. *Waste Management*, *73*, 123–129.

Kumar, K. S., Dahms, H.-U., Won, E.-J., Lee, J.-S., & Shin, K.-H. (2015). Microalgae – A promising tool for heavy metal remediation. *Ecotoxicology and Environmental Safety*, *113*, 329–352.

Kumar, M., Prasanna, R., Bidyarani, N., Babu, S., Mishra, B. K., Kumar, A., ... Singh, R. (2013). Evaluating the plant growth promoting ability of thermotolerant bacteria and cyanobacteria and their interactions with seed spice crops. *Scientia Horticulturae*, *164*, 94–101.

Kusvuran, A., & Kusvuran, S. (2019). Using of microbial fertilizer as biostimulant alleviates damage from drought stress in guar (*Cyamopsis Tetragonoloba* (L.) Taub.) seedlings. *International Letters of Natural Sciences, 76*, 147–157.

Lam, M. K., & Lee, K. T. (2012). Microalgae biofuels: A critical review of issues, problems and the way forward. *Biotechnology Advances, 30*, 673–690.

Latef, A. A. H. A., Srivastava, A. K., Saber, H., Alwaleed, E. A., & Tran, L.-S. P. (2017). *Sargassum muticum* and *Jania rubens* regulate amino acid metabolism to improve growth and alleviate salinity in chickpea. *Scientific Reports, 7*, 1–12.

Li, K., Liu, Q., Fang, F., Luo, R., Lu, Q., Zhou, W., ... Addy, M. (2019). Microalgae-based wastewater treatment for nutrients recovery: A review. *Bioresource Technology*, 121934–121949.

Li, S., Zhao, S., Yan, S., Qiu, Y., Song, C., Li, Y., & Kitamura, Y. (2019). Food processing wastewater purification by microalgae cultivation associated with high value-added compounds production – A review. *Chinese Journal of Chemical Engineering, 27*, 2845–2856.

Li, X., Xu, K., Fu, W., Wang, J., Zhu, Y., Li, C., & Zhou, X. (2014). Simultaneous in-situ excess sludge reduction and removal of organic carbon and nitrogen by a pilot-scale continuous aerobic–anaerobic coupled (CAAC) process for deeply treatment of soybean wastewater. *Biochemical Engineering Journal, 85*, 30–37.

Lu, Q., Zhou, W., Min, M., Ma, X., Chandra, C., Doan, Y. T., ... Griffith, R. (2015). Growing *Chlorella* sp. on meat processing wastewater for nutrient removal and biomass production. *Bioresource Technology, 198*, 189–197.

Lu, W., Wang, Z., Wang, X., & Yuan, Z. (2015). Cultivation of *Chlorella* sp. using raw dairy wastewater for nutrient removal and biodiesel production: Characteristics comparison of indoor bench-scale and outdoor pilot-scale cultures. *Bioresource Technology, 192*, 382–388.

Lu, Y., & Xu, J. (2015). Phytohormones in microalgae: A new opportunity for microalgal biotechnology? *Trends in Plant Science, 20*, 273–282.

Lum, K. K., Kim, J., & Lei, X. G. (2013). Dual potential of microalgae as a sustainable biofuel feedstock and animal feed. *Journal of Animal Science and Biotechnology, 4*, 53–59.

Luo, L.-Z., Shao, Y., Luo, S., Zeng, F.-J., & Tian, G.-M. (2019). Nutrient removal from piggery wastewater by *Desmodesmus* sp. CHX1 and its cultivation conditions optimization. *Environmental Technology, 40*, 2739–2746.

Manjunath, M., Kanchan, A., Ranjan, K., Venkatachalam, S., Prasanna, R., Ramakrishnan, B., ... Singh, B. (2016). Beneficial cyanobacteria and eubacteria synergistically enhance bioavailability of soil nutrients and yield of okra. *Heliyon, 2*, 66-93.

Mansori, M., Chernane, H., Latique, S., Benaliat, A., Hsissou, D., & El Kaoua, M. (2015). Seaweed extract effect on water deficit and antioxidative mechanisms in bean plants (*Phaseolus vulgaris* L.). *Journal of Applied Phycology, 27*, 1689–1698.

Marks, E. A. N., Montero, O., & Rad, C. (2019). The biostimulating effects of viable microalgal cells applied to a calcareous soil: Increases in bacterial biomass, phosphorus scavenging, and precipitation of carbonates. *Science of the Total Environment, 692*, 784–790.

Mata, T. M., Melo, A. C., Simões, M., & Caetano, N. S. (2012). Parametric study of a brewery effluent treatment by microalgae *Scenedesmus obliquus*. *Bioresource Technology, 107*, 151–158.

Meale, S. J., Chaves, A. V., He, M. L., & Mcallister, T. A. (2014). Dose–response of supplementing marine algae (*Schizochytrium* spp.) on production performance, fatty acid profiles, and wool parameters of growing lambs. *Journal of Animal Science, 92*, 2202–2213.

Mendes, L. B. B., & Vermelho, A. B. (2013). Allelopathy as a potential strategy to improve microalgae cultivation. *Biotechnology for Biofuels, 6*, 152–165.

Mercier, L., Lafitte, C., Borderies, G., Briand, X., Esquerré-Tugayé, M. T., & Fournier, J. (2001). The algal polysaccharide carrageenans can act as an elicitor of plant defence. *New Phytologist*, *149*, 43–51.

Michalak, I., Chojnacka, K., & Saeid, A. (2017). Plant growth biostimulants, dietary feed supplements and cosmetics formulated with supercritical CO_2 algal extracts. *Molecules*, *22*, 66–82.

Mobin, S., & Alam, F. (2017). Some promising microalgal species for commercial applications: A review. *Energy Procedia*, *110*, 510–517.

Mógor, Á. F., Ördög, V., Lima, G. P. P., Molnár, Z., & Mógor, G. (2018). Biostimulant properties of cyanobacterial hydrolysate related to polyamines. *Journal of Applied Phycology*, *30*, 453–460.

Molinuevo-Salces, B., Riaño, B., Hernández, D., & García-González, M. C. (2019). Microalgae and wastewater treatment: Advantages and disadvantages. In M. A. Alam, & Z. Wang (Eds.), *Microalgae biotechnology for development of biofuel and wastewater treatment*. Switzerland: Springer.

Moran, C. A., Currie, D., Keegan, J. D., & Knox, A. (2018). Tolerance of broilers to dietary supplementation with high levels of the DHA-rich microalga, *Aurantiochytrium Limacinum*: Effects on health and productivity. *Animals*, *8*, 180–190.

Mujtaba, G., & Lee, K. (2017). Treatment of real wastewater using co-culture of immobilized *Chlorella vulgaris* and suspended activated sludge. *Water Research*, *120*, 174–184.

Muniraj, I. K., Xiao, L., Hu, Z., Zhan, X., & Shi, J. (2013). Microbial lipid production from potato processing wastewater using oleaginous filamentous fungi *Aspergillus oryzae*. *Water Research*, *47*, 3477–3483.

Nisha, R. , Kaushik, A., & Kaushik, C. P. (2007). Effect of indigenous cyanobacterial application on structural stability and productivity of an organically poor semi-arid soil. *Geoderma*, *138*, 49–56.

Nur, M. M. A., & Buma, A. G. (2018). Opportunities and challenges of microalgal cultivation on wastewater, with special focus on palm oil mill effluent and the production of high value compounds. *Waste and Biomass Valorization*, 1–19.

Oh, S. T., Zheng, L., Kwon, H. J., Choo, Y. K., Lee, K. W., Kang, C. W., & An, B. K. (2015). Effects of dietary fermented *Chlorella vulgaris* (CBT (R)) on growth performance, relative organ weights, cecal microflora, tibia bone characteristics, and meat qualities in Pekin ducks. *Asian-Australasian Journal of Animal Sciences*, *28*, 95–101.

Ördög, V., Stirk, W. A., Van Staden, J., Novák, O., & Strnad, M. (2004). Endogenous cytokinins in three genera of microalgae from the Chlorophyta. *Journal of Phycology*, *40*, 88–95.

Osman, M. E. H., El-Sheekh, M. M., El-Naggar, A. H., & Gheda, S. F. (2010). Effect of two species of cyanobacteria as biofertilizers on some metabolic activities, growth, and yield of pea plant. *Biology and Fertility of Soils*, *46*, 861–875.

Pan, S., Jeevanandam, J., & Danquah, M. K. (2019). Benefits of algal extracts in sustainable agriculture. In A. Hallmann, & P. Rampelotto (Eds.), *Grand challenges in algae biotechnology*. Switzerland: Springer.

Pardee, K., Ellis, P., Bouthillier, M., Towers, G. H., & French, C. (2004). Plant virus inhibitors from marine algae. *Canadian Journal of Botany*, *82*, 304–309.

Park, J. H., Upadhaya, S. D., & Kim, I. H. (2015). Effect of dietary marine microalgae (*Schizochytrium*) powder on egg production, blood lipid profiles, egg quality, and fatty acid composition of egg yolk in layers. *Asian-Australasian Journal of Animal Sciences*, *28*, 391–397.

Pathak, J., Rajneesh., Maurya, P. K., Singh, S. P., Häder, D.-P., & Sinha, R. P. (2018). Cyanobacterial farming for environment friendly sustainable agriculture practices: Innovations and perspectives. *Frontiers in Environmental Science*, *6*, 7–19.

Pattanaik, B., & Lindberg, P. (2015). Terpenoids and their biosynthesis in cyanobacteria. *Life*, *5*, 269–293.

Pavela, R., & Benelli, G. (2016). Essential oils as ecofriendly biopesticides? Challenges and constraints. *Trends in Plant Science*, *21*, 1000–1007.

Peiretti, P. G., & Meineri, G. (2008). Effects of diets with increasing levels of *Spirulina platensis* on the performance and apparent digestibility in growing rabbits. *Livestock Science*, *118*, 173–177.

Pires, J. C. M., Alvim-Ferraz, M. C. M., Martins, F. G., & Simoes, M. (2012). Carbon dioxide capture from flue gases using microalgae: Engineering aspects and biorefinery concept. *Renewable & Sustainable Energy Reviews*, *16*, 3043–3053.

Porwal, H., Mane, A., & Velhal, S. (2015). Biodegradation of dairy effluent by using microbial isolates obtained from activated sludge. *Water Resources and Industry*, *9*, 1–15.

Posadas, E., Bochon, S., Coca, M., García-González, M., García-Encina, P., & Muñoz, R. (2014). Microalgae-based agro-industrial wastewater treatment: A preliminary screening of biodegradability. *Journal of Applied Phycology*, *26*, 2335–2345.

Prasanna, R., Triveni, S., Bidyarani, N., Babu, S., Yadav, K., Adak, A., ... Saxena, A. K. (2014). Evaluating the efficacy of cyanobacterial formulations and biofilmed inoculants for leguminous crops. *Archives of Agronomy and Soil Science*, *60*, 349–366.

Priya, M., Gurung, N., Mukherjee, K., & Bose, S. (2014). Microalgae in removal of heavy metal and organic pollutants from soil. In S. Das (Ed.), *Microbial Biodegradation and Bioremediation*. UK: Elsevier.

Qin, L., Wang, Z., Sun, Y., Shu, Q., Feng, P., Zhu, L., ... Yuan, Z. (2016). Microalgae consortia cultivation in dairy wastewater to improve the potential of nutrient removal and biodiesel feedstock production. *Environmental Science and Pollution Research*, *23*, 8379–8387.

Qiu, Y., Zu, Y., Song, C., Xie, M., Qi, Y., Kansha, Y., & Kitamura, Y. (2019). Soybean processing wastewater purification via *Chlorella* L166 and L38 with potential value-added ingredients production. *Bioresource Technology Reports*, *7*, 100195–100199.

Queiroz, M. I., Lopes, E. J., Zepka, L. Q., Bastos, R. G., & Goldbeck, R. (2007). The kinetics of the removal of nitrogen and organic matter from parboiled rice effluent by cyanobacteria in a stirred batch reactor. *Bioresource Technology*, *98*, 2163–2169.

Rajesh, K., Rohit, M. V., & Venkata Mohan, S. (2017). Microalgae-based carotenoids production. In R. P. Rastogi, D. Madamwar, & A. Pandey (Eds.), *Algal green chemistry*. Netherlands: Elsevier.

Ramjeawon, T. (2000). Cleaner production in Mauritian cane-sugar factories. *Journal of Cleaner Production*, *8*, 503–510.

Raposo, M. F. D. J., De Morais, A. M. M. B., & De Morais, R. M. S. C. (2015). Carotenoids from marine microalgae: A valuable natural source for the prevention of chronic diseases. *Marine Drugs*, *13*, 5128–5155.

Raposo, M. F. D. J., Oliveira, S. E., Castro, P. M., Bandarra, N. M., & Morais, R. M. (2010). On the utilization of microalgae for brewery effluent treatment and possible applications of the produced biomass. *Journal of the Institute of Brewing*, *116*, 285–292.

Rechter, S., König, T., Auerochs, S., Thulke, S., Walter, H., Dörnenburg, H., ... Marschall, M. (2006). Antiviral activity of *Arthrospira*-derived spirulan-like substances. *Antiviral Research*, *72*, 197–206.

Reddy, C. A., & Saravanan, R. S. (2013). Polymicrobial multi-functional approach for enhancement of crop productivity. In S. Sariaslani, & G. M. Gadd (Eds.), *Advances in applied microbiology*. USA: Academic Press.

Renuka, N., Guldhe, A., Prasanna, R., Singh, P., & Bux, F. (2018). Microalgae as multi-functional options in modern agriculture: Current trends, prospects and challenges. *Biotechnology Advances*, *36*, 1255–1273.

Renuka, N., Prasanna, R., Sood, A., Ahluwalia, A. S., Bansal, R., Babu, S., ... Nain, L. (2016). Exploring the efficacy of wastewater-grown microalgal biomass as a biofertilizer for wheat. *Environmental Science and Pollution Research*, *23*, 6608–6620.

Renuka, N., Prasanna, R., Sood, A., Bansal, R., Bidyarani, N., Singh, R., ... Ahluwalia, A. S. (2017). Wastewater grown microalgal biomass as inoculants for improving micronutrient availability in wheat. *Rhizosphere*, *3*, 150–159.

Riaño, B., Molinuevo, B., & García-González, M. (2011). Treatment of fish processing wastewater with microalgae-containing microbiota. *Bioresource Technology*, *102*, 10829–10833.

Ribeiro, T., Lordelo, M. M., Costa, P., Alves, S. P., Benevides, W. S., Bessa, R. J. B., ... Prates, J. A. M. (2014). Effect of reduced dietary protein and supplementation with a docosahexaenoic acid product on broiler performance and meat quality. *British Poultry Science*, *55*, 752–765.

Rodríguez, A. A., Stella, A. M., Storni, M. M., Zulpa, G., & Zaccaro, M. C. (2006). Effects of cyanobacterial extracellular products and gibberellic acid on salinity tolerance in *Oryza sativa* L. *Saline Systems*, *2*, 7–10.

Rodriguez-Garcia, A., Hosseini, S., Martinez-Chapa, S. O., & Cordell, G. A. (2017). Multi-target activities of selected alkaloids and terpenoids. *Mini-Reviews in Organic Chemistry*, *14*, 272–279.

Ronga, D., Biazzi, E., Parati, K., Carminati, D., Carminati, E., & Tava, A. (2019). Microalgal biostimulants and biofertilisers in crop productions. *Agronomy*, *9*, 192–213.

Ruiz, J., Arbib, Z., Álvarez-Díaz, P., Garrido-Pérez, C., Barragán, J., & Perales, J. (2013). Photobiotreatment model (PhBT): A kinetic model for microalgae biomass growth and nutrient removal in wastewater. *Environmental Technology*, *34*, 979–991.

Sakamoto, Y., Mori, K., Matsuo, Y., Mukojima, N., Watanabe, W., Sobaru, N., ... Watanuki, H. (2017). Breeding of a new potato variety 'Nagasaki Kogane' with high eating quality, high carotenoid content, and resistance to diseases and pests. *Breeding Science*, *67*, 320–326.

Sampaio, M., Gonçalves, M., & Marques, I. (2011). Anaerobic digestion challenge of raw olive mill wastewater. *Bioresource Technology*, *102*, 10810–10818.

Sarhan, T. Z., & Ismael, S. F. (2014). Effect of low temperature and seaweed extracts on flowering and yield of two cucumber cultivars (*Cucumis sativus* L.). *International Journal of Agricultural and Food Research*, *3*, 41–54.

Sarker, P. K., Kapuscinski, A. R., Bae, A. Y., Donaldson, E., Sitek, A. J., Fitzgerald, D. S., & Edelson, O. F. (2018). Towards sustainable aquafeeds: Evaluating substitution of fishmeal with lipid-extracted microalgal co-product (*Nannochloropsis oculata*) in diets of juvenile Nile tilapia (*Oreochromis niloticus*). *PLoS One*, *13*, 1–25.

Schulz, C., Gelbrecht, J., & Rennert, B. (2004). Constructed wetlands with free water surface for treatment of aquaculture effluents. *Journal of Applied Ichthyology*, *20*, 64–70.

Shahid, A., Malik, S., Zhu, H., Xu, J., Nawaz, M. Z., Nawaz, S., ... Mehmood, M. A. (2020). Cultivating microalgae in wastewater for biomass production, pollutant removal, and atmospheric carbon mitigation; a review. *Science of the Total Environment*, *704*, 135303–135319.

Sharma, R., Khokhar, M., Jat, R., & Khandelwal, S. (2012). Role of algae and cyanobacteria in sustainable agriculture system. *Wudpecker Journal of Agricultural Research*, *1*, 381–388.

Sheikhzadeh, N., Tayefi-Nasrabadi, H., Oushani, A. K., & Enferadi, M. H. N. (2012). Effects of *Haematococcus pluvialis* supplementation on antioxidant system and metabolism in rainbow trout (*Oncorhynchus mykiss*). *Fish Physiology and Biochemistry*, *38*, 413–419.

Singh, R., Parihar, P., Singh, M., Bajguz, A., Kumar, J., Singh, S., ... Prasad, S. M. (2017). Uncovering potential applications of cyanobacteria and algal metabolites in biology, agriculture and medicine: Current status and future prospects. *Frontiers in Microbiology*, *8*, 515–551.

Singh, S., & Datta, P. (2007). Outdoor evaluation of herbicide resistant strains of *Anabaena variabilis* as biofertilizer for rice plants. *Plant and Soil*, *296*, 95–102.

Sklyar, V., Epov, A., Gladchenko, M., Danilovich, D., & Kalyuzhnyi, S. (2003). Combined biologic (anaerobic-aerobic) and chemical treatment of starch industry wastewater. *Applied Biochemistry and Biotechnology*, *109*, 253–262.

Spruijt, J., Van Der Weide, R. & Van Krimpen, M. (2016). Opportunities for micro algae as ingredient in animal diets. ACRRES—Application Centre for Renewable Resources.

Sroka, E., Kamiński, W., & Bohdziewicz, J. (2004). Biological treatment of meat industry wastewater. *Desalination*, *162*, 85–91.

Stamatiadis, S., Evangelou, L., Yvin, J.-C., Tsadilas, C., Mina, J. M. G., & Cruz, F. (2015). Responses of winter wheat to *Ascophyllum nodosum* (L.) Le Jol. extract application under the effect of N fertilization and water supply. *Journal of Applied Phycology*, *27*, 589–600.

Stoneham, T. R., Kuhn, D. D., Taylor, D. P., Neilson, A. P., Smith, S. A., Gatlin, D. M., … O'Keefe, S. F. (2018). Production of omega-3 enriched tilapia through the dietary use of algae meal or fish oil: Improved nutrient value of fillet and offal. *PLoS One*, *13*, 1–14.

Swarnalakshmi, K., Prasanna, R., Kumar, A., Pattnaik, S., Chakravarty, K., Shivay, Y. S., … Saxena, A. K. (2013). Evaluating the influence of novel cyanobacterial biofilmed biofertilizers on soil fertility and plant nutrition in wheat. *European Journal of Soil Biology*, *55*, 107–116.

Sytar, O., Brestic, M., Rai, M., & Shao, H. (2012). Plant phenolic compounds for food, pharmaceutical and cosmetics production. *Journal of Medicinal Plants Research*, *6*, 2526–2539.

Till, B. E., Huntington, J. A., Posri, W., Early, R., Taylor-Pickard, J., & Sinclair, L. A. (2019). Influence of rate of inclusion of microalgae on the sensory characteristics and fatty acid composition of cheese and performance of dairy cows. *Journal of Dairy Science*, *102*, 10934–10946.

Toyomizu, M., Sato, K., Taroda, H., Kato, T., & Akiba, Y. (2001). Effects of dietary *Spirulina* on meat colour in muscle of broiler chickens. *British Poultry Science*, *42*, 197–202.

Umamaheswari, J., & Shanthakumar, S. (2016). Efficacy of microalgae for industrial wastewater treatment: A review on operating conditions, treatment efficiency and biomass productivity. *Reviews in Environmental Science and Bio/Technology*, *15*, 265–284.

Urrutia, O., Mendizabal, J. A., Insausti, K., Soret, B., Purroy, A., & Arana, A. (2016). Effects of addition of linseed and marine algae to the diet on adipose tissue development, fatty acid profile, lipogenic gene expression, and meat quality in lambs. *PLoS One*, *11*, 1–23.

Usman, A., Khalid, S., Usman, A., Hussain, Z., & Wang, Y. (2017). Algal polysaccharides, novel application, and outlook. In K. M. Zia, M. Zuber, & M. Ali (Eds.), *Algae based polymers, blends, and composites*. Netherlands: Elsevier.

Valente, L. M. P., Custodio, M., Batista, S., Fernandes, H., & Kiron, V. (2019). Defatted microalgae (*Nannochloropsis* sp.) from biorefinery as a potential feed protein source to replace fishmeal in European sea bass diets. *Fish Physiology and Biochemistry*, *45*, 1067–1081.

Van Emon, M. L., Loy, D. D., & Hansen, S. L. (2015). Determining the preference, in vitro digestibility, in situ disappearance, and grower period performance of steers fed a novel algae meal derived from heterotrophic microalgae. *Journal of Animal Science*, *93*, 3121–3129.

Vera, J., Castro, J., Gonzalez, A., & Moenne, A. (2011). Seaweed polysaccharides and derived oligosaccharides stimulate defense responses and protection against pathogens in plants. *Marine Drugs*, *9*, 2514–2525.

Vessey, J. K. (2003). Plant growth promoting rhizobacteria as biofertilizers. *Plant and Soil*, *255*, 571–586.

Vossen, E., Raes, K., Van Mullem, D., & De Smet, S. (2017). Production of docosahexaenoic acid (DHA) enriched loin and dry cured ham from pigs fed algae: Nutritional and sensory quality. *European Journal of Lipid Science and Technology*, *119*, 1600144–1600156.

Wang, H., Xiong, H., Hui, Z., & Zeng, X. (2012). Mixotrophic cultivation of *Chlorella pyrenoidosa* with diluted primary piggery wastewater to produce lipids. *Bioresource Technology, 104*, 215–220.

Wei, G., Jia, Q., Chen, X., Köllner, T. G., Bhattacharya, D., Wong, G. K.-S., ... Chen, F. (2019). Terpene biosynthesis in red algae is catalyzed by microbial type but not typical plant terpene synthases. *Plant Physiology, 179*, 382–390.

Win, T. T., Barone, G. D., Secundo, F., & Fu, P. (2018). Algal biofertilizers and plant growth stimulants for sustainable agriculture. *Industrial Biotechnology, 14*, 203–211.

Yaakob, Z., Ali, E., Zainal, A., Mohamad, M., & Takriff, M. S. (2014). An overview: Biomolecules from microalgae for animal feed and aquaculture. *Journal of Biological Research-Thessaloniki, 21*, 6–15.

Yan, L., & Kim, I. H. (2013). Effects of dietary omega-3 fatty acid-enriched microalgae supplementation on growth performance, blood profiles, meat quality, and fatty acid composition of meat in broilers. *Journal of Applied Animal Research, 41*, 392–397.

Yan, L., Lim, S. U., & Kim, I. H. (2012). Effect of fermented *Chlorella* supplementation on growth performance, nutrient digestibility, blood characteristics, fecal microbial and fecal noxious gas content in growing pigs. *Asian-Australasian Journal of Animal Sciences, 25*, 1742–1747.

Yu, W., Wen, G. L., Lin, H. Z., Yang, Y. K., Huang, X. L., Zhou, C. P., ... Li, T. (2018). Effects of dietary *Spirulina platensis* on growth performance, hematological and serum biochemical parameters, hepatic antioxidant status, immune responses and disease resistance of Coral trout *Plectropomus leopardus* (Lacepede, 1802). *Fish & Shellfish Immunology, 74*, 649–655.

Zhu, G.-F., Li, J.-Z., Wu, P., Jin, H.-Z., & Wang, Z. (2008). The performance and phase separated characteristics of an anaerobic baffled reactor treating soybean protein processing wastewater. *Bioresource Technology, 99*, 8027–8033.

Zhu, S., Qin, L., Feng, P., Shang, C., Wang, Z., & Yuan, Z. (2019). Treatment of low C/N ratio wastewater and biomass production using co-culture of *Chlorella vulgaris* and activated sludge in a batch photobioreactor. *Bioresource Technology, 274*, 313–320.

Zotte, A. D., Cullere, M., Sartori, A., Szendro, Z., Kovacs, M., Giaccone, V., & Dal Bosco, A. (2014). Dietary *Spirulina* (*Arthrospira platensis*) and *Thyme* (*Thymus vulgaris*) supplementation to growing rabbits: Effects on raw and cooked meat quality, nutrient true retention and oxidative stability. *Meat Science, 98*, 94–103.

Chapter 9

Microalgae and cyanobacteria as food: Legislative and safety aspects

Giorgos Markou[1], **Imene Chentir**[2,3] **and Ioannis Tzovenis**[4]
[1]Institute of Technology of Agricultural Products, Hellenic Agricultural Organization-Demeter (HAO-Demeter), Athens, Greece, [2]Higher School of Food Science and Agri-Food Industry, Algiers, Algeria, [3]Laboratory of Enzyme Engineering and Microbiology, National School of Engineering of Sfax, University of Sfax, Sfax, Tunisia, [4]MicroPhykos, Athens, Greece

9.1 Introduction

Food and feed security are forecast to be put under great pressure in the near future, due to the sharp rise in the world's population and the subsequent increase in protein demand. In the recent past, the global demand for animal and plant proteins has increased continuously. The global human population has more than doubled from 2.7 in 1950 to currently around 8 billion people. However, the increase in purchasing power of more people has led to an increase in meat production by a factor of 5, from 45 to 233 million tons per year (Van Der Spiegel, Noordam, & Van Der Fels-Klerx, 2013). The Food and Agriculture Organization of the United Nations (FAO) projects a global population of more than 9 billion people and meat production of 410 million tons per year by 2050 (Boland et al., 2013; Henchion, Hayes, Mullen, Fenelon, & Tiwari, 2017). Within this situation, the need for new and alternative sources of nutrients and in particular of proteins has become urgent.

The European Union (EU) has a protein deficit of about 70%–80% (De Visser, Schreuder, & Stoddard, 2014). The approximately 400,000 ha of soya grown in Europe today only represents around 3% of what Europe needs to produce vegetable proteins for animal feed. The EU faces a significant issue on protein deficiency which is also expressed by a EU Parliament motion[1] that

1. See the motion "The EU protein deficit: what solution for a long-standing problem?" [2010/2111(INI)], http://www.europarl.europa.eu/sides/getDoc.do?type = REPORT&reference = A7-2011-0026&language = EN.

Cultured Microalgae for the Food Industry. DOI: https://doi.org/10.1016/B978-0-12-821080-2.00003-4

called for putting more effort into research, development, and extension of cultivating protein-rich feed materials to increase the EU's own protein feed production (De Visser et al., 2014). However, employing terrestrial plants to replace the imported soya protein would require 10%–30% of EU arable land, which restricts the strategy of self-producing protein for feed in the EU (De Visser et al., 2014). In the frame of the EU's Strategies for Bioeconomy and Blue Growth, microalgal biomass is becoming increasingly important as a renewable and alternative source of protein but also as a source of diversity of commercial products and applications (e.g., food, feed, nutraceuticals, cosmetics, fertilizers, and bio-based products) and is gaining increasing interest.

The consumption of some microalgal and cyanobacterial species as foodstuff or food supplements has been common practice for centuries. They have been traditionally used as human food in many regions and countries, such as Mexico, Chad, China, Japan, Mongolia, Myanmar, Thailand, Vietnam, India, and other countries in South America. Examples of microalgal/cyanobacterial species that have been traditionally consumed include the cyanobacteria *Spirulina* (*Arthrospira platensis*, *Arthrospira maxima*), *Nostoc* (*N. commune*, *N. flagelliforme*, and *N. punctiforme*), *Aphanotheca sacrum*, *Anabaena flos-aquae* and *Aphanizomenon flos-aquae*, and the microalgal species *Chlorella*, *Spirogyra*, and *Oedogonium* (García, De Vicente, & Galán, 2017).

Microalgae, besides the basic nutrition that they can provide (proteins, lipids, and carbohydrates), are mainly considered and promoted as food supplements/functional foods/nutraceuticals because of the additional health benefits that they provide to humans and animals. Microalgae/cyanobacteria are effective sources of a variety of nutritional compounds, such as essential amino acids, essential fatty acids, vitamins, and minerals (in particular Fe), while they display antioxidant, antiinflammatory, anticancerous, or/and antimicrobial activity (Nethravathy, Mehar, Mudliar, & Shekh, 2019). The microalgal/cyanobacterial market has steadily grown in recent years (fivefold growth from 2000 to 2010) and it is expected to reach an annual turnover of around 300 M€ by 2022. Although the microalgal/cyanobacterial market is very small compared to the total food/feed market, it displays a steep growth trend with a promising dynamic with several new companies emerging every year. In Europe, more than 150 companies are currently active, with most of them located in France (García et al., 2017; Nethravathy et al., 2019).

The consumption of any alternative or even conventional food might have some risks, so the development and promotion of the microalgal/cyanobacterial market and their consumption must take into consideration the potential risks and harms to consumers (Wells et al., 2017). These risks and harms are related to intrinsic characteristics of the biomass (secondary metabolites such as toxic compounds, allergens, etc.) and to hazardous compounds that could contaminate the biomass during the cultivation or downstream processing of the harvested biomass. Thus the aim of this chapter is to give an overview and discuss some of the major aspects of safety of microalgal/cyanobacterial products consumption.

9.2 Regulatory framework and legislation in the European Union

In the EU, there are several categories of regulations that apply to the use and marketing of microalgal/cyanobacterial products (either as whole dried biomass or their extracts). An overview of the regulatory framework is shown in Table 9.1. The EU, through these regulations, establishes control standards and labeling and

TABLE 9.1 Overview of the EU regulations for food/feed

Regulatory topic	Description
1. General Food Law Regulation	Regulation (EC) No. 178/2002 of the European Parliament and of the Council of 28 January 2002 laying down the general principles and requirements of food law, establishing the European Food Safety Authority and laying down procedures in matters of food safety • Guarantee a high level of protection of human life and health and the protection of consumers' interests. Also guarantee fair practices in food trade, taking into account animal health and welfare, plant health, and the environment • Ensure free movement of food and feed manufactured and marketed in the Union, in accordance with the General Food Law Regulation • Facilitate global trade of safe feed and safe, wholesome food by taking into account international standards and agreements when developing Union legislation, except where this might undermine the high level of consumer protection pursued by the Union
2. Labeling and nutrition	• Food labeling legislation —(EU) No. 1169/2011 • Nutrition and Health Claims—(EC) No. 1924/2006 • Food supplements—Directive 2002/46/EC • Addition of vitamins and minerals—(EC) No. 1925/2006 • Food for specific groups—(EU) No. 609/2013 • *Trans* fat in food—(EC) No. 1925/2006
3. Biological and chemical safety	• Food hygiene—(EC) 852/2004 • Feed hygiene—(EC) Regulation 767/2009 • Feed additives—(EC Regulation 1831/2003) • Undesirable substances in animal feed—Council Directive 2002/32/EC. • Contaminants in foodstuff—(EC) No. 1881/2006
4. Novel food	(EU) 2015/2283 "Food business operators can place a novel food on the European Union market only after the Commission has processed an application for the authorization of a novel food, and has adopted an implementing act authorizing the placing on the market of a novel food and updating the Union list"

EU, European Union.

tries to guarantee safety in the areas of feed and food products (hygiene, health, contamination with unwanted substances, etc.). The main authority regarding food safety is the European Food Safety Authority (EFSA), which was established in 2002 by the European Parliament and Council for laying down procedures covering all matters on food and feed safety, health, and nutrition. For a complete overview of the EU food safety policy see Laky (2019).

The General Food Law Regulation establishes "general principles and requirements concerning food and feed law at Union level by developing an integrated approach to food safety 'from farm to fork'." Microalgae/cyanobacteria are subjected to the General Food Law Regulation as far as it concerns already marketable products, such as dried biomass of *Spirulina* (*A. platensis*, and *A. maxima*) or *Chlorella* sp. (*C. luteoviridis*, *C. pyrenoidosa*, and *C. vulgaris*), docosahexaenoic acid (DHA) oil from *Crypthecodinium cohnii*, and β-carotene from *Dunaliella salina*. *Spirulina* and *Chlorella* are commonly used as food supplements in the form of dried biomass (powder or compacted in tablets).

Marketable microalgal/cyanobacterial products are subject to all those food safety regulations (labeling and nutrition, and biological and chemical safety) that apply to all food products. The Nutrition and Health Claims Regulation rules on nutrition and health claims used by "food business operators when they want to highlight the particular beneficial effects of their products, in relation to health and nutrition, on the product label or in its advertising." The objective of the Nutrition and Health Claims Regulation is "to ensure that any claim made on a food's labeling, presentation or advertising in the EU is clear, accurate and based on scientific evidence." The Biological and Chemical Safety Regulation regulates all aspects of food and feed hygiene, rules for feed additives and the presence of undesirable contaminants in food and feed.

In addition, microalgae are subjected also to the regulations on novel foods. In principle, a novel food is defined as any "newly developed, innovative food, food produced using new technologies and production processes, as well as food which is or has been traditionally eaten outside of the EU" and that had not been consumed to a significant degree in the EU before May 15, 1997, when the regulations on novel foods came into force. If one food business operator wants to place a novel food in the EU market, it is required to submit an application for authorization of the product. The EFSA first conducts a scientific risk assessment, while the Commission manages the application files and processes the authorization for the novel food. Novel foods will only be authorized if they do not display any risk to public health, or any nutritionally disadvantages or misleading labeling compared to similar foods that they are intended to replace.

Besides the already marketable products, so far, there are a number of microalgal/cyanobacterial novel food products that have been approved by EU and are listed in the implementing regulation (EU) 2470/2017, where the conditions under which the novel food may be used, and specific labeling requirements are described. In this list the following microalgal/cyanobacterial products are listed: biomass of the diatom *Odontella aurita* to be used as a supplement (0.5%–1.5% maximum levels) in various foods such as flavored pasta, fish soups, marine terrines, etc.; dried biomass of the green microalga *Tetraselmis chuii* used in sauces, special salts, and condiments, etc.; and astaxanthin-rich oleoresin from the green alga *Haematococcus pluvialis* used as a food supplement. Moreover, the microalgal species *Euglena gracilis* and *Phaeodactylum tricornutum* are listed in the qualified presumption of safety list with the qualification "for production purposes only," which "implies the absence of viable cells of the production organism in the final product and can also be applied for food and feed products based on microbial biomass." In the category of microalgal novel foods are also included the oil rich in DHA and eicosapentaenoic acid from *Schizochytrium* sp. (from strains *S.* sp. ATCC PTA-9695, and *S.* sp. T18), used as a supplement in a variety of foods, including foods with medical purposes; DHA oil from the marine species *Ulkenia* sp. used in bakery products, cereal bars, or nonalcoholic beverages; (Koutsoumanis et al., 2019), which however are fungi-like heterokonts that cannot be considered as algae (Leyland, Leu, & Boussiba, 2017).

9.3 Risk and safety issues

9.3.1 Inherent to biomass risks and safety issues

Many alternative food or feed sources suffer from containing antinutritional factors (ANFs). ANFs are those compounds of biological origin contained in foods and feeds that reduce food intake and nutrient utilization, especially those of proteins, vitamins, and minerals. ANFs have adverse effects, such as toxicity, bitter taste, or they contribute to impaired gastrointestinal and metabolic performance preventing optimal exploitation of the nutrients contained in a food or feed (Gilani, Xiao, & Cockell, 2012). Examples of ANFs are glucosinolates, phytates, tannins, and polyphenols. Only a few works exist that assess ANFs of microalgae, which show that microalgal protein and mineral bioaccessibility or bioavailability do not differ or are superior to control foods (Neumann et al., 2018; Tibbetts et al., 2019; Uribe-Wandurraga, Igual, García-Segovia, & Martínez-Monzó, 2020). Moreover, many studies dealing with animal feeding have shown that the inclusion of microalgae in feed could improve animal growth and product quality, rendering them as a very promising alternative to food crops in feeds (Madeira et al., 2017). In addition, microalgae have shown positive effects also as health supplements (Koyande et al., 2019). Therefore it is unlikely that microalgae have adverse effects on health of animals or humans (Olvera-Novoa, Domínguez-Cen, Olivera-Castillo, & Martínez-Palacios, 1998). However, there are some reports in the literature pertaining to cases of microalgae and cyanobacteria that could have some health risks when consumed by humans, caused by various factors that are discussed below.

9.3.1.1 Allergenicity

Food allergy is a hypersensitivity reaction to normally harmless food components. It is an atopic disorder, which refers to immunoglobulin-E (IgE)-mediated reactions of the human immune system to food antigens, and specifically to some types of proteins or polysaccharides (Yu, Freeland, & Nadeau, 2016). The introduction of novel foods and ingredients, including microalgae and cyanobacteria, has therefore the potential to introduce allergens into the food chain (Van Putten, 2009). There are inadequate and even contradictory data on the allergenicity of microalgae/cyanobacteria, with some studies reporting that biomass of *Chlorella homosphaera* (Tiberg, Dreborg, & Björkstén, 1995) and *A. platensis* (Le, Knulst, & Röckmann, 2014), or the pigment phycocyanin from *A. platensis* (Petrus, Culerrier, Campistron, Barre, & Rougé, 2009) triggered allergic reactions. In most cases, patients who reacted to these products were also sensitized to many other allergens. These findings show that these products may be of clinical significance in certain groups of patients that would need to avoid their consumption. Nevertheless, some microalgae and cyanobacteria have been suggested as novel sources for the development of antiallergic agents (Fleurence & Levine, 2018). The interesting point is that among the candidates (such as *Porphyridium cruentum* and *D. salina*) for the development of antiallergic agents *Chlorella* sp. or *Arthrospira* sp. are also included because it has been found that they can mitigate anaphylactic shock and decrease histamine release (Vo, Ngo, & Kim, 2012; Yang, Lee, & Kim, 1997).

9.3.1.2 Nucleic acids

Typically, microalgae have lower nucleic acid (DNA and RNA) content when compared to other single-cell protein sources, such as bacteria or fungi. However, the total nucleic acid content of microalgae is about 4%–6% (Janczyk, Wolf, & Souffrant, 2005; Saleh, Hussein, Abdalla, El-Fouly, & Shaheen, 1985), which is much higher than that of most plant-based food sources (Jonas et al., 2001). The nucleic acid content is a noteworthy nutritional parameter, because the biochemical degradation of nucleic acids (mainly of purines) increases plasma uric acid causing gout and kidney stones (Ritala, Häkkinen, Toivari, & Wiebe, 2017). RNA content is much higher (0.7%–6.65%) than DNA (0.31%–0.86%), while RNA is more strongly associated with the risk factor for gout problems than DNA (Jonas et al., 2001). The recommended higher levels of daily nucleic acid intake are around 2 g/kg body weight (BW)/day for an adult (Altschul, 2013). Based on this recommendation the highest daily intake of microalgae is about 30–50 g/kg BW/day for an adult.

9.3.1.3 Toxicity

As was previously mentioned, there are several species of cyanobacteria that are traditionally consumed worldwide. Most are harvested from natural environments, where other toxic cyanobacteria or microalgae can potentially contaminate the biomass with toxins (see Section 9.3.2.2) raising, therefore the potential risk to human health (Caruana & Amzil, 2018; Johnson et al., 2008; Vichi, Lavorini, Funari, Scardala, & Testai, 2012). In addition, there are reports that indicate that species used as food supplements such as *Nostoc*, *A. sacrum*, *Ana. flos-aquae*, and *Aph. flos-aquae* potentially synthesize toxins themselves. There is evidence that *Aph. flos-aquae* carries toxin genes and produces significant levels of cylindrospermopsin and saxitoxins (Lyon-Colbert, Su, & Cude, 2018), while *Ana. flos-aquae* also is a potent anatoxin-a producer (Van Dolah, 2000). *N. commune* produces the neurotoxic amino acid β-*N*-methylamino-L-alanine (Johnson et al., 2008), however no detectable levels of microcystin were observed (Yang et al., 2011). However, the production of toxins by these cyanobacteria remains uncertain and more research is required to evaluate whether given cultivation and environmental conditions favor the synthesis of toxins or not. While cyanobacteria have these potential risks, toxins from other microalgae, especially of the Chlorellaceae family (to which *Chlorella* belongs), have never been reported so far.

9.3.1.4 Carcinogenicity

Some of the cyanobacterial toxins (such as microcystins) are potential genotoxic and carcinogenic agents but their carcinogenic mechanisms are not yet well understood. Their main mechanism of action is the inhibition of protein phosphatases leading to hyperphosphorylation of cellular proteins, which is considered to be a cause of tumor development (Žegura, Štraser, & Filipič, 2011). Besides cyanotoxins, no reports of the presence of carcinogens on other microalgal species have been found. In contrast, several recent studies report the potential anticarcinogenic effect and the ability to induce apoptotic cancer cell death of microalgal extracts (Abd El-Hack et al., 2019).

9.3.1.5 Pathogenicity

There are some microalgal species that can cause diseases themselves, for example, the saprophytic green alga *Prototheca* that causes protothecosis in dogs, cats, cattle, and to a lesser extent in humans (Consuelo Quinet Leimann, Cezar Fialho Monteiro, Lazéra, Ulloa Candanoza, & Wanke, 2004) or the green alga *Desmodesmus armatus* which caused several cases of soft-tissue infection (Fujimoto et al., 2018; Westblade et al., 2015).

9.3.1.6 Hypervitaminosis

Hypervitaminosis occurs when the storage levels of vitamins are abnormally high. Hypervitaminosis can lead to toxic symptoms and diverse health effects.

Microalgae are a very good source of vitamins (Brown, Mular, Miller, Farmer, & Trenerry, 1999; Fabregas & Herrero, 1990; Jalilian, Najafpour, & Khajouei, 2019; Sousa, Gouveia, Batista, Raymundo, & Bandarra, 2008), although their consumption as a food supplement is unlikely to lead to hypervitaminosis. According to EFSA (2016), the vitamins that can have a potential effect when they are taken up in high doses are vitamin A, vitamin D, vitamin B6, and vitamin B3 (nicotinic acid). Their upper levels for daily intake are set at 3 mg/kg BW/day, 50 μg/kg BW/d, 25 mg/kg BW/d and 8 mg/kg BW/d, respectively.

9.3.2 Risks related to the cultivation process

Besides the inherent characteristics of the biomass that pose a health risk, microalgae/cyanobacteria can display additional health risks associated with the contamination of the produced biomass with unwanted and hazardous compounds either during the cultivation stage or during processing of the harvested biomass. Such unwanted contaminants include heavy metals, pesticides, pathogens, other microalgal/cyanobacterial species, and a variety of organic and inorganic compounds. These compounds could contaminate biomass due of their presence in the various resources and materials used, such as water, fertilizers, or any other food contact material (FCM), like chemicals or machinery/equipment, packaging materials, etc. that could release and transfer their constituents to the biomass intended to be consumed by humans or animals. Regarding FCMs, standards of quality are regulated by legislations [see, e.g., (EC) No. 1935/2004, which sets out the general principles of safety and inertness for all FCMs]. The qualities of all these FCMs are more or less controllable and will not be discussed here further.

As far the use of water in the food production chain in the EU is concerned, its quality must meet the basic quality standards for drinking water (potable water), as set out by Council Directive 98/83/EC. However, microalgal/cyanobacterial biomass production is not a subject of these regulated quality standards and a variety of different water qualities could be used, as long as the produced biomass meets the food quality standards. Moreover, while accessibility of potable water is not always secure and there are plenty of cases in which access to potable water has a series of challenges, the use of water from natural water bodies might be unavoidable for the production of microalgae/cyanobacteria on a large scale. Moreover, some interesting microalgal species are from marine habitats and grow in seawater. The quality of natural seawater is not always controllable.

Water is a major resource for microalgae/cyanobacteria cultivation and is the main cause of contamination in the biomass, because most are either suspended in it, or they are transported through it. In a recent review conducted by the FAO and the World Health Organization, it was highlighted that there was a gap on existing guidance in areas relevant to food safety regarding specific situations in which water is used that needed to be addressed

(WHO, 2019). Since water pollution is a great concern that could have a negative effect on the quality of microalgal/cyanobacterial biomass, while at the same time there is a strategy for the transition to the circular (bio-)economy (European Commission, 2018), in this section the main risks arising during cultivation will be further discussed.

Open and uncovered cultivation systems are more susceptible to contamination by airborne contaminants because they do not provide any protection compared to closed (photobioreactors) or covered open pond systems (e.g., greenhouses). Closed or covered systems are more expensive, but they provide controllable growth conditions and have reliable production because of the reduced contamination risks (Apel & Weuster-Botz, 2015; Lam, Lee, Chen, & Chang, 2018). Nevertheless, large-scale cultivation of microalgae/cyanobacteria for food most often is conducted in open raceway pond systems.

9.3.2.1 Organic and inorganic contaminants

Microalgal/cyanobacterial cultures could be contaminated by a variety of organic and inorganic unwanted or even hazardous compounds, such as heavy metals, pesticides, polycyclic aromatic hydrocarbons, polychloro-biphenyls, bird droppings, debris, insects, and rodent hairs. In any case, marketable microalgal and cyanobacterial biomass is frequently analyzed to prove that specific contaminants are lower than the maximum levels set by regulatory agencies. The most important contaminants are heavy metals and pesticides.

Heavy metals, by definition, are elements with molecular densities higher than 5 g mL^{-1}, of which cadmium (Cd), lead (Pb), mercury (Hg), and inorganic arsenic (As) are the most potentially toxic (WHO, 2017). These heavy metal species have no beneficial effects in humans and their excessive intake has serious health effects (Abarikwu, 2013; Morais, E Costa, & De Lourdes Pereira, 2012). Heavy metals react with biomolecules, such as proteins, DNA, or enzymes, and form stable and highly biotoxic compounds, which can result in carcinogenesis, decreased reproductive ability, and damaging to the nervous, skeletal, circulatory, endocrine, and immune systems (Abarikwu, 2013; Li, Ma, Van Der Kuijp, Yuan, & Huang, 2014; Morais et al., 2012). However, not all heavy metals are hazardous, while some of them, such as Cu, Cr, Co, Mn, Ni, Se, Mo, and Zn, are essential and are required in very small quantities (i.e., micronutrients) because they are involved in physiological functions in animals and humans (Mertz, 2012). Heavy metals cannot be metabolized or degraded, and therefore they accumulate in the body. Consequently, most countries regulate their presence in foodstuffs. In the EU maximum levels of heavy metals in food and feed are regulated by Commission Regulation (EC) No. 1881/2006 and Directive 2002/32/EC, respectively.

Microalgae and cyanobacteria have high affinities and sorption capacities for heavy metals and are considered as hyperaccumulators, which means that they are capable of growing in media containing heavy metals and concentrating them at high levels in their cells (Kaplan, 2013; Leong & Chang, 2020). Sorption of heavy metals by microalgae/cyanobacteria is ruled by the cell wall parameters (electrostatic characteristics and organic matrices, mainly polysaccharides) (Rossi & De Philippis, 2016). This characteristic renders them highly susceptible for contamination with heavy metals. The sorption capacities of microalgae vary greatly depending on the cultivation conditions (heavy metal concentration, pH, salinity, microalgal species, etc.) and can vary significantly (Kaplan, 2013; Leong & Chang, 2020). Heavy metals are released into water bodies due to various industrial materials and activities, such as tanning, painting, electroplating, electrical devices, and electronics, and therefore natural water supplies are frequently severely contaminated by heavy metals (Mishra et al., 2019). Resources used in microalgal/cyanobacterial cultures, such as inorganic fertilizers (in particular phosphate fertilizers), could contribute also to heavy metal (in particular Cd and Zn) contamination (Mishra et al., 2019; Mortvedt, 1995).

In a recent study, *Chlorella* and *Spirulina*, registered in the EU as food supplements, were examined for heavy metal contamination and it was found that all studied supplements had Cd, Hg, and Ni levels below safety limits. However, some selected samples were characterized by a relatively high content of Al, Pb, and inorganic As (Rzymski et al., 2019). Nevertheless, taking into consideration that the commonly recommended daily dose of *Spirulina* or *Chlorella* is 3.0 g, the tolerable upper intake level and safety upper level/guidance level as set by the associated European guidance were not exceeded for any essential elements considered by the study (Rzymski et al., 2019).

Pesticides, also described as phytosanitary products, include a wide variety of organochlorine compounds, that are used throughout the globe on a large scale to combat mainly fungi and insects that destroy partially or fully crop plants and their products. Numerous studies and research have been carried out on detailing the harmful effects of pesticides on different sections of the environment, useful animal populations, and also on their combined effects on human health and the economy (Samsidar, Siddiquee, & Shaarani, 2018). Pesticides can enter microalgae/cyanobacteria cultivation ponds either through air transportation during the application or through water. Microalgal and cyanobacterial systems differ in their sensitivity to pesticide contamination. Although some strains can accumulate numerous pesticides, the microalgal and cyanobacterial responses to pesticides vary widely depending upon the concentration used, duration of exposure, and algal species tested (Hultberg, Bodin, Ardal, & Asp, 2016; Rath, 2012). The health risk depends on the strain metabolism and pesticide persistence in the culture medium (Jassby, 1988).

9.3.2.2 *Biological contamination*

Biological contamination of microalgal/cyanobacterial biomass is either associated with the presence of a wide range of coexisting small- and microorganisms, including bacteria, fungi, grazers, parasites, viruses, and toxic microalgae/cyanobacteria in the cultures (Lam et al., 2018; Wang, Zhang, Chen, Wang, & Liu, 2013) or biomass contamination at the processing stages (i.e., during harvesting, washing, drying, storage, or packaging, etc.), mainly depending on the hygiene practices and the water source used. The coexisting organisms could have severe effects on the cultures, ranging from reduction of the biomass yield, either due to the competition for available resources (nutrients, light, etc.) or the secretion of algicidal extracellular metabolites harmful to microalgae (Lam et al., 2018; Wang et al., 2013), to more serious issues related to the microalgal final products and the post-treatment they must receive before they can be used in any product, where it can present a serious potential health risk. Microalgal/cyanobacterial biomass production is generally carried out in outdoor open pond systems, where they can be easily contaminated by bacteria and toxic cyanobacteria. Bacterial contamination cannot be omitted from open mass cultures because of the nature of outdoor open systems. Contamination by pathogens such as *Listeria monocytogenes*, *Enterobacter sakazakii*, *Staphylococcus*, *Salmonella*, *Shigella*, and *Clostridium* and their toxins, as well as coliforms (*Escherichia coli*) are undesirable and need to be prevented through good pond management (Hoekstra, Volschenk, Collins, & Mcmaster, 2011; Vardaka, Kormas, Katsiapi, Genitsaris, & Moustaka-Gouni, 2016). These bacteria are common in 50% of foodborne illness cases (Becker, 1994), therefore microalgal/cyanobacterial biomass should meet all the requirements for food or feed quality standards regarding these pathogens by following the guidance and criteria of Good Hygiene and Manufacturing Practices and the Hazard Analysis and Critical Control Points. In recent years, the quality and safety of marketed microalgal supplements has been reconsidered due to alterations to the culture purity and the detection of cyanotoxins, for example, microcystin nodularin cylindrospermopsin anatoxin-a; anatoxin-a(s), lyngbyatoxin, or saxitoxin, which exhibit hepatotoxic, neurotoxic, cytotoxic, potentially carcinogenic, and dermatotoxic effects. Some co-occurring toxigenic species involved in open pond cultures are *Microcystis*, *Planktothrix*, *Nostoc*, and *Anabaena*, *Oscillatoria*, *Cylindrospermopsis*, *Aphanizomenon*, and *Lyngbya* (Manali, Arunraj, Kumar, & Ramya, 2017; Roy-Lachapelle, Solliec, Bouchard, & Sauvé, 2017; Scoglio, 2018; Vichi et al., 2012). Considering this biological contamination risk, monitoring and controlling processes should be carried during the cultivation process to avoid the cocultivation of wild toxic strains. Tests to exclude the presence of cyanotoxins should be carried out and recorded on such strains. To this end, the recent development of specific tools can be warmly welcomed; these are PCR primers for the detection of genes applied in the synthesis of cyanotoxins (Saker, Welker, & Vasconcelos, 2007).

9.4 Conclusion

Microalgal and cyanobacterial biomass is becoming increasingly important in the food/feed market. However, the EU considers microalgae/cyanobacteria (with the exception of a few products) as novel foods, and therefore they are subjected to the European regulatory and legislation frameworks. As novel foods it should be ensured that their consumption will not present any risks, which can be related either to the intrinsic characteristics of the biomass (secondary metabolites such as toxic compounds, allergens, etc.), to hazardous compounds that could contaminate the biomass during cultivation (organic and inorganic contaminants and biological contamination), or during the downstream processing of the harvested biomass. More research is needed to adequately cover this topic.

References

Abarikwu, S. O. (2013). *Lead, arsenic, cadmium, mercury: Occurrence, toxicity and diseases. Pollutant Diseases, Remediation and Recycling*. Springer.

Abd El-Hack, M. E., Abdelnour, S., Alagawany, M., Abdo, M., Sakr, M. A., Khafaga, A. F., ... Gebriel, M. G. (2019). Microalgae in modern cancer therapy: Current knowledge. *Biomedicine & Pharmacotherapy*, *111*, 42–50.

Altschul, A. M. (2013). *New protein foods: Animal protein supplies*. Academic Press.

Apel, A. C., & Weuster-Botz, D. (2015). Engineering solutions for open microalgae mass cultivation and realistic indoor simulation of outdoor environments. *Bioprocess and Biosystems Engineering*, *38*, 995–1008.

Becker, E. (1994). *Large-scale cultivation*. New York: Cambridge University Press.

Boland, M. J., Rae, A. N., Vereijken, J. M., Meuwissen, M. P., Fischer, A. R., Van Boekel, M. A., ... Hendriks, W. H. (2013). The future supply of animal-derived protein for human consumption. *Trends in Food Science & Technology*, 29, 62–73.

Brown, M., Mular, M., Miller, I., Farmer, C., & Trenerry, C. (1999). The vitamin content of microalgae used in aquaculture. *Journal of Applied Phycology*, *11*, 247–255.

Caruana, A. M. N., & Amzil, Z. (2018). Chapter 13 - Microalgae and toxins. In: I. A. Levine, & J. Fleurence (Eds.), *Microalgae in health and disease prevention*. Academic Press.

Consuelo Quinet Leimann, B., Cezar Fialho Monteiro, P., Lazéra, M., Ulloa Candanoza, E. R., & Wanke, B. (2004). Protothecosis. *Medical Mycology*, *42*, 95–106.

De Visser, C. L. M., Schreuder, R., & Stoddard, F. (2014). The EU's dependency on soya bean import for the animal feed industry and potential for EU produced alternatives. *OCL*, *21*, D407.

EFSA (2016). *European Food Safety Authority, Scientific Committee on FoodScientific Panel on dietetic products, nutrition and allergies, tolerable upper intake levels for vitamins and minerals*, p. 482.

European Commission. (2018). A sustainable bioeconomy for Europe: strengthening the connection between economy, society and the environment, Updated Bioeconomy Strategy. p. 107.

Fabregas, J., & Herrero, C. (1990). Vitamin content of four marine microalgae. Potential use as source of vitamins in nutrition. *Journal of Industrial Microbiology*, *5*, 259–263.

Fleurence, J., & Levine, I. A. (2018). Chapter 14 - Antiallergic and allergic properties. In: I. A. Levine, & J. Fleurence (Eds.), *Microalgae in health and disease prevention*. Academic Press.

Fujimoto, M., Inaba, Y., Takahashi, T., Nakanishi, G., Muraosa, Y., Yahiro, M., ... Murata, S. I. (2018). Image gallery: Granulomatous dermatitis due to infection with the chlorophyllic green alga Desmodesmus. *British Journal of Dermatology*, *179*, e167-e167.

García, J. L., De Vicente, M., & Galán, B. (2017). Microalgae, old sustainable food and fashion nutraceuticals. *Microbial Biotechnology*, *10*, 1017–1024.

Gilani, G. S., Xiao, C. W., & Cockell, K. A. (2012). Impact of antinutritional factors in food proteins on the digestibility of protein and the bioavailability of amino acids and on protein quality. *British Journal of Nutrition*, *108*, S315–S332.

Henchion, M., Hayes, M., Mullen, A. M., Fenelon, M., & Tiwari, B. (2017). Future protein supply and demand: strategies and factors influencing a sustainable equilibrium. *Foods*, *6*, 53.

Hoekstra, D. T., Volschenk, H., Collins, M., & Mcmaster, L. D. (2011). An investigation of Clostridium species present in nutraceutical preparations of Arthrospira platensis (Spirulina) for human consumption. *Journal of Applied Phycology*, *23*, 777–787.

Hultberg, M., Bodin, H., Ardal, E., & Asp, H. (2016). Effect of microalgal treatments on pesticides in water. *Environmental Technology*, *37*, 893–898.

Jalilian, N., Najafpour, G., & Khajouei, M. (2019). Enhanced vitamin B12 production using Chlorella vulgaris. *International Journal of Engineering*, *32*, 1–9.

Janczyk, P., Wolf, C., & Souffrant, W. B. (2005). Evaluation of nutritional value and safety of the green microalgae Chlorella vulgaris treated with novel processing methods. *Archiva Zootechnica*, *8*, 132–147.

Jassby, A. (1988). Some public health aspects of microalgal products. *Algae and human affairs/ edited by Carole A. Lembi, J. Robert Waaland; sponsored by the Phycological Society of America, Inc.*

Johnson, H. E., King, S. R., Banack, S. A., Webster, C., Callanaupa, W. J., & Cox, P. A. (2008). Cyanobacteria (Nostoc commune) used as a dietary item in the Peruvian highlands produce the neurotoxic amino acid BMAA. *Journal of Ethnopharmacology*, *118*, 159–165.

Jonas, D., Elmadfa, I., Engel, K.-H., Heller, K., Kozianowski, G., König, A., ... Kleiner, J. (2001). Safety considerations of DNA in food. *Annals of Nutrition and Metabolism*, *45*, 235–254.

Kaplan, D. (2013). Absorption and adsorption of heavy metals by microalgae. Handbook of microalgal culture: Applied phycology and biotechnology, 2, 602–611.

Koutsoumanis, K., Allende, A., Alvarez-Ordóñez, A., Bolton, D., Bover-Cid, S., Chemaly, M., ... Lindqvist, R. (2019). Update of the list of QPS-recommended biological agents intentionally added to food or feed as notified to EFSA 10: Suitability of taxonomic units notified to EFSA until March 2019. *EFSA Journal*, 17.

Koyande, A. K., Chew, K. W., Rambabu, K., Tao, Y., Chu, D.-T., & Show, P.-L. (2019). Microalgae: A potential alternative to health supplementation for humans. *Food Science and Human Wellness*, *8*, 16–24.

Laky, Z. (2019). Food Safety, Fact Sheets on the European Union - 2019.

Lam, T. P., Lee, T.-M., Chen, C.-Y., & Chang, J.-S. (2018). Strategies to control biological contaminants during microalgal cultivation in open ponds. *Bioresource Technology*, *252*, 180–187.

Le, T.-M., Knulst, A. C., & Röckmann, H. (2014). Anaphylaxis to Spirulina confirmed by skin prick test with ingredients of Spirulina tablets. *Food and Chemical Toxicology*, *74*, 309–310.

Leong, Y. K., & Chang, J.-S. (2020). Bioremediation of heavy metals using microalgae: Recent advances and mechanisms. *Bioresource Technology*, 122886.

Leyland, B., Leu, S., & Boussiba, S. (2017). Are Thraustochytrids algae? *Fungal Biology*, *121*, 835–840.

Li, Z., Ma, Z., Van Der Kuijp, T. J., Yuan, Z., & Huang, L. (2014). A review of soil heavy metal pollution from mines in China: Pollution and health risk assessment. *Science of the Total Environment*, *468*, 843–853.

Lyon-Colbert, A., Su, S., & Cude, C. (2018). A systematic literature review for evidence of Aphanizomenon flos-aquae toxigenicity in recreational waters and toxicity of dietary supplements: 2000–2017. *Toxins*, *10*, 254.

Madeira, M. S., Cardoso, C., Lopes, P. A., Coelho, D., Afonso, C., Bandarra, N. M., & Prates, J. A. M. (2017). Microalgae as feed ingredients for livestock production and meat quality: A review. *Livestock Science*, *205*, 111–121.

Manali, K. M., Arunraj, R., Kumar, T., & Ramya, M. (2017). Detection of microcystin producing cyanobacteria in Spirulina dietary supplements using multiplex HRM quantitative PCR. *Journal of Applied Phycology*, *29*, 1279–1286.

Mertz, W. (2012). *Trace elements in human and animal nutrition*. Elsevier.

Mishra, S., Bharagava, R. N., More, N., Yadav, A., Zainith, S., Mani, S., & Chowdhary, P. (2019). Heavy metal contamination: An alarming threat to environment and human health. In: R. C. Sobti, N. K. Arora, & R. Kothari (Eds.), *Environmental biotechnology: For sustainable future*. Singapore: Springer.

Morais, S., E Costa, F. G., & De Lourdes Pereira, M. (2012). *Heavy metals and human health*. Intech Open Access Publisher.

Mortvedt, J. J. (1995). Heavy metal contaminants in inorganic and organic fertilizers. *Fertilizer Research*, *43*, 55–61.

Nethravathy, M. U., Mehar, J. G., Mudliar, S. N., & Shekh, A. Y. (2019). Recent advances in microalgal bioactives for food, feed, and healthcare products: Commercial potential, market space, and sustainability. *Comprehensive Reviews in Food Science and Food Safety*, *18*, 1882–1897.

Neumann, U., Derwenskus, F., Gille, A., Louis, S., Schmid-Staiger, U., Briviba, K., & Bischoff, S. C. (2018). Bioavailability and safety of nutrients from the microalgae Chlorella vulgaris, Nannochloropsis oceanica and Phaeodactylum tricornutum in C57BL/6 mice. *Nutrients*, *10*, 965.

Olvera-Novoa, M., Domínguez-Cen, L., Olivera-Castillo, L., & Martínez-Palacios, C. A. (1998). Effect of the use of the microalga Spirulina maxima as fish meal replacement in diets for tilapia, Oreochromis mossambicus (Peters), fry. *Aquaculture Research*, *29*, 709–715.

Petrus, M., Culerrier, R., Campistron, M., Barre, A., & Rougé, P. (2009). ALLERGY Net. *Allergy*, *65*, 924–925.

Rath, B. (2012). Microalgal bioremediation: current practices and perspectives. *Journal of Biochemical Technology*, *3*, 299–304.

Ritala, A., Häkkinen, S. T., Toivari, M., & Wiebe, M. G. (2017). Single cell protein—State-of-the-art, industrial landscape and patents 2001–2016. *Frontiers in Microbiology*, 8.

Rossi, F., & De Philippis, R. (2016). Exocellular polysaccharides in microalgae and cyanobacteria: Chemical features, role and enzymes and genes involved in their biosynthesis. *The Physiology of Microalgae*. Springer.

Roy-Lachapelle, A., Solliec, M., Bouchard, M. F., & Sauvé, S. (2017). Detection of cyanotoxins in algae dietary supplements. *Toxins*, *9*, 76.

Rzymski, P., Budzulak, J., Niedzielski, P., Klimaszyk, P., Proch, J., Kozak, L., & Poniedziałek, B. (2019). Essential and toxic elements in commercial microalgal food supplements. *Journal of Applied Phycology*, *31*, 3567–3579.

Saker, M. L., Welker, M., & Vasconcelos, V. M. (2007). Multiplex PCR for the detection of toxigenic cyanobacteria in dietary supplements produced for human consumption. *Applied Microbiology and Biotechnology*, *73*, 1136–1142.

Saleh, A., Hussein, L., Abdalla, F., El-Fouly, M., & Shaheen, A. (1985). The nutritional quality of drum-dried algae produced in open door mass culture. *Zeitschrift für Ernährungswissenschaft*, *24*, 256–263.

Samsidar, A., Siddiquee, S., & Shaarani, S. M. (2018). A review of extraction, analytical and advanced methods for determination of pesticides in environment and foodstuffs. *Trends in Food Science & Technology*, *71*, 188–201.

Scoglio, S. (2018). Microcystins in water and in microalgae: Do microcystins as microalgae contaminants warrant the current public alarm? *Toxicology Reports*, *5*, 785–792.

Sousa, I., Gouveia, L., Batista, A. P., Raymundo, A., & Bandarra, N. M. (2008). Microalgae in novel food products. *Food Chemistry Research Developments*, 75–112.

Tibbetts, S. M., Patelakis, S. J., Whitney-Lalonde, C. G., Garrison, L. L., Wall, C. L., & Macquarrie, S. P. (2019). Nutrient composition and protein quality of microalgae meals produced from the marine prymnesiophyte Pavlova sp. 459 mass-cultivated in enclosed photobioreactors for potential use in salmonid aquafeeds. *Journal of Applied Phycology*, 1–20.

Tiberg, E., Dreborg, S., & Björkstén, B. (1995). Allergy to green algae (Chlorella) among children. *Journal of Allergy and Clinical Immunology*, *96*, 257–259.

Uribe-Wandurraga, Z. N., Igual, M., García-Segovia, P., & Martínez-Monzó, J. (2020). In vitro bioaccessibility of minerals from microalgae-enriched cookies. *Food & Function*.

Van Der Spiegel, M., Noordam, M. Y., & Van Der Fels-Klerx, H. J. (2013). Safety of novel protein sources (insects, microalgae, seaweed, duckweed, and rapeseed) and legislative aspects for their application in food and feed production. *Comprehensive Reviews in Food Science and Food Safety*, *12*, 662–678.

Van Dolah, F. M. (2000). Diversity of marine and freshwater algal toxins. Food *s*cience and *t*echnology. New York: Marcel Dekker-, 19–44.

Van Putten, M. C. (2009). *Novel foods and food allergies: an exploratory study of novel foods as allergy management strategy*. Wageningen University.

Vardaka, E., Kormas, K. A., Katsiapi, M., Genitsaris, S., & Moustaka-Gouni, M. (2016). Molecular diversity of bacteria in commercially available "Spirulina" food supplements. *PeerJ*, *4*, e1610.

Vichi, S., Lavorini, P., Funari, E., Scardala, S., & Testai, E. (2012). Contamination by Microcystis and microcystins of blue–green algae food supplements (BGAS) on the italian market and possible risk for the exposed population. *Food and Chemical Toxicology*, *50*, 4493–4499.

Vo, T.-S., Ngo, D.-H., & Kim, S.-K. (2012). Marine algae as a potential pharmaceutical source for anti-allergic therapeutics. *Process Biochemistry*, *47*, 386–394.

Wang, H., Zhang, W., Chen, L., Wang, J., & Liu, T. (2013). The contamination and control of biological pollutants in mass cultivation of microalgae. *Bioresource Technology*, *128*, 745–750.

Wells, M. L., Potin, P., Craigie, J. S., Raven, J. A., Merchant, S. S., Helliwell, K. E., ... Brawley, S. H. (2017). Algae as nutritional and functional food sources: Revisiting our understanding. *Journal of Applied Phycology*, *29*, 949–982.

Westblade, L. F., Ranganath, S., Dunne, W. M., JR, Burnham, C.-A. D., Fader, R., & Ford, B. A. (2015). Infection with a chlorophyllic eukaryote after a traumatic freshwater injury. *New England Journal of Medicine*, *372*, 982–984.

WHO. *Ten chemicals of major public health concern* (2017) http://www.who.int/ipcs/assessment/public_health/chemicals_phc/en/ Accessed 16.06.17.

WHO (2019). World Health Organization. Safety and quality of water used in food production and processing: meeting report.

Yang, H.-N., Lee, E.-H., & Kim, H.-M. (1997). Spirulina platensis inhibits anaphylactic reaction. *Life Sciences*, *61*, 1237–1244.

Yang, Y., Park, Y., Cassada, D. A., Snow, D. D., Rogers, D. G., & Lee, J. (2011). In vitro and in vivo safety assessment of edible blue-green algae, Nostoc commune var. sphaeroides Kützing and Spirulina plantensis. *Food and Chemical Toxicology*, *49*, 1560–1564.

Yu, W., Freeland, D. M. H., & Nadeau, K. C. (2016). Food allergy: Immune mechanisms, diagnosis and immunotherapy. *Nature Reviews Immunology*, *16*, 751.

Žegura, B., Štraser, A., & Filipič, M. (2011). Genotoxicity and potential carcinogenicity of cyanobacterial toxins – A review. *Mutation Research/Reviews in Mutation Research*, *727*, 16–41.

Chapter 10

Microalgae as structuring ingredients in food

Tom M.M. Bernaerts and Ann M. Van Loey
Laboratory of Food Technology (Member of Leuven Food Science and Nutrition Research Centre, LFoRCe), Department of Microbial and Molecular Systems (M^2S), KU Leuven, Leuven, Belgium

10.1 Introduction

The use of microalgae as functional food ingredients has gained increasing interest in the past few decades. Many recent studies have investigated the impact of microalgal biomass addition on the overall quality of enriched food products, as recently reviewed by several authors (Bernaerts, Gheysen, Foubert, Hendrickx, & Van Loey, 2019a; Caporgno & Mathys, 2018; Lafarga, 2019). While most of these studies have focused on the nutritional value, color, texture, and sensory properties of the enriched foods, few research studies have been published on the impact of microalgal biomass addition on the rheological properties of fluid and semisolid food products. However, microalgal biomasses are generally a rich source of different structural biopolymers such as proteins and polysaccharides. Therefore microalgae might also have an important contribution to the rheological quality aspects of enriched food products. This chapter addresses the impact of microalgal biomass on the viscosity and viscoelastic behavior of semisolid food products, in relation to different food processing operations.

10.2 Food products enriched with microalgal biomass: impact on food structure

The effect of microalgal biomass addition to (semisolid) food products on the rheological properties has been studied for cereal-based products (bread, pasta and cookie doughs), emulsions (o/w emulsions and mayonnaise), vegetable-based products (soups), and dairy products (yogurt and ice cream), as presented in Table 10.1. Most of these studies used biomass concentrations of 1%–2%, although lower (below 0.1%) and higher (up to 3%) concentrations have also been investigated.

Cultured Microalgae for the Food Industry. DOI: https://doi.org/10.1016/B978-0-12-821080-2.00012-5

TABLE 10.1 Research studies evaluating the rheological properties of different food products enriched with microalgal biomass.

Food category	Food product	Microalga species	Biomass concentration (%)	Rheology	References
Cereal-based products	Bread dough	*Chlorella vulgaris*	1	Viscoelastic (↘)	Nunes et al. (2020)
		C. vulgaris	~0.6, ~1.2, ~1.8	Viscoelastic (↗)	Graça et al. (2018)
		C. vulgaris	~2.4, ~3	Viscoelastic (↘)	Graça et al. (2018)
	Breadstick dough	*Arthrospira platensis* (*Spirulina*)	1.5	Viscoelastic (↗)	Uribe-Wandurraga et al. (2019)
		C. vulgaris	1.5	Viscoelastic (↗)	Uribe-Wandurraga et al. (2019)
	Bread sourdough	*Isochrysis galbana*	1.5	Viscosity (↗)	García-Segovia et al. (2017)
		Nannochloropsis gaditana	1.5	Viscosity (↗)	García-Segovia et al. (2017)
		Scenedesmus almeriensis	1.5	Viscosity (↗)	García-Segovia et al. (2017)
		Tetraselmis suecica	1.5	Viscosity (↗)	García-Segovia et al. (2017)
	Cookie dough	*A. platensis* (*Spirulina*)	0.5, 1, 1.5, 2	Viscoelastic (↗)	Uribe-Wandurraga et al. (2020a)
		C. vulgaris	0.5, 1, 1.5, 2	Viscoelastic (↗)	Uribe-Wandurraga et al. (2020a)
	Gluten-free bread dough	*Tetraselmis chuii*	1, 2, 3	Viscoelastic (↘)	Nunes et al. (2020a)

	Gluten-free pasta dough	*A. platensis* (*Spirulina*)	1, 2	Viscoelastic (↗)	Fradinho et al. (2020)
		A. platensis (*Spirulina*)	3	Viscoelastic (↘)	Fradinho et al. (2020)
Emulsified products	o/w emulsion	*C. vulgaris*	2	Viscosity (↗), Viscoelastic (↗)	Raymundo et al. (2005)
		A. platensis (*Spirulina*)	1	Viscosity, Viscoelastic	Uribe-Wandurraga et al. (2020a)
		C. vulgaris	1	Viscosity, Viscoelastic	Uribe-Wandurraga et al. (2020b)
		Dunaliella salina	1	Viscosity, Viscoelastic	
	Mayonnaise	*C. vulgaris*	~0.5, ~1.5, ~2.5	Viscosity (=)	Abd El-Razik and Mohamed (2013)
Fruit- and vegetable-based products	Broccoli soup	*A. platensis* (*Spirulina*)	0.5–2	Viscosity (↗)	Lafarga et al. (2019)
		Chlorella sp.	0.5–2	Viscosity (↗)	Lafarga et al. (2019)
		Tetraselmis sp.	0.5–2	Viscosity (↗)	Lafarga et al. (2019)
Dairy products	Yogurt	*A. platensis* (*Spirulina*)	0.1–1.5	Viscosity (=)	Sengupta and Bhowal (2017)
		A. platensis (*Spirulina*)	0.13–0.5	Viscosity (↘)	Khaledabad, Ghasempour, Kia, Bari, and Zarrin (2019)

(*Continued*)

TABLE 10.1 (Continued)

Food category	Food product	Microalga species	Biomass concentration (%)	Rheology	References
	Ice cream mix	*A. platensis* (*Spirulina*)	0.075–0.3	Viscosity (↘)	Malik et al. (2013)
		Diacronema vlkianum	0.1, 0.2, 0.3	Viscosity (↘)	Durmaz et al. (2020)
		Nannochloropsis oculata	0.1, 0.2, 0.3	Viscosity (↘)	Durmaz et al. (2020)
		Porphyridium cruentum	0.1, 0.2, 0.3	Viscosity (↗)	Durmaz et al. (2020)

In the preparation of bread doughs, replacement of wheat flour with biomass of different microalgal species did not lead to a conclusive effect on the viscosity and viscoelastic properties of the enriched bread doughs. For instance, Uribe-Wandurraga, Igual, García-Segovia, and Martínez-Monzó (2019) reported a slightly increased storage modulus after enriching breadstick dough with *Arthrospira platensis* and *Chlorella vulgaris* biomass. In contrast, a decrease in storage and loss modulus was observed by Nunes et al. (2020b) for bread doughs enriched with *C. vulgaris* at a concentration of 1% indicating loss of network structure, although the loss factor (and therefore the degree of viscoelasticity) was not altered by the microalgae addition. Graça, Fradinho, Sousa, and Raymundo (2018) showed that the effect of *C. vulgaris* biomass on the viscoelastic behavior is dependent on the concentration. The authors suggested that concentrations up to ~1.8% caused a strengthening effect of the dough structure, presumably related to a reinforcement of the protein matrix. In contrast, a destructuring effect was observed for concentrations of ~2.4% and ~3.0%, which was appointed to disruption of the gluten matrix due to segregation of the microalgal particles. García-Segovia, Pagán-Moreno, Lara, and Martínez-Monzó (2017) reported increased viscosities of sourdoughs enriched with microalgal biomass at a concentration of 1.5%, but clearly showed differences in the structuring potential of different microalgal species. While the viscosity was almost doubled by addition of *Isochrysis galbana*, only a small viscosity increase was observed for *Scenedesmus almeriensis*, *Nannochloropsis gaditana*, or *Tetraselmis suecica*. It should be noted that the differences (increases and decreases) in viscosity and viscoelastic behavior observed in all the aforementioned studies after addition of microalgal biomass were rather limited, and that the rheological properties of the doughs were not representative for the texture of the final bread, since the firmness of the bread was not substantially affected by the addition of microalgae (García-Segovia et al., 2017; Graça et al., 2018; Nunes et al., 2020b).

Fradinho et al. (2020) investigated the effect of *A. platensis* incorporation on the rheological properties of gluten-free fresh pasta dough. While addition of 1% and 2% of *A. platensis* biomass resulted in increased storage modulus, incorporation of 3% biomass reduced the elasticity of the pasta doughs, as evidenced by a lower storage modulus. Hence, this observation confirms the importance of the concentration of microalgal biomass as reported by Graça et al. (2018), suggesting that the ideal incorporation concentration of microalgae in cereal-based doughs is below 2%. However, the differences in viscoelastic properties of the pasta doughs were not reflected in different textural properties of the cooked pasta (Fradinho et al., 2020).

A more pronounced change in rheological properties was observed for o/w emulsions enriched with 2% of *C. vulgaris* biomass, representing mayonnaises and salad dressings, showing a drastic increase in elastic properties (evidenced by a higher storage modulus and a frequency independence) as well as in zero shear viscosity. The authors attributed this to the high amounts of proteins and carbohydrates in *C. vulgaris* biomass, reinforcing the emulsion structure through the formation of physical entanglements (Raymundo, Gouveia, Batista, Empis, & Sousa, 2005). Similarly, addition of *Arthrospira* sp., *Chlorella* sp., or *Tetraselmis* sp. to broccoli soups resulted in increased viscosity (Lafarga et al., 2019).

The diversity in structuring potential of different microalgal species was obvious from a recent study by Durmaz et al. (2020), who investigated the rheological properties of an ice cream formulation after the addition of *Diacronema vlkianum*, *Nannochloropsis oculata*, and *Porphyridium cruentum*. Only addition of *P. cruentum* biomass resulted in a substantial viscosity increase compared to the control ice cream mix, which was already observed at a concentration of 0.1%. Given that no food processing was applied to the microalgal biomasses, that is, assuming that the microalgal cells were not disrupted, the improved viscosity was presumably attributed to the sulfated extracellular polysaccharides present in the red microalga. In contrast, addition of *D. vlkianum* or *N. oculata* led to a decrease in viscosity, regardless of their high protein contents (Durmaz et al., 2020). Hence, depending on the food application, the type of structural biopolymers and their molecular structure most likely determine the structuring potential of the microalgal biomasses. This was also observed in other studies using *A. platensis*. Despite its high protein content, no viscosity increase was observed after addition of this cyanobacterium to yogurt (Sengupta & Bhowal, 2017) or ice cream mix (Malik, Kempanna, & Paul, 2013). However, it should be noted that the latter authors used the biomass to replace the commercial stabilizer sodium alginate, and therefore concluded that *A. platensis* can only be used to partially replace sodium alginate in ice cream, due to the lower water-holding capacity of the cyanobacterium compared to sodium alginate (Malik et al., 2013).

Although *A. platensis* did not lead to enhanced rheological properties of dairy products, a strong gel reinforcement of gelatin-maltodextrin and κ-carrageenan gels was observed by addition of this cyanobacterium (Batista et al., 2011; Firoozmand & Rousseau, 2014). Hence, it can be concluded that the type of food product is another important factor determining the structuring potential of microalgal biomass. In addition, food processing in all the aforementioned studies was limited to incubation or fermentation processes at relatively low temperatures (between 8°C and 40°C). The use of more intense processing techniques could lead to more pronounced changes in rheological properties, as will be discussed later in the chapter.

10.3 Rheological properties of microalgal biomass in aqueous model systems

As discussed in the previous section, evaluation of the rheological properties of microalgal biomass in various food products does not lead to a conclusive answer on the structuring capacity of microalgae, which could partially be attributed to the complexity of the different food matrices. Investigation of the rheological properties in simple model systems such as aqueous suspensions is therefore very useful to assess the intrinsic structuring potential of microalgal biomasses of different microalgal species. An overview of the studies available in literature is shown in Table 10.2. Most studies focused on the viscosity (and/or flow behavior) of microalgal suspensions. Given that most of these experiments were established in the context of optimization of cultivation and downstream processing of microalgae, a broad range of biomass concentrations has been studied (Bernaerts et al., 2019a). On the one hand, dilute systems (0.004%–0.5%) have been analyzed representing concentrations in open raceway ponds (typically up to 0.5 $g\,L^{-1}$) or closed photobioreactors (typically up to 5 $g\,L^{-1}$) (Zhang et al., 2013). On the other hand, rheological properties of highly concentrated microalgal slurries during downstream processing, obtained after dewatering of microalgal cultures, have been studied in concentrations up to 25% (Chen et al., 2018; Mettu et al., 2019; Yap, Martin, & Scales, 2016). At low concentrations, microalgal suspensions typically show Newtonian flow behavior often characterized by low viscosities. Above a certain critical concentration, non-Newtonian behavior is observed (usually pseudoplastic flow behavior). The value of this critical concentration depends on the microalgal species, and is influenced by several factors such as particle size, particle deformability, and extracellular compounds, among others (Wileman, Ozkan, & Berberoglu, 2012). Comparison of the critical concentration of different microalgal species can therefore serve as a first indication of the structuring potential of the microalgal biomass, apart from the viscosity values and degree of non-Newtonian behavior at a specific biomass concentration.

One of the most promising microalgal species for inducing thickening in food products is *P. cruentum*. In comparison to several other microalgal species, *P. cruentum* displayed the highest viscosities in aqueous suspensions at 8% w/w. In addition, suspensions of this red microalga showed the most pronounced shear-thinning behavior, characterized by flow behavior indices of 0.20–0.23 (Bernaerts et al., 2017). The structuring capacity of *P. cruentum* biomass is attributed to its extracellular polysaccharides, which are characterized as high-molecular-weight polymers with a weight average molar mass between 2.4×10^5 and $4 \times 10^6\,g\,mol^{-1}$ (Bernaerts, Kyomugasho, et al., 2018; Geresh, Adin, Yarmolinsky, & Karpasas, 2002; Patel et al., 2013). They possess unique building blocks composed of xylose, glucose, galactose, and glucuronic acid, although different ratios of these monosaccharides have been reported in the literature (Bernaerts, Kyomugasho, et al., 2018; Geresh & Arad, 1991; Geresh et al., 2002, 2009; Gloaguen et al., 2004; Heaney-Kieras & Chapman, 1976; Patel et al., 2013; Percival & Foyle, 1979; Roussel et al., 2015; Soanen, Da Silva, Gardarin, Michaud, & Laroche, 2015). In addition to the glucuronic acid residues, the presence of sulfate groups esterified to glucose or galactose units (ranging from 1% to 15.1%) gives anionic characteristics to the polysaccharides (Bernaerts, Kyomugasho,

TABLE 10.2 Research studies evaluating the rheological properties of microalgal biomass suspensions.

References	Microalgal species	Biomass concentration	Flow behavior: shear rate (s^{-1})	Linear viscoelastic behavior: shear strain (%), angular frequency (rad s^{-1})	Processing of microalgal biomass
Adesanya, Vadillo, and Mackley (2012)	*Scenedesmus obliquus*	6.8%–15% (w/v) $\phi = 0.015–0.15$	/	0.01%–100%, 50–3000 Hz	/
Bernaerts et al. (2017)	*Arthrospira platensis, Chlorella vulgaris, Nannochloropsis* sp., *Odontella aurita, Phaeodactylum tricornutum, Porphyridium cruentum, Schizochytrium* sp.	8% (w/w)	0.1–100 s^{-1}	0.01%–1000%, 0.1–10 rad s^{-1}	Untreated HPH (100 MPa, 1 pass) HPH + thermal treatment ($^{10°C}PU_{90°C} = 3$ min) HPH + thermal treatment ($^{10°C}F_{121.1°C} = 5$ min)
Bernaerts, Panozzo, et al. (2018)	*C. vulgaris, P. cruentum*	8% (w/w)	0.1–100 s^{-1}	0.01%–1000%, 0.1–10 rad s^{-1}	Untreated Thermal treatment (95°C, 15 min) Thermal treatment + HPH (20 MPa, 1 pass) Thermal treatment + HPH (100 MPa, 1 pass)

Buchmann, Bloch, and Mathys (2018)	*A. platensis*	5–100 g L^{-1}	0.5–500 s^{-1}	/	/
Cagney, Zhang, Bransgrove, Allen, and Balabani (2017)	*Chlorella* sp., *P. tricornutum, Tetraselmis chuii*	$\phi = 0.05–0.2$	20–200 s^{-1}	/	/
Chang et al. (2014)	*Schizochytrium* sp.	47.5–130.7 g L^{-1}	1–100 s^{-1}	/	/
Chen, Chen, and Gong (1997)	*Haematococcus lacustris*	7–64 g L^{-1}	~15–~123 s^{-1}	/	/
Chen et al. (2018)	*Chlorella pyrenoidosa*	10%–20% (w/w)	21.5–1000 s^{-1}	/	/
Martínez-Sanz et al. (2020)	*A. platensis, Nannochloropsis gaditana, Scenedesmus almeriensis*	1%–25% (w/w)	0.1–100 s^{-1}	0.01%–1000%, 0.1–10 rad s^{-1}	Untreated Ultrasound (24 kHz, 400 W, 5 min)
Mettu et al. (2019)	*Nannochloropsis* sp.	12%–24% (w/w)	0.005–300 s^{-1}	/	Untreated Incubation (40°C, 48 h) + HPH (120 MPa, one pass)
Schneider and Gerber (2014)	*Nannochloropsis salina*	9.7%–24% (w/w)	0.076–600 s^{-1}	/	Untreated Thermal treatment (120°C, 2 + 2.25 h)

(*Continued*)

TABLE 10.2 (Continued)

References	Microalgal species	Biomass concentration	Flow behavior: shear rate (s^{-1})	Linear viscoelastic behavior: shear strain (%), angular frequency (rad s^{-1})	Processing of microalgal biomass
Wileman et al. (2012)	*C. vulgaris, Nannochloris* sp., *P. tricornutum*	0.5–80 kg m^{-3}	5–280 s^{-1}	/	/
Yap et al. (2016)	*Nannochloropsis* sp.	25% (w/w)	0.001–5000 s^{-1}	0.1–10 Hz	Untreated HPH (100 MPa, 1 pass)
Zhang et al. (2013)	*Chlorella* sp.	$\phi = 1.18–4.77$	5–100 s^{-1}	/	/
Zhang et al. (2018)	*C. pyrenoidosa*	5%–15% (w/w)	21.5–1000 s^{-1}	/	/

HPH, High-pressure homogenization.

et al., 2018; de Jesus Raposo, de Morais, & de Morais, 2014; Geresh & Arad, 1991; Heaney-Kieras & Chapman, 1976; Patel et al., 2013; Percival & Foyle, 1979). However, despite the large abundance of negatively charged residues in *Porphyridium* sp. polysaccharides, addition of different cations (Na^+, K^+, Ca^{2+}, or Zn^{2+}) only results in screening of electrostatic repulsions (already observed at low cation concentrations), but does not lead to network formation through cation-mediated gelation as known for several conventional hydrocolloids such as carrageenans, alginates, and pectins (Bernaerts, Kyomugasho, et al., 2018; Eteshola, Karpasas, Arad, & Gottlieb, 1998; Liberman, Ochbaum, Arad, & Bitton, 2016; Saha & Bhattacharya, 2010). Nevertheless, in a biomass suspension, extracellular polysaccharides are responsible for intercellular interactions of neighboring cells, which result in the formation of large and stable cell clusters. In addition, solubilization of these viscosity-increasing polysaccharides causes shear-thinning flow behavior of the continuous serum phase. Hence, it is obvious that extracellular polysaccharides have an important contribution to the structuring capacity of *P. cruentum* biomass suspensions, yielding samples that hardly pour with viscosities of up to 100 Pa s (i.e., 100,000 mPa s) (Bernaerts et al., 2017; Bernaerts, Panozzo, et al., 2018).

Extracellular polysaccharides have also been reported for different strains of *Chlorella* (Bernaerts et al., 2019a; Ferreira et al., 2020), but generally show lower thickening properties than extracellular polysaccharides of *P. cruentum*. Shear-thinning behavior has been reported for 0.5% solutions of extracellular polysaccharides isolated from *Chlorella* sp., but with viscosities below 25 mPa s (Yalcin, Hicsasmaz, Boz, & Bozoglu, 1994). Even though the extracellular polysaccharides of *C. vulgaris* have been shown to establish the microstructural properties of biomass suspensions, creating some cell clusters by intercellular interactions, their contribution to the viscosity of the biomass suspensions is not as large as observed for *P. cruentum*. The latter was obvious from much lower values for consistency coefficient (approximately six times lower) and higher flow behavior indices (between 0.35 and 0.51) for biomass suspensions at a concentration of 8% w/w (Bernaerts et al., 2017; Bernaerts, Panozzo, et al., 2018). According to Wileman et al. (2012), the critical concentration of *C. vulgaris* biomass suspensions would be between 4% and 6%, indicating that a relatively high biomass concentration is required to create cell aggregates, which result in shear-thinning behavior upon flow. While this critical concentration corresponds to a volume fraction of approximately 0.115, an apparent yield stress would only occur at a volume fraction of $\sim$0.25 (Cagney et al., 2017; Souliès, Pruvost, Legrand, Castelain, & Burghelea, 2013). Indeed, although some authors observed a yield stress behavior for *Chlorella* sp. suspensions, they emphasized that the values were rather low (Bernaerts et al., 2017; Wu & Shi, 2008).

In contrast to the high critical concentration of *C. vulgaris* biomass, suspensions of *A. platensis* already showed deviation from Newtonian behavior at biomass concentrations of 0.5%. This shear-thinning behavior was attributed to the shape and size of the cells of this cyanobacterium. Compared to spherical cells, the filaments of *A. platensis* show stronger interactions with the medium at rest, but align to the flow when shearing forces are applied (Buchmann et al., 2018; Torzillo, 1997). In spite of this lower critical concentration, the viscosity increase at higher concentrations was much lower compared to *C. vulgaris* suspensions, with viscosities in 8% suspensions being approximately 10 times lower for *A. platensis* compared to the green microalga (Bernaerts et al., 2017).

No pronounced shear-thinning behavior has been observed for *Phaeodactylum tricornutum* suspensions (flow behavior indices between 0.83 and 1), despite the higher aspect ratio of the fusiform morphotypes compared to spherical microalgae cells, and the fact that nonspherical particles generally orient themselves to the flow (Bernaerts et al., 2017; Cagney et al., 2017; Wileman et al., 2012). In addition, the viscosities observed were low, probably due to the absence of extracellular polysaccharides in the fusiform morphotypes. Since the ovoid type is the only morphotype of *P. tricornutum* that produces extracellular polysaccharides (Desbois, Walton, & Smith, 2010), it would be interesting to compare its rheological properties in biomass suspensions to the other morphotypes, which, to the best of our knowledge, has not been reported so far.

The microalga species with the highest critical concentration reported so far is *Nannochloropsis* sp., presumably due to the small size of the spherical cells, and also to the rigid and uncharged cell wall lacking extracellular polysaccharides (Bernaerts, Gheysen, et al., 2018; Scholz et al., 2014). Schneider, Fortin, Span, and Gerber (2016) concluded that the critical concentration for *Nannochloropsis salina* ranged between 15.8% and 17.7%. In addition, even though shear-thinning flow behavior and increased viscosities were observed for *Nannochloropsis* sp. suspensions at higher concentrations (up to 25%), viscosities were relatively low and no substantial yield stress was observed, indicating that *Nannochloropsis* sp. cells do not tend to aggregate and show typical behavior of rigid spherical particles (Mettu et al., 2019; Schneider et al., 2016; Yap et al., 2016).

In contrast to the flow behavior, the linear viscoelastic behavior of microalgal suspensions has been very poorly investigated. However, small oscillatory deformations within the linear viscoelastic region allow characterization of the native structure of the sample without disturbing the network structure, and these analyses are very useful for evaluating the intrinsic structuring potential of microalgae. As a matter of fact, large variability in the linear viscoelastic behavior was observed by Bernaerts et al. (2017) when comparing seven microalgal species. Only three of the microalgal species presented a substantial network structure, and were characterized as weak gels ($G' > G''$ and $G' \propto \omega^{0.1}$). As it was shown in a subsequent study that these three microalgae (*P. cruentum*, *C. vulgaris*, and *Odontella aurita*) were the extracellular polysaccharide-producing microalgae, it can be concluded that extracellular polysaccharides are responsible for the organized structure in nonprocessed microalgal suspensions, aggregating the intact cells and resulting in stable cell clusters. Higher absolute values of G' resulted in a stronger network structure in *P. cruentum* suspensions compared to the other two microalgae, presumably related to the higher amounts and different molecular structure of the extracellular polysaccharides (Bernaerts et al., 2017; Bernaerts, Gheysen, et al., 2018). In contrast, no intrinsic structuring potential was shown for *A. platensis*, *P. tricornutum*, *Nannochloropsis* sp., *Schizochytrium* sp., and *Scenedesmus obliquus*, with most of them lacking extracellular polysaccharides (Adesanya et al., 2012; Bernaerts et al., 2017, 2019a). A strong frequency dependence of *S. obliquus* suspensions in concentrations between 3% and 9% indicated a liquid-like behavior, and hence a low network structure for this microalga (Adesanya et al., 2012).

10.4 Food processing operations for targeted functionalization of microalgal biomass

Whereas the rheological properties of unprocessed microalgal suspensions can generally be attributed to particle properties (size, shape, and deformability) and the presence of extracellular polymers, processing techniques can be used to modify the microstructure of the microalgal biomass, and therefore to target specific rheological properties. In conventional food products, mechanical processing is often used as a functionalization step, for example, for physical stabilization of oil droplets in milk products or for creating the desired structural properties of vegetable-based products, the latter generally being related to the rheological properties (Lopez-Sanchez et al., 2011). In the context of microalgal processing, mechanical techniques have often been used for disruption of microalgal cells. As cell disruption results in the release of intracellular compounds, allowing interactions between liberated biopolymers to occur, this process often impacts the rheological properties as well.

Regarding microalgal cell disruption methods, high-pressure homogenization is considered one of the most effective techniques. Safi et al. (2014) compared this technique with manual grinding, ultrasonication, and chemical treatments, and concluded that high-pressure homogenization was more efficient both for microalgae with fragile cell walls (*A. platensis* and *P. cruentum*) and for more rigid cell-walled species (*N. oculata*, *Haematococcus pluvialis*, and *C. vulgaris*). It should be noted that the authors applied very intense homogenization conditions (two passes at 270 MPa) for all the aforementioned microalgal species. The degree of cell disruption is obviously determined by the pressure level and the number of passes. Spiden et al. (2013) showed for instance that one pass at 40 MPa was sufficient for nearly complete cell disruption in the fragile cell-walled *T. suecica*, whereas multiple passes at pressures >100 MPa were required for disrupting *Chlorella* sp.

and *Nannochloropsis* sp. cells. The latter was confirmed by Bernaerts, Gheysen, Foubert, Hendrickx, and Van Loey (2019b), concluding nearly complete disruption of *Nannochloropsis* sp. cells after four passes at 100 MPa, while the number of passes could be reduced to two when operating at an ultra-high pressure of 250 MPa. Hence, the use of moderate high-pressure homogenization conditions generally applied in the food industry (typically far below 100 MPa) will not be sufficient to fully disrupt microalgal cells with rigid cell walls.

The impact of high-pressure homogenization on the rheological properties of microalgal suspensions has so far only been studied at 100 and 120 MPa (Bernaerts et al., 2017; Mettu et al., 2019; Yap et al., 2016). High-pressure homogenization for a single pass at 100 MPa variously affected the rheological properties of seven microalgal species in 8% suspensions. For *A. platensis* and *C. vulgaris*, a (slight but significant) viscosity increase was observed, due to the release of intracellular compounds as a result of cell disruption. In contrast, decreased viscosity and gel strength were observed for *P. cruentum* after high-pressure homogenization at 100 MPa. The authors attributed this to breakage of the extracellular polysaccharide network under high shear forces, leading to reduced serum viscosity and destabilization of the cell clusters in the original suspension. Decreased viscosity after high-pressure homogenization was also seen for *O. aurita*, but this was attributed to a drastic particle size reduction, given the somewhat larger cells (around 50 μm) of this diatom (Wiltshire & Dürselen, 2004). Finally, the impact of high-pressure homogenization on the rheological properties was negligible for the other microalgal species. For *Nannochloropsis* sp., this was attributed to the very low degree of cell disruption occurring at a single pass at 100 MPa, barely altering the microstructure (Bernaerts et al., 2017). In contrast, Mettu et al. (2019) reported a drastic increase in viscosity after high-pressure homogenization for one pass at 120 MPa. However, this mechanical treatment was performed on *Nannochloropsis* sp. cells that were weakened during a preceding incubation treatment (48 h, 40°C), which were presumably more susceptible to cell disruption in the subsequent high-pressure homogenization process (Mettu et al., 2019).

While several other food processing techniques can be used for microalgal cell disruption or for changing microstructural properties, no studies were found on their impact on the rheological behavior of microalgal suspensions. Although microwave treatments have been shown to disrupt cell membranes and/or cell walls of microalgae due to the heat and pressure build-up inside the cells (Günerken et al., 2015), intracellular compounds are released to a lesser extent than with high-pressure homogenization (Cho et al., 2012; Heo et al., 2017). The same conclusion has been drawn for extrusion processes (Wang et al., 2018). In contrast to the aforementioned techniques, pulsed electric field (PEF) treatment generally induces membrane permeabilization without distorting the microalgal cell wall ('t Lam et al., 2017). While Carullo et al. (2018) clearly showed the shrinkage of *C. vulgaris* cells after PEF treatment, indicating the release of intracellular compounds through the electroporated membranes, most of the high-molecular-weight polymers are assumed to be retained inside the cell wall. Consequently, aggregate formation in PEF-treated suspensions is less expected compared to those subjected to high-pressure homogenization, and possible viscosity increases are therefore supposed to be less pronounced (Bernaerts et al., 2019a).

On the contrary, thermal processing, commonly used as a preservation step to guarantee the safety of food products, led in many cases to altered rheological properties of microalgal suspensions due to solubilization and/or other heat-induced reactions of several structural biopolymers. Bernaerts, Panozzo, et al. (2018) showed a drastic increase in viscosity and gel strength (up to seven times) after thermally treating suspensions of *P. cruentum* and *C. vulgaris* for 15 min at 95°C, and this structural build-up was attributed to the formation of strong aggregates. However, while gelation of extracellular polysaccharides was identified as the mechanism in *P. cruentum* suspensions, the formation of aggregates in *C. vulgaris* was likely due to protein solubilization and/or denaturation (Bernaerts, Panozzo, et al., 2018). Another study showed that the intensity of the thermal treatment largely influences the extent of structural build-up. In this context, an intense sterilization process resulted in a higher viscosity and gel strength for different microalgal species compared to a milder pasteurization treatment (Bernaerts et al., 2017). This was also observed by González-Fernández, Sialve, Bernet, and Steyer (2012), who reported cell disruption of *Scenedesmus* sp. to occur when heated at 90°C, resulting in the release of intracellular material and hence aggregate formation, whereas an intact cell wall was observed after a thermal treatment at 70°C. However, thermal processing did not increase the viscosity for all microalgae. For instance, *Nannochloropsis* sp. suspensions showed unaltered or low viscosities after various heat treatments, including a pasteurization ($^{10°C}PU_{90°C} = 3$ min), a sterilization ($^{10°C}F_{121.1°C} = 5$ min), and an intense preheating (2.25 h, 120°C). Apart from some protein solubilization, the microstructural properties of *Nannochloropsis* sp. suspensions were not drastically changed for any of these thermal treatments (Bernaerts et al., 2017; Schneider & Gerber, 2014).

Combining different processing techniques might actually be the best strategy to maximize the structuring potential of microalgae. Thermal processing of *C. vulgaris* suspensions was proven more effective when preceded by a high-pressure homogenization step, as this cell disruption treatment caused more intracellular components to be available for heat-induced interactions during the subsequent thermal treatment (Bernaerts, Panozzo, et al., 2018). In addition, it is important to note that different processing sequences also lead to distinct microstructures, as schematically represented for *C. vulgaris* in Fig. 10.1. These different microstructural properties not only alter the rheological properties, they are also relevant from a nutritional perspective. It was recently proven that cell disruption upon high-pressure homogenization

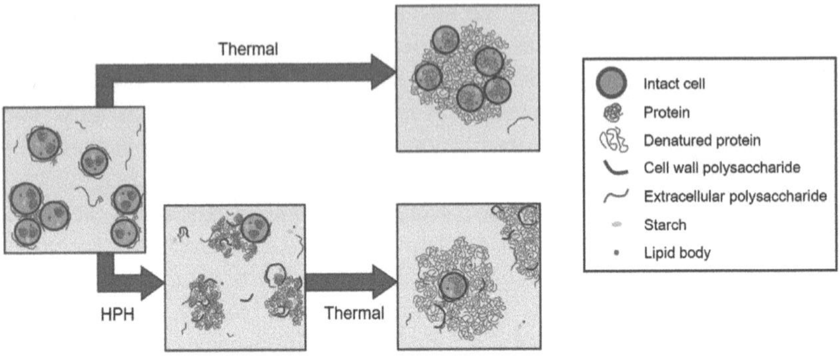

FIGURE 10.1 Schematic overview of microstructural changes of *Chlorella vulgaris* suspensions upon different processing sequences, based on Bernaerts et al. (2017), Bernaerts, Panozzo, et al. (2018). *HPH*, High-pressure homogenization.

improved in vitro bioaccessibility of carotenoids and ω3-LC-PUFA in the microalga *Nannochloropsis* sp. (Bernaerts et al., 2020). Hence, as *C. vulgaris* also possesses a (semi)rigid cell wall, combination of mechanical and thermal processing will not only result in the highest rheological properties, but presumably also lead to the highest nutrient bioaccessibility.

10.5 Conclusion

Besides the use of microalgae as functional food ingredients for nutritional enrichment of foods, microalgal biomass of particular species could also improve the rheological properties of the enriched food products. Despite the little research studies on rheological properties of microalgal biomasses as food ingredients, compilation of the literature data available so far indicates the structuring potential of certain microalgae species. In general, the structuring capacity of nonprocessed (intact) biomass is low, given that most structural biopolymers are located inside the individual cells and cannot interact to create polymer networks. Only for microalgal species rich in extracellular polymers, high viscosities can be observed by a combination of aggregate formation and polymer solubilization into the continuous phase, as shown for *P. cruentum* suspensions. However, (food) processing can be used to modify the microstructure and therefore the rheological properties of the microalgal suspensions. On the one hand, this has been shown for high-pressure homogenization, inducing cell disruption resulting in improved rheological properties. On the other hand, the use of thermal processes promotes heat-induced solubilization and gelation reactions, beneficial for creating structural build-up. However, there is a need for more research on the impact of (other) food processing operations on the rheological and microstructural properties of microalgal biomass suspensions. Even though the number of research studies on the rheological properties of microalgae-enriched foods has been largely increasing in recent years, investigation of food processing operations during preparation of these enriched food products will be essential in evaluating the overall potential of microalgae as structuring food ingredients.

Acknowledgments

Tom M.M. Bernaerts is a postdoctoral researcher funded by Research Foundation Flanders (FWO) (Grant no. 1252221N). Financial support was also obtained from the Research Fund KU Leuven (KP/14/004).

References

Abd El-Razik, M. M., & Mohamed, A. G. (2013). Utilization of acid casein curd enriched with *Chlorella vulgaris* biomass as substitute of egg in mayonnaise production. *World Applied Sciences Journal*, *26*, 917–925. Available from https://doi.org/10.5829/idosi.wasj.2013.26.07.13523.

Adesanya, V. O., Vadillo, D. C., & Mackley, M. R. (2012). The rheological characterization of algae suspensions for the production of biofuels. *Journal of Rheology (New York)*, *56*, 925–939. Available from https://doi.org/10.1122/1.4717494.

Batista, A. P., Nunes, M. C., Raymundo, A., Gouveia, L., Sousa, I., Cordobés, F., ... Franco, J. M. (2011). Microalgae biomass interaction in biopolymer gelled systems. *Food Hydrocolloids*, *25*, 817–825. Available from https://doi.org/10.1016/j.foodhyd.2010.09.018.

Bernaerts, T. M. M., Gheysen, L., Foubert, I., Hendrickx, M. E., & Van Loey, A. M. (2019a). The potential of microalgae and their biopolymers as structuring ingredients in food: A review. *Biotechnology advances*, *37*, 107419. Available from https://doi.org/10.1016/j.biotechadv.2019.107419.

Bernaerts, T. M. M., Gheysen, L., Foubert, I., Hendrickx, M. E., & Van Loey, A. M. (2019b). Evaluating microalgal cell disruption upon ultra high pressure homogenization. *Algal Research*, *42*, 101616. Available from https://doi.org/10.1016/j.algal.2019.101616.

Bernaerts, T. M. M., Gheysen, L., Kyomugasho, C., Jamsazzadeh Kermani, Z., Vandionant, S., Foubert, I., ... Van Loey, A. M. (2018). Comparison of microalgal biomasses as functional food ingredients: Focus on the composition of cell wall related polysaccharides. *Algal Research*, *32*, 150–161. Available from https://doi.org/10.1016/j.algal.2018.03.017.

Bernaerts, T. M. M., Kyomugasho, C., Van Looveren, N., Gheysen, L., Foubert, I., Hendrickx, M. E., & Van Loey, A. M. (2018). Molecular and rheological characterization of different cell wall fractions of *Porphyridium cruentum*. *Carbohydrate Polymers*, *195*, 542–550. Available from https://doi.org/10.1016/j.carbpol.2018.05.001.

Bernaerts, T. M. M., Panozzo, A., Doumen, V., Foubert, I., Gheysen, L., Goiris, K., ... Van Loey, A. M. (2017). Microalgal biomass as a (multi)functional ingredient in food products: Rheological properties of microalgal suspensions as affected by mechanical and thermal processing. *Algal Research*, *25*, 452–463. Available from https://doi.org/10.1016/j.algal.2017.05.014.

Bernaerts, T. M. M., Panozzo, A., Verhaegen, K. A. F., Gheysen, L., Foubert, I., Moldenaers, P., ... Van Loey, A. M. (2018). Impact of different sequences of mechanical and thermal processing on the rheological properties of *Porphyridium cruentum* and *Chlorella vulgaris* as functional food ingredients. *Food & Function*, *9*, 2433–2446. Available from https://doi.org/10.1039/c8fo00261d.

Bernaerts, T. M. M., Verstreken, H., Dejonghe, C., Gheysen, L., Foubert, I., Grauwet, T., & Van Loey, A. M. (2020). Cell disruption of *Nannochloropsis* sp. improves in vitro bioaccessibility of carotenoids and ω3-LC-PUFA. *Journal of Functional Foods*, *65*, 103770. Available from https://doi.org/10.1016/j.jff.2019.103770.

Buchmann, L., Bloch, R., & Mathys, A. (2018). Comprehensive pulsed electric field (PEF) system analysis for microalgae processing. *Bioresource Technology*, *265*, 268–274. Available from https://doi.org/10.1016/j.biortech.2018.06.010.

Cagney, N., Zhang, T., Bransgrove, R., Allen, M. J., & Balabani, S. (2017). Effects of cell motility and morphology on the rheology of algae suspensions. *Journal of Applied Phycology*, *29*, 1145–1157. Available from https://doi.org/10.1007/s10811-016-1033-y.

Caporgno, M. P., & Mathys, A. (2018). Trends in microalgae incorporation into innovative food products with potential health benefits. *Frontiers in Nutrition*, *5*, 58. Available from https://doi.org/10.3389/fnut.2018.00058.

Carullo, D., Abera, B. D., Casazza, A. A., Donsì, F., Perego, P., Ferrari, G., & Pataro, G. (2018). Effect of pulsed electric fields and high pressure homogenization on the aqueous extraction of intracellular compounds from the microalgae *Chlorella vulgaris*. *Algal Research*, *31*, 60–69. Available from https://doi.org/10.1016/j.algal.2018.01.017.

Chang, G., Wu, J., Jiang, C., Tian, G., Wu, Q., Chang, M., & Wang, X. (2014). The relationship of oxygen uptake rate and kLa with rheological properties in high cell density cultivation of docosahexaenoic acid by *Schizochytrium* sp. S31. *Bioresource Technology*, *152*, 234–240. Available from https://doi.org/10.1016/j.biortech.2013.11.002.

Chen, F., Chen, H., & Gong, X. (1997). Mixotrophic and heterotrophic growth of *Haematococcus lacustris* and rheological behaviour of the cell suspensions. *Bioresource Technology*, *62*, 19–24. Available from https://doi.org/10.1016/S0960-8524(97)00115-6.

Chen, H., Fu, Q., Liao, Q., Zhang, H., Huang, Y., Xia, A., & Zhu, X. (2018). Rheological properties of microalgae slurry for application in hydrothermal pretreatment systems. *Bioresource Technology*, *249*, 599–604. Available from https://doi.org/10.1016/j.biortech.2017.10.051.

Cho, S., Choi, W., Oh, S., Lee, C., Seo, Y., Kim, J., … Lee, H. (2012). Enhancement of lipid extraction from marine microalga, *Scenedesmus* associated with high-pressure homogenization process. *Journal of Biomedicine & Biotechnology*, *2012*. Available from https://doi.org/10.1155/2012/359432.

de Jesus Raposo, M. F., de Morais, A. M. M. B., & de Morais, R. M. S. C. (2014). Influence of sulphate on the composition and antibacterial and antiviral properties of the exopolysaccharide from *Porphyridium cruentum*. *Life Sciences*, *101*, 56–63. Available from https://doi.org/10.1016/j.lfs.2014.02.013.

Desbois, A. P., Walton, M., & Smith, V. J. (2010). Differential antibacterial activities of fusiform and oval morphotypes of *Phaeodactylum tricornutum* (Bacillariophyceae). *Journal of the Marine Biological Association of the United Kingdom*, *90*, 769–774. Available from https://doi.org/10.1017/S0025315409991366.

Durmaz, Y., Kilicli, M., Toker, O. S., Konar, N., Palabiyik, I., & Tamtürk, F. (2020). Using spray-dried microalgae in ice cream formulation as a natural colorant: Effect on physicochemical and functional properties. *Algal Research*, *47*, 101811. Available from https://doi.org/10.1016/j.algal.2020.101811.

Eteshola, E., Karpasas, M., Arad, S., & Gottlieb, M. (1998). Red microalga exopolysaccharides: 2. Study of the rheology, morphology and thermal gelation of aqueous preparations. *Acta Polymerica*, *49*, 549–556, https://doi.org/10.1002/(SICI)1521-4044(199810)49::10/11 < 549:AID-APOL549 > 3.0.CO;2-T.

Ferreira, A. S., Ferreira, S. S., Correia, A., Vilanova, M., Silva, T. H., Coimbra, M. A., & Nunes, C. (2020). Reserve, structural and extracellular polysaccharides of *Chlorella vulgaris*: A holistic approach. *Algal Research*, *45*, 101757. Available from https://doi.org/10.1016/j.algal.2019.101757.

Firoozmand, H., & Rousseau, D. (2014). Tailoring the morphology and rheology of phase-separated biopolymer gels using microbial cells as structure modifiers. *Food Hydrocolloids*, *42*, 204–214. Available from https://doi.org/10.1016/j.foodhyd.2014.04.040.

Fradinho, P., Niccolai, A., Soares, R., Rodolfi, L., Biondi, N., Tredici, M. R., … Raymundo, A. (2020). Effect of *Arthrospira platensis* (spirulina) incorporation on the rheological and bioactive properties of gluten-free fresh pasta. *Algal Research*, *45*, 101743. Available from https://doi.org/10.1016/j.algal.2019.101743.

García-Segovia, P., Pagán-Moreno, M. J., Lara, I. F., & Martínez-Monzó, J. (2017). Effect of microalgae incorporation on physicochemical and textural properties in wheat bread formulation. *Food Science and Technology International*, *23*, 437–447. Available from https://doi.org/10.1177/1082013217700259.

Geresh, S., Adin, I., Yarmolinsky, E., & Karpasas, M. (2002). Characterization of the extracellular polysaccharide of *Porphyridium* sp.: Molecular weight determination and rheological properties. *Carbohydrate Polymers*, *50*, 183–189. Available from https://doi.org/10.1016/S0144-8617(02)00019-X.

Geresh, S., & Arad, S. (1991). The extracellular polysaccharides of the red microalgae: Chemistry and rheology. *Bioresource Technology*, *38*, 195–201. Available from https://doi.org/10.1016/0960-8524(91)90154-C.

Geresh, S., Arad, S., Levy-Ontman, O., Zhang, W., Tekoah, Y., & Glaser, R. (2009). Isolation and characterization of poly- and oligosaccharides from the red microalga *Porphyridium* sp. *Carbohydrate Research*, *344*, 343–349. Available from https://doi.org/10.1016/j.carres.2008.11.012.

Gloaguen, V., Ruiz, G., Morvan, H., Mouradi-Givernaud, A., Maes, E., Krausz, P., & Strecker, G. (2004). The extracellular polysaccharide of *Porphyridium* sp.: An NMR study of lithium-resistant oligosaccharidic fragments. *Carbohydrate Research*, *339*, 97–103. Available from https://doi.org/10.1016/j.carres.2003.09.020.

González-Fernández, C., Sialve, B., Bernet, N., & Steyer, J. P. (2012). Thermal pretreatment to improve methane production of *Scenedesmus* biomass. *Biomass and Bioenergy*, *40*, 105–111. Available from https://doi.org/10.1016/j.biombioe.2012.02.008.

Graça, C., Fradinho, P., Sousa, I., & Raymundo, A. (2018). Impact of *Chlorella vulgaris* on the rheology of wheat flour dough and bread texture. *LWT Food Science and Technology*, *89*, 466–474. Available from https://doi.org/10.1016/j.lwt.2017.11.024.

Günerken, E., D'Hondt, E., Eppink, M. H. M., Garcia-Gonzalez, L., Elst, K., & Wijffels, R. H. (2015). Cell disruption for microalgae biorefineries. *Biotechnology Advances*, *33*, 243–260. Available from https://doi.org/10.1016/j.biotechadv.2015.01.008.

Heaney-Kieras, J., & Chapman, D. J. (1976). Structural studies on the extracellular polysaccharide of the red alga *Porphyridium cruentum*. *Carbohydrate Research*, *52*, 169–177. Available from https://doi.org/10.1016/S0008-6215(00)85957-1.

Heo, Y. M., Lee, H., Lee, C., Kang, J., Ahn, J., Lee, Y. M., … Kim, J. (2017). An integrative process for obtaining lipids and glucose from *Chlorella vulgaris* biomass with a single treatment of cell disruption. *Algal Research*, *27*, 286–294. Available from https://doi.org/10.1016/j.algal.2017.09.022.

Khaledabad, M. A., Ghasempour, Z., Kia, E. M., Bari, M. R., & Zarrin, R. (2019). Probiotic yoghurt functionalised with microalgae and Zedo gum: Chemical, microbiological, rheological and sensory characteristics. *International Journal of Dairy Technology*, *70*, 1–9. Available from https://doi.org/10.1111/1471-0307.12625.

Lafarga, T. (2019). Effect of microalgal biomass incorporation into foods: Nutritional and sensorial attributes of the end products. *Algal Research, 41*, 101566. Available from https://doi.org/10.1016/j.algal.2019.101566.

Lafarga, T., Acién-Fernández, F. G., Castellari, M., Villaró, S., Bobo, G., & Aguiló-Aguayo, I. (2019). Effect of microalgae incorporation on the physicochemical, nutritional, and sensorial properties of an innovative broccoli soup. *LWT Food Science and Technology, 111*, 167–174. Available from https://doi.org/10.1016/j.lwt.2019.05.037.

Liberman, G. N., Ochbaum, G., Arad, S., & Bitton, R. (2016). The sulfated polysaccharide from a marine red microalga as a platform for the incorporation of zinc ions. *Carbohydrate Polymers, 152*, 658–664. Available from https://doi.org/10.1016/j.carbpol.2016.07.025.

Lopez-Sanchez, P., Nijsse, J., Blonk, H. C. G., Bialek, L., Schumm, S., & Langton, M. (2011). Effect of mechanical and thermal treatments on the microstructure and rheological properties of carrot, broccoli and tomato dispersions. *Journal of the Science of Food and Agriculture, 91*, 207–217. Available from https://doi.org/10.1002/jsfa.4168.

Malik, P., Kempanna, C., & Paul, A. (2013). Quality characteristics of ice cream enriched with Spirulina powder. *International Journal of Food and Nutritional Science*, 2, 44–50.

Martínez-Sanz, M., Garrido-Fernández, A., Mijlkovic, A., Krona, A., Martínez-Abad, A., Coll-Marqués, J. M., … Lopez-Sanchez, P. (2020). Composition and rheological properties of microalgae suspensions: Impact of ultrasound processing. *Algal Research, 49*101960. Available from https://doi.org/10.1016/j.algal.2020.101960.

Mettu, S., Yao, S., Law, S. Q. K., Sun, Z., Scales, P. J., Ashokkumar, M., & Martin, G. J. O. (2019). Rheological properties of concentrated slurries of harvested, incubated and ruptured Nannochloropsis sp. cells. *BMC Chemical Engineering, 1*, 1–11. Available from https://doi.org/10.1186/s42480-019-0011-y.

Nunes, M. C., Fernandes, I., Vasco, I., Sousa, I., & Raymundo, A. (2020a). Tetraselmis chuii as a sustainable and healthy ingredient to produce gluten-free bread: Impact on structure, colour and bioactivity. *Foods, 9*, 579–593. Available from https://doi.org/10.3390/foods9050579.

Nunes, M. C., Graça, C., Vlaisavljević, S., Tenreiro, A., Sousa, I., & Raymundo, A. (2020b). Microalgal cell disruption: Effect on the bioactivity and rheology of wheat bread. *Algal Research, 45*, 101749. Available from https://doi.org/10.1016/j.algal.2019.101749.

Patel, A. K., Laroche, C., Marcati, A., Ursu, A. V., Jubeau, S., Marchal, L., … Michaud, P. (2013). Separation and fractionation of exopolysaccharides from *Porphyridium cruentum*. *Bioresource Technology, 145*, 345–350. Available from https://doi.org/10.1016/j.biortech.2012.12.038.

Percival, E., & Foyle, R. A. J. (1979). The extracellular polysaccharides of *Porphyridium cruentum* and *Porphyridium aerugineum*. *Carbohydrate Research, 72*, 165–176. Available from https://doi.org/10.1016/S0008-6215(00)83932-4.

Raymundo, A., Gouveia, L., Batista, A. P., Empis, J., & Sousa, I. (2005). Fat mimetic capacity of *Chlorella vulgaris* biomass in oil-in-water food emulsions stabilized by pea protein. *Food Research International, 38*, 961–965. Available from https://doi.org/10.1016/j.foodres.2005.02.016.

Roussel, M., Villay, A., Delbac, F., Michaud, P., Laroche, C., Roriz, D., … Diogon, M. (2015). Antimicrosporidian activity of sulphated polysaccharides from algae and their potential to control honeybee nosemosis. *Carbohydrate Polymers, 133*, 213–220. Available from https://doi.org/10.1016/j.carbpol.2015.07.022.

Safi, C., Ursu, A. V., Laroche, C., Zebib, B., Merah, O., Pontalier, P., & Vaca-Garcia, C. (2014). Aqueous extraction of proteins from microalgae: Effect of different cell disruption methods. *Algal Research, 3*, 61–65. Available from https://doi.org/10.1016/j.algal.2013.12.004.

Saha, D., & Bhattacharya, S. (2010). Hydrocolloids as thickening and gelling agents in food: A critical review. *Journal of Food Science and Technology*, *47*, 587–597. Available from https://doi.org/10.1007/s13197-010-0162-6.

Schneider, N., Fortin, T. J., Span, R., & Gerber, M. (2016). Thermophysical properties of the marine microalgae *Nannochloropsis salina*. *Fuel Processing Technology*, *152*, 390–398. Available from https://doi.org/10.1016/j.fuproc.2016.06.039.

Schneider, N., & Gerber, M. (2014). Correlation between viscosity, temperature and total solid content of algal biomass. *Bioresource Technology*, *170*, 293–302. Available from https://doi.org/10.1016/j.biortech.2014.07.107.

Scholz, M. J., Weiss, T. L., Jinkerson, R. E., Jing, J., Roth, R., Goodenough, U., ... Gerken, H. G. (2014). Ultrastructure and composition of the *Nannochloropsis gaditana* cell wall. *Eukaryotic Cell*, *13*, 1450–1464. Available from https://doi.org/10.1128/EC.00183-14.

Sengupta, S., & Bhowal, J. (2017). Optimization of ingredient and processing parameter for the production of *Spirulina platensis* incorporated soy yogurt using response surface methodology. *Journal of Microbiology, Biotechnology and Food Sciences*, *6*, 1081–1085. Available from https://doi.org/10.15414/jmbfs.2017.6.4.1081-1085.

Soanen, N., Da Silva, E., Gardarin, C., Michaud, P., & Laroche, C. (2015). Improvement of exopolysaccharide production by *Porphyridium marinum*. *Bioresource Technology*, *213*, 231–238. Available from https://doi.org/10.1016/j.biortech.2016.02.075.

Souliès, A., Pruvost, J., Legrand, J., Castelain, C., & Burghelea, T. I. (2013). Rheological properties of suspensions of the green microalga *Chlorella vulgaris* at various volume fractions. *Rheologica Acta*, *52*, 589–605. Available from https://doi.org/10.1007/s00397-013-0700-z.

Spiden, E. M., Yap, B. H. J., Hill, D. R. A., Kentish, S. E., Scales, P. J., & Martin, G. J. O. (2013). Quantitative evaluation of the ease of rupture of industrially promising microalgae by high pressure homogenization. *Bioresource Technology*, *140*, 165–171. Available from https://doi.org/10.1016/j.biortech.2013.04.074.

't Lam, G. P., van der Kolk, J. A., Chordia, A., Vermuë, M. H., Olivieri, G., Eppink, M. H. M., & Wijffels, R. H. (2017). Mild and selective protein release of cell wall deficient microalgae with pulsed electric field. *ACS Sustainable Chemistry & Engineering*, *5*, 6046–6053. Available from https://doi.org/10.1021/acssuschemeng.7b00892.

Torzillo, G. (1997). *Tubular bioreactors*. Spirulina platensis *(Arthrospira): Physiology, cell-biology and biotechnology* (pp. 101–116). CRC Press.

Uribe-Wandurraga, Z. N., Igual, M., García-Segovia, P., & Martínez-Monzó, J. (2019). Effect of microalgae addition on mineral content, colour and mechanical properties of breadsticks. *Food & Function*, *10*, 4685–4692. Available from https://doi.org/10.1039/c9fo00286c.

Uribe-Wandurraga, Z. N., Igual, M., Reino-Moyón, J., García-Segovia, P., & Martínez-Monzó, J. (2020a). Effect of Microalgae (Arthrospira platensis and Chlorella vulgaris) Addition on 3D Printed Cookies. *Food Biophysics*. Available from https://doi.org/10.1007/s11483-020-09642-y.

Uribe-Wandurraga, Z. N., Martínez-Sanchez, I., Savall, C., García-Segovia, P., & Martínez-Monzó, J. (2020b). Microalgae fortification of low-fat oil-in-water food emulsions: an evaluation of the physicochemical and rheological properties. *Journal of Food Science and Technology*. Available from https://doi.org/10.1007/s13197-020-04828-1.

Wang, M., He, C., Chen, S., Wen, S., Wu, X., Zhang, D., ... Wei, C. (2018). Microalgal cell disruption via extrusion for the production of intracellular valuables. *Energy*, *142*, 339–345. Available from https://doi.org/10.1016/j.energy.2017.10.061.

Wileman, A., Ozkan, A., & Berberoglu, H. (2012). Rheological properties of algae slurries for minimizing harvesting energy requirements in biofuel production. *Bioresource Technology*, *104*, 432–439. Available from https://doi.org/10.1016/j.biortech.2011.11.027.

Wiltshire, K. H., & Dürselen, C. (2004). Revision and quality analyses of the Helgoland Reede long-term phytoplankton data archive. *Helgoland Marine Research*, 252–268. Available from https://doi.org/10.1007/s10152-004-0192-4.

Wu, Z., & Shi, X. (2008). Rheological properties of *Chlorella pyrenoidosa* culture grown heterotrophically in a fermentor. *Journal of Applied Phycology*, *20*, 279–282. Available from https://doi.org/10.1007/s10811-007-9244-x.

Yalcin, I., Hicsasmaz, Z., Boz, B., & Bozoglu, F. (1994). Characterization of the extracellular polysaccharide from freshwater microalgae *Chlorella* sp. *LWT Food Science and Technology*, *27*, 158–165. Available from https://doi.org/10.1006/fstl.1994.1032.

Yap, B. H. J., Martin, G. J. O., & Scales, P. J. (2016). Rheological manipulation of flocculated algal slurries to achieve high solids processing. *Algal Research*, *14*, 1–8. Available from https://doi.org/10.1016/j.algal.2015.12.007.

Zhang, H., Liao, Q., Fu, Q., Chen, H., Huang, Y., Xia, A., ... Li, J. (2018). Rheological properties of microalgae slurry under subcritical conditions for hydrothermal hydrolysis systems. *Algal Research*, *33*, 78–83. Available from https://doi.org/10.1016/j.algal.2018.04.026.

Zhang, X., Jiang, Z., Chen, L., Chou, A., Yan, H., Zuo, Y. Y., & Zhang, X. (2013). Influence of cell properties on rheological characterization of microalgae suspensions. *Bioresource Technology*, *139*, 209–213. Available from https://doi.org/10.1016/j.biortech.2013.03.195.

Chapter 11

Biological activities of natural products from microalgae

Joseph Samuel Palmer[1,2], Dolores Camacho Muñoz[1] and Christine Edwards[1]

[1]*School of Pharmacy and Life Sciences, Robert Gordon University, Aberdeen, United Kingdom,* [2]*ScotBio, BioCity Scotland, Newhouse, United Kingdom*

11.1 Introduction

11.1.1 What are bioactive compounds?

Bioactive molecules can be defined as compounds that affect physiological or cellular activities of the humans or animals that have consumed them. The types of bioactives and the biochemical changes they impart on an organism are various, ranging from eye, skin, and liver tissue protection provided by some microalgal pigments, antiinflammatory and anticancer properties of some lipids and vitamins, through to the health-debilitating actions of neurotoxins. Potential health effects of these compounds are dependent on the method of application or ingestion. Many bioactives are obtained through the diet, or more recently through food supplements called nutraceuticals. Their uptake is dependent upon the digestion process, which affects their stability and bioavailability and consequently their scope for affecting health (Carbonell-Capella et al., 2014). Bioactive compounds are also administered therapeutically as pharmaceuticals or topically through cosmeceuticals.

Many bioactive compounds are refined from natural sources or are chemically synthesized based upon the chemistries of naturally occurring bioactives. Various natural bioactives are already incorporated into commercial products, including phytosterols, carotenoids, polyunsaturated fatty acids (PUFAs), and polyphenols (Gallego et al., 2018). Increasing consciousness about our health and well-being, better knowledge of and demand for healthier foods and health-benefiting products, plus an increased preference for naturally sourced goods is fueling consumer demand for bioactive-containing foods and therapeutics. This is evidenced by the surge in popularity of health food shops, changing consumer preferences to perceived healthier goods

Cultured Microalgae for the Food Industry. DOI: https://doi.org/10.1016/B978-0-12-821080-2.00004-6

(such as vegetable oils over animal fats) and the rapid growth of the sports nutrition market. There is also a continuing drive in research to identify bioactive compounds with therapeutic uses in preventative and palliative medications. Increasing diagnosis rates of cancer, cardiovascular disease, diabetes, neurological, and age-related diseases in developed countries, plus heightened concerns about global issues such as antimicrobial resistance and emerging viral pathogens have further necessitated the identification of new bioactive molecules. For example, the polysaccharide calcium-spirulan, derived from the cyanobacteria *Arthrospira platensis*, demonstrates broad-acting antiviral activity and could be utilized as a pre-exposure prophylaxis, in combination with conventional treatments and vaccines. Consideration must also be given for safe and sustainable methods of manufacturing bioactive components for inclusion in foods, pharmaceuticals, nutraceuticals, and cosmeceuticals, to align with greater consumer and public concerns over product quality and the use of petroleum and animal-derived products. Opportunely, nature provides both an arsenal of effective, novel bioactive chemicals to greatly improve public health and the means for their sustainable production. In this regard, the diverse and largely untapped pool of metabolomic components in microalgae is an area of key scientific interest, while the unique physiologies and growth strategies of these organisms can provide a means to produce natural, high-quality chemicals to meet the growing demand.

11.1.2 Microalgae: a key resource for bioactives

Importantly, microalgae not only provide a natural source for many high-quality bioactive compounds, but also represent a vast pool for new bioactive components for elucidation and utilization. As outlined by Gallego et al. (2018), the wide diversity both in terms of microalgal taxonomy and metabolic strategies provides a substantial resource for bioactive compound discovery, with considerable research efforts currently focused on strain screening. Furthermore, microalgae can often be isolated and cultured in photobioreactors (PBR), with controlled conditions allowing growth optimization and increased target compound production, quality, and traceability (Wang, Lan, & Horsman, 2012). Microalgal-derived bioactives are being used as sustainable, vegan, and environmentally friendly alternatives for compounds usually sourced from animals and petrochemicals, such as PUFAs and carotenoids (Doughman, Krupanidhi, & Sanjeevi, 2007). PBRs have minimal land-use and resource requirements and can be installed in many locations where other agricultural practices are unsuitable, providing a regular source of nutrients and bioactives to malnourished communities across the globe (Koyande et al., 2019). The production of bioactives from microalgae becomes more economically favorable when other products such as biofuels, protein, or other bioactives are coextracted from the same microalgae biomass in a microalgal biorefinery approach (Koyande et al., 2019; Zhu, 2015).

This chapter highlights the range of bioactives, both health-promoting and deleterious that can be produced by various microalgae. The review looks at both key bioactive chemical groups, such as fatty acids (FAs), pigments, and vitamins that are microalgae components, as well as introducing more niche and unusual molecules, such as glycocarotenoids and alkaloids that may provide unique commercial and pharmaceutical opportunities in the future. The bioactive mechanisms of these compounds are discussed, and their production in the context of a microalgal biorefinery setting is also addressed. Microalgal toxins, another key group of bioactives associated with this group of organisms are discussed regarding food production and safety.

11.2 Lipids

11.2.1 Polyunsaturated fatty acids

FAs are aliphatic hydrocarbon chains with a carboxyl [C(=O)OH] group at one end and a methyl (CH_3) group at the omega (ω) end of the molecule. Naturally occurring FAs typically have an even number of carbon atoms (typically 4–22 carbon atoms), with the aliphatic chains either saturated or unsaturated, based on the presence of single or double covalent bonds between carbons atoms, respectively. FAs can have either single (monounsaturated) or multiple double bonds (polyunsaturated), with FAs listed with the convenient notation of number of carbon atoms:number of double bonds, such as palmitic acid (C16:0) or linolenic acid (C18:3) (Hasenhuettl, 2016). An important structural aspect of unsaturated FAs is the position of the first double bond from the ω end. PUFAs with the first double bond three carbon atoms away from the ω end are termed omega-3 (ω-3) FAs. ω-3 FAs cannot be produced in the human body and are classed as essential fatty acids (EFAs), such as eicosapentaenoic acid (EPA, C20:5 ω-3) and docosahexaenoic acid (DHA, C22:6 ω-3). Similarly, omega-6 (ω-6) FAs are also essential and can only be obtained by humans through the diet. ω-6 compounds include γ-linolenic acid (GLA, C18:3 ω-6) and arachidonic acid (ARA, C20:4 ω-6). Stereoisomerism also occurs in FAs, with *cis* and *trans* geometric isomers common for many PUFAs. This isomerization results in markedly different biological activities, and isomers are therefore named differently, such as oleic acid (*cis*-9-octadecanoic acid) and elaidic acid (*trans*-9-octadecanoic acid). Although FAs can be found in living organisms as free FA, they are typically found esterified in nonpolar glycerides (esterification between FA and glycerol), polar phospholipids, and glycolipids, or potentially with other bioactive compounds (Cuellar-Bermudez et al., 2015).

PUFAs have numerous nutraceutical and pharmaceutical applications. FAs regulate many physiological functions and health conditions including cardiovascular disease, type II diabetes, and inflammatory disease, with potential use in the treatment of atherosclerosis, cancer, rheumatoid arthritis, psoriasis, and age-related diseases such as Alzheimer disease and macular degeneration (Calder, 2015a; D'Eliseo et al., 2016; Patil et al., 2007). The association between PUFAs and cardiovascular disease is well researched, with current data providing strong concordant evidence that ω-3 PUFAs reduce the risk of cardiac death and can also help lower plasma triglycerides, resting heart rate, blood pressure, and may also improve myocardial filling and efficiency and improve vascular function (Mozaffarian & Wu, 2011). Furthermore, EPA and DHA have been shown to exert antineoplastic activity on a range of solid and hematological tumors by inducing apoptotic cell death (D'Eliseo et al., 2016). In vitro and in vivo antitumor activity is exhibited both alone and in combination with conventional treatments, in a dose- and time-dependent manner. Furthermore, a highly selective effect is observed, with little or no toxicity on healthy cells, as overviewed by D'Eliseo et al. (2016). Higher levels of circulating ω-3 FAs, specifically docosapentaenoic acid (DPA) and DHA, have been associated with decreased risk of incident inflammatory arthritis (Gan et al., 2017). Frommer et al. (2015) found that dose-dependent secretion of the proinflammatory cytokine IL-6, the chemokines IL-8 and MCP-1, and matrix-degrading enzymes pro-MMP1 and MMP3 was enhanced with increasing concentrations of both saturated and unsaturated free FAs in rheumatoid arthritis synovial fibroblasts. The various antiinflammatory mechanisms of ω-3 FAs are detailed by (Calder, 2015a) including alteration of cell membrane phospholipid FA composition, disturbing of lipid rafts, and inhibition of proinflammatory transcription factor kappa B, reducing expression of inflammatory genes. ω-3 FAs are also important for eye health, demonstrating antiinflammatory action in the lacrimal gland, preventing apoptosis of secretory epithelial cells (Calder, 2015b). Supplementation of ω-3 FAs can protect the tear film and cornea by clearing meibomitis (Roncone, Bartlett, & Eperjesi, 2010). Recent research indicates that a high ratio of ω-6 to ω-3 FAs is common in Western diets due to excessive consumption of ω-6-rich but ω-3-deficient foods. High ω-6/ω-3 ratios promote the onset of cardiovascular disease, cancer, inflammatory, and autoimmune diseases, whereas increased levels of ω-3 PUFAs (a lower ω-6/ω-3 ratio) show suppressive effects (Simopoulos, 2008). In a complex series of biochemical reactions, PUFAs are catalyzed into a myriad of derivatives which play a key role as mediators of the local symptoms of inflammation including vasoconstriction, vasodilatation, coagulation, pain, and fever (Simopoulos, 2008).

Current international guidelines recommend for the general population to consume at least 250 mg day^{-1} of long-chain ω-3 PUFAs or at least two servings of oily fish per week (Mozaffarian & Wu, 2011). Although oily fish is a key source of long-chain ω-3 PUFAs, concerns over the sustainability of global fish stocks, destructive fishing methods, and fish contamination have warranted identification of alternate-ω-3 sources (Cuellar-Bermudez et al., 2015; Ryckebosch et al., 2012). Microalgae have been touted as a sustainable, vegan source of PUFAs, and are in fact the origin of EPA and DHA in aquatic ecosystems (Doughman et al., 2007). Numerous species of microalgae have comparable or superior ω-3 PUFA contents to oily-fish species on a weight for weight basis and are potential candidates for the large-scale production of ω-3 PUFAs (Adarme-Vega et al., 2012). Indeed, most DHA is industrially produced from the fermentation of microbes such as heterotrophic microalgae, whereas ARA and EPA are often extracted from fungal species (Khozin-Goldberg, Leu, & Boussiba, 2016). Microalgae belonging to the red algal lineage (Rhodophyta) can synthesize ARA and EPA PUFAs, but not C22-length PUFAs. This is also the case for Glaucophyta, with the acquisition of enzymes necessary for C18 PUFA elongation and desaturation retained from early nonphotosynthetic, eukaryotic ancestors which went through primary endosymbiosis with cyanobacterium (Khozin-Goldberg et al., 2016). Picoplanktonic microalgae from phylum Prasinophyceae can synthesize EPA and DHA, whereas unicellular microalgae from the core green lineage have FA profiles similar to vascular plants (Khozin-Goldberg et al., 2016). Methods for lipid extraction are well developed for biofuel generation in microalgal biorefineries, including environmentally friendly techniques such as supercritical and subcritical fluid, pressurized liquid and microwave-assisted extraction. While "fuel only" production models have proved commercially unviable (Adarme-Vega et al., 2012), similar processing techniques could be adopted for the extraction and purification of specialty PUFAs as coproducts along with other bioactive molecules (Ryckebosch et al., 2012).

11.2.2 Sterols

Sterols are a group of large, amphiphilic compounds belonging to the triterpene family of lipids. Algal sterols are structurally and functionally analogous to cholesterol, which is biosynthesized by all animal cells and acquired endogenously through the diet, whereas some sterols (i.e., phytosterols) are synthesized de novo by fungi, plants, and microalgae (Luo, Su, & Zhang, 2015). The algal sterols group includes both sterols and stanols, with their structures broadly categorized by a four-ring steroid nucleus, an aliphatic side chain, double bonds, and a hydroxyl (OH) group where conjugates with FA and sugar moieties are often bonded, as demonstrated by the sterols ergosterol and cholesterol (Fig. 11.1). Stanols are saturated sterols, having no double bonds in the sterol ring structure. Microalgal sterols are biosynthesized from isopentenyl diphosphate and dimethylallyl diphosphate to squalene. When consumed by animals, they provide therapeutic action against hypercholesterolemia, helping lower blood low-density lipoprotein-cholesterol (Ahmed, Zhou, & Schenk, 2015; Luo et al., 2015). Microalgal sterols

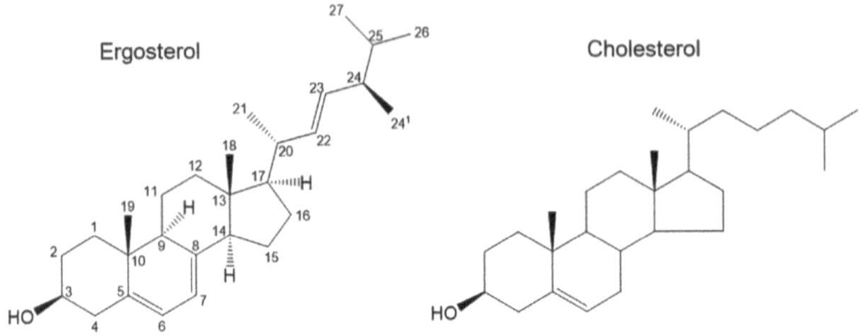

FIGURE 11.1 Chemical structure of ergosterol and cholesterol.

have been shown to downregulate intestinal acyl-CoA:cholesterol acyltransferase 2 and hepatic 3-hydroxy-3-methylglutaryl-CoA reductase genes involved in cholesterol gastrointestinal absorption and synthesis, respectively (Chen et al., 2014), thus reducing the onset of cardiovascular diseases. Algal sterols show antiinflammatory, antiatherogenicity, anticancer, antioxidant, and immunomodulatory activities, with research also showing neuroprotective functions (Ahmed et al., 2015; Luo et al., 2015). Additionally, many phytosterols function as precursors for other important bioactive molecules such as vitamin D, cortisone, and hormones (Luo et al., 2015).

Ahmed et al. (2015) profiled and quantified sterols in 10 microalgal strains, with *Pavlova lutheri* having both the highest concentration (26 mg g^{-1} DW) and number of different sterols (11). They also found different dominant sterols between genera, with other researchers showing different microalgal classes are also characterized by specific sterols (Luo et al., 2015). *Chlorella vulgaris* and *Dunalliella tertiolecta* contain ergosterol, which inhibits highly proinflammatory cytokine (TNF-α) production and cyclooxygenase-2 (COX-2) expression, suppressing the lipopolysaccharide-induced inflammatory response, while other algal sterols showed effective antiinflammatory activities and antitumor promotion in mice (Yasukawa et al., 1996). Algal sterols have been shown to have antibacterial activity, and sterol-containing extracts from microalgae *Isochrysis galbana* could inhibit multidrug-resistant *Mycobacterium tuberculosis* at lower doses than various tuberculosis drugs (Prakash, Sasikala, & Aldous, 2010).

11.2.3 Tocopherols and tocotrienols

Tocopherols and tocotrienols are a series of chemically related, lipid-soluble compounds, synthesized in photosynthetic organisms (Goiris, Muylaert, & De Cooman, 2015). Alpha-tocopherol (α-tocopherol) is one of the four most widespread homologs of tocopherol, the major form of vitamin E, and is the most active antioxidant tocopherol and the form that the liver preferentially incorporates into lipoproteins that are released into the bloodstream for distribution to peripheral tissues (Harris, Devaraj, & Jialal, 2002). Tocopherols are comprised of a chromanol ring and a 16-carbon aliphatic chain. α-Tocopherol is trimethylated at the 5, 7, and 8 positions of the chromanol ring (Das Gupta & Suh, 2016), with the different tocopherol homologs differing in their 5 and 7 groups (Fig. 11.2). Tocopherols and tocotrienols have the same basic chemical structure, although tocotrienols have three conjugated double bonds in the aliphatic chain, whereas the chain is saturated in tocopherols (Goiris et al., 2015) (Fig. 11.2). Unmethylated carbons at the 5- and 7-positions are electrophilic centers that trap reactive oxygen and nitrogen species, giving tocopherols their strong antioxidant properties. In microalgae, these antioxidant compounds are present in the lipid bilayers of cell membranes, and production is stimulated by UV, salt stress, and nitrogen limitation (Goiris et al., 2015). Tocopherols have received considerable attention within the food sciences for their association with lower risks of coronary heart disease and cancer (Mendiola et al., 2008). Tocopherols are thought to lower the risk of coronary disease by decreasing lipid peroxidation, inhibiting platelet adhesion and aggregation and smooth muscle cell proliferation, exerting antiinflammatory effects on monocytes and by improving endothelial function (Harris et al., 2002). However, although α-tocopherol is thought to be chemopreventive, recent research indicates that at high doses it may lose its effectiveness (Das Gupta & Suh, 2016).

Quantification of α-tocopherol contents in over 130 microalgae and cyanobacterial (blue-green algae) strains revealed that (1) α-tocopherol contents are maximal in the stationary phase of growth, (2) α-tocopherol was significantly higher in nitrogen-depleted growth media, and (3) α-tocopherol contents were characteristic of phylogenetic class (Mudimu et al., 2017). The study found that α-tocopherol content varied considerably, with generally lower concentrations in Rhodophyta and Cyanobacteria. Under optimal conditions, a maximum concentration of 1.4 mg g^{-1} DW was found in *Nannochloropsis oculate*.

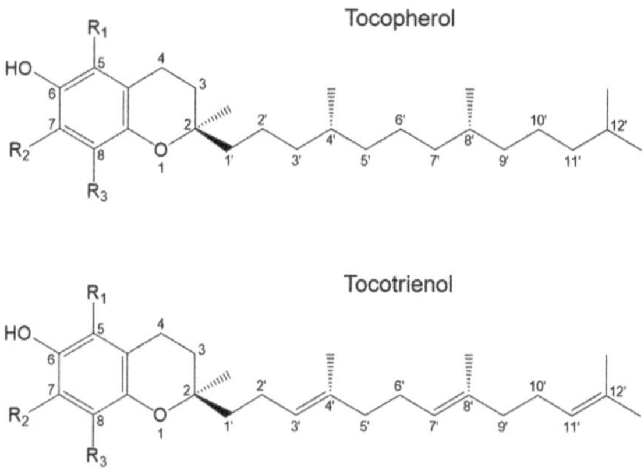

	R_1	R_2	R_3
α-Tocopherol	CH_3	CH_3	CH_3
β-Tocopherol	CH_3	H	CH_3
γ-Tocopherol	H	CH_3	CH_3
δ-Tocopherol	H	H	CH_3

	R_1	R_2	R_3
α-Tocotrienol	CH_3	CH_3	CH_3
β-Tocotrienol	CH_3	H	CH_3
γ-Tocotrienol	H	CH_3	CH_3
δ-Tocotrienol	H	H	CH_3

FIGURE 11.2 Chemical structure of tocopherols and tocotrienols.

11.3 Pigments

As photosynthetic organisms, microalgae have evolved to contain a diverse suite of photosynthetic pigments involved directly and indirectly in the capture of solar energy. Pigments are either directly involved in light harvesting by transferring light energy to the photosynthetic electron transport chain, indirectly as photoprotective compounds which filter, quench, or scavenge excess light, or as both (Mulders et al., 2014). Mulders et al. (2014) provide an excellent overview of the biosynthesis, function, and diversity of pigments in microalgae, including the barriers and opportunities of pigment production from microalgae. Photosynthetic pigments can be grouped into three distinct groups based on their chemistries, chlorophylls, carotenoids, and phycobiliproteins, all exhibiting bioactive capability, which is discussed in the following section. While the chemistries of these compounds differ within and between these groups, all compounds exhibit a large, conjugated system of double bonds, whereby delocalized electrons absorb light of specific wavelengths. Nonabsorbed light is reflected, giving rise to the vivid and distinct colors of these pigments (Mulders et al., 2014). Pigments can be further categorized into primary and secondary pigments, with primary pigments functionally and structurally linked to the photosynthetic apparatus. Additionally, the bioactivity of flavonoids is also discussed in this section.

11.3.1 Carotenoids

Carotenoids are the pigments responsible for the red, orange, and yellow pigmentation of plant foliage, as well as the coloration seen in feathers, crustacean shells, fish flesh, and skin (Del Campo, García-González, & Guerrero, 2007). There are over 1100 different carotenoid molecules, the chemical structures of which are mainly derived from a 40-carbon polyene chain (lycopene), made up of eight isoprene units. The ends of the lycopene molecule can then be enzymatically converted into terminal cyclic groups, which importantly, may or may not be complimented by oxygen-containing functional groups (Del Campo et al., 2007; Tanaka, Shnimizu, & Moriwaki, 2012). This variation distinguishes between the two main carotenoid groups: the pure hydrocarbons, carotenes, and their oxygenated derivatives, the xanthophylls (Higuera-Ciapara, Felix-Valenzuela, & Goycoolea, 2006). In the latter group, oxygen-containing functional groups can be present as hydroxyl groups (such as zeaxanthin), oxi-groups (=O) (such as echinenone and canthaxanthin), or in a combination of both (such as astaxanthin). Various other chemical modifications can also differentiate carotenoid compounds, including ketolation, desaturation, stereoisomer, esterification, and *cis/trans* isomerization, as shown in Fig. 11.3. The chemistry and biological source of a carotenoid will affect the form it is in. For example, astaxanthin may be esterified in one or both hydroxyl groups with different FAs, creating

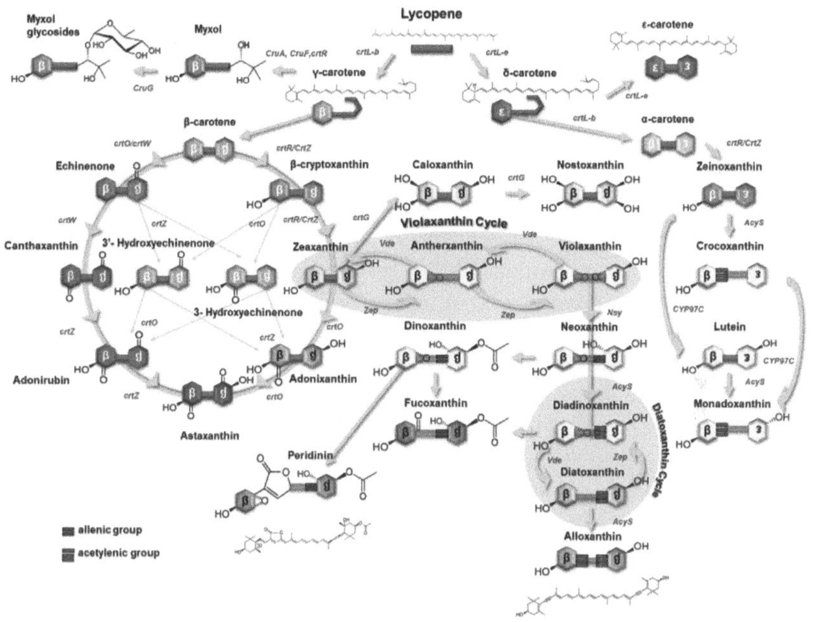

FIGURE 11.3 Schematic representation of cartogenesis of common microalgal carotenoids.

more lipophilic mono- or diesters. Carotenoids may also form chemical complexes with proteins (carotenoproteins), lipoproteins (carotenolipoproteins), sugar moieties (carotenoid glycosides), or be free compounds (Takaichi & Mochimaru, 2007).

Carotenoids perform various biological functions, depending on their chemical structure. Carotenoids with terminal β-groups (such as β-carotene) can act as provitamin A, precursors to vitamin A (retinol). This vitamin is important for vision, immunological function, reproduction, and embryonic development of vertebrates. Recent meta-analysis on the association of total and individual carotenoid levels in human serum in relation to metabolic syndrome found that total and individual carotenoids serum levels all showed an inverse association with metabolic syndrome, with β-carotene having the strongest putative protective effect, followed by α-carotene and β-cryptoxanthin (Beydoun et al., 2018). The role of carotenoids in cardiometabolic health is emphasized, as they play an essential role in the control of adipogenesis (formation of fat cells) and the production of adipokines and inflammatory mediators that affect the distribution of central adiposity and the resultant insulin resistance (Beydoun et al., 2018). In plants and algae, carotenoids with oxygen on both end groups act as membrane stabilizers, giving structural support in both cell membranes and photosynthetic complexes (Havaux, 1998). Free carotenoids occupy the hydrophobic region of membrane bilayers, with xanthophylls rigidifying membranes, including the thylakoid membrane, and β-carotene and echinenone have a neutral or fluidizing effect (Zakar et al., 2016).

Carotenoids which do not have vitamin activity contribute to in vivo antioxidant activity in humans, explained by the biological functions they provide in plants and microalgae. In addition to helping capture light energy, many compounds have evolved to play key roles in scavenging reactive oxidative species (ROS) in times of excess light, hence protecting chlorophyll and other pigments from photodamage (Mulders et al., 2014; Novoveská et al., 2019). This antioxidant ability provides the basis for the protective action against oxidative stress in many organisms, including animals. In mammals, carotenoids are mainly accumulated in the liver and combined into lipoproteins for release into the blood circulation (Watson & Preedy, 2012). They provide an antioxidant defense system in the presence of high concentrations of free radicals, with studies reporting that β-carotene, lycopene, lutein, and β-cryptoxanthin protect against lipid peroxidation in rat livers. Lutein and zeaxanthin are antioxidants found in the human retina and macula, providing a protective effect against photoinduced damage to the lens and retina (Roberts & Dennison, 2015). This is achieved by quenching of radicals and the absorption of blue light, simultaneously. Resultantly, deficiencies in lutein and zeaxanthin are associated with age-related degenerative diseases such as macular degeneration and cataracts (Koushan et al., 2013). Furthermore, in vivo studies on human skin have shown that carotenoids are vital components of the antioxidative protective system of skin and could serve as marker substances for the overall antioxidative status (Darvin et al., 2011). Antioxidants can decrease mutagenesis, and therefore carcinogenesis, both by decreasing oxidative damage of DNA and by decreasing oxidant-stimulated cell division. Preclinical studies have shown that some carotenoids have potent antitumor effects both in vitro and in vivo, indicating potential preventive and/or therapeutic roles for the compounds. Some carotenoids demonstrate antiproliferative activity (prohibiting the growth of tumor cells) (Gagez et al., 2012). β-Carotene, lycopene, lutein, zeaxanthin, β-cryptoxanthin, fucoxanthin, canthaxanthin, and astaxanthin have been shown to demonstrate chemopreventive and/or anticarcinogenic activity in several tissues (Tanaka et al., 2012). Reported antiproliferative mechanisms include the ability to induce apoptosis, cell cycle arrest, inhibit the growth factor signal transduction (IGF), and stimulate the expression of antioncogenes (Ugocsai et al., 2005). Cha, Koo, and Lee (2008) demonstrated that xanthophylls from *Chlorella* (predominantly violaxanthin, antheraxanthin, lutein, and zeaxanthin) inhibited cancerous cell growth (Cha et al., 2008). Schwartz et al. (1988) showed that feeding *Spirulina* extract to hamsters ($n = 20$) gave complete chemoprotection when exposed to carcinogens, compared to 100% tumor development in a control group (Schwartz et al., 1988). Some carotenoids and carotenoid derivatives have been found to reverse multidrug resistance in cancer cells by inhibiting the activity of ATP-dependent multidrug transporters, which belong to the superfamily of ATP-binding cassette (ABC) proteins. Such proteins actively pump xenobiotic compounds like chemotherapeutic drugs out of tumorous cells, reducing the efficacy of the treatment. Carotenoids, especially xanthophylls, have been shown to competitively inhibit these proteins, and thus reverse the multidrug resistance (Eid, El-Readi, & Wink, 2012). Such carotenoids could therefore be used in conjunction with chemotherapeutic drugs to synergistically increase their cytotoxicity, therefore improving their effectiveness (Eid et al., 2012). For example, some epoxy-β-carotenes, which are present in some microalgae, including cyanobacteria, have been found to reverse epirubicin resistance by inhibiting multidrug resistance protein 1 (P-gp) in some tumor cell lines (Gagez et al., 2012; Yuan, Gong, & Chen, 1997).

Pertinent reviews of carotenoid composition and biosynthesis in microalgae are provided by Arroyo-Currás et al. (2016) and Christaki et al. (2012), along with their potential applications. Some species of microalgae are already commercially exploited for carotenoid production, including *Haematococcus pluvialis* (astaxanthin), *Dunaliella salina* (β-carotene), *Scenedesmus* sp. (lutein), and diatoms such as *Mallomonas* sp. (fucoxanthin) (Ambati et al., 2019; Novoveská et al., 2019). As these carotenoids can be extracted from other natural sources such as plants and bacteria, valorization relies upon the carotenoid being present in naturally high concentrations and/or grown in conditions that promote overproduction of the compound to be commercially competitive. While there are well-established markets for these bioactives, demand for more uncommon carotenoids is likely to increase as their therapeutic applications are elucidated and suitable natural sources are identified. In this regard, microalgae are the best source of many carotenoids which are common or unique to microalgae, such as some keto-carotenoids and carotenoid glycosides. For example, echinenone, 3'-hydroxyechinenone, myxol-2'-fucoside, and oscillaxanthin are common to cyanobacteria, such as *Arthrospira* sp. (Takaichi & Mochimaru, 2007), while diadinoxanthin, diatoxanthin, and dinoxanthin are only found in certain microalgal classes. A recent study demonstrated significant antioxidant and antihyperglycemic activity of myxoxanthophyll, and other cyanobacterial pigments (Ghosh et al., 2016), supporting the use of raw pigment extracts or purified carotenoids in lieu of synthetic diabetes therapies with unwanted side effects.

11.3.2 Phycobiliproteins

Phycobiliproteins (PBPs) are a group of water-soluble pigments also only found in cyanobacteria and algae including *Rhodophytes*, *Cryptomonads*, and *Glaucocystophytes* (Glazer, 1994). PBPs are formed of two components; a linear tetrapyrroles unit, phycobilin, also containing a large conjugated system of double bonds, which are peptide linked to a cysteinyl residue of a protein. Four main phycobilins have been identified, phycocyanobilin (blue), phycobiliviolin (purple), phycoerythrobilin (red), and phycourobilin (yellow), giving the PBPs their vivid and distinguishable colors (Glazer, 1994; Mulders et al., 2014). PBPs form part of light-harvesting supramolecular complexes called phycobilisomes in cyanobacteria and red algae, in which hundreds of PBPs can be aggregated on the outer surface of the thylakoid membranes. Phycocyanobilin and phycoerythrobilin are the most common PBPs in microalgae, with phycocyanobilin being the main pigment unit in phycocyanin (deep blue) and allophycocyanin (light blue), and phycoerythrobilin in phycoerythrin (orange-red). The absorbance characteristics of individual PBPs are influenced by the environment and interactions between other chromophores, which are strongly dictated by the protein structure, which also affects the fluorescence of the molecules. This latter property has enabled the use of PBPs in the fields of molecular biology, immunology, and cytology as fluorescent markers (Sekar & Chandramohan, 2008; Yu et al., 2017).

There has been considerable interest in the role of PBPs as bioactives, particularly as anticancer agents. Documented pharmacological properties include potent antioxidant, antiinflammatory, antiviral, antifungal, antihyperglycemic, neuroprotective, and hepatoprotective activities (Sekar & Chandramohan, 2008). PBPs have significant antioxidative scavenging action against ROS and show dose-dependent decreases in lipid peroxidation, which exerts a protective effect against hemolysis and provides the basis of antiinflammatory and antitumor activity exhibited by these molecules (Romay & González, 2000; Yu et al., 2017). Animal studies have demonstrated C-phycocyanin from *Arthrospira platensis* (Spirulina) increased survival of mice after being irradiated by a lethal dose of ^{60}Co ray, helped the recovery of T-cells and increased the number of antibody-forming cells in normal and immunocompromised mice, as well as other beneficial immunomodulatory functions (Yu et al., 2017). Rimbau et al. (1999) demonstrated the neuroprotective role of C-phycocyanin in rats pre-exposed to neuronal damage, indicating a possible role in treating oxidative stress-induced neuronal injury in neurodegenerative diseases, such as Alzheimer and Parkinson diseases (Rimbau et al., 1999). Some studies have also demonstrated antifungal and antiviral activity. Both crude and purified C-phycocyanin from *A. plantensis* showed inhibition against *Candida albicans*, *Aspergillus*, *Penicillium*, and *Rhizopus* fungal species and good activity against Newcastle, avian adenovirus, hepatitis A and polio viruses (Murugan, 2011), while allophycocyanin from *Arthrospira* prevented plaque formation and delayed viral RNA synthesis of enterovirus 71 in rhabdomyosarcoma cells, reducing the cytopathic effects and apoptosis of host cells (Shih et al., 2003).

For over 30 years, a growing body of research has demonstrated antitumor activity of phycocyanin and phycoerythrin using in vitro studies on human tumor cell lines and in vivo animal models, as reviewed by Yu et al. (2017). Promising results from animal studies using PBPs-containing extracts on tumor-bearing animals (Schwartz et al., 1988) have since been demonstrated with various human cancer cell lines. Phycocyanin can function as a chemotherapeutic agent by generation of ROS via activation of caspase-3 (Khan et al., 2006; Pardhasaradhi et al., 2003), in addition to mechanisms such as cytochrome *c* release, decreases in membrane potential and poly (ADP-ribose) polymerase cleavage (Roy et al., 2007), resulting in apoptosis of selected hepatocellular carcinoma, leukemia, and rat histiocytic tumor cell lines. Phycocyanin has also been shown to inhibit and reduce the expression of COX-2 and B-cell lymphoma 2 (Bcl-2), respectively, both of which suppress apoptosis in various tumor cells (Pardhasaradhi et al., 2003; Roy et al., 2007). C-phycocyanin from *Arthronema africanum* showed time- and concentration-dependent inhibition of Graffi myeloid tumor cells and increased survival time of tumor-bearing hamsters (Gardeva et al., 2014), while phycoerythrin from red algae could inhibit the growth of human oral epithelial, mice ascitic hepatoma, and Bcap-3 human breast cancer cell lines (Cai et al., 2014). Similarly, B-phycoerythrin from *Porphyridium cruentum* can suppress the proliferation of Graffi myeloid tumor cells in a dose-dependent manner (Minkova et al., 2011).

Phycocyanin has also been used to overcome the multidrug resistance phenotype of cancer cells. Downregulation of multidrug resistance-1 (MDR1) poly glycoprotein in human hepatocarcinoma cells was achieved using C-phycocyanin derived from *A. platensis* (Nishanth et al., 2010). The authors concluded that reductions of NF-κB and AP-1, in a dose-dependent manner, downregulate MDR1, in addition to inactivation of the signal transduction pathways involving Akt, ERK, JNK, and p38 genes. This resulted in a fivefold increase in the sensitivity of hepatocarcinoma cells to the chemotherapy drug, doxorubicin (Nishanth et al., 2010). PBPs also have application in photodynamic therapy of cancers, whereby PBPs are used as photosensitizers, inducing apoptosis in tumor cells. Purified phycoerythrin and phycocyanin from *Porphyra yezoensis* Ueda inhibited hepatocellular carcinoma cell growth after laser irradiation by 80% and 59%, respectively, and $100\ mg\ L^{-1}$ purified phycoerythrin and phycocyanin induced the apoptosis of the cells by 31.5% and 32.5%, respectively, 8 h after photodynamic therapy (Cai et al., 2014).

11.3.3 Chlorophyll and derivatives

Chlorophylls are tetrapyrroles classified as porphyrins. Like the iron-containing heme group in hemoglobin, chlorophyll has one, centrally chelated magnesium ion inside a porphyrin head. This consists of four pyrrole rings (nitrogen arranged in a carbon ring) with the porphyrin bound to a hydrocarbon, phytyl tail (Fig. 11.4). Five structurally characterized families of chlorophyll are documented, denoted *a*, *b*, *c*, *d*, and *f*, with *a* and *b* the dominant forms in most photosynthesizers (Queiroz Zepka, Jacob-Lopes, & Roca, 2019). Type *c* is found in microalgal phyla including *Cryptophyta*,

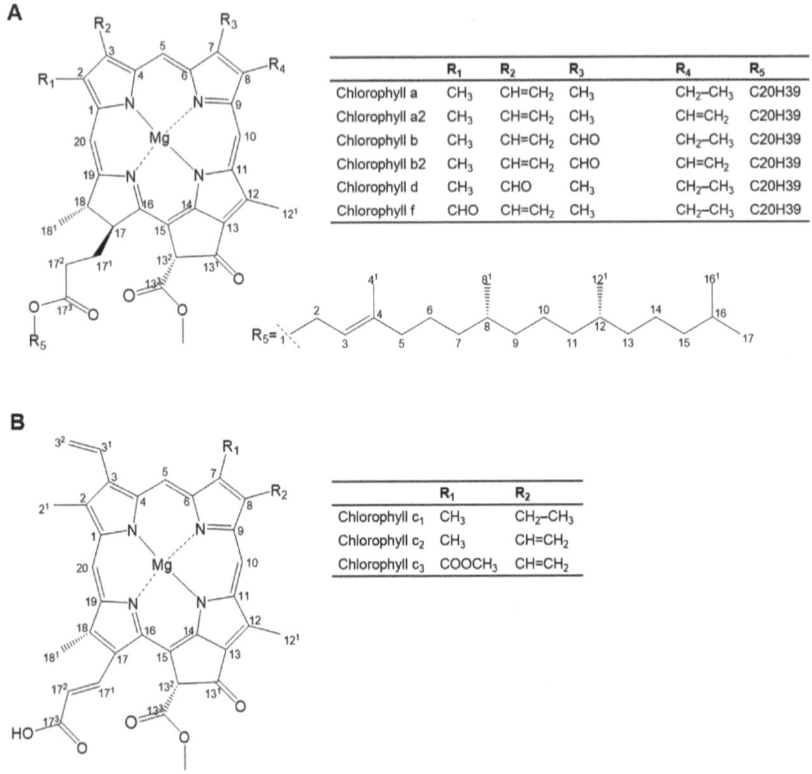

	R_1	R_2	R_3	R_4	R_5
Chlorophyll a	CH_3	$CH{=}CH_2$	CH_3	$CH_2{-}CH_3$	C20H39
Chlorophyll a2	CH_3	$CH{=}CH_2$	CH_3	$CH{=}CH_2$	C20H39
Chlorophyll b	CH_3	$CH{=}CH_2$	CHO	$CH_2{-}CH_3$	C20H39
Chlorophyll b2	CH_3	$CH{=}CH_2$	CHO	$CH{=}CH_2$	C20H39
Chlorophyll d	CH_3	CHO	CH_3	$CH_2{-}CH_3$	C20H39
Chlorophyll f	CHO	$CH{=}CH_2$	CH_3	$CH_2{-}CH_3$	C20H39

	R_1	R_2
Chlorophyll c_1	CH_3	$CH_2{-}CH_3$
Chlorophyll c_2	CH_3	$CH{=}CH_2$
Chlorophyll c_3	$COOCH_3$	$CH{=}CH_2$

FIGURE 11.4 Chemical structure of (A) chlorophyll a, a2, b, b2, d and f and (B) chlorophyll c1, c2 and c3.

Haptophyta, *Bacillariophyta*, *Chlorophyta*, and *Cyanophyta*, and differs from other chlorophyll types in that it does not have a phytyl tail or a reduced ring D (Fig. 11.4B). Chlorophyll types *a* and *b* differ only in that a methyl group at the C7 position in *a* is replaced by a formyl group in *b* (Fig. 11.4A), with the molecules cycling between one another (Pérez-Gálvez & Roca, 2017). Chlorophylls *d* and *f* (Fig. 11.4A) are common to cyanobacteria, both absorbing far-red light, which allows these organisms to grow in deep water where visible light is depleted and infrared radiation is enhanced (Nürnberg et al., 2018). From chlorophyll *a/b*, various catabolic reactions can occur, known collectively as the pheophorbide *a* oxygenase (PAO)/phyllobilin pathway. Chlorophyll *a/b* is enzymatically dechelated and dephytylated, resulting in pheophorbide *a* through the intermediates of chlorophyllide and phaeophytin, respectively (Hörtensteiner, 1999). During senescence/ripening in higher plants, pheophorbide is enzymatically modified to primary or not, fluorescent chlorophyll catabolites, which are then tautomerized to not fluorescent chlorophyll catabolites (NCCs) or dioxobilin-NCC (Pérez-Gálvez & Roca, 2017).

In microalgae, only limited examples of pheophorbide catabolism have been demonstrated, such as the production of red pigments, red chlorophyll carbolites in *Chlorella protothecoides* and *Isochrysis galbana* (Bale, Airs, & Llewellyn, 2011; Engel et al., 1991). Additionally, chlorophyll can also be oxidized into a range of derivatives.

While the economic potential for chlorophyll is predominantly as a colorant, increasing focus is on their potential health benefits. The chemical structures of the molecules, including the chelated metal, phytyl chain, and hydroxyl and formyl groups influence the bioactivity of chlorophylls and their carbolites. Chlorophyll and its derivatives show antioxidant capability, which is due to the porphyrin structure, the central metal ion, and side groups. Lanfer-Marquez, Barros, and Sinnecker (2005) found higher antioxidant activities in chlorophyll *b*-derived compounds compared to *a*-derivatives, using the β-carotene bleaching method, illustrating the functionality of the aldehyde group. Studies using mouse models have shown that chlorophyllin, a semisynthetic copper derivative of chlorophyll, attenuated intestinal and hepatic inflammation and ameliorated liver fibrosis by downregulation of inflammatory cytokines and fibrotic proteins, respectively, in response to xenobiotic CCl_4 (Zheng et al., 2018). Chemoprotective and antimutagenic roles of chlorophyll and related compounds have also been demonstrated. Chlorophylls can complex with carcinogens, reducing the bioavailability and uptake of carcinogens into the gut (Dashwood et al., 1998; Negishi, Rai, & Hayatsu, 1997).

11.3.4 Flavonoids

Flavonoids are a highly diverse group of over 6000 different naturally occurring polyphenolic compounds; a major group of pigments in higher plants, algae, and microalgae, boasting a variety of nutraceutical, pharmaceutical, and cosmetic applications (Goiris et al., 2014). Flavonoids are characterized by the flavan nucleus (C6C3C6), shown in Fig. 11.5, formally considered as benzo-γ-pyrone (chromone) derivatives (Heim, Tagliaferro, & Bobilya, 2002; Panche, Diwan, & Chandra, 2016). Flavonoid categorization is based on oxidation of the "C" ring of the basic flavan structure (Fig. 11.5). Major classes include flavones, flavanones, flavanes, flavonols, catechins, and anthocyanidins (Havsteen, 2002), as shown in Fig. 11.5. Different compounds within a class will have different substitution and conjugation patterns on the A and B rings, including hydrogenation, hydroxylation, methylation, sulfation, and glycosylation (addition of carbohydrate), with further differentiation between structural and geometric isomers and tautomers (Amić et al., 2003; Heim et al., 2002). Flavonoids may exist as monomers or are polymerized into dimeric or oligomeric forms. These polymeric flavonoids are called condensed tannins, whereas hydrolyzable tannins also contain gallic acid esterified to a carbohydrate

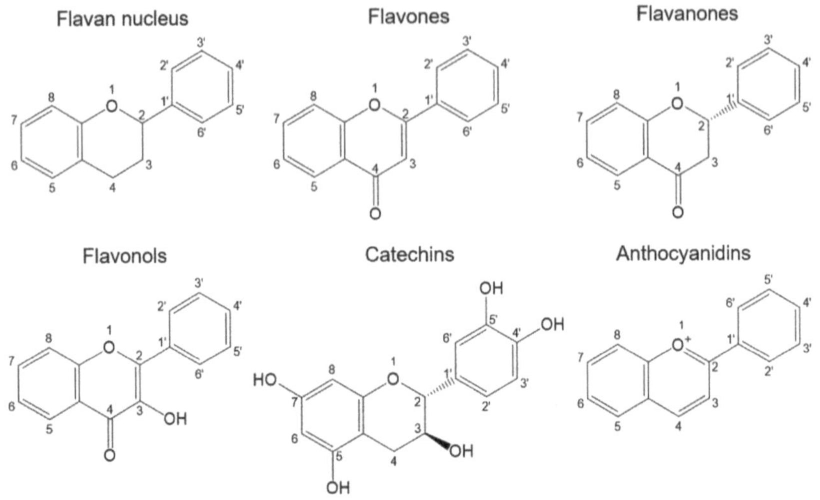

FIGURE 11.5 Chemical structure of the flavan nucleus, flavones, flavanones, flavonols, catechins and anthocyanidins.

(Amić et al., 2003). Many flavonoids occur naturally as glycosylated derivatives, such as anthocyanins, esterified to sugars such as glucose, galactose, and lignin (Amić et al., 2003; Heim et al., 2002). The bioactivity of a flavonoid and their metabolites is dependent on the chemical structure and orientation of the numerous moieties on the molecule. After consumption, flavonoids are subsequently metabolized by gut microbiota, in the intestinal walls and in the liver, forming various potentially bioactive breakdown products (Panche et al., 2016; Thilakarathna & Rupasinghe, 2013). These metabolites impart a host of beneficial health effects, namely due to their significant antioxidant and chelating abilities. Neuroprotective, cardioprotective, chemoprotective, antibacterial, antidiabetic, antiviral, and antiinflammatory effects have been found (Abotaleb et al., 2018; Mullie et al., 2008). Flavonoid metabolites help prevent lipid peroxidation of PUFA in cell membranes, which can lead to atherosclerosis and cancer (Amić et al., 2003). The radical-scavenging ability of a flavonoid is mainly due to the high reactivities of hydroxyl substituents (Vinayagam & Xu, 2015). The iron-chelating ability of some flavonoids is thought to inhibit the Fenton reaction, an important source of free radicals causing lipid peroxidation, DNA damage, and mutagenesis, and reduces the activities of enzymes such as xanthine oxidase, cyclo-oxygenase, and lipoxygenase by chelation/reduction of iron in their active site (Arroyo-Currás et al., 2016). Abotaleb et al. (2018) concluded that the chemoprotective ability of flavonoids is also due to their interference with multiple signal transduction pathways, limiting proliferation, angiogenesis, and metastasis of tumors or by increasing apoptosis.

11.4 Other health-promoting bioactive compounds

11.4.1 Alkaloids

Alkaloids are a broad group of roughly 2000 chemicals, defined as compounds containing a nitrogen atom(s) in a cyclic ring. Key groups in microalgae are phenylethylamine (PEA), indole, and halogenated-indole alkaloids (Güven, Percot, & Sezik, 2010). Phenylethylamines (PEAs) consist in their simplest form of a benzene ring with an ethylamine side chain (Fig. 11.6) and are formed through the decarboxylation of phenylalanine (Güven et al., 2010). Their reported bioactivity includes as hormones, stimulants, hallucinogens, entactogens, anorectics, bronchodilators, and antidepressants (Güven et al., 2010). These molecules can act as neurostimulators by passing through the blood–brain barrier and increasing the release of dopamine and serotonin (Rajakumar, 2018). The psychoactive PEAs act as a natural antidepressant in human and animal studies, and consumption of PEA-containing food supplements has been encouraged (Güven et al., 2010; Rajakumar, 2018). There are limited data on the abundance of PEA in microalgae, although 4-hydroxyphenylethylamine has been identified in the green alga *Scenedesmus acutus* (Rolle et al., 1977), while concentrations of 1.4 mg g^{-1} of PEA having been reported in cyanobacteria *Aphanizomenon flos-aquae*. The dried biomass of this species is sold under the brand "Phycomin EX" by Bio Cell Co., Ltd. (Cappello, 2009; Rajakumar, 2018; Scoglio et al., 2017).

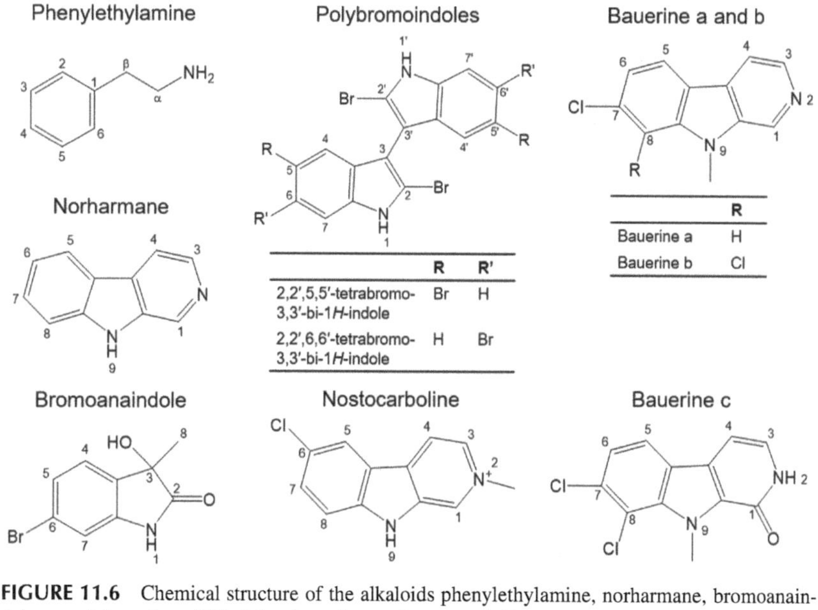

FIGURE 11.6 Chemical structure of the alkaloids phenylethylamine, norharmane, bromoanaindolone, polybrominated biindoles, bauerines and nostocarboline.

Indole alkaloids are tryptophan-derived compounds with benzylpyrrole groups. Norharmane is one such example, and contains a complete β-carboline moiety (Fig. 11.6). It has been identified as an exometabolite of the cyanobacteria *Anabaena cylindrica*, *Anabaena inaequalis*, *Anabaena siamensis*, *Chroococcus minutus*, *Nostoc carneum*, *Nostoc commune*, *Nostoc harveyana*, and *Phormidium foveolarum* (Volk, 2008). Norharmane bioactivity includes the inhibition of monoamine oxidase, indoleamine 2,3-dioxygenase, nitric oxide synthase, and acetylcholinesterase enzymes, increase of insulin secretion from human islets of Langerhans, apoptosis in human neuroblastoma SH-SY5Y cells, and possible comutagenesis with other compounds (Volk, 2008). Other examples include hapalindole-type molecules, namely the related hapalindoles, fischerindoles, welwitindolinones, and ambiguines. These compounds, derived from cyanobacteria *Fischerella* sp. and *Hapalosiphon* sp., boast a plethora of biochemical activities including antifungal, antituberculosis, and anticancer properties (Hillwig, Zhu, & Liu, 2014). Halogenated-indole alkaloids have also been identified in microalgae, such as bromoanaindolone (Fig. 11.6) isolated from cyanobacterium *Anabaena constricta*, which exhibits antibacterial and anticyanobacterial activity (Volk et al., 2009), antiinflammatory polybrominated biindoles (Fig. 11.6) from *Rivularia firma* (Hodder & Capon, 1991), three chlorinated carboline derivatives, antiviral bauerines (Fig. 11.6) from *Dichotrix baueriana* (Larsen, Moore, & Patterson, 1994), and a carboline alkaloid, nostocarboline, a cholinesterase inhibitor from *Nostoc* str. 78-12A with potential for the treatment of Alzheimer disease (Becher, Beuchat, Gademann, & Jüttner, 2005; Volk et al., 2009; Fig. 11.6).

11.4.2 Polysaccharides

Polysaccharides are biological polymers ubiquitous to all algae. Polysaccharides are formed of two or more monosaccharide (sugar) monomers that can be decomposed by hydrolysis. Homopolysaccharides contain identical monomers, whereas heteropolysaccharides contain different monomers, joining to form either linear or branched polysaccharides. In microalgae these molecules provide a range of functions including storage, cell development, signaling and differentiation, structure, and protection from environmental stresses through tough cell-wall complexes, often comprising a high percentage of the overall algal dry matter (Senni et al., 2011; Tannin-Spitz et al., 2005). Some water-soluble polysaccharides are secreted by microalgae into the surrounding environment, termed exopolysaccharides (EPSs), playing important roles in biofilm formation as well as many other biological functions (Raposo, De Morais, & Bernardo de Morais, 2013). In terms of bioactive properties, sulfated EPS/polysaccharides (SPS) and β-glucans are of key interest. Sulfated polysaccharide molecules are normally characterized by high molecular weights, anionic, and acidic properties due to the presence of carboxyl and sulfate groups (Raposo, De Morais, & Bernardo de Morais, 2013). β-Glucans are a group of polysaccharides formed of D-glucose monomers typically joined by β-glycosidic bonds, often found in the cell walls of microbes. β-Glucans can be soluble or insoluble in water, exhibit both branched or linear structures with sugar or nonsugar branches, and can also be sulfated.

Both types of molecule demonstrate a plethora of bioactive properties, making them suitable candidates for medical, pharmaceutical, and biotechnological applications (Senni et al., 2011). In microalgae, a large body of research has focused on the bioactive properties of SPS from the red microalga *Porphyridium* and *Rhodella*. Water-soluble polysaccharides from *Porphyridium* show significant antioxidant activity again ROS, and can inhibit lipid autoxidation in a dose-dependent manner (Tannin-Spitz et al., 2005). Other studies have shown antiglycemic (Huang et al., 2006), antiinflammatory (Matsui et al., 2003), and in vitro and in vivo antiproliferative action against Graffi myeloid cells with SPS from these microalgae (Gardeva et al., 2009). Similar properties have also been documented with *Arthrospira*, *Chlorella*, *Gyrodinium*, and *Phaeodactylum* SPS. Neuroprotective effects have also been noted in SPS isolated from the aforementioned microalgae and *Cylindrotheca*, *Isochrysis*, and *Aphanocapsa* sp. These may have a range of therapeutic applications, from prevention of the onset of Alzheimer disease to improving mood and memory (Rajakumar, 2018). One of the most relevant bioactive functions of SPS is their significant antiviral properties. An overview of the EC_{50}/ED_{50} of SPS from various marine microalgae against different viral strains is given by (Raposo, De Morais, & Bernardo de Morais, 2013). Calcium-Spirulan, an SPS from *A. platensis*, was found to inhibit the in vitro replication of several enveloped viruses including herpes simplex type I (HSV-1), human cytomegalovirus, measles virus, mumps virus, influenza A virus, and HIV-1 virus (Hayashi, Hayashi, & Kojima, 1996; Mader et al., 2016). *Porphyridium* SPS could inhibit HSV-1, HSV-2, varicella zoster virus, hepatitis B virus, viral hemorrhagic septicemia virus, vaccinia virus, and others (Raposo, De Morais, & De Morais, 2014). It is suggested that a main antiviral mechanism of SPS is interaction of the sulfate groups with glycoproteins on the viral membrane, affecting their ability to infect host cells.

β-Glucans also show medicinal potential, including immuno-stimulating, hepatoprotective, antiproliferative, antioxidative, and antiviral effects (Rodríguez-Zavala et al., 2010; Schulze et al., 2016). Examples include paramylon (β-1,3-glucan) and lentinan, and further addition of sulfate or carboxymethyl groups can improve the bioactivity of some β-glucans. Species such as *Eugenia gracilis* have been shown to produce up to 93 mg g^{-1} dry weight (DW) of paramylon under certain growth conditions, which is higher than the concentrations in other plant and algal sources (Rodríguez-Zavala et al., 2010). A screen of over 40 microalgal species showed that *Scenedesmus ovalternus* SAG 52.80 and *Porphyridium purpureum* SAG 1380-1d strains could accumulate β-glucans to over 20% DW, and enhanced irradiance and nitration starvation could increase yields for some strains (Schulze et al., 2016).

11.5 Microalgal toxins

There are clearly many benefits from microalgae, and production will most certainly increase the chances of meeting the United Nation's Sustainable Goals (UNSGs) 2 and 3, with zero hunger along with good health and well-being by 2030.

Since microalgal production is achieved by a broad range of means, from harvesting natural biomass such as *A. flos-aquae* from the Upper Klamath Lake in Oregon, United States, to highly controlled food-grade biomass production systems, and there is a clear need for consideration of product contamination. Whilst there is always the possibility of pathogen and heavy metal contamination, we must consider the fact that several microalgae naturally produce toxins with greater potency in acute toxicity tests than strychnine. In addition, many of these toxins cause chronic symptoms following prolonged exposure to low doses and therefore must be considered and ideally tested if the biomass or product(s) is to be used for animal/human health or nutrition. Where microalgae are being produced for use as biofertilizers, which are subject to less rigorous safely tests, there is the possibility that low concentrations of toxins can bioaccumulate in plants. While most biomass production is either PBR or outdoor ponds/raceways, contamination is most likely from toxic freshwater algae, although where natural seawater is being used, especially from coastal sources, pretreatment with UV is common practice.

This section describes those toxins most likely to contaminate cultured microalgae as supported by recent evidence despite a few studies published on cyanotoxins in food supplements which would represent a direct route of exposure.

11.5.1 Hepatotoxins

11.5.1.1 Microcystins and nodularins

Microcystins (MCs) are a large family of cyclic heptapeptides (>200 variants characterized; Meriluoto, Spoof, & Codd, 2017) produced by many genera of cyanobacteria and constitute the most commonly occurring cyanobacterial toxins (Table 11.1). MCs have a general structure of cyclo-(D-Ala^1-X^2-D-MeAsp^3-Z^4-Adda^5-D-Glu^6-Mdha^7), where X and Z represent variable L amino acids and Adda is the unusual amino acid (2S,3S,8S,9S,)-3-amino-9-methoxy-2,6,8-trimethyl-10-phenyldeca-4,6-dienoic acid (Fig. 11.7). As well as variations in L amino acids, other variants are the result of demethylation at D-MeAsp or Mdha, methyl esterification at D-Glu, geometric changes, and modifications of Adda. The nomenclature is based on the variable amino acids present in positions 2 and 4, for example, microcystin-LR

TABLE 11.1 Commonly occurring cyanobacterial toxins and reported detection levels in microalgal products.

Cyanotoxin	Genera of origin	Mode of action (MOA) and toxicity data	WHO TDI ($\mu g\ kg^{-1}$ bw)	Detection in commercial microalgal products ($\mu g\ g^{-1}$ dw)	References
Microcystins (cyclic heptapeptides)	*Microcystis, Anabaena, Nostoc Planktothrix, Oscillatoria Anabaenopsis, Haphalosiphon Aphanocapsa, Dolichospernum Fischerella, Gleotrichia Leptolyngbya, Phormidium Pseudoanabanena, Rivularia, Snowella, Synechococcus, Synechocystsis, Tolypothrix Woronichinia, Scytonema*	MOA: Protein phosphatase inhibition (PP1, 2A, 2B, 3, 4, and 5) Acute toxicity: liver toxin (LD_{50}: 50–300 $\mu g\ kg^{-1}$ bw) Chronic toxicity: tumor promoter	0.04	0.008[a] MC-LR (UKL and *Spirulina*) 0.002[a] MC-LR (UKL)	Sánchez-Parra et al. (2020)
				≈0.01 – ≈42[a,f] MCs (A. *flos-aquae*) ≈0.01 – ≈54[b,f] MCs (A. *flos-aquae*) ≈0.01 – ≈60[c,f] MCs (A. *flos-aquae*)	Miller et al. (2020)
				0.21–0.84[a] MCs (Spirulina) 0.8-8.2[a] MCs (*A. flos-aquae*)	Roy-Lachapelle et al. (2017)
				<0.01–5.8[a] MCs (*A. flos-aquae*)	Heussner et al. (2012)
				<1.2–5.20[a] MCs (*A. flos-aquae*)	Vichi et al. (2012)
				0.014–0.163[a] MCs (*Spirulina*)	Jiang (2008)

(*Continued*)

TABLE 11.1 (Continued)

Cyanotoxin	Genera of origin	Mode of action (MOA) and toxicity data	WHO TDI ($\mu g\ kg^{-1}$ bw)	Detection in commercial microalgal products ($\mu g\ g^{-1}$ dw)	References
				$0.048-0.051^{b}$ MCs (BGA) $0.028-0.078^{b}$ MCs (*Spirulina*) $0.01-0.036^{b}$ MCs (*Chlorella*)	Yu et al. (2002)
				$<0.1-35.7^{a,b}$ MCs (BGA) $<0.3-49^{c}$ MCs (BGA)	Lawrence et al. (2001)
				$0.30-10.42^{b}$ MCs (UKL) $0.15-0.52^{b}$ MCs (*Spirulina*)	Gilroy et al. (2000)
Nodularins (cyclic pentapeptides)	*Nodularia, Nostoc*	MOA: Protein phosphatase inhibition (PP1, 2A, and 3) Acute toxicity: liver toxin (LD_{50}: 30–50 $\mu g\ kg^{-1}$ bw) Chronic toxicity: tumor initiator	n.d.	Not detected to date	Van Apeldoorn et al. (2007)

Cylindrospermopsins (tricyclic guanidine alkaloids)	*Aphanizomenom, Chrysosporum, Dolichospernum, Lyngbya, Microseria, Phormidium, Raphidopsis, Umezakia*	MOA: Protein synthesis inhibitor Acute toxicity: Organ and tissue damage particularly liver Chronic toxicity: carcinogen(?) 2100 μg kg^{-1} bw in 24 h 200 μg kg^{-1} bw in 5–6 days	0.06–0.3	Not detected to date	Van Apeldoorn et al. (2007)
Saxitoxin(s) (carbamate alkaloids)	*Aphanizomenom, Cylindrospermum, Dolichospernum, Geitlerinema, Microcystis, Phormidium, Raphidopsis, Scytonema*	MOA: Neurotoxin—blocks voltage-gated sodium channels 10–65 μg kg^{-1} bw	0.05		Van Apeldoorn et al. (2007)
Anatoxin-a (bicyclic alkaloids)	*Aphanizomenom, Arthrospira, Cuspidothrix, Dolichospernum, Geitlerinema, Microcoleus, Microcystis, Oscillatoria, Phormidium, Planktothrix, Raphidopsis, Tychonema*	MOA: Neurotoxin binding to neuronal nicotinic acetylcholine receptors. 200–250 μg kg^{-1} bw	0.1	0.034[a] ATX-a (*Chlorella*) 0.025[a] ATX-a (UKL and *Spirulina*) 0.002[a] ATX-a (UKL)	Sánchez-Parra et al. (2020)
				0.34[a] DihydroATX-a (*Spirulina*) 1.6 and 2.7[a] epoxyATX-a (*Spirulina*) 0.15 and 0.40[a] ATX-a (*A. flos-aquae*) and 7.2[a] dihydroATX-a (*A. flos-aquae*)	Roy-Lachapelle et al. (2017)

(*Continued*)

TABLE 11.1 (Continued)

Cyanotoxin	Genera of origin	Mode of action (MOA) and toxicity data	WHO TDI ($\mu g\ kg^{-1}$ bw)	Detection in commercial microalgal products ($\mu g\ g^{-1}$ dw)	References
				0.07 and 0.92[a] epoxyATX-a (*A. flos-aquae*)	
				0.011[d,e] ATX-a (*Spirulina*)	Rellán et al. (2009)
				9–10[a] dihydrohomoATX-a (BGA) 18–19[a] epoxyATX-a (BGA)	Draisci et al. (2001)
Anatoxin-a(s) (phosphorylated cyclic *N*-hydroxyguanine)	*Dolichospernum*	MOA: Neurotoxin Acetyl cholinesterase inhibitor 20 $\mu g\ kg^{-1}$ bw	n.d.	Not detected to date	

AEG, N-(2-aminoethyl)glycine; *A. flos-aquae, Aphanizomenon flos-aquae*; *BGA*, blue-green algae; *BMAA*, β-methylamino-L-alanine; *bw*, body weight; *DAB*, 2,4-diaminobutyric acid; *dw*, dry weight; LD_{50}, lethal dose (50%) by intraperitoneal injection (mouse or rat); *n.d.*, not determined; *TDI*, tolerable daily intake; *UKL*, Upper Klamath algae.
[a] *Determined by liquid chromatography-mass spectrometry.*
[b] *Determined by ELISA.*
[c] *Determined by phosphatase enzyme assay.*
[d] *Determined by liquid chromatography-fluorescence detection.*
[e] *Determined by gas chromatography-mass spectrometry.*
[f] *Data were estimated from data in the corresponding publication.*

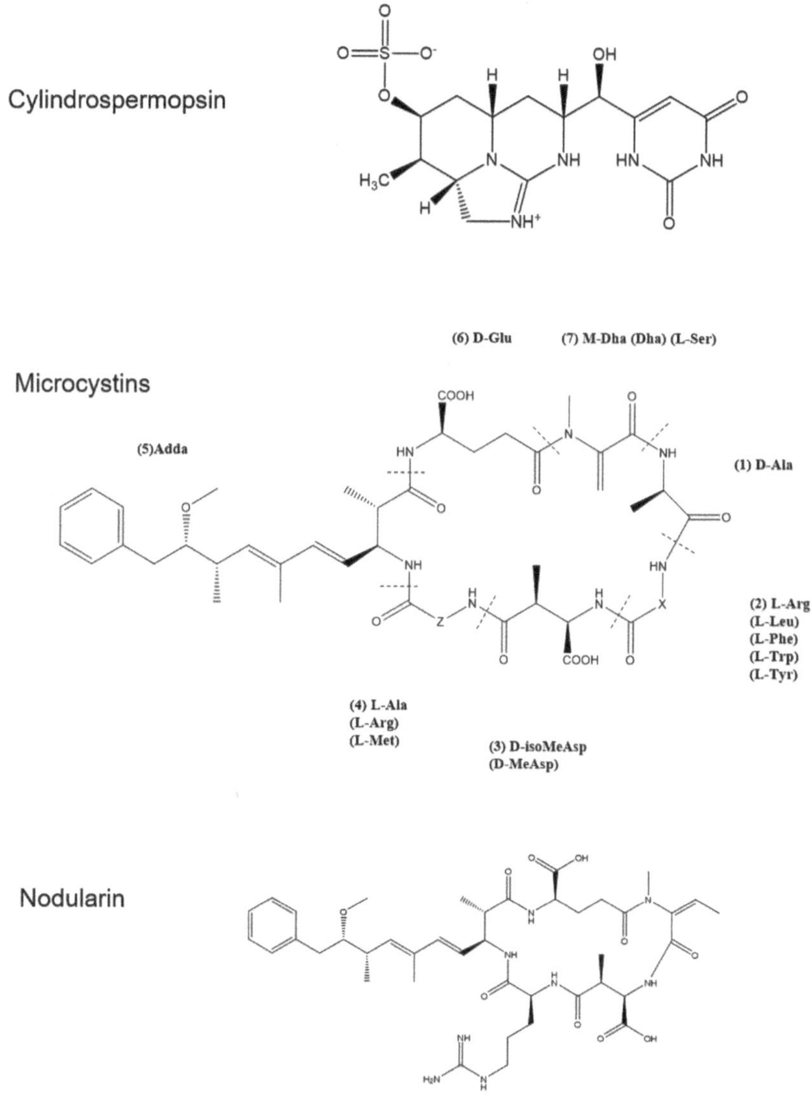

FIGURE 11.7 Chemical structure of Cylindrospermopsin and Microcystins and Nodularin hepatoxins.

(MC-LR) consists of leucine (L) at position 2 and arginine (R) at position 4. MCs are very stable and not destroyed by processes such as boiling or autoclaving so they would not be removed from any microalgal biomass used for feed or fertilizer.

The occurrence of MCs is a serious health hazard and acute doses have been responsible for illness and fatalities in animals and humans, leading the

World Health Organization (WHO) to set a guideline limit of 1 μg L^{-1} MC-LR as the maximum allowable concentration in drinking water. Their effect is most prominent in the liver as the toxins are rapidly transported into cells via transmembrane organic anion transporter peptides (OATP), specifically OATP1B and OAT1B3 (bile acid transport system), hence they are commonly known as hepatotoxins. Essential cellular processes such as control of the cell cycle, metabolic, and transcriptional regulation along with several DNA repair pathways are controlled by protein phosphorylation/dephosphorylation by kinases/phosphatases, respectively. These toxins are potent inhibitors of eukaryotic protein phosphatases resulting in hyperphosphorylation and loss of cell regulation as observed by deformation of cell structure, tumor promotion, and lipid peroxidation. The amino acid methyl-dehydroalanine (Mdha) at position 7 in the MC structure covalently binds to protein phosphatase residues such as cysteine-273 in PP1 and is thus irreversible. In June 2006, eight countries met at the International Agency for Research on Cancer (IARC) to review data on the potential carcinogenicity of ingested cyanobacterial peptide toxins, microcystin-LR (MC-LR) and nodularin (NOD). This group concluded that MC-LR is "possibly carcinogenic to humans" and classified it accordingly (group 2B; Grosse, 2006). Epidemiological studies have linked chronic exposure of MCs via drinking water supply to incidents of primary liver cancer (Svirčev et al., 2014; Ueno et al., 1996), supporting the need for efficient monitoring and management practices in the provision of safe water. In addition, more recent studies have demonstrated multiple and complex mechanisms of toxicity, including damage to other organs such as the kidneys, reproductive system, brain, and heart, and these are well reviewed by McLellan and Manderville (2017).

Over the last few decades, the main focus of research and regulation has been in aquatic environments used for recreation and drinking water supply. However, more recently studies have explored the risk to public health of MC and other cyanotoxins accumulating in the food chain and in microalgal supplements which have been extensively reviewed by Testai et al. (2017).

Based on toxicity studies, the WHO has calculated a tolerable daily intake (TDI) of MCs as 0.04 μg kg^{-1} body weight for an adult, that is, consumption of 2.4 μg MCs per day for a 60 kg adult. While concentrations in algal health food supplements obtained from wild biomass frequently exceed safe levels, there are growing reports of low concentrations of MCs in *Spirulina* and *Chlorella* products, some of which are close to TDI values (Table 11.1). Some countries, such as the United States, have set a maximum acceptable concentration of MCs at 1 μg g^{-1} in microalgal supplements.

NODs are a group of 10 related pentapeptides, where the structure of NOD-R is cyclo-(D-MeAsp^1-L-Arg^2-Adda^3-D-Glu^4-Mdhb^5) where Mdhb represents N-methyldehydrobutyrine (Fig. 11.7). NODs are produced by just a few species, principally *Nodularia spumigena* which is normally found in brackish water habitats. NODs have a similar mode of action to MCs in inhibiting several protein phosphatases. They are distinct from MCs in that they do not irreversibly bind to the protein phosphatases and they are tumor initiators rather than promoters. IARC classified NOD as a group 3 compound as available data were limited. NODs have not been detected in food supplements, as expected given the niche habitat.

11.5.2 Cylindrospermopsins

The most common toxin of this small group is cylindrospermopsin (CYN), the tricyclic guanidine alkaloid coupled to a hydroxymethyl uracil (Fig. 11.7) and several structurally modified variants have also been identified (CYNs: 7-deoxyCYN and 7-epiCYN). CYNs are protein synthesis inhibitors leading to multiple types of toxicity (hepatoxic, cytotoxic, nephrotoxic, and genotoxic). The effects of this group of toxins were first experienced in 1979 by a community in Palm Island, Australia, where over 100 people developed hepatoenteritis following consumption of drinking water contaminated with a bloom *Raphidopsis raciborskii* (syn. *Cylindrospermosis raciborskii*) which had been treated with copper sulfate, liberating CYN from the cells. Whilst this organism/toxin was originally considered as a tropical problem, this is no longer the case as the number of CYN-producing genera has increased and it has been detected in temperate ecosystems. The few studies to date have not detected CYNs in microalgal supplements (Heussner et al., 2012; Roy-Lachapelle et al., 2017).

11.5.3 Neurotoxins

11.5.3.1 Saxitoxins

Saxitoxin (STX) is a tricyclic alkaloid with unique hydropurine rings (Fig. 11.8). There are approximately 50 chemical variants which include nonsulfated compounds such as STX, which have single or double sulfation as in gonyautoxins and C-toxins, respectively. Other variants produced by strains of the genus *Lyngbya* are decarbamoyl derivatives. Whilst STXs are produced by several genera of cyanobacteria (Table 11.1), they are also produced by marine dinoflagellates, where they bioaccumulate in shellfish causing paralytic shellfish poisoning (PSP). Symptoms of PSP include tingling, numbness, perspiration, vomiting, and diarrhea, while if the dose is severe this will be followed by loss of motor coordination, paralysis, and death. STXs disrupt signal transmission between neurons and muscles by blocking voltage-gated sodium channels, preventing conductance of signal along the neuron leading to paralysis and death. In the few studies reported, STXs have not been detected in microalgal food supplements (Heussner et al., 2012; Roy-Lachapelle et al., 2017), however many studies do not perform analysis for these toxins as the methodology is more challenging.

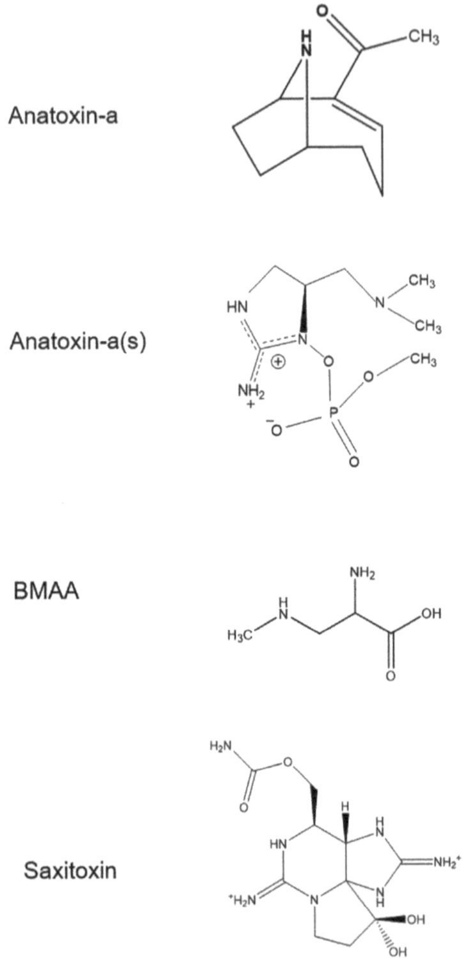

FIGURE 11.8 Chemical structure of the microalgal neurotoxins Anatoxin-a, Anatoxin-a(s), β-N-methylamino-L-alanine (BMAA) and Saxitoxin.

11.5.3.2 Anatoxins

Anatoxin-a (ATX-a), a bicyclic alkaloid (Fig. 11.8), along with the related homoanatoxin-a (HTX; propionyl instead of acetyl group), are potent binders to acetyl choline receptors, triggering muscle contraction, but as the complex cannot be broken down by acetyl cholinesterase as normal, the muscle cells are overstimulated until they arrest. ATX-a also affects other nicotinic cholinergic receptors in the cardiovascular system resulting in increased blood pressure and heart rate.

These toxins are produced by a range of planktonic and benthic cyanobacteria (Table 11.1), and on occasions some strains produce MCs at the same time as providing enhanced synergistic toxicity. It is important to note that when analyzing for ATX-a by preferred methods such as liquid chromatography-mass spectrometry, it must be differentiated by the isobaric phenylalanine [both mass charge (*m/z*) 166], which has been misidentified as ATX-a in previous poisoning incidents (Furey et al., 2005). Anatoxins can be readily modified in alkaline pH and UV light, resulting in less toxic transformation products such as dihydro and epoxy derivatives which have been detected in *Spirulina* tablets in several studies (Table 11.1). One product contained MCs as well as ATX transformation products. Interestingly, a recent study detected ATX-a in *Chlorella* tablets at a low concentration which was well below recommended the TDI (Table 11.1; Sánchez-Parra, Boutarfa, and Aboal, 2020). No transformation products were reported, but as with most of these studies there is no consistent use of validated methods.

11.5.3.3 Anatoxin-a(S)

Anatoxin-a(s) [ATX-a(s)] is a hydroxyguanidine methyl phosphate ester (Fig. 11.8), a natural organophosphate, and a potent inhibitor of acetylcholinesterase with a mechanism akin to those of carbamate and organophosphorus insecticides. ATX-a(s) binds to the active site of acetylcholinesterase and prevents degradation of acetylcholine in the peripheral nervous system. Persistent stimulation results in tremors, convulsions, and salivation, followed by death. It has been so far only reported by a single genus, *Dolichospermum* (formerly *Anabaena*) and is rarely detected (Devic et al., 2002). The limited occurrence of ATX-a(s) is reflected by lack of methods and reports in the literature and is typically not considered.

11.5.3.4 β-N-methylamino-L-alanine

β-N-methylamino-L-alanine (BMAA, Fig. 11.8) and its common structural isomers 2,4-diaminobutyric acid (DAB) and N-2-aminoethylglycine (AEG) are nonproteinaceous diamino acids of low molecular weight which are considered by some to be cyanotoxins associated with multiple neurodegenerative diseases. However, this is highly controversial as many reports of their detection are inconsistent, largely due to poor analytical methods (Manolidi et al., 2019). In addition to controversy over their occurrence, there is also an ongoing debate on the mode of action and, given the current uncertainty, they are not recognized as cyanotoxins by the WHO. Reports of these compounds in microalgal supplements seem to be method dependent and until such a time that there are robust, validated methods available it is not possible to consider these compounds further.

11.6 Conclusions

Microalgae are undoubtedly versatile cell factories producing a wide range of bioactive metabolites which can be exploited in many applications from bulk biomass in feed applications to highly purified high-value compounds. Many of these compounds have shown significant and varied bioactivity against a range of diseases and the successful exploitation of microalgae for therapeutics may, in part, help address a number of health issues in the developed and developing world. As we explore diverse strains along with improved production at scale, microalgae will offer multiple solutions for the food–energy–water nexus. However, we must remain vigilant over the potential for production/contamination by toxic metabolites. The potency of some of these compounds means that suitable toxin-screening procedures should be utilized to ensure the safety of microalgae-derived feeds and compounds.

References

Abotaleb, M., et al. (2018). Flavonoids in cancer and apoptosis. *Cancers, 11*(1), 28. Available from: https://doi.org/10.3390/cancers11010028, Multidisciplinary Digital Publishing Institute.

Adarme-Vega, T., et al. (2012). Microalgal biofactories: a promising approach towards sustainable omega-3 fatty acid production. *Microbial Cell Factories, 11*(1), 96. Available from: https://doi.org/10.1186/1475-2859-11-96, BioMed Central.

Ahmed, F., Zhou, W., & Schenk, P. M. (2015). *Pavlova lutheri* is a high-level producer of phytosterols. *Algal Research, 10*, 210–217. Available from: https://doi.org/10.1016/j.algal.2015.05.013, Elsevier.

Ambati, R. R., et al. (2019). Industrial potential of carotenoid pigments from microalgae: current trends and future prospects. *Critical Reviews in Food Science and Nutrition*, 1880–1902. Available from: https://doi.org/10.1080/10408398.2018.1432561, Taylor and Francis Inc.

Amić, D., et al. (2003). Structure-radical scavenging activity relationships of flavonoids. *Croatica chemica acta, 76*(1), 55–61, Hrvatsko kemijsko društvo.

Arroyo-Currás, N., et al. (2016). Substituent inductive effects on the electrochemical oxidation of flavonoids studied by square wave voltammetry and ab initio calculations. *Molecules, 21* (11), 1422. Available from: https://doi.org/10.3390/molecules21111422. (*Basel, Switzerland*) Multidisciplinary Digital Publishing Institute.

Bale, N. J., Airs, R. L., & Llewellyn, C. A. (2011). Type I and Type II chlorophyll-a transformation products associated with algal senescence. *Organic Geochemistry, 42*(5), 451–464. Available from: https://doi.org/10.1016/j.orggeochem.2011.03.016, Pergamon.

Becher, P. G., Beuchat, J., Gademann, K., & Jüttner, F. (2005). Nostocarboline: isolation and synthesis of a new cholinesterase inhibitor from *Nostoc* 78-12A. *Journal of Natural Products. American Chemical Society, 68*(12), 1793–1795. Available from: https://doi.org/10.1021/np050312l.

Beydoun, M. A., et al. (2018). Carotenoids, vitamin A, and their association with the metabolic syndrome: a systematic review and meta-analysis. *Nutrition Reviews, 77*(1), 32–45, Oxford University Press.

Cai, C., et al. (2014). Purification and photodynamic bioactivity of phycoerythrin and phycocyanin from *Porphyra yezoensis* Ueda. *Journal of Ocean University of China, 13*(3), 479–484. Available from: https://doi.org/10.1007/s11802-014-2148-x, Science Press.

Calder, P. C. (2015a). Functional roles of fatty acids and their effects on human health. *Journal of Parenteral and Enteral Nutrition, 39*(1_suppl), 18S–32S. Available from: https://doi.org/10.1177/0148607115595980, John Wiley & Sons, Ltd.

Calder, P. C. (2015b). Marine omega-3 fatty acids and inflammatory processes: effects, mechanisms and clinical relevance. *Biochimica et Biophysica Acta (BBA) - Molecular and Cell Biology of Lipids, 1851*(4), 469–484. Available from: https://doi.org/10.1016/J.BBALIP.2014.08.010, Elsevier.

Cappello, J.V. (2009) 'Blue-green algae composition'. Google Patents.

Carbonell-Capella, J. M., et al. (2014). Analytical methods for determining bioavailability and bioaccessibility of bioactive compounds from fruits and vegetables: a review. *Comprehensive Reviews in Food Science and Food Safety, 13*(2), 155–171. Available from: https://doi.org/10.1111/1541-4337.12049, John Wiley & Sons, Ltd.

Cha, K. H., Koo, S. Y., & Lee, D.-U. (2008). Antiproliferative effects of carotenoids extracted from *Chlorella ellipsoidea* and *Chlorella vulgaris* on human colon cancer cells. *Journal of Agricultural and Food Chemistry. American Chemical Society, 56*(22), 10521–10526.

Chen, J., et al. (2014). Algal sterols are as effective as β-sitosterol in reducing plasma cholesterol concentration. *Journal of Agricultural and Food Chemistry, 62*(3), 675–681. Available from: https://doi.org/10.1021/jf404955n, American Chemical Society.

Christaki, E., et al. (2012). Functional properties of carotenoids originating from algae. *Journal of Science of Food and Agriculture, 93*(1), 5–11. Available from: https://doi.org/10.1002/jsfa.5902.

Cuellar-Bermudez, S. P., et al. (2015). Extraction and purification of high-value metabolites from microalgae: essential lipids, astaxanthin and phycobiliproteins. *Microbial Biotechnology, 8*(2), 190–209. Available from: https://doi.org/10.1111/1751-7915.12167, John Wiley & Sons, Ltd (10.1111).

Darvin, M. E., et al. (2011). The role of carotenoids in human skin. *Molecules, 16*(12), 10491–10506. (Basel, Switzerland), Molecular Diversity Preservation International.

Das Gupta, S., & Suh, N. (2016). Tocopherols in cancer: an update. *Molecular Nutrition & Food Research, 60*(6), 1354–1363. Available from: https://doi.org/10.1002/mnfr.201500847, John Wiley & Sons, Ltd.

Dashwood, R., et al. (1998). Chemopreventive properties of chlorophylls towards aflatoxin B: 1 a review of the antimutagenicity and anticarcinogenicity data in rainbow trout. *Mutation Research.*

D'Eliseo, D., et al. (2016). Omega-3 fatty acids and cancer cell cytotoxicity: implications for multi-targeted cancer therapy. *Journal of Clinical Medicine, 5*(2), 15. Available from: https://doi.org/10.3390/jcm5020015, Multidisciplinary Digital Publishing Institute.

Del Campo, J. A., García-González, M., & Guerrero, M. G. (2007). Outdoor cultivation of microalgae for carotenoid production: current state and perspectives. *Applied Microbiology and Biotechnology, 74*(6), 1163–1174, Springer.

Devic, E., et al. (2002). Detection of anatoxin-a(s) in environmental samples of cyanobacteria by using a biosensor with engineered acetylcholinesterases. *Applied and Environmental Microbiology, 68*(8), 4102–4106. Available from: https://doi.org/10.1128/AEM.68.8.4102-4106.2002, American Society for Microbiology.

Doughman, S. D., Krupanidhi, S., & Sanjeevi, C. B. (2007). Omega-3 fatty acids for nutrition and medicine: considering microalgae oil as a vegetarian source of EPA and DHA. *Current Diabetes Reviews, 3*(3), 198–203. Available from: https://doi.org/10.2174/157339907781368968.

Draisci, R., et al. (2001). Identification of anatoxins in blue-green algae food supplements using liquid chromatography-tandem mass spectrometry. *Food Additives and Contaminants, 18*(6), 525–531. Available from: https://doi.org/10.1080/02652030118558.

Eid, S. Y., El-Readi, M. Z., & Wink, M. (2012). Carotenoids reverse multidrug resistance in cancer cells by interfering with ABC-transporters. *Phytomedicine: International Journal of Phytotherapy and Phytopharmacology, 19*(11), 977–987, Elsevier.

Engel, N., et al. (1991). Chlorophyll catabolism in *Chlorella protothecoides* isolation and structure elucidation of a red bilin derivative. *FEBS Letters*, *293*(1–2), 131–133. Available from: https://doi.org/10.1016/0014-5793(91)81168-8, John Wiley & Sons, Ltd.

Frommer, K. W., et al. (2015). Free fatty acids: potential proinflammatory mediators in rheumatic diseases. *Annals of the Rheumatic Diseases*, *74*(1), 303–310. Available from: https://doi.org/10.1136/ANNRHEUMDIS-2013-203755, BMJ Publishing Group Ltd.

Furey, A., et al. (2005). Strategies to avoid the mis-identification of anatoxin-a using mass spectrometry in the forensic investigation of acute neurotoxic poisoning, in*Journal of Chromatography. A*, 91–97. Available from: https://doi.org/10.1016/j.chroma.2005.05.040, Elsevier.

Gagez, A.-L., et al. (2012). Epoxycarotenoids and cancer. Review. *Current Bioactive Compounds*, *8*(2), 109–141, Bentham Science Publishers.

Gallego, R., et al. (2018). Green extraction of bioactive compounds from microalgae. *Journal of Analysis and Testing.*, *2*(2), 109–123. Available from: https://doi.org/10.1007/s41664-018-0061-9, Springer Singapore.

Gan, R. W., et al. (2017). The association between omega-3 fatty acid biomarkers and inflammatory arthritis in an anti-citrullinated protein antibody positive population. *Rheumatology*, *56* (12), 2229–2236. Available from: https://doi.org/10.1093/rheumatology/kex360, Oxford University Press.

Gardeva, E. G., et al. (2009). Cancer protective action of polysaccharide, derived from red microalga *Porphyridium cruentum*—a biological background. *Biotechnology and Biotechnological Equipment*, *23*, 783–787. Available from: https://doi.org/10.1080/13102818.2009.10818540, Taylor & Francis.

Gardeva, E. G., et al. (2014). Antitumor activity of C-phycocyanin from *Arthronema africanum* (Cyanophyceae). *Brazilian Archives of Biology and Technology*, *57*(5), 675–684. Available from: https://doi.org/10.1590/S1516-89132014005000018, Instituto de Tecnologia do Parana.

Ghosh, T., et al. (2016). Cyanobacterial pigments as natural anti-hyperglycemic agents: an in vitro study. *Frontiers in Marine Science*, *3*(AUG), 146. Available from: https://doi.org/10.3389/fmars.2016.00146, Frontiers Media S. A.

Gilroy, D. J., et al. (2000). Assessing potential health risks from microcystin toxins in blue-green algae dietary supplements. *Environmental Health Perspectives*, *108*(5), 435–439. Available from: https://doi.org/10.1289/ehp.00108435, Public Health Services, US Dept of Health and Human Services.

Glazer, A. N. (1994). Phycobiliproteins—a family of valuable, widely used fluorophores. *Journal of Applied Phycology*, *6*(2), 105–112. Available from: https://doi.org/10.1007/BF02186064, Kluwer Academic Publishers.

Goiris, K., et al. (2014). Detection of flavonoids in microalgae from different evolutionary lineages. *Journal of Phycology*, *50*(3), 483–492. Available from: https://doi.org/10.1111/jpy.12180, Edited by M. Posewitz. John Wiley & Sons, Ltd (10.1111).

Goiris, K., Muylaert, K., & De Cooman, L. (2015). *Microalgae as a novel source of antioxidants for nutritional applications. Handbook of Marine Microalgae* (pp. 269–280). Academic Press, 10.1016/B978-0-12-800776-1.00017-0.

Grosse, Y (2006). Carcinogenicity of nitrate, nitrite, and cyanobacterial peptide toxins. *Lancet Oncology*, *7*(8), 628–629.

Güven, K. C., Percot, A., & Sezik, E. (2010). Alkaloids in marine algae. *Marine Drugs*, 269–284. Available from: https://doi.org/10.3390/md8020269, Multidisciplinary Digital Publishing Institute.

Harris, A., Devaraj, S., & Jialal, I. (2002). Oxidative stress, alpha-tocopherol therapy, and atherosclerosis. *Current Atherosclerosis Reports*, *4*(5), 373–380. Available from: https://doi.org/10.1007/s11883-002-0075-6, Current Medicine Group.

Hasenhuettl, G. L. (2016). *Fats and fatty oils. Kirk-Othmer Encyclopaedia of Chemical Technology* (pp. 1–37). Hoboken, NJ, USA: John Wiley & Sons, Inc, 10.1002/0471238961.0601201908011905.a01.pub3.

Havaux, M. (1998). Carotenoids as membrane stabilizers in chloroplasts. *Trends in Plant Science, 3*(4), 147–151, Elsevier.

Havsteen, B. H. (2002). The biochemistry and medical significance of the flavonoids. *Pharmacology & Therapeutics, 96*(2–3), 67–202. Available from: https://doi.org/10.1016/S0163-7258(02)00298-X, Pergamon.

Hayashi, K., Hayashi, T., & Kojima, I. (1996). A natural sulfated polysaccharide, calcium spirulan, isolated from spirulina platensis: in vitro and ex vivo evaluation of anti-herpes simplex virus and anti-human immunodeficiency virus activities. *AIDS Research and Human Retroviruses, 12*(15), 1463–1471. Available from: https://doi.org/10.1089/aid.1996.12.1463.

Heim, K. E., Tagliaferro, A. R., & Bobilya, D. J. (2002). Flavonoid antioxidants: chemistry, metabolism and structure-activity relationships. *The Journal of Nutritional Biochemistry, 13*(10), 572–584. Available from: https://doi.org/10.1016/S0955-2863(02)00208-5, Elsevier.

Heussner, A. H., et al. (2012). Toxin content and cytotoxicity of algal dietary supplements. *Toxicology and Applied Pharmacology, 265*(2), 263–271. Available from: https://doi.org/10.1016/j.taap.2012.10.005, Academic Press.

Higuera-Ciapara, I., Felix-Valenzuela, L., & Goycoolea, F. M. (2006). Astaxanthin: a review of its chemistry and applications. *Critical Reviews in Food Science and Nutrition, 46*(2), 185–196, Taylor & Francis.

Hillwig, M. L., Zhu, Q., & Liu, X. (2014). Biosynthesis of ambiguine indole alkaloids in cyanobacterium Fischerella ambigua. *ACS Chemical Biology, 9*(2), 372–377. Available from: https://doi.org/10.1021/cb400681n, American Chemical Society.

Hodder, A. R., & Capon, R. J. (1991). A new brominated biindole from an Australian cyanobacterium, Rivularia firma. *Journal of Natural Products, 54*(6), 1661–1663, ACS Publications.

Hörtensteiner, S. (1999). Chlorophyll breakdown in higher plants and algae. *Cellular and Molecular Life Sciences*, 330–347. Available from: https://doi.org/10.1007/s000180050434, Springer.

Huang, J., et al. (2006). Reduction in the blood glucose level of exopolysaccharide of porphyridium cruentum in alloxan-induced diabetic mice. *Journal of Fujian Normal University*, 3, Natural Science Edition.

Jiang, Y (2008). Detection of the hepatotoxic microcystins in 36 kinds of cyanobacteria Spirulina food products in China. *Food Additives and Contaminants - Part A Chemistry, Analysis, Control, Exposure and Risk Assessment, 25*(7), 885–894.

Khan, M., et al. (2006). C-phycocyanin ameliorates doxorubicin-induced oxidative stress and apoptosis in adult rat cardiomyocytes. *Journal of Cardiovascular Pharmacology, 47*(1), 9–20. Available from: https://doi.org/10.1097/01.fjc.0000191520.48404.27.

Khozin-Goldberg, I., Leu, S., & Boussiba, S. (2016). Microalgae as a source for VLC-PUFA production. *Subcellular Biochemistry, 86*, 471–510. Available from: https://doi.org/10.1007/978-3-319-25979-6_19, Springer New York.

Koushan, K., et al. (2013). The role of lutein in eye-related disease. *Nutrients., 5*(5), 1823–1839, Multidisciplinary Digital Publishing Institute.

Koyande, A. K., et al. (2019). Microalgae: a potential alternative to health supplementation for humans. *Food Science and Human Wellness, 8*(1), 16–24. Available from: https://doi.org/10.1016/j.fshw.2019.03.001, Elsevier BV.

Lanfer-Marquez, U. M., Barros, R. M. C., & Sinnecker, P. (2005). Antioxidant activity of chlorophylls and their derivatives, in*Food Research International*, 885–891. Available from: https://doi.org/10.1016/j.foodres.2005.02.012.

Larsen, L. K., Moore, R. E., & Patterson, G. M. L. (1994). β-Carbolines from the blue-green alga *Dichothrix baueriana*. *Journal of Natural Products*, *57*(3), 419–421, ACS Publications.

Lawrence, J. F., et al. (2001). Comparison of liquid chromatography/mass spectrometry, ELISA, and phosphatase assay for the determination of microcystins in blue-green algae products. *Journal of AOAC International*, *84*(4), 1035–1044, AOAC International.

Luo, X., Su, P., & Zhang, W. (2015). Advances in microalgae-derived phytosterols for functional food and pharmaceutical applications. *Marine Drugs*, *13*(7), 4231–4254. Available from: https://doi.org/10.3390/md13074231, Multidisciplinary Digital Publishing Institute, AG.

Mader, J., et al. (2016). Calcium spirulan derived from *Spirulina platensis* inhibits herpes simplex virus 1 attachment to human keratinocytes and protects against herpes labialis. *Journal of Allergy and Clinical Immunology*, *137*(1), 197–203. Available from: https://doi.org/10.1016/J.JACI.2015.07.027, Mosby.

Manolidi, K., et al. (2019). Neurotoxin BMAA and its isomeric amino acids in cyanobacteria and cyanobacteria-based food supplements. *Journal of Hazardous Materials*, *365*, 346–365. Available from: https://doi.org/10.1016/j.jhazmat.2018.10.084, Elsevier B.V.

Matsui, M. S., et al. (2003). Sulfated polysaccharides from red microalgae have antiinflammatory properties in vitro and in vivo. *Applied Biochemistry and Biotechnology - Part A Enzyme Engineering and Biotechnology*, *104*(1), 13–22. Available from: https://doi.org/10.1385/ABAB:104:1:13, Springer.

McLellan, N. L., & Manderville, R. A. (2017). Toxic mechanisms of microcystins in mammals. *Toxicology Research*, *6*(4), 391–405. Available from: https://doi.org/10.1039/c7tx00043j, Royal Society of Chemistry.

Mendiola, J. A., et al. (2008). Enrichment of vitamin E from *Spirulina platensis* microalga by SFE. *The Journal of Supercritical Fluids*, *43*(3), 484–489. Available from: https://doi.org/10.1016/J.SUPFLU.2007.07.021, Elsevier.

Meriluoto, J., Spoof, L., & Codd, G. A. (2017). *Handbook of Cyanobacterial Monitoring and Cyanotoxin Analysis*. John Wiley & Sons.

Miller, T. R., et al. (2020). Microcystin toxins at potentially hazardous levels in algal dietary supplements revealed by a combination of bioassay, immunoassay, and mass spectrometric methods. *Journal of Agricultural and Food Chemistry*, *68*(30), 8016–8025. Available from: https://doi.org/10.1021/acs.jafc.0c02024, American Chemical Society.

Minkova, K. M., et al. (2011). Antitumor activity of B-phycoerythrin from *Porphyridium cruentum*. *Journal of Pharmacy Research*, *4*(5), 1480–1482.

Mozaffarian, D., & Wu, J. H. Y. (2011). Omega-3 fatty acids and cardiovascular disease. *Journal of the American College of Cardiology*, *58*(20), 2047–2067. Available from: https://doi.org/10.1016/j.jacc.2011.06.063.

Mudimu, O., et al. (2017). Screening of microalgae and cyanobacteria strains for α-tocopherol content at different growth phases and the influence of nitrate reduction on α-tocopherol production. *Journal of Applied Phycology*, *29*(6), 2867–2875. Available from: https://doi.org/10.1007/s10811-017-1188-1, Springer Netherlands.

Mulders, K. J. M., et al. (2014). Phototrophic pigment production with microalgae: biological constraints and opportunities. *Journal of Phycology*, *50*(2), 229–242. Available from: https://doi.org/10.1111/jpy.12173, Edited by R. Bassi. Blackwell Publishing Inc.

Mullie, P., et al. (2008). Estimation of daily human intake of food flavonoids. *International Journal of Food Sciences and Nutrition, 59*(4), 291–298. Available from: https://doi.org/10.1080/09687630701539293, Taylor & Francis.

Murugan, T. (2011). Screening for antifungal and antiviral activity of C-phycocyanin from *Spirulina platensis. Association of Pharmaceutical Innovators.*

Negishi, T., Rai, H., & Hayatsu, H. (1997). Antigenotoxicity activity of natural chlorophylls. *Mutation Research – Fundamental and Molecular Mechanisms of Mutagenesis, 376*(1–2), 97–100. Available from: https://doi.org/10.1016/S0027-5107(97)00030-4, Elsevier.

Nishanth, R. P., et al. (2010). C-Phycocyanin inhibits MDR1 through reactive oxygen species and cyclooxygenase-2 mediated pathways in human hepatocellular carcinoma cell line. *European Journal of Pharmacology, 649*(1–3), 74–83. Available from: https://doi.org/10.1016/j.ejphar.2010.09.011, Elsevier.

Novoveská, L., et al. (2019). Microalgal carotenoids: a review of production, current markets, regulations, and future direction. *Marine Drugs*. Available from: https://doi.org/10.3390/md17110640, Multidisciplinary Digital Publishing Institute, AG.

Nürnberg, D. J., et al. (2018). Photochemistry beyond the red limit in chlorophyll f-containing photosystems. *Science (New York, N.Y.), 360*(6394), 1210–1213. Available from: https://doi.org/10.1126/science.aar8313, American Association for the Advancement of Science.

Panche, A. N., Diwan, A. D., & Chandra, S. R. (2016). Flavonoids: an overview. *Journal of Nutritional Science, 5*, e47. Available from: https://doi.org/10.1017/jns.2016.41, Cambridge University Press.

Pardhasaradhi, B. V. V., et al. (2003). Phycocyanin-mediated apoptosis in AK-5 tumor cells involves down-regulation of Bcl-2 and generation of ROS. *Molecular Cancer Therapeutics, 2*(11), 1165–1170, American Association for Cancer Research.

Patil, V., et al. (2007). Fatty acid composition of 12 microalgae for possible use in aquaculture feed. *Aquaculture International, 15*(1), 1–9. Available from: https://doi.org/10.1007/s10499-006-9060-3, Kluwer Academic Publishers.

Pérez-Gálvez, A., & Roca, M. (2017). Phyllobilins: a new group of bioactive compounds. *Studies in Natural Products Chemistry*, 159–191. Available from: https://doi.org/10.1016/B978-0-444-63931-8.00004-7, Elsevier B.V.

Prakash, S., Sasikala, S. L., & Aldous, V. H. J. (2010). Isolation and identification of MDR–*Mycobacterium tuberculosis* and screening of partially characterised antimycobacterial compounds from chosen marine micro algae. *Asian Pacific Journal of Tropical Medicine, 3*(8), 655–661. Available from: https://doi.org/10.1016/S1995-7645(10)60158-7, Elsevier.

Queiroz Zepka, L., Jacob-Lopes, E., & Roca, M. (2019). Catabolism and bioactive properties of chlorophylls. *Current Opinion in Food Science*, 94–100. Available from: https://doi.org/10.1016/j.cofs.2019.04.004, Elsevier Ltd.

Rajakumar, P. D. (2018). Psychoactive properties of microalgae, in*Microalgae in Health and Disease Prevention*, 325–334. Available from: https://doi.org/10.1016/b978-0-12-811405-6.00016-5, Elsevier.

Raposo, M., De Morais, A. M. M. B., & De Morais, R. M. S. C. (2014). Influence of sulphate on the composition and antibacterial and antiviral properties of the exopolysaccharide from *Porphyridium cruentum. Life Sciences, 101*(1–2), 56–63. Available from: https://doi.org/10.1016/j.lfs.2014.02.013, Elsevier Inc.

Raposo, M., De Morais, R., & Bernardo de Morais, A. (2013). Bioactivity and applications of sulphated polysaccharides from marine microalgae. *Marine Drugs, 11*(1), 233–252. Available from: https://doi.org/10.3390/md11010233, Multidisciplinary Digital Publishing Institute.

Rellán, S., et al. (2009). First detection of anatoxin-a in human and animal dietary supplements containing cyanobacteria. *Food and Chemical Toxicology*, *47*(9), 2189–2195. Available from: https://doi.org/10.1016/j.fct.2009.06.004, Pergamon.

Rimbau, V., et al. (1999). Protective effects of C-phycocyanin against kainic acid-induced neuronal damage in rat hippocampus. *Neuroscience Letters*, *276*(2), 75–78. Available from: https://doi.org/10.1016/S0304-3940(99)00792-2, Elsevier Ireland Ltd.

Roberts, J. E., & Dennison, J. (2015). The photobiology of lutein and zeaxanthin in the eye. *Journal of ophthalmology*, *2015*, Hindawi.

Rodríguez-Zavala, J. S., et al. (2010). Increased synthesis of α-tocopherol, paramylon and tyrosine by *Euglena gracilis* under conditions of high biomass production. *Journal of Applied Microbiology*, *109*(6), 2160–2172. Available from: https://doi.org/10.1111/j.1365-2672.2010.04848.x, John Wiley & Sons, Ltd.

Rolle, I., et al. (1977). Amines in unicellular green algae: 2. Amines in *Scenedesmus acutus*. *Analytical Biochemistry*, *77*(1), 103–109. Available from: https://doi.org/10.1016/0003-2697(77)90294-9, Academic Press.

Romay, C., & González, R. (2000). Phycocyanin is an antioxidant protector of human erythrocytes against lysis by peroxyl radicals. *Journal of Pharmacy and Pharmacology*, *52*(4), 367–368. Available from: https://doi.org/10.1211/0022357001774093, Wiley-Blackwell.

Roncone, M., Bartlett, H., & Eperjesi, F. (2010). Essential fatty acids for dry eye: a review. *Contact Lens and Anterior Eye*, *33*(2), 49–54. Available from: https://doi.org/10.1016/J.CLAE.2009.11.002, Elsevier.

Roy, K. R., et al. (2007). Alteration of mitochondrial membrane potential by *Spirulina platensis* C-phycocyanin induces apoptosis in the doxorubicinresistant human hepatocellular-carcinoma cell line HepG2. *Biotechnology and Applied Biochemistry*, *47*(3), 159. Available from: https://doi.org/10.1042/ba20060206, Wiley.

Roy-Lachapelle, A., et al. (2017). Detection of cyanotoxins in algae dietary supplements. *Toxins*, *9*(3), 76. Available from: https://doi.org/10.3390/toxins9030076, Multidisciplinary Digital Publishing Institute, AG.

Ryckebosch, E., et al. (2012). Microalgae as an alternative source of omega-3 long chain polyunsaturated fatty acids. *Lipid Technology*, *24*(6), 128–130. Available from: https://doi.org/10.1002/lite.201200197, John Wiley & Sons, Ltd.

Sánchez-Parra, E., Boutarfa, S., & Aboal, M. (2020). Are cyanotoxins the only toxic compound potentially present in microalgae supplements? Results from a study of ecological and non-ecological products. *Toxins.*, *12*(9), 552. Available from: https://doi.org/10.3390/toxins12090552, NLM (Medline).

Schulze, C., et al. (2016). Screening of microalgae for primary metabolites including β-glucans and the influence of nitrate starvation and irradiance on β-glucan production. *Journal of Applied Phycology*, *28*(5), 2719–2725. Available from: https://doi.org/10.1007/s10811-016-0812-9, Springer Netherlands.

Schwartz, J., et al. (1988). Prevention of experimental oral cancer by extracts of *Spirulina–Dunaliella* algae. *Nutrition and Cancer*, Taylor & Francis.

Scoglio, S. *et al.* (2017) 'Neuro-modulating preparations for treatment of neurological and mood disorders'. Google Patents.

Sekar, S., & Chandramohan, M. (2008). Phycobiliproteins as a commodity: trends in applied research, patents and commercialization. *Journal of Applied Phycology*, 113–136. Available from: https://doi.org/10.1007/s10811-007-9188-1, Springer.

Senni, K., et al. (2011). Marine polysaccharides: a source of bioactive molecules for cell therapy and tissue engineering. *Marine Drugs*, *9*(9), 1664–1681. Available from: https://doi.org/10.3390/md9091664, Multidisciplinary Digital Publishing Institute.

Shih, S. R., et al. (2003). Inhibition of enterovirus 71-induced apoptosis by allophycocyanin isolated from a blue-green alga Spirulina platensis. *Journal of Medical Virology*, *70*(1), 119–125. Available from: https://doi.org/10.1002/jmv.10363, John Wiley & Sons, Ltd.

Simopoulos, A. P. (2008). The importance of the omega-6/omega-3 fatty acid ratio in cardiovascular disease and other chronic diseases. *Experimental Biology and Medicine*, *233*(6), 674–688. Available from: https://doi.org/10.3181/0711-MR-311, SAGE PublicationsSage UK: London, England.

Svirčev, Z., et al. (2014). Epidemiology of cancers in serbia and possible connection with cyanobacterial blooms. *Journal of Environmental Science and Health - Part C Environmental Carcinogenesis and Ecotoxicology Reviews*, *32*(4), 319–337. Available from: https://doi.org/10.1080/10590501.2014.967053, Taylor and Francis Inc.

Takaichi, S., & Mochimaru, M. (2007). Carotenoids and carotenogenesis in cyanobacteria: unique ketocarotenoids and carotenoid glycosides. *Cellular and Molecular Life Sciences*, 2607–2619. Available from: https://doi.org/10.1007/s00018-007-7190-z, Springer.

Tanaka, T., Shnimizu, M., & Moriwaki, H. (2012). Cancer chemoprevention by carotenoids. *Molecules (Basel, Switzerland)*, *17*(3), 3202–3242, Molecular Diversity Preservation International.

Tannin-Spitz, T., et al. (2005). Antioxidant activity of the polysaccharide of the red microalga *Porphyridium* sp. *Journal of Applied Phycology*, *17*(3), 215–222. Available from: https://doi.org/10.1007/s10811-005-0679-7, Springer.

Testai, E., et al. (2017). Review and analysis of occurrence, exposure and toxicity of cyanobacteria toxins in food. *EFSA Supporting Publications*, *13*(2). Available from: https://doi.org/10.2903/sp.efsa.2016.en-998, Wiley-Blackwell.

Thilakarathna, S., & Rupasinghe, H. (2013). Flavonoid bioavailability and attempts for bioavailability enhancement. *Nutrients*, *5*(9), 3367–3387. Available from: https://doi.org/10.3390/nu5093367, Multidisciplinary Digital Publishing Institute, AG.

Ueno, Y., et al. (1996). Detection of microcystins, a blue-green algal hepatotoxin, in drinking water sampled in Haimen and Fusui, endemic areas of primary liver cancer in China, by highly sensitive immunoassay. *Carcinogenesis*, *17*(6), 1317–1321. Available from: https://doi.org/10.1093/carcin/17.6.1317, Oxford Academic.

Ugocsai, K., et al. (2005). Effects of selected flavonoids and carotenoids on drug accumulation and apoptosis induction in multidrug-resistant colon cancer cells expressing MDR1/LRP, in vivo *International Institute of Anticancer Research*, *19*(2), 433–438.

Van Apeldoorn, M. E., et al. (2007). Toxins of cyanobacteria. *Molecular Nutrition and Food Research*, 7–60. Available from: https://doi.org/10.1002/mnfr.200600185, John Wiley & Sons, Ltd.

Vichi, S., et al. (2012). Contamination by microcystis and microcystins of blue-green algae food supplements (BGAS) on the Italian market and possible risk for the exposed population. *Food and Chemical Toxicology*, *50*(12), 4493–4499. Available from: https://doi.org/10.1016/j.fct.2012.09.029, Pergamon.

Vinayagam, R., & Xu, B. (2015). Antidiabetic properties of dietary flavonoids: a cellular mechanism review. *Nutrition & Metabolism*, *12*(1), 60. Available from: https://doi.org/10.1186/s12986-015-0057-7, BioMed Central.

Volk, R. B. (2008). Screening of microalgae for species excreting norharmane, a manifold biologically active indole alkaloid. *Microbiological Research*, *163*(3), 307–313. Available from: https://doi.org/10.1016/j.micres.2006.06.002, Elsevier GmbH.

Volk, R. B., et al. (2009). Bromoanaindolone, a novel antimicrobial exometabolite from the cyanobacterium *Anabaena constricta. Natural Product Research, 23*(7), 607–612. Available from: https://doi.org/10.1080/14786410802114068, Taylor & Francis Group.

Wang, B., Lan, C. Q., & Horsman, M. (2012). Closed photobioreactors for production of microalgal biomasses. *Biotechnology Advances*, 904–912. Available from: https://doi.org/10.1016/j.biotechadv.2012.01.019, Elsevier.

Watson, R. R., & Preedy, V. R. (2012). *Bioactive Food as Dietary Interventions for Liver and Gastrointestinal Disease: Bioactive Foods in Chronic Disease States*. Academic Press.

Yasukawa, K., et al. (1996). Inhibitory effects of sterols isolated from *Chlorella vulgaris* on 12-o-tetradecanoylphorbol-13-acetate-induced inflammation and tumor promotion in mouse skin. *Biological & Pharmaceutical Bulletin, 19*(4), 573–576. Available from: https://doi.org/10.1248/bpb.19.573, Pharmaceutical Society of Japan.

Yu, F. Y., et al. (2002). Development of a sensitive ELISA for the determination of microcystins in algae. *Journal of Agricultural and Food Chemistry, 50*(15), 4176–4182. Available from: https://doi.org/10.1021/jf0202483, American Chemical Society.

Yu, P., et al. (2017). Purification and bioactivities of phycocyanin. *Critical Reviews in Food Science and Nutrition, 57*(18), 3840–3849. Available from: https://doi.org/10.1080/10408398.2016.1167668, Taylor and Francis Inc.

Yuan, J.-P., Gong, X.-D., & Chen, F. (1997). Separation and analysis of carotenoids and chlorophylls in *Haematococcus lacustris* by high-performance liquid chromatography photodiode array detection. *Journal of Agricultural and Food Chemistry, 45*(5), 1952–1956, American Chemical Society Publications.

Zakar, T., et al. (2016). Carotenoids assist in cyanobacterial photosystem II assembly and function. *Frontiers in Plant Science, 7*, 295, Frontiers.

Zheng, H., et al. (2018). Chlorophyllin modulates gut microbiota and inhibits intestinal inflammation to ameliorate hepatic fibrosis in mice. *Frontiers in Physiology, 9*, 1671. Available from: https://doi.org/10.3389/fphys.2018.01671, Frontiers Media SA.

Zhu, L. (2015). Biorefinery as a promising approach to promote microalgae industry: an innovative framework. *Renewable and Sustainable Energy Reviews, 41*, 1376–1384. Available from: https://doi.org/10.1016/J.RSER.2014.09.040, Pergamon.

Chapter 12

Bioaccessibility and bioavailability of bioactive compounds delivered from microalgae

María Janeth Rodríguez-Roque[1], Rogelio Sánchez-Vega[2], Ingrid Aguiló-Aguayo[3], Alan Eduardo Medina-Antillón[1], Mayra Cristina Soto-Caballero[1], Nora Aideé Salas-Salazar[1] and Carlos Guillermo Valdivia-Nájar[4]

[1]Faculty of Agrotechnological Science, Autonomous University of Chihuahua, Chihuahua, Mexico, [2]Faculty of Zootechnics and Ecology, Autonomous University of Chihuahua, Chihuahua, Mexico, [3]Institut de Recerca i Tecnologia Agroalimentàries (IRTA), XaRTA-Postharvest, Parc Científic i Tecnològic Agroalimentari de Lleida, Parc de Gardeny, Edifici Fruitcentre, Lleida, Spain, [4]Department of Food Technology, Centro de Investigación y Asistencia en Tecnología y Diseño del Estado de Jalisco (CIATEJ), Zapopan, Mexico

12.1 Introduction

There is currently great interest in microalgae production as they are considered a rich source of highly nutritional and functional ingredients. Microalgae synthesize bioactive molecules including vitamins, carotenoids, peridinin, phycocyanin, polyunsaturated fatty acids (PUFAs), and phytosterols, among others (Le Goff et al., 2019). These compounds have a wide array of physiological and pharmacological activities, such as antioxidant, anticancer, antiviral, cardioprotective, hepatoprotective, and immunomodulatory actions (El-Baz, Abdel-Jaleel, Saleh, & Hussein, 2018). Epidemiological evidence suggests that the intake of natural foods containing bioactive molecules improves human health, wellness, and nutritional status (Granado-Lorencio et al., 2009).

The health benefits of these constituents have increased the demand for microalgae production. In this context, the isolation of bioactive compounds from microalgae has gained considerable relevance due to the high productivity of microalgae (Rasmussen & Morrisey, 2007). Additionally, bioactive

Cultured Microalgae for the Food Industry. DOI: https://doi.org/10.1016/B978-0-12-821080-2.00006-X

compounds from microalgae have important commercial applications as they are used as food additives, nutraceuticals, and in the pharmaceutical industry (Se-Kwon & Kyong-Hwa, 2011).

Several studies have focused on analyzing the bioactive constituents content of microalgae, however, information regarding their bioaccessibility and bioavailability remains scarce. A compound can be considered as "bioactive" only when its physiological response has been proved (Carbonell-Capella, Buniowska, Barba, Esteve, & Frígola, 2014). For this reason, there is great interest in both bioaccessibility and bioavailability of microalgae-derived molecules.

Bioavailability is defined as the fraction of ingested compound that reaches the relevant sites in the human body in its active form and an appropriate concentration (Rein et al., 2013). Bioavailability comprises two other terms, bioaccessibility and bioactivity (Fernández-García, Carvajal-Lérida, & Pérez-Gálvez, 2009), which cover the following steps: digestion, absorption, metabolism, and distribution, as shown in Fig. 12.1. Overall, bioavailability and bioaccessibility are the most commonly used terms. Bioaccessibility represents the fraction of a compound that is released from the food matrix after digestion to be solubilized in the gut for uptake (Bernaerts, Gheysen, Foubert, Hendrickx, & Van Loey, 2019). On the other hand, bioactivity is when compounds are transported into the systemic circulation to reach a target tissue, interact with other biomolecules, and generate a cascade of biochemical and physiological effects (Cilla, Bosch, Barberá, & Alegría, 2018).

Although a great amount of bioactive compounds can be found in microalgae, only a small proportion will be absorbed, metabolized, and utilized to exert their biological function. For this reason, it is important to take into consideration both the bioaccessibility and bioavailability of bioactive constituents. The analysis of these two parameters allows a better understanding of the in vivo biological activity of bioactive constituents.

To this end, different in vivo and in vitro methodologies have been developed for analyzing the bioaccessibility and bioavailability of bioactive compounds. In vivo studies are usually carried out in animals or humans. Human intervention studies are the reference standard methodology for obtaining specific information regarding the bioavailability of bioactive compounds,

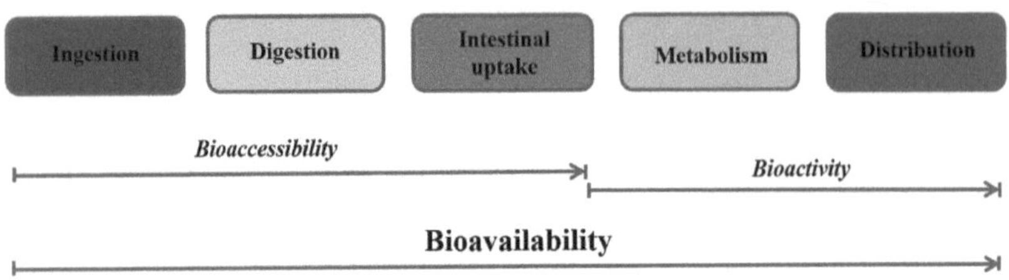

FIGURE 12.1 Bioavailability and bioaccessibility steps.

however, they are expensive and long-term studies with difficulties in data interpretation (Rodríguez-Roque, 2014). Animal studies can also provide important data related to the bioavailability of compounds with biological activity, they are less expensive than human trials and the analysis of results is also difficult due to the differences in metabolism between animals and humans (Van Buggenhout et al., 2010). In vitro studies are usually used as the first step in analyzing the bioaccessibility of bioactive compounds for predictive purposes (Cardoso, Afonso, Lourenço, Costa, & Nunes, 2015; Failla & Chitchumroonchokchai, 2005). They are cost-effective, reproducible, and fast methodologies, however, they cannot mimic in vivo processes which comprise feedback mechanisms, mucosa cell activity, hormonal and nervous control, immune system, and peristaltic movements (Guerra et al., 2012). The main advantages and disadvantages of these methodologies are presented in Table 12.1.

In the following sections, the bioaccessibility and bioavailability of the main bioactive compounds from microalgae are presented and discussed.

12.2 Bioaccessibility and bioavailability of bioactive compounds from microalgae

A wide range of studies have evaluated the bioactive compounds contained in microalgae. It is clear that these microorganisms are rich sources of biologically active molecules with a high potential for commercial purposes in pharmacological, cosmetic, nutraceutical, and functional food industries. For this reason, they are receiving a great deal of attention and diverse microalgae species have been studied. Despite this context, information regarding the bioaccessibility, bioactivity, and bioavailability of these molecules with biological activity remains scarce. It is important to focus attention not only on the type and quantity of bioactive compounds contained in microalgae, but also on the portions of these compounds that are released, absorbed, metabolized, and distributed within the human body to reach the target tissues (Fig. 12.2).

12.3 Vitamins

Vitamins are important bioactive compounds as they are precursors of enzyme cofactors and play an important role in the scavenging of reactive oxygen species. Among the most important vitamins synthesized by microalgae are the following: provitamin A, vitamin B (B_1, B_2, B_3, B_5, B_6, B_8, B_9, and B_{12}), vitamin C, vitamin D_3, vitamin E, and vitamin K_1 (Barbosa et al., 2005; Carballo-Cárdenas, Tuan, Janssen, & Wijffels, 2003; Durmaz, 2007; Edelmann, Aalto, Chamlagain, Kariluoto, & Piironen, 2019; Galasso et al., 2019; Levasseur, Perré, & Pozzobon, 2020; Tarento et al., 2018). The vitamin content of microalgae depends on the specie and environmental

TABLE 12.1 The main advantages and disadvantages of in vivo and in vitro methodologies used for analyzing bioaccessibility and bioavailability.

Methodology type	Main assessment	Advantages	Disadvantages
In vivo studies – Human intervention studies	• Bioavailability	• Relevant data related to bioavailability • Reference standard • High scientific evidence	• Limitations in experimental design • Difficult data interpretation • Expensive • Ethical constraints • Long-term study
In vivo studies – Animal studies	• Bioavailability • Bioactivity	• Recovery of important data related to bioavailability	• Less expensive than human intervention studies • Difficult data interpretation due to differences between human and animal metabolism • Ethical constraints
In vitro studies – Gastrointestinal digestion – Cell culture	• Bioaccessibility • Bioactivity	• Cost-effective methodology • Reproducible • Fast	• Data interpretation for predictive purposes only • It does not mimic the in vivo process

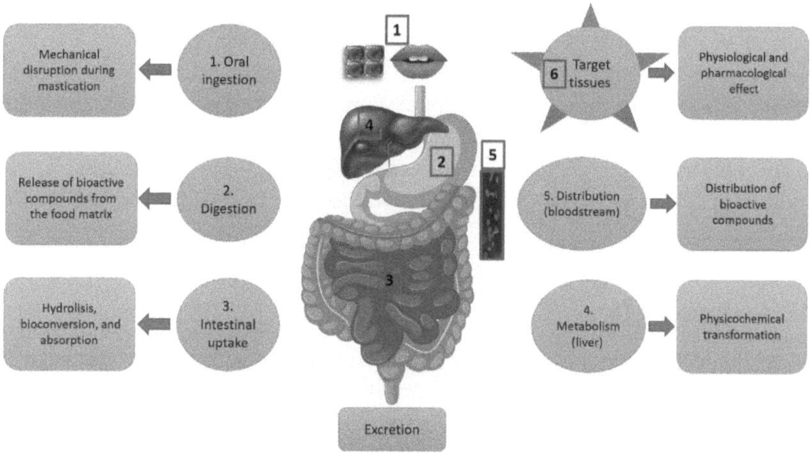

FIGURE 12.2 General scheme of bioactive compounds bioavailability.

conditions (light, nutrient availability, harvesting stage, among others) as highlighted in other chapters in this book.

Each vitamin has a specific metabolism pathway for being bioavailable, and thus exerting its biological activity. For instance, vitamin D_3 must be metabolized by both liver and kidney to exert a physiological effect on the human body. Once ingested, vitamin D_3 is converted to 25-hydroxyvitamin D in the liver and hydroxylated in the kidney to the active form 1,25-dihydroxyvitamin D (Pilz et al., 2018). The concentration of vitamin D_3 from the microalgae *Nannochloropsis oceanica* was increased by exposure to artificial ultraviolet B radiation at $6\ kJ \cdot m^{-2} \cdot day^{-1}$. However, there is no information on the percentage of this compound that is converted into active form, as well as the amount that is bioaccessible and bioavailable.

In the case of vitamin C, the intestinal uptake occurs by means of different mechanisms (Ball, 2006): (1) via a sodium-dependent active transporter (SVCT1) in the ileum and jejunum for ascorbic acid; (2) via facilitated diffusion, with the aid of glucose transporters, in the duodenum and jejunum for dehydroascorbic acid (dehydroascorbic acid is transported into cells with the aid of glucose transporters and is quickly recycled to vitamin C); and (3) via sodium-dependent transporter (SVCT2) to transport vitamin C into target tissues. Similarly to vitamin D_3, information regarding bioaccessibility and bioavailability of vitamins (B, C, D_3, E, and K_1) produced by microalgae was not found.

Research on the bioaccessibility, bioactivity, and bioavailability of vitamins contained in microalgae will contribute to elucidating the effect of these compounds on health. Taking into consideration that microalgae are able to synthesized the active form of some vitamins it is necessary to extend the knowledge in this area. In the literature, only a small number of studies have focused on this topic. For instance, it was reported that the active isomer of vitamin K_1 was produced by cyanobacteria, while the chemical synthesis of this vitamin produced up to 20% of their inactive form (Tarento et al., 2018). Additionally, microalgae consumption modified the intestinal microbiota, leading to an increase in vitamin absorption. For instance, the consumption of *Arthrospira* improved the absorption of vitamin B_1 by increasing the concentration of *Lactobacillus* in the gut (Ambrosi, Reinehr, Bertolin, Costa, & Colla, 2008).

12.4 Pigments

Microalgae are important sources of natural pigments with a wide range of commercial applications, mainly as nutritional supplements or natural colorants (Spolaore, Joannis-Cassan, Duran, & Isambert, 2006). The main pigments contained in microalgae are carotenoids and chlorophylls but also other pigments (i.e., peridinin and phycocyanin) can be found. Pigments naturally encapsulate within cells and thus resist harsh processing conditions. Cell wall integrity can significantly limit the availability of these compounds, therefore disruption of the microalgal cell wall is necessary as a pretreatment for allowing the liberation of cellular content (Nunes et al., 2020).

12.4.1 Carotenoids

One of the most important pigments contained in microalgae is carotenoids. These are a widespread family of fat-soluble compounds synthesized by algae and also by plants and photosynthetic bacteria. Although there are more than 700 carotenoids identified, only 50 can be absorbed by the human body and just six reach the blood (α-carotene, β-carotene, β-cryptoxanthin, lycopene, lutein, and zeaxanthin) (Maiani et al., 2009). These compounds in human tissues come from the diet, as humans are not able to synthesize carotenoids de novo (Granado-Lorencio et al., 2009).

Once carotenoids are released from microalgae, they must be solubilized in oil droplets and transferred to bile salt mixed micelles to become bioavailable. Mixed micelles are generated during digestion of cholesterol esters, phospholipids, and triglycerides for producing free cholesterol, fatty acids, and mono- and di-acylglycerides (Harrison, 2012). Thus the formation of micelles is an important factor related to carotenoid bioavailability. In addition to the formation of micelles, the absorption of carotenoids depends on other factors, such as the presence of other bioactive compounds, the presence of dietary fiber, the coingested food, nutrient status of the host, and food processing, among others (Van Het Hof, West, Weststrate, & Hautvast, 2000). In this sense, it has been established that carotenoid solubility into micelles depends on their molecular structure. Fiber can interact with biliary fluids, producing an insoluble gel which reduces the carotenoid micellization during gastric digestion (Rodríguez-Roque, Rojas-Graü, Elez-Martínez, & Martín-Belloso, 2014). In the case of carotenoid interaction with other bioactive compounds or food constituents, xanthophylls reduce the carotenoid bioaccessibility due to their being more liposoluble.

Regarding carotenoids from microalgae, *Chlorella* spp., *Dunaliella salina*, and *Haematococcus pluvialis* have been widely studied for production of carotenoids under stress conditions, reaching up to 14% dry weight of β-carotene and 2%–3% of astaxanthin (Se-Kwon & Kyong-Hwa, 2011). The bioaccessibility of the main carotenoids from microalgae has been reported through in vitro gastrointestinal digestion (Table 12.2). For instance, the in vitro bioaccessibility of lutein and zeaxanthin from lyophilized biomass and lutein-rich extracts in olive oil produced from the microalgae *Scenedesmus almeriensis* (CCAP 276/24) was analyzed (Granado-Lorencio et al., 2009). In this study, more than 75% of microalgae carotenoids remained in the digesta but their bioaccessibility was low (<1%) for both lyophilized biomass and oil extracts enriched at low doses. Carotenoid isomerization was low during in vitro digestion, while it increased during the micellar phase. The incorporation of lutein into the aqueous micellar phase was improved when olive oil was used as the vehicle. In this sense, the relative incorporations of lutein, zeaxanthin, α-carotene, and β-carotene into micelles were 35%–40%, 45%–60%, 50%, and 25%, respectively (Granado-Lorencio et al., 2009).

In vitro bioaccessibility of β-carotene ranged from 0% to 2.6% for *Chlorella* sp. and around 1.7% for zeaxanthin (Cha et al., 2012; Gille et al., 2016). β-Carotene bioaccessibility from *Chlamydomonas reinhardtii* reached up to 10% and up to 29% in *Phaeodactylum tricornutum* (Gille et al., 2019). The high carotenoid bioaccessibility of *Chlamydomonas reinhardtii* and *Phaeodactylum tricornutum* could be related to the fact that these microalgae possess a less rigid cell wall, in comparison to *Chlorella* sp.

It was reported that cell disruption increased the in vitro bioaccessibility of carotenoids from microalgae. Indeed, the in vitro bioaccessibility of carotenoids from *Nannochlopsis* sp. was improved when the cell was disrupted. The authors of that study reported incomplete lipid digestion and thus, a low digestibility of β-carotene, antheraxanthin, and zeaxanthin (between 1% and 6%) in nondisrupted cell microalgae. On the other hand, the carotenoid bioaccessibility increased by threefold when the microalga was high-pressure homogenized due to disruption of cells (Bernaerts et al., 2020).

12.4.2 Peridinin

This pigment is a dinoflagellate-derived compound that has shown antiproliferation of tumor cell activity. Only one publication focused on evaluating the bioactivity of this compound was found in the literature. The study evaluated the influence of peridinin isolated from the dinoflagellate *Heterocpsa*

TABLE 12.2 In vitro bioaccessibility of microalgae bioactive compounds.

Bioactive compounds	Microalgae	Bioaccessibility (%)	References
Antheraxanthin	*Chlorella* sp.	0%	Cha, Koo, Song, and Pan (2012)
	Nannochloropsis sp.	1%–9% (depending on microalgae processing)	Bernaerts et al. (2020)
β-Carotene	*Chlamydomonas reinhardtii*	10%	Gille, Trautmann, Posten, and Briviba (2016)
	Chlorella sp.	0%–2.6% (depending on extraction technique and microalgae processing)	Cha et al. (2012), Gille, Neumann, Louis, Bischoff, and Briviba (2018)
	Phaeodactylum tricornutum	19%–76% (depending on extraction technique and microalgae processing)	Gille et al. (2018), Gille, Hollenbach, Trautmann, Posten, and Briviba (2019)
	Nannochloropsis sp.	5%–16% (depending on microalgae processing)	Bernaerts et al. (2020)
Fucoxanthin	*Phaeodactylum tricornutum*	20%–62% (depending on extraction technique and microalgae processing)	Gille et al. (2018, 2019), Kim et al. (2016)
Lutein	*Chlamydomonas reinhardtii*	20%	Gille et al. (2016)
	Chlorella sp.	0%–26% (depending on microalgae processing)	Cha et al. (2011), Gille et al. (2016, 2018)
	Scenedesmus almeriensis	<1%	Granado-Lorencio et al. (2009)
Zeaxanthin	*Chlorella* sp.	1.7%–18% (depending on microalgae processing)	Cha et al. (2012)

(*Continued*)

TABLE 12.2 (Continued)

Bioactive compounds	Microalgae	Bioaccessibility (%)	References
	Phaeodactylum tricornutum	0%–30% (depending on extraction technique and microalgae processing)	Gille et al. (2018, 2019)
	Scenedesmus almeriensis	<1%	Granado-Lorencio et al. (2009)
	Nannochloropsis sp.	5%–15% (depending on microalgae processing)	Bernaerts et al. (2020)
Chlorophylls	*Chlorella* sp.	77%–84% (depending on extraction technique and microalgae processing)	Rego, Costa, Pereira, and Redondo (2015)
Fatty acids	*Isochrysis galbana*	8%–13%	Bonfanti et al. (2018)
Eicosapentaenoic acid	*Nannochloropsis* sp.	14%–29% (depending on microalgae processing)	Bernaerts et al. (2020)

triquetra on the DLD-1 human colon cancer cells and demonstrated that peridinin induced the apoptosis of this cell line (Sugawara et al., 2007).

12.4.3 Phycocyanin

Phycocyanin is a phycobiliprotein, a blue pigment produced by cyanobacteria including *Arthrospira* (*Spirulina*). It regulates cholesterol metabolism and lipid peroxidation, protects against fibrosis, and has an antioxidant effect (reducing the superoxide anion, nitric oxide, and thiobarbituric reactive substances) (Coué et al., 2019). However, up to the best of the authors' knowledge, there is no information on bioaccessibility and bioavailability of phycocyanin.

12.5 Polyunsaturated fatty acids

Microalgae-derived PUFAs have been associated with antimicrobial activity, in particular to long-chain unsaturated fatty acids (C_{16}–C_{20}) (Zheng et al., 2005).

The biological importance of these compounds has led to its approval for use as a food additive. Docosahexaenoic acid (DHA) and eicosapentaenoic acid (EPA) extracted from *Schizochytrium*, as well as oil with high EPA content extracted from *Phaeodactylum tricornutum*, oil rich in PUFAs, and DHA obtained from *Ulkenia* sp. have been authorized by different international commissions (Zanella & Vianello, 2020). Oils produced from unicellular organisms, including PUFAs, are considered as novel and functional food ingredients (Astiasarán & Ansorena, 2009).

Similarly to carotenoids, PUFAs must be incorporated into mixed micelles to become bioaccessible. Therefore their solubilization into micelles, as well as the disruption of the cell walls, determines the extent to which these compounds are both bioaccessible and bioavailable. Bernaerts et al. (2020) evaluated the influence of cell integrity of *Nannochloropsis* sp. on the in vitro lipid digestibility and bioaccessibility of omega-3 long-chain PUFA. Cell disruption by high-pressure homogenization of a *Nannochloropsis* sp. suspension resulted in nearly complete lipid digestibility (56%–62% free fatty acid). The bioaccessibility of EPA was increased threefold (between 27% and 29%) when applying high-pressure homogenization of *Nannochloropsis* sp. in comparison to when untreated (EPA bioaccessibility $< 15\%$), as shown in Table 12.2 (Bernaerts et al., 2020).

Dawczynski et al. (2017) evaluated the clinical effect of daily intake of foods enriched with microalgae oil as a source of DHA (C22:6n-3). Thirty-eight patients diagnosed with rheumatoid arthritis participated in the study. Treatment with DHA resulted in a significant reduction of clinical and biochemical signs of inflammation. Therefore microalgae oil, a rich source of PUFAs, could be an alternative treatment for rheumatoid arthritis.

12.6 Peptides

Bioactive peptides are short sequences of amino acids that are encrypted within the sequence of their parent protein and are liberated upon proteolysis. They are promising ingredients due to their biological activity and health benefits, showing anticancer, antioxidant, and antihypertensive activities, as well as preventing cardiovascular disease (Fan, Bai, Zhu, Yang, & Zhang, 2014; Kim & Kang, 2011; Nazih & Bard, 2018). Generally, they are formed by 2–20 amino acid residues. Species, such as *Chlorella vulgaris*, *Dunaliella salina*, *Arthrospira maxima*, *Navicula incerta*, and *Chlamydomonas reinhardtii* contain between 40% and 60% of protein on a dry weight basis. Overall, the protein content of microalgae is higher than that of other conventional protein sources (milk 26%, and meat 43% dry matter) (Kim & Kang, 2011). Additionally, Shein, Fang, Wu, and Lin (2010) demonstrated that microalgae protein was an excellent source of anticarcinogenic peptides with better antioxidant activity toward peroxyl radicals than Trolox.

The bioaccessibility of proteins from microalgae is affected by the cell wall, microalgae strain, and the presence of other compounds. The cell wall of microalgae must be disrupted to afford the interaction of proteins and digestive enzymes (Becker, 2007). In this sense, Wild, Steingaß, and Rodehutscord (2018) reported that cell wall disruption increased the in vitro protein digestibility in different microalgal species. These authors obtained an in vitro protein digestibility of 54% in noncell wall-disrupted *Nannochloropsis*, against 84% for cell wall-disrupted *Chlorella*.

Recently, research was carried out to evaluated the in vitro digestibility of some microalgal biomasses belonging to species approved as food: *Arthrospira platensis* F&M-C256, *Aphanizomenon flos-aquae*, *Nostoc sphaeroides* F&M-C117, *Chlorella sorokiniana* F&M-M49, *Chlorella sorokiniana* IAM C-212, *Chlorella vulgaris* Allma, *Tetraselmis suecica* F&M-M33, *Porphyridium purpureum* F&M-M46, *Phaeodactylum tricornutum* F&M-M40, *Tisochrysis lutea* F&M-M36, and *Nannochloropsis oceanica* F&M-M24 (Niccolai, Zittelli, Rodolfi, Biondi, & Tredici, 2019). The authors of that study reported that *A. platensis* F&M-C256, *C. sorokiniana* F&M-M49, and *C. vulgaris* Allma showed higher crude protein content and in vitro digestibility. On the contrary, marine species were less digestible. The differences in digestibility were associated with the complexity of the cell wall as exopolysaccharides may have limited the action of digestive enzymes.

Human gastric cancer cell lines were treated with microalgae-derived peptides from *Chlorella vulgaris* (36 $\mu g\ mL^{-1}$) for different time periods (up to 24 h). Through flow cytometry, it was observed that cells in the G1 phase decreased, and the peptide fraction arrested the cells in a post-G1 phase. These results suggest that microalgae-derived peptides induce cell death and inhibit the growth of human gastric cancer (Shein et al., 2010).

12.7 Phytosterols

Phytosterols are structural membrane components of microalgae, thus they are contained in all species in the range of 0.4%–5.1% dry weight biomass (Luo, Su, & Zhang, 2015). The microalgae *Isochrysis galbana*, *Nannochloropsis* sp., *Phaeodactylum tricornutum*, *Diacronema lutheri* (syn. *Pavlova lutheri*), *Tetraselmis* sp., and *Nannochloropsis* sp. are the main phytosterol producers (Le Goff et al., 2019; Ryckebosch, Bruneel, Termote-Verhalle, Muylaert, & Foubert, 2014).

They regulate the membrane fluidity and permeability but also have functional activity as hormonal precursors from signal transduction in organisms, antiinflammatory, antihypercholesterolemic, antioxidant, anticancer, antidiabetic, and immunomodulatory actions (Luo et al., 2015; Stirk et al., 2013). β-Sitosterol, campesterol, and stigmasterol are involved in the prevention of cardiometabolic diseases and cancers, in the reduction of inflammatory markers by means of regulating C-Jun N-terminal kinases and nuclear factor-kappa B signaling pathways, as well as suppressing the production of proinflammatory mediators (Le Goff et al., 2019).

Phytosterols derived from microalgae reduce the gastrointestinal absorption of dietary cholesterol that is endogenously produced. *Schizochytrium*-derived sterols showed a cholesterol-lowering ability by means of downregulation of both intestinal gene ACAT2 expression (cholesterol absorption in the gut) (Chen, Jiao, Jiang, Bi, & Chen, 2014) and hepatic 3-hydroxy-33-methylglutaryl-CoA (HMG-CoA) reductase (cholesterol synthesis) (Luo et al., 2015). *Schizochytrium* sterol extract at doses of 0.06 and 0.30 g kg^{-1} diet decrease plasma cholesterol concentrations by 19.5% and 34%, respectively, in hamsters fed with a high-cholesterol diet (Chen et al., 2014). Moreover, a lipid-rich extract from the blue-green alga *Nostoc commune* var. *sphaeroides* Kützing (*N. commune*), containing campesterol, β-sitosterol, and clionasterol, inhibited the synthesis of cholesterol on human hepatoma cell lines by reducing the mRNA expression of 3-hydroxy-3-methylglutaryl-CoA reductase (HMGR) and low-density lipoprotein receptor (Rasmussen et al., 2008).

Regarding the antiinflammatory effect of microalgae phytosterols, some in vitro and in vivo studies have evaluated the functionality of these compounds. Ergosterol from *Chlorella vulgaris* and *Dunalliella tertiolecta* reduces the in vitro LPS-induced response of macrophages by avoiding the production of cytokine (TNF-α) and COX-2 expression in sheep and mice (Caroprese, Albenzio, Ciliberti, Francavilla, & Sevi, 2012; Yasukawa et al., 1996). In addition, stigmasterol isolated from the microalgae *Navicula incerta* has been shown to have an anticancer activity. It produced significant toxicity on human hepatocarcinoma (HepG2) cells and induced apoptosis via the upregulation of the proapoptotic genes *Bax* and *p53* and downregulation of the antiapoptotic gene *Bcl-2* (Kim, Li, Kang, Ryu, & Kim, 2014).

12.8 Other compounds

Phycotoxins and biotoxins are produced by at least 95 dinoflagellates according to the Taxonomic Reference List of Toxic Plankton Algae of the Intergovermental Oceanographic Commission. These compounds have interesting applications in human or veterinary medicine as analgesic, antitumor, anticholesterol, cytotoxic, antiinfective, immunosuppressive, and neurological disease therapeutics (Table 12.3) (Assunção, Guedes, & Malcata, 2017; Gallardo-Rodríguez et al., 2012).

Other interesting compounds from microalgae are polysaccharides. In this sense, polysaccharides from *Spirulina platensis* inhibited the proliferation of ascitic hepatoma cells and reduced lung metastasis of B16-BL6 melanoma cells in mice (Lisheng, Baojiang, Jihong, Guangquan, & Botang, 1991). The polysaccharide "calcium spirulan" (a sulfated polysaccharide chelating calcium) was isolated from *Spirulina platensis* and reduced the lung metastasis

TABLE 12.3 Health benefits of some biotoxins isolated from microalgae dinoflagellates.

Health benefits	Biotoxin
Pain treatment	Saxitoxin Tetrodotoxin
Prevention of neurodegenerative disease	Yessotoxin Gambierol Gymnodimine
Antitumor	Yessotoxin
Wounds treatment	Gonyautoxin

of B16-BL6 melanoma cells by inhibiting tumor invasion (Mishima et al., 1998).

12.9 Final remarks

Nowadays, microalgae production has gained increased attention as they are considered an important source of sustainable food and bioactive ingredients. They have high photosynthetic efficiency with high biomass productivity when compared to traditional crops. One of the most important features of microalgae is the wide range of biologically active molecules that some species are able to synthesize.

Vitamins, carotenoids, chlorophylls, PUFAs, and proteins (that can be used for generating bioactive peptides) are among the main bioactive constituents from microalgae. It is clear that all these compounds exert a physiological or pharmaceutical effect on the human body once ingested. Several studies in the literature have focused their research in analyzing the main source of bioactive compounds, the quantity, type, and the method for improving the concentration of a given compound. However, information regarding bioavailability, bioaccessibility, or bioactivity of all these interesting compounds remains scarce. Other important tasks to be considered in future studies are the synergistic interactions among bioactive compounds from microalgae. These topics are an unlimited area of basic and applied research and represent a great challenge for researchers.

Acknowledgments

I. Aguiló-Aguayo thanks the National Programme for the Promotion of Talent and Its Employability of the Spanish Ministry of Economy, Industry, and Competitiveness and the European Social Fund for the Postdoctoral Senior Grant "Ramon y Cajal" (RYC-2016–19949).

References

Ambrosi, M. A., Reinehr, C. O., Bertolin, T. E., Costa, J. A. V., & Colla, L. M. (2008). Propriedades de saúde de Spirulina spp. *Revista de Ciências Farmacêuticas Básica e Aplicada*, *29*, 109–117.

Assunção, J., Guedes, A. C., & Malcata, F. X. (2017). Biotechnological and pharmacological applications of biotoxins and other bioactive molecules from dinoflagellates. *Marine Drugs*, *15*, 393.

Astiasarán, I., & Ansorena, D. (2009). Algal oils. In R. Moreau, & A. Kamal-Eldin (Eds.), *Gourmet and health-promoting specialty oils* (pp. 491–513). Academic Press.

Ball, G. F. M. (2006). *Vitamins in foods: Analysis, bioavailability, and stability. Food science and technology* (Vol. 156, p. 785). Boca Raton, Florida: CRC/Taylor & Francis.

Barbosa, M. J., Zijffers, J. W., Nisworo, A., Vaes, W., van Schoonhoven, J., & Wijffels, R. H. (2005). Optimization of biomass, vitamins, and carotenoid yield on light energy in a flat-panel reactor using the A-stat technique. *Biotechnology and Bioengineering*, *890*(2), 233–242.

Becker, E. W. (2007). Micro-algae as a source of protein. *Biotechnology Advances*, *25*, 207–210.

Bernaerts, T. M. M., Gheysen, L., Foubert, I., Hendrickx, M. E., & Van Loey, A. M. (2019). The potential of microalgae and their biopolymers as structuring ingredients in food: A review. *Biotechnology Advances*, *37*(8), 107419.

Bernaerts, T. M. M., Verstreken, H., Dejonghe, C., Gheysen, L., Foubert, I., Grauwet, T., & Van Loy, A. M. (2020). Cell disruption of *Nannochloropsis sp.* improves *in vitro* bioaccessibility of carotenoids and ω3-LC-PUFA. *Journal of Functional Foods*, *65*, 103770.

Bonfanti, C., Cardoso, C., Afonso, C., Matos, J., Garcia, T., Tanni, S., & Bandarra, N. M. (2018). Potential of microalga *Isochrysis galbana*: Bioactivity and bioaccessibility. *Algal Research*, *29*, 242–248.

Carballo-Cárdenas, E. C., Tuan, P. M., Janssen, M., & Wijffels, R. H. (2003). Vitamin E (α-tocopherol) production by the marine microalgae *Dunaliella tertiolecta and Tetraselmis suecica* in batch cultivation. *Biomolecular Engineering*, *200*(4), 139–147.

Carbonell-Capella, J. M., Buniowska, M., Barba, F. J., Esteve, M. J., & Frígola, A. (2014). Analytical methods for determining bioavailability and bioaccessibility of bioactive compounds from fruits and vegetables: A review. *Comprehensive Review in Food Science and Food Safety*, *13*, 155–171.

Cardoso, C., Afonso, C., Lourenço, H., Costa, S., & Nunes, M. L. (2015). Bioaccessibility assessment methodologies and their consequences for the risk benefit evaluation of food. *Trends Food Science and Technology*, *41*, 5–23.

Caroprese, M., Albenzio, M., Ciliberti, M. G., Francavilla, M., & Sevi, A. (2012). A mixture of phytosterols from *Dunaliella tertiolecta* affects proliferation of peripheral blood mononuclear cells and cytokine production in sheep. *Veterinary Immunology Immunopathology*, *150*, 27–35.

Cha, K. H., Koo, S. Y., Song, D., & Pan, C. (2012). Effect of microfluidization on bioaccessibility of carotenoids from *Chlorella ellipsoidea* during simulated digestion. *Journal of Agricultural and Food Chemistry*, *60*, 9437–9442.

Cha, K. H., Lee, J. Y., Song, D., Kim, S. M., Lee, D., Jeon, J., & Pan, C. (2011). Effect of microfluidization on in vitro micellization and intestinal cell uptake of lutein from *Chlorella vulgaris*. *Journal of Agricultural and Food Chemistry*, *59*, 8670–8674.

Chen, J., Jiao, R., Jiang, Y., Bi, Y., & Chen, Z. Y. (2014). Algal Sterols are as effective as β-sitosterol in reducing plasma cholesterol concentration. *Journal of Agricultural and Food Chemistry*, *62*, 675–681.

Cilla, A., Bosch, L., Barberá, R., & Alegría, A. (2018). Effect of processing on the bioaccessibility of bioactive compounds—A review focusing on carotenoids, minerals, ascorbic acid, tocopherols and polyphenols. *Journal of Food Composition and Analysis*, *68*, 3–15.

Coué, M., Tesse, A., Falewée, J., Aguesse, A., Croyal, M., Fizanne, L., ... Ouguerram, K. (2019). Spirulina liquid extract protects against fibrosis related to non-alcoholic steatohepatitis and increases ursodeoxycholic acid. *Nutrients*, *11*(1), 194.

Dawczynski, C., Dittrich, M., Neumann, T., Goetze, K., Welzel, A., Oelzner, P., ... Jahreis, G. (2017). Docosahexaenoic acid in the treatment of rheumatoid arthritis: A double-blind, placebo-controlled, randomized cross-over study with microalgae vs. sunflower oil. *Clinical Nutrition*, *37*(2), 494–504.

Durmaz, Y. (2007). Vitamin E (α-tocopherol) production by the marine microalgae *Nannochloropsis oculata* (Eustigmatophyceae) in nitrogen limitation. *Aquaculture*, *2720*(1), 717–722.

Edelmann, M., Aalto, S., Chamlagain, B., Kariluoto, S., & Piironen, V. (2019). Riboflavin, niacin, folate and vitamin B_{12} in commercial microalgae powders. *Journal of Food Composition and Analysis*, *82*, 103226.

El-Baz, F., Abdel-Jaleel, G. A., Saleh, D. O., & Hussein, R. A. (2018). Protective and therapeutic potentials of *Dunaliella salina* on aging-associated cardiac dysfunction in rats. *Asian Pacific Journal of Tropical Biomedicine*, *8*, 403.

Failla, M. L., & Chitchumroonchokchai, C. (2005). *In vitro* models as tools for screening the relative bioavailabilities of provitamin A carotenoids in foods. *HarvestPlus Technical Monograph*, *3*, 32.

Fan, X., Bai, L., Zhu, L., Yang, L., & Zhang, X. (2014). Marine algae-derived bioactive peptides for human nutrition and health. *Journal of Agricultural and Food Chemistry*, *62*, 9211–9222.

Fernández-García, E., Carvajal-Lérida, I., & Pérez-Gálvez, A. (2009). *In vitro* bioaccessibility assessment as a prediction tool of nutrient efficiency. *Nutrition Research*, *29*, 751–760.

Galasso, C., Gentile, A., Orefice, I., Ianora, A., Bruno, A., Noonan, D. M., ... Brunet, C. (2019). Microalgal derivatives as potential nutraceutical and food supplements for human health: A focus on cancer prevention and interception. *Nutrients*, *110*(6), 1226.

Gallardo-Rodríguez, J. J., Sánchez-Mirón, A., García-Camacho, F., López-rosales, L., Chisti, Y., & Molina-Grima, E. (2012). Bioactives from microalgal dinoflagellates. *Biotechnology Advances*, *30*, 1673–1684.

Gille, A., Hollenbach, R., Trautmann, A., Posten, C., & Briviba, K. (2019). Effect of sonication on bioaccessibility and cellular uptake of carotenoids from preparations of photoautotrophic *Phaeodactylum tricornutum*. *Food Research International*, *118*, 40–48.

Gille, A., Neumann, U., Louis, S., Bischoff, S. C., & Briviba, K. (2018). Microalgae as a potential source of carotenoids: Comparative results of an in vitro digestion method and a feeding experiment with C57BL/6J mice. *Journal of Functional Foods*, *49*, 285–294.

Gille, A., Trautmann, A., Posten, C., & Briviba, K. (2016). Bioaccessibility of carotenoids from *Chlorella vulgaris* and *Chlamydomonas reinhardtii*. *International Journal of Food Sciences and Nutrition*, *67*(5), 507–513.

Granado-Lorencio, F., Herrero-Barbudo, C., Acién-Fernández, G., Molina-Grima, E., Fernández-Sevilla, J. M., Pérez-Sacristán, B., & Blanco-Navarro, I. (2009). *In vitro* bioaccesibility of lutein and zeaxanthin from the microalgae *Scenedesmus almeriensis. Food Chemistry, 114*, 747–752.

Guerra, A., Etienne-Mesmin, L., Livrelli, V., Denis, S., Blanquet-Diot, S., & Alric, M. (2012). Relevance and challenges in modeling human gastric and small intestinal digestion. *Trends in Biotechnology, 30*, 591–600.

Harrison, E. H. (2012). Mechanisms involved in the intestinal absorption of dietary vitamin A and provitamin A carotenoids. *Biochimica et Biophysica Acta, 1821*, 70–77.

Kim, J. H., Kim, S. M., Cha, K. H., Mok, I., Koo, S. Y., Pan, C., & Lee, J. K. (2016). Evaluation of the anti-obesity effect of the microalga *Phaeodactylum tricornutum. Applied Biological Chemistry, 59*, 283–290.

Kim, S.-K., & Kang, K.-H. (2011). Medicinal effects of peptides from marine microalgae. In S.-K. Kim (Ed.), *Advances in Food and Nutrition Research* (Vol 64, pp. 313–323). San Diego, CA, USA: Academic Press.

Kim, Y. S., Li, X. F., Kang, K. H., Ryu, B., & Kim, S. K. (2014). Stigmasterol isolated from marine microalgae *Navicula incerta* induces apoptosis in human hepatoma HepG2 cells. *BMB Report, 47*, 433.

Le Goff, M., Le Ferrec, E., Mayer, C., Mimouni, V., Lagadic-Gossmann, D., Schoefs, B., & Ulmann, L. (2019). Microalgal carotenoids and phytosterols regulate biochemical mechanisms involved in human health and disease prevention. *Biochimie, 167*, 106–118.

Levasseur, W., Perré, P., & Pozzobon, V. (2020). A review of high value-added molecules production by microalgae in light of the classification. *Biology Advances, 41*, 107545. Available from https://doi.org/10.1016/j.biotechadv.2020.107545.

Lisheng, L., Baojiang, G., Jihong, R., Guangquan, Q., & Botang, W. (1991). Inhibitive effect and mechanism of polysaccharide of Spirulina platensis on transplanted tumor in mice. *Marine Science, 5*, 33–38.

Luo, X., Su, P., & Zhang, W. (2015). Advances in microalgae-derived phytosterols for functional food and pharmaceutical applications. *Marine Drugs, 13*, 4231–4254.

Maiani, G., Castón, M. J. P., Catasta, G., Toti, E., Cambrodón, I. G., Bysted, A., ... Schlemmer, U. (2009). Carotenoids: Actual knowledge on food sources, intakes, stability and bioavailability and their protective role in humans. *Molecular Nutrition and Food Research, 53* (SUPPL. 2), 194–218.

Mishima, T., Murata, J., Toyoshima, M., Fujii, K., Nakajima, M., Hayashi, T., ... Saiki, I. (1998). Inhibition of tumor invasion and metastasis by calcium spirulan (Ca-SP), a novel sulfated poly- saccharide derived from a blue-green alga, *Spirulina platensis. Clinical & Experimental Metastasis, 16*, 541–550.

Nazih, H., & Bard, J.-M. (2018). Microalgae in human health: Interest as a functional food. In Levine, & Fleurence (Eds.), *Microalgae in Health and Disease Prevention* (pp. 211–226). Academic Press.

Niccolai, A., Zittelli, G. C., Rodolfi, L., Biondi, N., & Tredici, M. (2019). Microalgae of interest as food source: Biochemical composition and digestibility. *Algal Research, 42*, 101617.

Nunes, M., Graça, C., Vlaisavljević, S., Tenreiro, A., Sousa, I., & Raymundo, A. (2020). Microalgal cell disruption: Effect on the bioactivity and rheology of wheat bread. *Algal Research, 45*, 101749.

Pilz, S., März, W., Cashman, K. D., Kiely, M. E., Whiting, S. J., Holick, M. F., ... Zittermann, A. (2018). Rationale and plan for vitamin D food fortification: A review and guidance paper. *Frontiers in Endocronology, 9*, 373.

Rasmussen, H. E., Blobaum, K. R., Park, Y. K., Ehlers, S. J., Lu, F., & Lee, J. Y. (2008). Lipid extract of *Nostoc commune var. sphaeroides Kützing*, a blue-green alga, inhibits the activation of sterol regulatory element binding proteins in HepG2 Cells. *Journal of Nutrition*, *138*, 476–481.

Rasmussen, R. S., & Morrisey, M. T. (2007). Marine biotechnology for production of food ingredients. *Advances in Food and Nutrition Research*, *52*, 237–292.

Rego, D., Costa, L., Pereira, M. T., & Redondo, L. M. (2015). Cell membrane permeabilization studies of *Chlorella sp.* by pulsed electric fields. *IEEE Transaction on Plasma Science*, *43*, 3483–3488.

Rein, M. J., Renouf, M., Cruz-Hernandez, C., Actis-Goretta, L., Thakkar, S. K., & da Silva Pinto, M. (2013). Bioavailability of bioactive food compounds: A challenging journey to bioefficacy. *British Journal of Clinical Pharmacology*, *75*(3), 588–602.

Rodríguez-Roque, M.J. 2014. *In vitro* bioaccessibility of health related compounds from beverages based on fruit juice, milk or soymilk: Influence of food matrix and processing (thesis dissertation), University of Lleida, 2014.

Rodríguez-Roque, M. J., Rojas-Graü, M. A., Elez-Martínez, P., & Martín-Belloso, O. (2014). *In vitro* bioaccessibility of health-related compounds as affected by the formulation of fruit juice- and milk-based beverages. *Food Research International*, *62*, 771–778.

Ryckebosch, E., Bruneel, C., Termote-Verhalle, R., Muylaert, K., & Foubert, I. (2014). Influence of extraction solvent system on extractability of lipid components from different microalgae species. *Algal Research*, *3*, 36–43. Available from https://doi.org/10.1016/j.algal.2013.11.001.

Se-Kwon, K., & Kyong-Hwa, K. (2011). Medicinal effects of peptides from marine microalgae. Chapter 25. *Advances in Food and Nutrition Research*, *64*.

Shein, I. C., Fang, T. J., Wu, T. K., & Lin, P. H. (2010). Anticancer and antioxidant activities of the peptide fraction from algae protein waste. *Journal of Agricultural and Food Chemistry*, *58*, 1202–1207.

Spolaore, P., Joannis-Cassan, C., Duran, E., & Isambert, A. (2006). Commercial applications of microalgae. *Journal of Bioscience and Bioengineering*, *101*, 87–96.

Stirk, W. A., Bálint, P., Tarkowská, D., Novák, O., Strnad, M., Ördög, V., & van Staden, J. (2013). Hormone profiles in microalgae: Gibberellins and brassinosteroids. *Plant Physiology and Biochemistry*, *70*, 348–353.

Sugawara, T., Yamashita, K., Sakai, S., Asai, A., Nagao, A., Shiraishi, T., ... Hirata, T. (2007). Induction of apoptosis in DLD-1 human colon cancer cells by peridinin isolated from the Dinoflagellate, *Heterocapsa triquetra*. *Bioscience, Biotechnology, and Biochemistry*, *71*, 1069–1072.

Tarento, T. D. C., McClure, D. D., Vasiljevski, E., Schindeler, A., Dehghani, F., & Kavanagh, J. M. (2018). Microalgae as a source of vitamin K_1. *Algal Research*, *36*, 77–87.

Van Buggenhout, S., Alminger, M., Lemmens, L., Colle, I., Knockaert, G., Moetlants, K., ... Hendrickx, M. (2010). *In vitro* approaches to estimate the effect of food processing on carotenoid bioavailability need thorough understanding of process induced microstructural changes. *Trends in Food Science and Technology*, *21*, 607–618.

Van Het Hof, K. H., West, C. E., Weststrate, J. A., & Hautvast, J. G. A. J. (2000). Dietary factors that affect the bioavailability of carotenoids. *Journal of Nutrition*, *130*(3), 503–506.

Wild, K. J., Steingaß, H., & Rodehutscord, M. (2018). Variability in nutrient composition and in vitro crude protein digestibility of 16 microalgae products. *Journal of Animal Physiology and Animal Nutrition*, *102*(5), 1306–1319.

Yasukawa, K., Akihisa, T., Kanno, H., Kaminaga, T., Izumida, M., Sakoh, T., ... Takido, M. (1996). Inhibitory effects of sterols isolated from *Chlorella vulgaris* on 12-O-tetradecanoylphorbol-13-acetate-Induced inflammation and tumor promotion in mouse skin. *Biological and Pharmaceutical Bulletin*, *19*, 573–576.

Zanella, L., & Vianello, F. (2020). Microalgae of the genus *Nannochloropsis*: Chemical composition and functional implications for human nutrition. *Journal of Functional Foods*, *68*, 103919.

Zheng, C. J., Yoo, J.-S., Lee, T.-G., Cho, H.-Y., Kim, Y.-H., & Kim, W.-G. (2005). Fatty acid synthesis is a target for antibacterial activity of unsaturated fatty acids. *FEBS Letters*, *579* (23), 5157–5162.

Chapter 13

Sustainability of microalgae cultivation

Benoit Guieysse and Maxence Plouviez
School of Food & Advanced Technology, Massey University, Palmerston North, New Zealand

13.1 Introduction

Microalgae biotechnology can be defined as the industrial use of microalgae to generate products and/or services such as food, fuels, pharmaceuticals, and wastewater treatment (Xu, Weathers, Xiong, & Liu, 2009). Because artificial illumination remains economically prohibitive in most applications, these processes are nearly all based on microalgae cultivation outdoors using sunlight as the source of energy. "Dark" heterotrophic cultivation also remains infrequent and conceptually similar to "dark" microbial fermentation from a sustainability standpoint: it is therefore not discussed further.

Outdoor microalgae cultivation is often considered as intrinsically "sustainable" because it is based on the use of sunlight to convert carbon dioxide (CO_2) and nutrients into biomass (Ahmad, Yasin, Derek, & Lim, 2011). However, engineering this natural process at an industrial scale requires land, energy, water, nutrients, and materials that must be processed and transported (Ketzer, Skarka, & Rösch, 2018). It is therefore critical to assess how efficiently these resources are used to optimize and benchmark microalgae biotechnologies against alternatives. This assessment is especially critical to prevent the misallocation of resources and unintended environmental mistakes (e.g., *Jatropha*; Kant & Wu, 2011). Numerous studies have thus attempted to evaluate resources use and environmental impacts during microalgae cultivation (Tables 13.1–13.3). Focusing on microalgae cultivation in open ponds, this chapter provides an overview of the state of the art in the field and discusses the most important impacts and contributors during cultivation in open ponds. Refer to Béchet, Plouviez, Chambonnière, & Guieysse (2017) for a comprehensive technical methodology.

Cultured Microalgae for the Food Industry. DOI: https://doi.org/10.1016/B978-0-12-821080-2.00013-7

TABLE 13.1 Life cycle analyses of biofuel production from microalgae.

Product	System[a]	Species	Productivity $g \cdot m^{-2} \cdot d^{-1}$	Approach	Environmental impact categories[b]	CO_2eq emissions[c]	References
Biodiesel	R	*Chlorella vulgaris*	25	Cradle-to-grave	AbD, Ac, Eu, GWP, Ozone, HumTox, Land, MarTox, Rad, Photo	0.059–0.134 kg CO_2 eq. MJ^{-1}	Lardon et al. (2009)
Biodiesel	PBR	*Nannochloropsis* sp.	25	Well-to-pump	GWP	−0.075 kg CO_2 eq. MJ^{-1}	Batan, Quinn, Willson, and Bradley (2010)
Biodiesel	R	n.s.	30	Cradle-to-grave	GWP	−0.028 to 0.018 kg CO_2 eq. t km^{-1}	Campbell, Beer, and Batten (2011)
Biomethane	R	*C. vulgaris*	25	Well-to-wheel	AbD, Ac, Eu, GWP, Hum Tox, Land, Ozone, Photo, Rad	0.083[d] kg CO_2 eq. MJ^{-1}	Collet et al. (2011)
Biomass	PBR	*Alexandrium minute; Karlodinium veneficum; Heterosigma akashiwo*		Cradle-to-gate	AbD, Ac, Eu, FreTox, GWP, HumTox, MarTox, Ozone, Photo, TeTox	22.54–23.94 kg CO_2 eq. kg $biomass^{-1}$	Sevigné Itoiz et al. (2012)
Biofuel	R	*Nannochloropsis* sp.	25–50	Well-to-wheel	GWP	−0.042 to 0.497 kg CO_2 eq. MJ^{-1}	Quinn, Smith, Downes, and Quinn (2014)
Bio-oil	R	*Scenedesmus dimorphus*	6.5	Well-to-pump	GWP	−0.011 to 0.210 kg CO_2 eq. MJ^{-1}	Bennion, Ginosar, Moses, Agblevor, and Quinn (2015)

Biodiesel and coproducts	R	n.s.	30	Well-to-wheel	GWP	0.060–1.320 kg CO_2 eq. MJ^{-1}	Dutta, Neto, and Coelho (2016)
Biodiesel	R	*Spirulina platensis*; *Nannochloropsis* sp.	14–27	Cradle-to-gate	Ac, Aland, CC, FD, FreEu, FreTox, HumTox, MarEu, MarTox, MinD, NLT, Ozone, Photo, PMF, Rad, TeTox, ULand, WD	19.3–26.4 kg CO_2 eq. kg $biodiesel^{-1}$	Foteinis, Antoniadis-Gavriil, and Tsoutsos (2018)
Hydrocarbons	R	*Botryococcus braunii*	9	Well-to-use	GWP, Land, W	−0.9 kg CO_2 eq. kg $hydrocarbons^{-1}$	Chaudry, Bahri, and Moheimani (2019)
Biodiesel and biogas	R	*Scenedesmus dimorphus*	6–15	Cradle-to-gate	Ac, Eu, FreTox, GWP	0.80–10.7 kg CO_2 eq. kg $biomass^{-1}$	Mediboyina, Banuvalli, Chauhan, and Mudliar (2020)

[a] *R, raceway pond or high-rate algal pond (HRAP); PBR, photobioreactor.*
[b] *AbD, abiotic depletion; Ac, acidification; Aland, agricultural land occupation; AqTox; aquatic toxicity; Carci, carcinogen; CC, climate change; EcoQ, ecosystem quality; En, energy use; Eu, eutrophication; F, fossil fuel use; FD, fossil fuel depletion; FreEu, freshwater eutrophication; FreTox, freshwater toxicity; GWP, global warming potential; HumH, human health; HumTox, human toxicity; Land, land use; MarEu, marine eutrophication; MarTox, marine toxicity; MD, metals depletion; MinEx, mineral extraction; MinD, minerals resources depletion; Mo, metal ores; N, nuclear energy; NLT, natural land transformation; NonCarci; noncarcinogen; NOx, nitrogen oxides; NRen, nonrenewable energy; NRr, nonrenewable resources; Ozone, ozone depletion; Photo, photochemical oxidation; PMF, particulate matter formation; Rad, ionizing radiation; Rinorg, respiratory inorganics; Rorg, respiratory organics; TeAc, terrestrial acidification; TeTox, terrestrial toxicity; Uland, urban land occupation; W, water use; WD, water depletion; WR, water resources.*
[c] *The ranges listed represent the lowest and highest emissions reported during outdoor cultivations and without differentiation of the scenarios being studied.*
[d] *Collet et al. (2011) estimated the GWP of algal-based methane was 1.42 higher than the GWP of algal biodiesel. We therefore estimated the GWP of algae-based methane to 0.083 kg CO_2eq. MJ^{-1} based on the GWP of algal biodiesel of 0.059 kg CO_2eq. MJ^{-1} computed by Lardon et al. (2009).*

TABLE 13.2 Life cycle analyses of microalgae-based food and nutraceuticals. The cradle-to-gate approach was used in all the studies cited.

Product	System[a]	Species	Productivity $g \cdot m^{-2} \cdot d^{-1}$	Environmental impact categories[b]	CO_2eq emissions[c]	References
Biomass	PBR	*Nannochloropsis* sp.	4.5–15.1	GWP	0.09–1.76 kg $CO_2 \cdot$ MJexergy-Dry algal biomass^{-1}	Taelman, De Meester, Roef, Michiels, and Dewulf (2013)
Astaxanthin	PBR	*Haematococcus pluvialis*		AbD, Ac, Eu, FreTox, GWP, HumTox, MarTox, Ozone, Photo, TeTox	0.3–0.48 kg CO_2eq $\cdot$ g asthaxanthin^{-1}	Pérez-López, González-García, Jeffryes et al. (2014)
Bioactive	PBR, PBR + R	*Tetraselmis suecica*		AbD, Ac, Eu, GWP, Ozone, Photo	56.7 kg CO_2eq $\cdot$ kg algal biomass^{-1}	Pérez-López, González-García, Ulloa et al. (2014)
Biomass	R, PBR, F	*Chlorella vulgaris*, *Arthrospira platensis*		Ac, AqTox, Carci, Land, MinEx, NRen, GWP, Rorg, Rinorg, NonCarci, Ozone, TeTox, Rad	14.7–245 kg CO_2eq. kg protein powder^{-1}	Smetana, Sandmann, Rohn, Pleissner, and Heinz (2017)
Phycocyanin	R	*A. platensis*	15	AbD, Ac, Eu, FreTox, GWP, HumTox, MarTox, Ozone, Photo, TeTox	1180–326,000 kg CO_2eq $\cdot$ kg phycocyanin^{-1}	Papadaki, Kyriakopoulou, Tzovenis, and Krokida (2017)
Biomass		*C. vulgaris*	1.2–2.6 g L^{-1}	Ac, Aland, CC, FD, FreEu, FreTox, HumTox, MarEu, MD, NLT, Ozone, Photo, PMF, Rad, TeTox, ULand, WD	1.05–3.07 kg CO_2eq. kg biomass^{-1}	D'Imporzano et al. (2018)

ß-Carotene	R	*Dunaliella salina*	$1.25-2.55 \times 10^5$ kg h^{-1}	Ac, Eu, FreTox, GWP, HumTox, TeTox	265–520 kg CO_2eq. kg ß-carotene^{-1}	Espada et al., 2019
Bioplastic	R, PBR, PBR + R	*Scenedesmus acutus*	12.5–25	GWP	−0.315 to 0.656 kg CO_2eq. kg bioplastic feedstock^{-1}	Beckstrom, Wilson, Crocker, and Quinn (2020)

[a] *R, raceway pond or high-rate algal pond (HRAP); PBR, photobioreactor.*
[b] *AbD, abiotic depletion; Ac, acidification; Aland, agricultural land occupation; AqTox; aquatic toxicity; Carci, carcinogen; CC, climate change; EcoQ, ecosystem quality; En, energy use; Eu, eutrophication; F, fossil fuel use; FD, fossil fuel depletion; FreEu, freshwater eutrophication; FreTox, freshwater toxicity; GWP, global warming potential; HumH, human health; HumTox, human toxicity; Land, land use; MarEu, marine eutrophication; MarTox, marine toxicity; MD, metals depletion; MinEx, mineral extraction; MinD, minerals resources depletion; Mo, metal ores; N, nuclear energy; NLT, natural land transformation; NonCarci; noncarcinogen; NOx, nitrogen oxides; NRen, nonrenewable energy; NRr, nonrenewable resources; Ozone, ozone depletion; Photo, photochemical oxidation; PMF, particulate matter formation; Rad, ionizing radiation; Rinorg, respiratory inorganics; Rorg, respiratory organics; TeAc, terrestrial acidification; TeTox, terrestrial toxicity; Uland, urban land occupation; W, water use; WD, water depletion; WR, water resources.*
[c] *The ranges listed represent the lowest and highest emissions reported during outdoor cultivations and without differentiation of the scenarios being studied.*

TABLE 13.3 Life cycle analyses of microalgae and bioproduct production combined with wastewater treatment.

Product	System[a]	Species	Productivity $g \cdot m^{-2} \cdot d^{-1}$	Approach	Environmental impact categories[b]	CO_2eq emissions[c]	References
Biomass	R		31.4	Cradle-to-gate	En, Eu, GWP, Land, W	−24,400 to 23,800 kg $CO_2eq \cdot 317\ GJ^{-1}$	Clarens, Resurreccion, White, and Colosi (2010)
Biodiesel	R			Well-to-pump	GWP	−20.9 to 135.7 kg CO_2eq. 1000 MJ algae $biodiesel^{-1}$	Sander and Murthy (2010)
Bioelectricity	R	Marine algae		Well-to-wheel	En, GWP, W	0.290–0.570 kg $CO_2eq \cdot$ vehicle km $traveled^{-1}$	Clarens, Nassau, Resurreccion, White, and Colosi (2011)
Bio-jet fuel	R			Cradle-to-wake	GWP	35.2–131.9 kg $CO_2eq \cdot GJ^{-1}$	Fortier, Roberts, Stagg-Williams, and Sturm (2014)
Biodiesel	R	*Scenedesmus obliquus*	30	Well-to-wheel	GWP	0.110–0.150 kg CO_2eq. MJ $fuel^{-1}$	Mata, Mendes, Caetano, and Martins (2014)
Biofuel: liquid, electric	PBR		15–34.6	Well-to-wheel	Eu, F, GWP, W	−7.50 to 1.50 kg $CO_2eq \cdot km^{-1}$	Mu et al. (2014)

Biofuel and animal feed	R, PBR	*Staurosira* sp.; *Desmodesmus* sp.	23	Cradle-to-gate	CC, EcoQ, HumH, NRr, WD		Beal et al. (2015)
Wastewater treatment	R	Microalgal-bacterial flocs	22	Cradle-to-gate	AbD, F, FreEu, Land, MinD, MarEu, Mo, N, WR	0.71–26.0 kg CO_2eq · kg MaB-floc TSS^{-1}	Sfez, Van Den Hende, Taelman, De Meester, and Dewulf (2015)
Oil	R	*Scenedesmus obliquus*	11	Cradle-to-gate	Aland, CC, FD, FreEu, FreTox, HumTox, MarEu, MarTox, MD, TeAc, TeTox, WD	1.00 kg CO_2eq · MJ^{-1}	Jez et al. (2017)
Biofuel	R		24	Well-to-wheel	Eu, F, GWP, NOx	0.28–0.36 kg CO_2 eq · MJ^{-1}	Mu et al. (2017)
Biofuel	R	*Chlorella sp.*	5–35	Cradle-to-gate	GWP	−2800 to 30,000 kg CO_2 eq · 50,700 MJ^{-1} $year^{-1}$	Roostaei and Zhang (2017)
Biodiesel	R	*Chlorella vulgaris*	25	Cradle-to-grave	Aland, CC, FD, FreEu, FreTox, HumTox, MarEu, MarTox, MD, NLT, Ozone, Photo, PMF, Rad, TeAc, TeTox, Uland, WD	0.306–1.64 kg CO_2eq. kg $biomass^{-1}$	Collotta, Champagne, Mabee, and Tomasoni (2018)
Biogas	R		18	Cradle-to-gate	AbD, Ac, GWP, Ozone, Photo, Eu	0.106–0.254 kg CO_2eq · m^{-3} of raw wastewater	Colzi Lopes, Valente, Iribarren, and González-Fernández (2018)

(*Continued*)

TABLE 13.3 (Continued)

Product	System[a]	Species	Productivity $g \cdot m^{-2} \cdot d^{-1}$	Approach	Environmental impact categories[b]	CO_2eq emissions[c]	References
Biogas or biofertilizer	R		12		CC, FD, FreEu, HumTox, MarEu, MD, Ozone, Photo, PMF, TeAc, TeTox	−0.190 to 0.285 kg $CO_2eq \cdot m^{-3}$ of treated WW	Arashiro et al. (2018)

[a] *R, raceway pond or high-rate algal pond (HRAP); PBR, photobioreactor.*
[b] *AbD, abiotic depletion; Ac, acidification; Aland, agricultural land occupation; AqTox; aquatic toxicity; Carci, carcinogen; CC, climate change; EcoQ, ecosystem quality; En, energy use; Eu, eutrophication; F, fossil fuel use; FD, fossil fuel depletion; FreEu, freshwater eutrophication; FreTox, freshwater toxicity; GWP, global warming potential; HumH, human health; HumTox, human toxicity; Land, land use; MarEu, marine eutrophication; MarTox, marine toxicity; MD, metals depletion; MinEx, mineral extraction; MinD, minerals resources depletion; Mo, metal ores; N, nuclear energy; NLT, natural land transformation; NonCarci; noncarcinogen; NOx, nitrogen oxides; NRen, nonrenewable energy; NRr, nonrenewable resources; Ozone, ozone depletion; Photo, photochemical oxidation; PMF, particulate matter formation; Rad, ionizing radiation; Rinorg, respiratory inorganics; Rorg, respiratory organics; TeAc, terrestrial acidification; TeTox, terrestrial toxicity; Uland, urban land occupation; W, water use; WD, water depletion; WR, water resources.*
[c] *The ranges listed represent the lowest and highest emissions reported during outdoor cultivations and without differentiation of the scenarios being studied.*

13.2 Life cycle assessment of microalgae cultivation

Life cycle assessment (LCA) aims to quantify the environmental impacts associated with a product or service. For the relatively "immature" microalgae biotechnology industry, LCAs are especially useful to benchmark a product/process against alternatives or identify "hot spots" that can be targeted to efficiently reduce environmental impacts. For example, Lardon, Hélias, Sialve, Steyer, and Bernard (2009) performed one of the earliest comprehensive LCAs of the production and combustion of microalgae biodiesel in order to identify "obstacles and limitations" to improve the sustainability of the process. Assuming a *Chlorella vulgaris* biomass productivity of 25 g dry biomass $\cdot$ m^{-2} $\cdot$ d^{-1}, their analysis evidenced that if microalgal cultivation would reduce land and pesticide use in comparison to conventional energetic crops, it would significantly contribute to the use of fertilizers. The authors concluded that, at its level of maturity, the production of algal biodiesel was not "environmentally competitive" in comparison to other crops, but environmental performance could be improved via the selection of species with high lipid and productivity yields or the recovery of bioproducts (e.g., energy and nutrients from oil cake).

As illustrated by the study described in Box 13.1, an LCA typically involves four main steps. First, the goal and scope must be defined (e.g., comparison versus hotspot identification), which is critical to determine the functional unit and system boundary. Data are then collected and mathematical models are developed to determine the amounts of water, energy, and other key resources needed to produce, process, transport materials, equipment, and products, including the wastes generated (the extent of this exercise is ring-fenced by the system boundary). This modeling enables to compute impact data for each selected impact categories. Finally, the output data can be normalized, weighted, and interpreted. Subjectivity cannot be avoided, especially when defining the functional unit and system boundary, when allocating effects (e.g., when the same process can have two different products or services such as, for example, treating wastewater and generating biofuel feedstock), and when normalizing data and weighting effects. The flexibility and inherent subjectivity of the LCA methodology generate a large variability in how the results from these studies are generated and interpreted. As can be seen in Table 13.1 focusing on LCAs of microalgae biodiesels, common system boundaries include cradle-to-grave or cradle-to-gate, and different production systems have been considered. Functional units vary from a "unit of freight transported using microalgae biodiesel" to a "unit of biomass or biodiesel generated at the gate." This variability reflects differences in goals as, for example, LCAs seeking to compare microalgae biofuels against alternative fuels are likely to use large boundaries and energy-based functional units, whereas LCAs seeking to compare different production systems for the same outputs can afford a narrower scope. These

BOX 13.1 Example of LCA focusing on microalgal biofuel production.

The study by Collet et al. (2011) illustrates a typical LCA of microalgal biofuel production based on the four main steps of: (1) definition of goal and scope; (2) life cycle inventory; (3) life cycle impact assessment; and (4) interpretation of results. The virtual system assessed was designed based on laboratory studies (Ras, Lardon, Bruno, Bernet, & Steyer, 2011) and included 100 ha of open ponds combined with a 3000 m^3 digester to grow and convert the microalgae biomass into biomethane.

Goal and scope: Collet et al. (2011) compared the environmental impacts generated by the production and use of 1 MJ of algae-based methane against algae-based biodiesel and traditional diesel. The functional unit for each impact was based on 1 MJ of energy produced. The "cradle-to-grave" system boundary included the construction, operation, and decommissioning of all facilities needed to produce, harvest, and anaerobically digest the algae, as well as the facilities needed for methane combustion. The extraction and shipping of resources (e.g., concrete, fertilizers, etc.) was also considered.

Life cycle inventory: Inventory data were largely derived from academic resources, communications with industrial producers, and inventories described in the Ecoinvent database. For example, harvesting was described as a two-step process including biomass settling and centrifugation and the settlers were designed based on laboratory data. Water evaporation losses were determined based on local climatic data.

Life cycle impact assessments: The impact assessment CML method developed by the Institute of Environmental Science (CML) at Leiden University, The Netherlands (Guinée, 2002), was used to assess the potential environmental impacts of the production of 1 MJ produced by burning algal fuel through an internal combustion engine. After grouping the different stages of the process into four categories (energy, infrastructure, combustion, and fertilizers), the authors analyzed their contributions toward nine impacts: abiotic depletion (Abd), potential acidification (Ac), eutrophication (Eu), global warming potential (GWP), human toxicity (HumTox), land competition (Land), ozone layer depletion (Ozone), photochemical oxidation (Photo), and ionizing radiation (Rad). In accordance with ISO guidelines, the authors used the substitution method to account for the impacts of by-products.

Interpretations of results and conclusions: When comparing to other fuels, Collet et al. (2011) found that algal methane was the worst performing fuel with regards to abiotic depletion (1.25 times higher than algal biodiesel), ionizing radiation (2.5 times higher than algal biodiesel), human toxicity (1.7 times higher than algal biodiesel), and global warming (1.4 times higher than algal biodiesel). Electricity consumption was the most important contributor to the impacts selected, representing more than 50% of the Abd, Ac, GWP, HumTox, Photo, and Rad impacts. Consequently, Collet et al. (2011) proposed that the facility should generate its own renewable electricity using, for example, solar panels or wind turbines. The authors also acknowledged that their study focused on a simplified process where methane was the only product, and that combining lipid recovery from biomass and methane production from biomass residues could improve economics and reduce environmental impacts.

differences are even more pronounced when looking at LCA studies of algal-based food products due to the large range of products considered (Table 13.2), and microalgae-based wastewater treatment as multiple products and services can be delivered (Table 13.3).

The lack of maturity of the microalgae industry causes considerable uncertainty in the estimation of biomass productivity, which magnifies the uncertainty in the effects assessed. While more realistic estimates are increasingly used (Moody, McGinty, & Quinn, 2014), many studies are still based on unverified high-productivities estimates, especially when growth is predicted in closed photobioreactors (PBRs) (Box 13.2). Uncertainty and high-productivity bias in turn generate a high uncertainty and positive bias in the conclusions of many studies. Due to the large variability in LCA

BOX 13.2 Closed photobioreactors versus raceway ponds.

The advantages and disadvantages of closed photobioreactors versus open ponds are well discussed in the literature, including in the context of environmental impact assessment (Jerney & Spilling, 2018; Ketzer et al., 2018; Tredici, 2010; Xu et al., 2009). On the positive side, closed systems reduce the risks of contamination, which is a clear advantage for food production, especially when the algae cultivated are slow growers. Closed systems also enable good control of temperature, pH, mass transfer (e.g., supply of CO_2 and removal of O_2), meaning near optimum nutrient conditions can thus be maintained. Closed systems prevent evaporation and can be designed to enable very high cell densities in the broth (due to efficient mixing), which can reduce downstream processing costs. Finally, vertical closed systems theoretically enable higher daytime productivities by reducing light limitation during peak hours. On the negative side, these systems are prone to fouling, which reduces light transmission efficiency, and can experience very high temperature (due the lack of "natural" evaporating cooling and the high area/volume geometry). In addition, high-density cultivation (at high cell retention times) increase the relative impact of biomass losses at night due to respiration. This respiration often offsets the gain in daytime productivity.

In practice, the considerable costs of constructing and operating closed systems (especially to prevent fouling and control temperature) can only be justified for the production of high-value algae at small-scale (Jerney & Spilling, 2018). In their analysis of the energy life-cycle for the production of oil-rich *Nannochloropsis* sp. biomass, Jorquera, Kiperstok, Sales, Embiruçu, and Ghirardi (2010) determined that the cultivation of *Nannochloropsis* sp. in horizontal tubular photobioreactors would not be economically feasible based on its net energy ratio of 0.07 for oil production and 0.20 for biomass production.

With regards to sustainability, both Resurreccion et al. (2012) and Sander and Murthy (2010) concluded that algae cultivation in closed photobioreactors would be less sustainable than cultivation in open systems. Stephenson et al. (2010) specifically reported that the global warming potential associated with the cultivation of *Chlorella vulgaris* would be 16 times higher in photobioreactors than in open ponds.

frameworks used, the inherent subjectivity of these exercises, and the high uncertainty associated with the results, it is difficult to extract overarching conclusions from the literature. Focusing on the cultivation step, several high-level conclusions can nevertheless be drawn from past LCAs of microalgae cultivation (see references in Tables 13.1–13.3):

- The main environmental impacts are associated with energy and fertilizers (nutrients) use;
- Outdoor cultivation is more sustainable in raceway ponds than in PBRs (Box 13.2);
- Using wastewater to supply water and nutrients reduces environmental impacts but has major limitations (Box 13.3);
- The need for new buildings and equipment (i.e., piping, etc.) generates significant environmental impacts;
- Various authors have acknowledged that microalgae cultivation is not sustainable for biofuel production and proposed a biorefinery approach (e.g., covalorization of the sludge after oil extraction). However, there are no full-scale systems in operation yet;
- Although commercial microalgae cultivation is currently almost exclusively focusing on food products for human consumption (e.g., *Arthrospira platensis*, *C. vulgaris*, *Dunaliella salina*, *Haematococcus pluvialis*), most past LCAs have focused on biofuels and/or microalgae-based wastewater treatment.

Various researchers have proposed guidelines to normalize LCAs and facilitate comparison between studies (Bradley, Maga, & Antón, 2015; Collet, Hélias, Lardon, Steyer, & Bernard, 2015; Ketzer et al., 2018). As suggested by Ketzer et al. (2018), it is especially critical to consider impacts such as land, energy, and water use in the current context of climate change and natural resources depletion. Because the use of closed PBRs is infrequent and only realistically feasible for the small-scale production of high-value metabolites (Box 13.2), the following section focuses on microalgae cultivation in open ponds, as done during the commercial cultivation of the four species named above.

13.3 Land use and location

High estimates of areal productivity (e.g., >20 g dry biomass $\cdot m^{-2} \cdot d^{-1}$, Tables 13.1–13.3) have often been used in past LCAs of microalgae biotechnologies. However, full-scale long-term productivities from the mature *A. platensis* (*Spirulina*) industry are around 10–20 t dry biomass $\cdot ha^{-1}$ $year^{-1}$ in open raceways (Vonshak, 1997). The upper value of this range corresponds to an average productivity of 6.6 g dry biomass $\cdot m^{-2} \cdot d^{-1}$ over 300 days of production each year, which requires efficient pond management. Higher values may be obtained by using other species or improving

BOX 13.3 Algae cultivation during wastewater treatment.

Given the costs and environmental issues associated with the use of water and fertilizers during algae cultivation, there is considerable interest in combining algae-based wastewater treatment with algae cultivation (see Table 13.3 for references). From an LCA perspective, this approach assumes that a large fraction of the environmental impact associated with "conventional" algae cultivation can be allocated to the wastewater treatment function. Caution is nevertheless required as algae-based wastewater treatment cannot be directly compared to conventional wastewater treatment in terms of, in particular, treatment efficiency or land and energy use (as further explained below).

In this process, algal photosynthesis provides the oxygen required by heterotrophic bacteria to aerobically biodegrade organic pollutants (and even oxidize ammonia via nitrification), which provides great savings as oxygen supply via mechanical aeration can consume 0.14–1.08 kWh of electricity per m^3 of water treated (Tchobanoglous, Burton, Stensel, & Metcalf & Eddy, Inc., 2003). The additional phototrophic biomass generated from CO_2 fixation increases nutrient removal capacity (mainly by assimilation into biomass) and chemical energy output as biomass. To illustrate this potential, Posadas et al. (2017) described a case study for the treatment of primary-settled domestic wastewater using raceway ponds. Assuming a yearly average algal biomass productivity of $20\,g \cdot m^{-2} \cdot d^{-1}$, the authors estimated land requirement at 6.5 m^2 per capita.

The availability of affordable land near the wastewater source can however be limited (Roostaei & Zhang, 2017) and the biomass generated is poorly suitable for biodiesel generation due its low and/or variable lipid content (caused by the inherent variability of wastewater flow and characteristics). In addition, the presence of pathogens or hazardous pollutants can prevent biomass valorization as food or even fertilizers (as the application of biosolids containing heavy metals or harmful microorganisms onto land is highly constrained). Consequently, many researchers now propose that biomass should be anaerobically digested, and that the biogas thus generated could even be purified using algal photosynthesis (to remove the CO_2 from the biogas). Unfortunately, biomass digestion causes the release of nutrients into aqueous streams that must then be treated or disposed of (Posadas et al., 2017), meaning that it may be more efficient to simply dry and burn the biomass for onsite energy production. Finally, the microalgae-based wastewater treatment must meet increasingly stringent discharge requirements including for pathogens and micropollutants, although recent research in this area shows interesting potential (Chambonnière, 2019; Norvill, Shilton, & Guieysse, 2016).

These practical issues may explain the limited number of full-scale applications currently operating (Hernández, Riaño, Coca, & García-González, 2013; Peng & Colosi, 2015). Regardless, the considerable environmental credential of wastewater-based microalgae cultivation drives extensive research currently focusing on developing new biorefinery capacity using microalgae (Javed et al., 2019; Pancha, Chokshi, & Mishra, 2019).

pond management, design, and location and, for example, Moody et al. (2014) predicted maximal annual average rates of 13–15 g dry biomass $\cdot m^{-2} \cdot d^{-1}$ (47–55 t dry biomass $\cdot ha^{-1}$ $year^{-1}$ over full-year operation) for *Nannochloropsis* grown in a PBR submerged in a pool to control temperature. There are however many practical limitations to how fast microalgae can grow in light-limited full-scale systems exposed to variable environmental conditions (Béchet, Shilton, & Guieysse, 2013). Exposure to suboptimal temperatures is a typical example of such limitation because temperature control is costly and requires energy and/or water (see Section 13.5). High productivities are thus often forecast under "sunny" climatic conditions that also generate high risks of temperature inhibition and are poorly suitable for water-cooling (Béchet et al., 2017; Béchet, Shilton, & Guieysse, 2016). The trade-offs between climate, temperature risks, and water use are extensively discussed by Béchet et al. (2016), who provide models and methodology for optimizing location, design, and operation based on climatic conditions and microalgal biology (light response and temperature dependence). Considerable gain in land use efficiency can thus be achieved through intelligent dynamic operation of mean cell retention time and water depth (Béchet et al., 2016).

Conservative estimates of microalgal productivity (e.g., 47–55 t dry biomass $\cdot ha^{-1}$ $year^{-1}$ over full-year operation) evidence efficient land use compared to even the most productive terrestrial crops (e.g., up to 73–87 t dry biomass $\cdot ha^{-1}$ $year^{-1}$ for sugarcane; Dismukes, Carrieri, Bennette, Ananyev, & Posewitz, 2008). It is also possible to harvest and use the entire microalgal biomass photosynthesized, which further improves land-use efficiency for food production. For example, productivities up to 19 t dry biomass $\cdot ha^{-1}$ $year^{-1}$ are reportedly achieved by Earthrise Nutritionals (United States) during commercial *Spirulina* production as whole-cell food (Vonshak, 1997), against 2.9–3.2 t dry biomass $\cdot ha^{-1}$ $year^{-1}$ for soybean seed (Matsuo, Yamada, Takada, Fukami, & Hajika, 2018), and of 7 t dry biomass $\cdot ha^{-1}$ $year^{-1}$ for corn grain (Dismukes et al., 2008). Microalgae cultivation also does not require arable land and, as such, it does not necessarily compete with "traditional" food production. Microalgae cultivation nevertheless still impacts land use because even marginal lands have ecological and economical functions. In addition, auxiliary land use (for inoculum and safety factors) can be significant.

13.4 Energy use

Nearly all commercial large-scale microalgae cultures are mixed to optimize cell exposure to light in dense (i.e., opaque) cultures. In raceway ponds, this mixing is typically provided using paddle wheels circulating the fluid at an average horizontal velocity of around $0.2\ m \cdot s^{-1}$. As a typical example, Béchet et al. (2017) estimated this mixing requirement to $0.083\ W\ m^{-2}$, or

2.6 MJ m^{-2} $year^{-1}$ for a 0.25 m deep (water depth) lined pond yielding 4.2 kg dry biomass $\cdot m^{-2} \cdot yr^{-1}$ (average cell retention time of 10 days, paddle wheel efficiency of 20%). Assuming a biomass heat value of 21 MJ kg^{-1}, the energy required for mixing represents only 3% of the chemical energy output. This energy can also be supplied using "clean" electricity rather than fossil fuels (as often needed to operate heavy machinery in traditional farming).

Energy is also needed for harvesting and processing, and this energy demand can be high depending on the final product delivered. For example, drying a microalgae "cake" containing 20% solids (which can be obtained by a combination of pressing and filtration, or by centrifugation) requires evaporating 4 kg of water per kg of final dry products (assuming the final moisture content is negligible). At an energy efficiency of around 50% for spray drying, this would represent nearly 90% of the chemical energy output as dry biomass in the example described above, although dryers that are more energy efficient can be used. While large facilities would typically use electricity onsite, there are examples of small *Spirulina* farms working exclusively on off-grid solar power for electricity and heat (drying) generation, with minimal additional land use. Large-scale dryers can be powered using fossil fuels, but this reduces environmental performance. Electricity-driven modern dryers and evaporators with high-energy efficiencies should therefore be preferred, but only if electricity is itself produced from renewable sources.

Energy is finally required to produce, transport, and use the chemicals, materials, and equipment needed to cultivate, process, and deliver the microalgae products to final consumers (including all necessary steps of waste management). This demand is highly case-specific and can be computed based on available databases and established methodologies for the system boundary selected. Energy requirements for supplying inputs and final algal products are fundamentally not different to those for other industrial bioprocesses.

13.5 Freshwater use and freshwater footprint

During large-scale microalgae cultivation in open ponds, freshwater is mainly used to prepare and refresh the culture medium and compensate for evaporation losses (leaks from well-lined ponds can be neglected). As in the case of energy use, water is used during the production, processing, and transport of chemicals, energy, materials, and equipment needed to produce, process, and deliver the final product to the end-user. The computation of this indirect off-site water use is case specific and not discussed further.

Evaporation from free surface ponds can be estimated using mathematical models (Béchet, Sialve, Steyer, Shilton, & Guieysse, 2018) and was predicted to range from 0.45 m^3 m^{-2} $year^{-1}$ under tropical climatic conditions, up to 2.1 m^3 m^{-2} $year^{-1}$ under arid climatic conditions, during *C. vulgaris* cultivation in a 100 m^2 pond (Béchet et al., 2017). This demand can be reduced by operating ponds inside greenhouses but this will reduce evaporative-cooling and, therefore, increase the risks of culture crash at high temperature.

The amount of freshwater used during processing depends on the level of water recycling following harvest and there is a lack of data on what is realistically achievable (Guieysse, Béchet, & Shilton, 2013). Complete recycling without treatment can entail significant productivity losses while treating water prior to recycling can be costly and have additional environmental impacts. In the example described above, Béchet et al. (2017) estimated process water demand (i.e., excluding evaporation) as 10% of the amount of water processed during each harvest cycle (e.g., 90% water recycle efficiency), which amounted to approximately 0.9 m^3 m^{-2} $year^{-1}$ in their base-case scenario earlier detailed (temperate weather). This corresponds to a daily refreshment rate of 1% of the total pond volume, a value recommended during *Spirulina* cultivation (Jourdan, 2018). Based on Béchet et al. (2017), predictions for evaporation and *C. vulgaris* productivity in 0.25 m ponds operated at 10 d HRT in central California of 1.2 m^3 m^{-2} $year^{-1}$ and 4.2 kg dry biomass $\cdot$ $m^{-2} \cdot d^{-1}$, respectively, the combined freshwater demand for evaporation and process water can thus be estimated to around 0.5 m^3 kg dry $biomass^{-1}$, or a net demand of 0.46 m^3 kg dry $biomass^{-1}$ when accounting for rainwater precipitation in the pond (see Béchet et al., 2017 for details). This water demand is similar to the freshwater demand for corn grain (0.56 m^3 kg dry $biomass^{-1}$, Dismukes et al., 2008). For *Spirulina* commercially grown in southern California (1.9 kg dry biomass $\cdot$ $m^{-2} \cdot d^{-1}$), the freshwater demand can be estimated at 1.6 m^3 kg dry $biomass^{-1}$ based on the evaporation rate of 2.1 m^3 m^{-2} $year^{-1}$ predicted by Béchet et al. (2017) under similar climatic conditions. While it may be difficult to practically increase the process water recycling above 90%, the net freshwater demand can be significantly reduced by actively controlling pond depth and HRT over time (Béchet et al., 2016), or selecting a better climatic location (Béchet et al., 2017). A considerable reduction in freshwater demand can also be achieved by cultivating microalgae in seawater and brackish water (the specific case of wastewater is discussed in Box 13.3).

In the literature, the water footprint (WF) is generally computed as the sum of the green, blue, and gray WFs (Chapagain & Hoekstra, 2011). Based on the Water Footprint Network (2020), the green WF is "water from precipitation that is stored in the root zone of the soil and evaporated, transpired or incorporated by plants," the blue WF is "water that has been sourced from surface or groundwater resources and is either evaporated, incorporated into a product or taken from one body of water and returned to another, or returned at a different time," and the gray WF is "the amount of fresh water required to assimilate pollutants to meet specific water quality standards." In the case of microalgae cultivation, evaporation losses can be considered as a combination of green and blue WF, as rainwater is often not enough to sustain these losses. The gray WF is very case specific as it will depend on the amount and composition of the process water discharged, but also local regulations and the sensitivity of the recipient ecosystem. Consequently, the economic and ecological value of water is highly dependent on location, and caution is required when using the WF as a metric to compare the impact of freshwater use at different locations. To illustrate this point, Guieysse et al. (2013) predicted that sustaining the production of 1 GJ of biomass energy per m^2 of pond area would require a yearly freshwater demand equivalent to 602 years of rainfall in an arid climate, against 8.7 years in a tropical climate despite the WF being slightly lower under the arid climate due to higher biomass productivity.

13.6 Carbon footprint

The production of microalgal biomass is generally asserted to be carbon neutral because atmospheric CO_2 is recycled during photosynthesis (Ahmad et al., 2011). However, carbon neutrality cannot be easily achieved considering the materials and energy needed to manage wastes and grow, process, and transport the microalgal biomass. Energy consumption represents one of the most important contributors to the carbon footprint of microalgae production systems (e.g., see Collet et al., 2011; Jez et al., 2017; Lardon et al., 2009; Resurreccion, Colosi, White, & Clarens, 2012). Indirect CO_2 emissions (caused by the use of material and fossil fuels to generate the electricity used on site) vary greatly between countries (Béchet et al., 2017). In addition, the greenhouse gases nitrous oxide (N_2O) and methane (CH_4) can be emitted during the production of microalgae. These gases have global warming potential (GWP) 298 and 34 times greater than CO_2 (Myhre et al., 2013), so their release can increase the impact of algae cultivation on climate change.

13.6.1 N_2O emissions

While indirect "off-site" N_2O emissions related to, for example, fertilizer production and transport are included in the global warming impact categories of LCAs, few LCAs have considered potential N_2O emissions directly associated with microalgae cultivation (Bauer, Grotz, Connelly, & Colosi, 2016; D'Imporzano, Veronesi, Salati, & Adani, 2018). Bauer et al. (2016) specifically focused on these emissions and categorized "low" and "high" N_2O flux reported in the literature. These authors then estimated that while a "low" N_2O production of 0.72 $\mu g\ N_2O \cdot g$ microalgae^{-1} day^{-1} would increase the baseline value of the GWP of microalgae biodiesel (41.3 g $CO_2eq \cdot MJ^{-1}$) by less than 1%, a "high" N_2O production of 530 $\mu g\ N_2O \cdot g$ microalgae^{-1} day^{-1} would increase this value by 25%. There are, however, very few data to support such an assessment and as preliminary data, Plouviez, Shilton, Packer, and Guieysse (2017) reported that 0.1%–0.4% of the N load could be emitted as N_2O-N during *C. vulgaris* cultivation in outdoor 50 L PBRs with supplied nitrate as the N source. In a similar approach, Plouviez, Chambonnière, Shilton, Packer, and Guieysse (2019) documented that during microalgae-based wastewater treatment in a 1 m^3 high-rate algal pond (HRAP), N_2O emissions could represent 0.13%–0.57% of the influent N load. Indirect N_2O emissions may also occur via ammonia volatilization at high pH, as the subsequent redeposition of this ammonia onto land can trigger N_2O synthesis via nitrification (Bartram et al., 2019), or when wastewater is discharged (as the ammonia or nitrate present can trigger N_2O emissions via nitrification or denitrification, respectively; Bartram et al., 2019). Plouviez et al. (2019) thus showed that indirect N_2O emissions (0.10%–0.85% of the influent N load) during microalgae-based wastewater treatment in a HRAP could significantly increase the total N_2O emissions from the system. Consequently, both direct and indirect N_2O emissions should be considered during the environmental assessments of microalgae cultivation.

13.6.2 CH_4 emissions

Ferron, Ho, Johnson, and Huntley (2012) measured positive CH_4 emissions of $7.4 - 31.5\ \mu mol\ m^{-2}\ d^{-1}$ (average of $19.9 \pm 5.6\ \mu mol\ m^{-2}\ d^{-1}$) in a 24 h period during the cultivation of the marine diatom *Staurosira* sp. To the best of our knowledge, this study is the only one that has documented emissions of the greenhouse gas CH_4 during microalgae cultivation. This pioneer study evidences the need to further evaluate the long-term occurrence of CH_4 emissions during microalgal cultivation.

13.7 Nutrient use and eutrophication

In comparison to many alternatives, cultivating microalgae in ponds enables efficient nutrient use because the whole cell can be harvested and used. In addition, there is little risk of nutrient losses via runoff, which also prevents eutrophication. The process water generated must however be properly managed to prevent eutrophication issues via improper discharge, and this waste management will also generate impacts (e.g., energy use).

13.8 Conclusions

Despite considerable variability and the lack of long-term full-scale data, microalgae cultivation for food production generally appears to be more sustainable than terrestrial crop farming in terms of land use efficiency and, when process water is properly managed, eutrophication potential.

Microalgae can also be cultivated on marginal land (i.e., areas of low natural plant productivity) but this practice can have a high impact on stressed local freshwater resources in arid climates. These environmental credentials can be considerably improved if seawater or brackish water can be used as process water for cultivation. The sustainability of microalgae cultivation for energy production can be challenged with regards to the use of fertilizers, freshwater, and energy. There is therefore considerable interest in microalgae-based wastewater treatment for enhanced biogas generation, but this entails additional costs and issues (Posadas, Muñoz, & Guieysse, 2017) for a low-value fuel career that can readily be produced from massive amounts of readily available green wastes (Shilton & Guieysse, 2010). It is however critical to remember that no truly generic conclusions can be made and that a thorough LCA must still be conducted for each new microalgae-based application.

Acknowledgment

This work was supported by Massey University, New Zealand.

References

Ahmad, A. L., Yasin, N. H. M., Derek, C. J. C., & Lim, J. K. (2011). Microalgae as a sustainable energy source for biodiesel production : A review. *Renewable and Sustainable Energy Reviews*, *15*, 584–593.

Arashiro, L. T., Montero, N., Ferrer, I., Acién, F. G., Gómez, C., & Garfí, M. (2018). Life cycle assessment of high rate algal ponds for wastewater treatment and resource recovery. *The Science of the Total Environment*, *622–623*, 1118–1130.

Bartram D., Short M.D., Ebie Y., Farkaš J., Gueguen C., Peters G.M., ... Masuda, S. (2019). Wastewater treatment and discharge. In *2019 refinement to the 2006 IPCC guidelines for national greenhouse gas inventories*.

Batan, L., Quinn, J., Willson, B., & Bradley, T. (2010). Net energy and greenhouse gas emission evaluation of biodiesel derived from microalgae. *Environmental Science & Technology*, *44* (20), 7975–7980.

Bauer, S. K., Grotz, L. S., Connelly, E. B., & Colosi, L. M. (2016). Reevaluation of the global warming impacts of algae-derived biofuels to account for possible contributions of nitrous oxide. *Bioresource Technology*, *218*, 196–201.

Beal, C. M., Gerber, L. N., Sills, D. L., Huntley, M. E., Machesky, S. C., Walsh, M. J., ... Greene, C. H. (2015). Algal biofuel production for fuels and feed in a 100-Ha facility: A comprehensive techno-economic analysis and life cycle assessment. *Algal Research*, *10*, 266–279.

Béchet, Q., Plouviez, M., Chambonnière, P., & Guieysse, B. (2017). Environmental impacts of full-scale algae cultivation. *Microalgae-based biofuels bioproduct from feed cultivation to end-products*, 505–525.

Béchet, Q., Shilton, A., & Guieysse, B. (2013). Modeling the effects of light and temperature on algae growth: State of the art and critical assessment for productivity prediction during outdoor cultivation. *Biotechnology Advances*, *31*(8), 1648–1663.

Béchet, Q., Shilton, A., & Guieysse, B. (2016). Maximizing productivity and reducing environmental impacts of full-scale algal production through optimization of open pond depth and hydraulic retention time. *Environmental Science & Technology*, *50*(7), 4102–4110.

Béchet, Q., Sialve, B., Steyer, J. P., Shilton, A., & Guieysse, B. (2018). Comparative assessment of evaporation models in algal ponds. *Algal Research*, *35*(August), 283–291.

Beckstrom, B. D., Wilson, M. H., Crocker, M., & Quinn, J. C. (2020). Bioplastic feedstock production from microalgae with fuel co-products: A techno-economic and life cycle impact assessment. *Algal Research*, *46*(January), 101769.

Bennion, E. P., Ginosar, D. M., Moses, J., Agblevor, F., & Quinn, J. C. (2015). Lifecycle assessment of microalgae to biofuel: Comparison of thermochemical processing pathways. *Applied Energy*, *154*, 1062–1071.

Bradley, T., Maga, D., & Antón, S. (2015). Unified approach to life cycle assessment between three unique algae biofuel facilities. *Applied Energy*, *154*, 1052–1061.

Campbell, P. K., Beer, T., & Batten, D. (2011). Life cycle assessment of biodiesel production from microalgae in ponds. *Bioresource Technology*, *102*(1), 50–56.

Chambonnière, P. (2019). *Understanding the mechanisms involved in Escherichia coli decay during wastewater treatment in high rate algal ponds* (Ph.D. thesis). Massey University.

Chapagain, A. K., & Hoekstra, A. Y. (2011). The blue, green and grey water footprint of rice from production and consumption perspectives. *Ecological Economics: The Journal of the International Society for Ecological Economics*, *70*(4), 749–758.

Chaudry, S., Bahri, P. A., & Moheimani, N. R. (2019). Life cycle analysis of milking of microalgae for renewable hydrocarbon production. *Computers & Chemical Engineering*, *121*, 510–522.

Clarens, A. F., Nassau, H., Resurreccion, E. P., White, M. A., & Colosi, L. M. (2011). Environmental impacts of algae-derived biodiesel and bioelectricity for transportation. *Environmental Science & Technology*, *45*(17), 7554–7560.

Clarens, A. F., Resurreccion, E. P., White, M. A., & Colosi, L. M. (2010). Environmental life cycle comparison of algae to other bioenergy feedstocks. *Environmental Science & Technology*, *45*(2), 834.

Collet, P., Hélias Arnaud, A., Lardon, L., Ras, M., Goy, R. A., & Steyer, J. P. (2011). Life-cycle assessment of microalgae culture coupled to biogas production. *Bioresource Technology*, *102*(1), 207–214.

Collet, P., Hélias, A., Lardon, L., Steyer, J. P., & Bernard, O. (2015). Recommendations for life cycle assessment of algal fuels. *Applied Energy*, *154*, 1089–1102.

Collotta, M., Champagne, P., Mabee, W., & Tomasoni, G. (2018). Wastewater and waste CO_2 for sustainable biofuels from microalgae. *Algal Research*, *29*(November 2016), 12–21.

Colzi Lopes, A., Valente, A., Iribarren, D., & González-Fernández, C. (2018). Energy balance and life cycle assessment of a microalgae-based wastewater treatment plant: A focus on alternative biogas uses. *Bioresource Technology*, *270*(July), 138–146.

D'Imporzano, G., Veronesi, D., Salati, S., & Adani, F. (2018). Carbon and nutrient recovery in the cultivation of *Chlorella vulgaris*: A life cycle assessment approach to comparing environmental performance. *Journal of Cleaner Production*, *194*, 685–694.

Dismukes, G. C., Carrieri, D., Bennette, N., Ananyev, G. M., & Posewitz, M. C. (2008). Aquatic phototrophs: Efficient alternatives to land-based crops for biofuels. *Current Opinion in Biotechnology*, *19*(3), 235–240.

Dutta, S., Neto, F., & Coelho, M. C. (2016). Microalgae biofuels: A comparative study on techno-economic analysis & life-cycle assessment. *Algal Research*, *20*, 44–52.

Espada, J. J., Pérez-Antolín, D., Vicente, G., Bautista, L. F., Morales, V., & Rodríguez, R. (2019). Environmental and techno-economic evaluation of β-carotene production from *Dunaliella Salina*. A biorefinery approach. *Biofuels, Bioproducts and Biorefining*, *14*(1), 43–54.

Ferron, S., Ho, D. T., Johnson, Z. I., & Huntley, M. E. (2012). Air – water fluxes of N_2O and CH_4 during microalgae (*Staurosira Sp.*) cultivation in an open raceway pond. *Environmental Science & Technology*, *46*(19), 10842–10848.

Fortier, M. O. P., Roberts, G. W., Stagg-Williams, S. M., & Sturm, B. S. M. (2014). Life cycle assessment of bio-jet fuel from hydrothermal liquefaction of microalgae. *Applied Energy*, *122*(July 2011), 73–82.

Foteinis, S., Antoniadis-Gavriil, A., & Tsoutsos, T. (2018). Life cycle assessment of algae-to-biodiesel shallow pond production systems in the Mediterranean: Influence of species, pond type, by(co)-product valorisation and electricity mix. *Biofuels, Bioproducts and Biorefining*, *12*(4), 542–558.

Guieysse, B., Béchet, Q., & Shilton, A. (2013). Variability and uncertainty in water demand and water footprint assessments of fresh algae cultivation based on case studies from five climatic regions. *Bioresource Technology*, *128*, 317–323.

Guinée, J. B. (2002). *Handbook on life cycle assessment: Operational guide to the ISO standards*. New York: Springer.

Hernández, D., Riaño, B., Coca, M., & García-González, M. C. (2013). Treatment of agro-industrial wastewater using microalgae-bacteria consortium combined with anaerobic digestion of the produced biomass. *Bioresource Technology*, *135*, 598–603.

Javed F., Aslam M., Rashid, N. Shamair, Z., Khan, A. L., Yasin, M. et al. (2019) Microalgae-based biofuels, resource recovery and wastewater treatment: A pathway towards sustainable biorefinery. Fuel Nov:255:e 115826

Jerney, J., & Spilling, K. (2018). Large scale cultivation of microalgae: Open and closed systems. In K. Spilling (Ed.), *Biofuels from algae. Methods in molecular biology* (vol 1980, pp. 1–8). New York, NY: Humana.

Jez, S., Spinelli, D., Fierro, A., Dibenedetto, A., Aresta, M., Busi, E., & Basosi, R. (2017). Comparative life cycle assessment study on environmental impact of oil production from micro-algae and terrestrial oilseed crops. *Bioresource Technology*, *239*, 266–275.

Jorquera, O., Kiperstok, A., Sales, E. A., Embiruçu, M., & Ghirardi, M. L. (2010). Comparative energy life-cycle analyses of microalgal biomass production in open ponds and photobioreactors. *Bioresource Technology*, *101*(4), 1406–1413.

Jourdan, J.P. (2018). *Grow your own Spirulina*. http://spirulinefrance.free.fr/Resources/Manuel%20du%202%20fevrier%202018.pdf

Kant, P., & Wu, S. (2011). The extraordinary collapse of Jatropha as a global biofuel. *Environmental Science & Technology*, *45*(17), 7114–7115.

Ketzer, F., Skarka, J., & Rösch, C. (2018). Critical review of microalgae LCA studies for bioenergy production. *BioEnergy Research*, *11*(1), 95–105.

Lardon, L., Hélias, A., Sialve, B., Steyer, J. P., & Bernard, O. (2009). Life-cycle assessment of biodiesel production from microalgae. *Environmental Science & Technology*, *43*(17), 6475–6481.

Mata, T. M., Mendes, A. M., Caetano, N. S., & Martins, A. A. (2014). Sustainability and economic evaluation of microalgae grown in brewery wastewater. *Bioresource Technology*, *168*, 151–158.

Matsuo, N., Yamada, T., Takada, Y., Fukami, K., & Hajika, M. (2018). Effect of plant density on growth and yield of new soybean genotypes grown under early planting condition in southwestern Japan. *Plant Production Science*, *21*(1), 16–25.

Mediboyina, M. K., Banuvalli, B. K., Chauhan, V. S., & Mudliar, S. N. (2020). Comparative life cycle assessment of autotrophic cultivation of *Scenedesmus dimorphus* in raceway pond coupled to biodiesel and biogas production. *Bioprocess and Biosystems Engineering*, *43*(2), 233–247.

Moody, J. W., McGinty, C. M., & Quinn, J. C. (2014). Global evaluation of biofuel potential from microalgae. *Proceedings of the National Academy of Sciences of the United States of America*, *111*(23), 8691–8696.

Mu, D., Min, M., Krohn, B., Mullins, K. A., Ruan, R., & Hill, J. (2014). Life cycle environmental impacts of wastewater-based algal biofuels. *Environmental Science & Technology*, *48*(19), 11696–11704.

Mu, D., Ruan, R., Addy, M., Mack, S., Chen, P., & Zhou, Y. (2017). Life cycle assessment and nutrient analysis of various processing pathways in algal biofuel production. *Bioresource Technology*, *230*, 33–42.

Myhre, G., Shindell, D., Bréon, F.-M., Collins, W., Fuglestvedt, J., Huang, J., ... Mendoza, B. (2013). Anthropogenic and natural radiative forcing. In *Climate change 2013: The physical science basis. Contribution of Working Group I to the Fifth Assessment Report of the Intergovernmental Panel on Climate Change* (pp. 659–740). Cambridge, United Kingdom and New York, NY, USA: Cambridge University Press.

Norvill, Z. N., Shilton, A., & Guieysse, B. (2016). Emerging contaminant degradation and removal in algal wastewater treatment ponds: Identifying the research gaps. *Journal of Hazardous Materials*, *313*, 291–309.

Pancha I., Chokshi K., & Mishra S. (2019) Industrial wastewater-based microalgal biorefinery: A dual strategy to remediate waste and produce microalgal bioproducts. In: Gupta S., Bux F. (eds) Application of Microalgae in Wastewater Treatment. Springer, Cham.

Papadaki, S., Kyriakopoulou, K., Tzovenis, I., & Krokida, M. (2017). Environmental impact of phycocyanin recovery from *Spirulina platensis* cyanobacterium. *Innovative Food Science and Emerging Technologies*, *44*, 217–223.

Peng, S., & Colosi, L. M. (2015). Anaerobic digestion of algae biomass to produce energy during wastewater treatment. *Water Environment Research: A Research Publication of the Water Environment Federation*, *88*(1), 29–39.

Pérez-López, P., González-García, S., Jeffryes, C., Agathos, S. N., McHugh, E., Walsh, D., ... Moreira, M. T. (2014). Life cycle assessment of the production of the red antioxidant carotenoid astaxanthin by microalgae: From lab to pilot scale. *Journal of Cleaner Production*, *64*, 332–344.

Pérez-López, P., González-García, S., Ulloa, R. G., Sineiro, J., Feijoo, G., & Moreira, M. T. (2014). Life cycle assessment of the production of bioactive compounds from *Tetraselmis suecica* at pilot scale. *Journal of Cleaner Production*, *64*, 323–331.

Plouviez, M., Chambonnière, P., Shilton, A., Packer, M. A., & Guieysse, B. (2019). Nitrous oxide (N_2O) emissions during real domestic wastewater treatment in an outdoor pilot-scale high rate algae pond. *Algal Research*, *44*, 101670.

Plouviez, M., Shilton, A., Packer, M. A., & Guieysse, B. (2017). N_2O emissions during microalgae outdoor cultivation in 50 L column photobioreactors. *Algal Research*, *26*, 348–355.

Posadas, E., Muñoz, R., & Guieysse, B. (2017). Integrating nutrient removal and solid management restricts the feasibility of algal biofuel generation via wastewater treatment. *Algal Research*, *22*, 39–46.

Quinn, J. C., Smith, T. G., Downes, C. M., & Quinn, C. (2014). Microalgae to biofuels lifecycle assessment—Multiple pathway evaluation. *Algal Research*, *4*(1), 116–122.

Ras, M., Lardon, L., Bruno, S., Bernet, N., & Steyer, J. P. (2011). Experimental study on a coupled process of production and anaerobic digestion of *Chlorella vulgaris*. *Bioresource Technology*, *102*(1), 200–206.

Resurreccion, E. P., Colosi, L. M., White, M. A., & Clarens, A. F. (2012). Comparison of algae cultivation methods for bioenergy production using a combined life cycle assessment and life cycle costing approach. *Bioresource Technology, 126*, 298–306.

Roostaei, J., & Zhang, Y. (2017). Spatially explicit life cycle assessment: Opportunities and challenges of wastewater-based algal biofuels in the United States. *Algal Research, 24*, 395–402.

Sander, K., & Murthy, G. S. (2010). Life cycle analysis of algae biodiesel. *The International Journal of Life Cycle Assessment, 15*(7), 704–714.

Sevigné Itoiz, E., Fuentes-Grünewald, C., Gasol, C. M., Garcés, E., Alacid, E., Rossi, S., & Rieradevall, J. (2012). Energy balance and environmental impact analysis of marine microalgal biomass production for biodiesel generation in a photobioreactor pilot plant. *Biomass and Bioenergy, 39*, 324–335.

Sfez, S., Van Den Hende, S., Taelman, S. E., De Meester, S., & Dewulf, J. (2015). Environmental sustainability assessment of a microalgae raceway pond treating aquaculture wastewater: From up-scaling to system integration. *Bioresource Technology, 190*, 321–331.

Shilton, A., & Guieysse, B. (2010). Sustainable sunlight to biogas is via marginal organics. *Current Opinion in Biotechnology, 21*(3), 287–291.

Smetana, S., Sandmann, M., Rohn, S., Pleissner, D., & Heinz, V. (2017). Autotrophic and heterotrophic microalgae and cyanobacteria cultivation for food and feed: Life cycle assessment. *Bioresource Technology, 245*, 162–170. (August).

Stephenson, A. L., Kazamia, E., Dennis, J. S., Howe, C. J., Scott, S. A., & Smith, A. G. (2010). Life-cycle assessment of potential algal biodiesel production in the United Kingdom: A comparison of raceways and air-lift tubular bioreactors. *Energy &Fuels, 24*(7), 4062–4077.

Taelman, S. E., De Meester, S., Roef, L., Michiels, M., & Dewulf, J. (2013). The environmental sustainability of microalgae as feed for aquaculture: A life cycle perspective. *Bioresource Technology, 150*, 513–522.

Tchobanoglous, G., Burton, F. L., & Stensel, H.Metcalf & Eddy, Inc.. (2003). (4th Edition). *Wastewater engineering – Treatment and reuse*, (Vol. 4). NY: McGraw-Hill.

Tredici, M. R. (2010). Photobiology of microalgae mass cultures: Understanding the tools for the next green revolution. *Biofuels, 1*(1), 143–162.

Vonshak, A. (1997). Spirulina platensis *(Arthrospira): Physiology, cell-biology and biotechnology*. Verlag: Taylor & Francis Ltd.

Water Footprint Network (2020). *What is a water footprint*? https://waterfootprint.org/en/water-footprint/what-is-water-footprint/.

Xu, L., Weathers, P. J., Xiong, X. R., & Liu, C. Z. (2009). Microalgal bioreactors: Challenges and opportunities. *Engineering in Life Sciences, 9*(3), 178–189.

Chapter 14

Consumer acceptance and attitudes toward microalgae and microalgal-derived products as food

Silvia Villaró[1,2], Inmaculada Viñas[1] and Tomás Lafarga[2]
[1]Food Technology Department, University of Lleida, Lleida, Spain, [2]Department of Chemical Engineering, University of Almería, Almería, Spain

14.1 Introduction

Marine resources are gaining increased importance in the context of the European bioeconomy, where microalgae are especially attractive because of their high content of high-quality proteins, carbohydrates, lipids, and other bioactive molecules (Lafarga, 2019). Microalgae show potential for being used to enhance or improve the nutritional value of traditional foods and theoretically could supplement the nutritional and energy needs of the world's increasing population. For this reason, together with an increasing need for producing environmentally sustainable foods, great efforts are being made by governmental authorities, food scientists, and food processors to incorporate microalgae into our diets and promote their production. However, despite the huge range of nutritional, technological, and bioactive benefits that microalgae offer, microalgal biomass is not yet a common ingredient. Indeed, most of the microalgal biomass currently being produced for food applications is commercialized as dietary supplements and sold as dry powders, tablets, or capsules and not as an ingredient for traditional preparations.

The utilization of microalgae as a food ingredient faces some challenges that still need to be overcome. Some of the most commonly mentioned drawbacks of microalgae biomass are their deep (generally) green color, the "fishy" taste and odor of microalgae-containing foods, and its powdery consistency (Chacón-Lee & González-Mariño, 2010). Other issues that are limiting the development of microalgae-enriched foods include low production capacities, high production costs, and regulatory and legislative issues, all of which are discussed in this

Cultured Microalgae for the Food Industry. DOI: https://doi.org/10.1016/B978-0-12-821080-2.00001-0

book. However, over the last decade a relatively large number of foods containing microalgae have been launched into the market and are currently being commercialized with success. It is true that most of these products used microalgae mainly as a coloring agent or as a marketing strategy (Lafarga, 2019), but still microalgae-containing products have had some commercial success.

Furthermore, different scientific publications developed microalgae-containing foods that, after a sensorial analysis, showed excellent acceptability—some of them, even higher than the controls. For example, Lucas, de Morais, Santos, and Costa (2018) developed a snack containing *Spirulina* sp. LEB 18 at a concentration of 2.6% (Fig. 14.1) and the statistical analysis of the sensorial evaluation revealed no differences between the microalgae-containing snack and the control (except for the color attribute). Similar results were observed by Lafarga, Mayre, et al. (2019) who produced wheat crackers enriched in *Tetraselmis* sp. or *Nannochloropsis* sp. at a concentration of 2.5% and observed higher overall acceptance scores in the microalgae-containing crackers when compared with the wheat-only controls.

Sensorial quality is of key importance in food acceptance as consumers demand foods with certain sensorial characteristics. However, not only the characteristics of the food itself but also consumer characteristics, namely genetic, age, group, gender, and physiological and psychological state, as well as its environment (religion, cultural habits, family, price, or convenience, among other factors) influence consumers' decisions to accept or reject a food product (Costell, Tárrega, & Bayarri, 2010). Large food brands have gone "green" over the last couple of years and products such as matcha tea, *Spirulina*, or kale are top trends in the food industry. As green-colored foods are "trendy," is the green color of microalgae really a drawback? Does microalgae incorporation into traditional foods affect their quality? Are microalgae-containing foods well accepted? What are the main factors

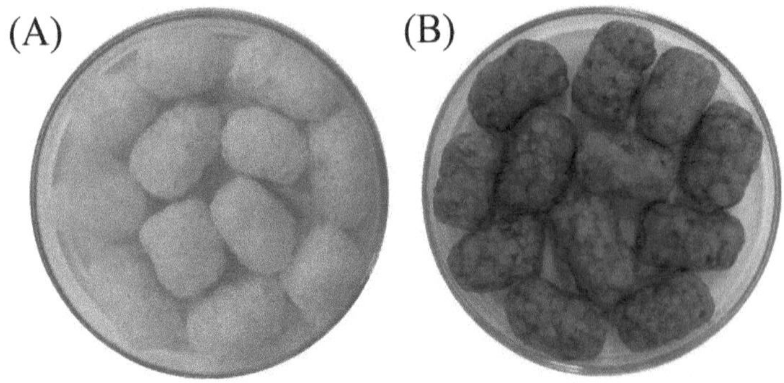

FIGURE 14.1 Control and *Spirulina*-enriched extruded snacks. (A) Control snacks and (B) snacks enriched with *Spirulina* sp. LEB 18 at a concentration of 2.6%. *Reprinted from Lucas, B. F., de Morais, M.G., Santos, T.D., & Costa, J.A.V. (2018). Spirulina for snack enrichment: Nutritional, physical and sensory evaluations.* LWT, 90, *270–276.*

affecting the acceptance of food? The aim of this chapter is to answer all of these questions and to identify which are the key aspects to consider when developing innovative microalgae-containing foods.

14.2 Cues affecting food acceptance and consumption

While eating, our brain receives different sensory inputs namely visual, gustatory, olfactory, or tactile that definitely affect the acceptance or rejection of a product. Acceptance or rejection of a certain food product is also influenced by: (1) an affective component, which generates a positive/negative response toward the product; (2) a cognition component which depends on knowledge and opinions about the product; and (3) a behavioral component, which involves intentions or actions defining the willingness of a consumer to do something in a certain situation (Costell et al., 2010). The process by which consumers accept or reject food is of a complex multidimensional nature and its structure is both dynamic and variable (Costell et al., 2010).

Taste, aroma, and texture, the main sensorial attributes of food, are key factors affecting the acceptance or rejection of food (Fig. 14.2). Aroma is

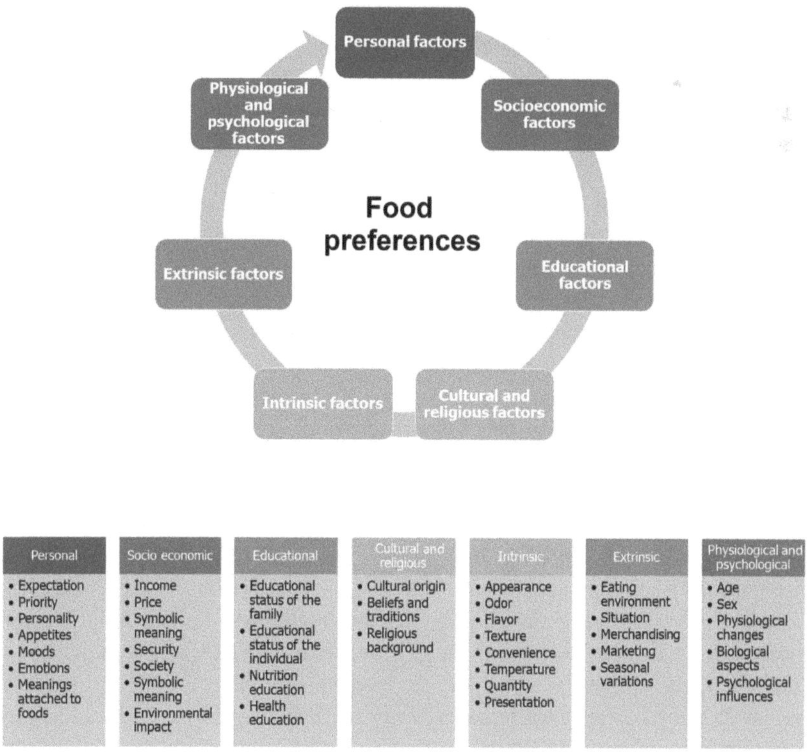

FIGURE 14.2 Summary of the main factors affecting acceptance/rejection of a product.

defined as an odor, sensed through the nose and retronasally (and also through the back of the mouth), while taste is the sense experienced by the tongue (which describes sensations of saltiness, sweetness, sourness, bitterness, or umami). Several factors alter aroma release from food, modulating the final perception. The most important factors include the chemical properties of the volatile compounds present in the food and the physicochemical properties of the food matrix (Ployon, Morzel, & Canon, 2017). The combination of both aroma and taste is defined as flavor, which is the major determinant of food acceptance. Not only flavor but also texture plays a key role. Flavor may be influenced by tactile, thermal, painful, and/or kinesthetic effects, as well as visual and auditory cues, although these are not intrinsic to it (Spence, Levitan, Shankar, & Zampini, 2010). Texture is a multiparameter attribute and a sensory property (thus, only a human being can perceive and describe it) that derives from the structure of the food (Szczesniak, 2002). Factors affecting textural acceptance can be divided into: (1) those that depend on the consumer; (2) those residing in the food; and (3) those that depend on the eating situation (Szczesniak, 2002). The perception of texture as well as the expectation of texture affect acceptance. Werthmann et al. (2015) conducted a repeated-measures-design where one feature (color, texture, or taste) of a well-liked and well-known yogurt (strawberry-raspberry-flavored) was manipulated while the other features were kept constant. The authors observed that color and taste manipulations of the yogurt did not affect young children's intake, while food texture led to reduced food intake of yogurt, indicative of fussy eating. In a different study, Poelman, Delahunty, and de Graaf (2013) observed that cooking time of *Brassica* vegetables, which leads to decreased firmness and associated texture attributes, significantly affected acceptance. However, the authors highlighted that it is likely that the combination of flavor and firmness attributes rather than texture alone determines acceptance for vegetables as cooking time affected texture but altered the flavor profile at the same time.

The perception transmitted by visual appearance of the package also plays a key role as the first contact between the consumers and the product. Seeing food can change feeding behaviors and food consumption volumes. The Napping procedure, derived from the French word *nappe* (tablecloth), consists of placing a group of products on a two-dimensional perceptual surface according to their similarities, and the assessment of the resemblance between the two products by a subject or judge (Pagès, 2005). Carrillo, Varela, & Fiszman, 2012 evaluated the influence of sensory and nonsensory characteristics on the overall acceptance and perceived healthiness of functional biscuits under three different scenarios: blind, informed, and expected conditions. In that study, the authors showed that the overall acceptance of the biscuits was higher in the blind test, while seeing only the package showed a trend toward a higher perception of healthiness. These results demonstrate that nonsensory factors could influence the first buy, and sensory characteristics could determine loyalty and repeat consumption of the products. Not only visual appearance of the product or the products' package affect acceptance but also eating environments. For example, García-Segovia, Harrington, & Seo, 2015 evaluated whether there is an interaction between eating environments and food acceptance and consumption. In that study, the authors offered an identical roasted chicken at different table settings, from a plastic tray to a gourmet table, both in testing booths and realistic contexts. In that study, the authors demonstrated that both, small scales (such as table setting) and large scales (like eating location) can affect food acceptance and intake. Similar results were observed by Zellner, Loss, Zearfoss, & Remolina, 2014 after plating a meal containing starch, vegetables, meat, and sauce more or less attractively. The authors concluded that attractive food presentation increases liking of food flavor. Moreover, portion size, number, and size are other visual cues that can affect food intake and increase the consumption of healthy foods in both children and adults (Wadhera & Capaldi-Phillips, 2014). Color plays an important role in food acceptance. Results reported so far on the effect of color on taste are contradictory, as although several studies demonstrated a significant effect of increasing the intensity of color on taste across different drinks, some others failed to demonstrate such an effect or demonstrated complex interactions that are difficult to interpret (Spence et al., 2010). In turn, a large number of studies claimed that flavor intensity is indeed influenced by color (Spence et al., 2010).

The affective component of the consumers' response to a food summarizes the general response (sensations, feelings, and emotions) that the consumer experiences in response to consuming a food item. As highlighted previously, a cognitive component can also affect acceptance or rejection of a product. This component is related to the information that the consumer has about a product and to her/his attitudes and beliefs (Costell et al., 2010). The relative dominance of affective or cognitive components toward a product influences eating behavior (Dubé & Cantin, 2000). Would consumers better accept microalgae as food if they knew (or better understood) the health benefits associated with their consumption? Consumer perceptions of risks and benefits are dynamic processes to the extent that attitudes toward consuming new flavors or ingredients are in continuous evolution (Costa-Font, Gil, & Traill, 2008). These dynamic processes can be motivated by, for example, improving the amount of evidence that demonstrates the health outcomes of consuming microalgae or increasing consumer knowledge on the health benefits of consuming microalgae. For this reason, further studies elucidating the potential health benefits of consuming algae as well as further efforts made by scientists and food processors to transmit the healthy attributes of microalgae will certainly have an effect on the acceptance of the microalgae-containing foods currently being produced and of those that will certainly be developed in the future. The importance of this cognition component depends on, for example, a cultural factor. Indeed, Japanese consumers consider functional foods a distinct class product and the importance of their health-promoting properties exceeds the importance of their sensory attributes (Lafarga & Hayes, 2017). Western cultures have a different attitude toward functional foods and the acceptance of a product depends largely on other attributes, especially taste (Verbeke, 2006). For example, the Napping procedure carried out by Carrillo, Varela, & Fiszman, 2012 in Spain demonstrated that participants were not willing to compromise sensory characteristics of functional biscuits for health.

14.3 Evolution of microalgae from a simple food to a top-trend ingredient

The European Union (EU) adopted a strategy entitled "Innovating for sustainable growth: A bioeconomy for Europe" in 2012, aiming to promote sustainable productions and limit their negative impacts on the environment (EC, 2012). The Europe 2020 Strategy called for a bioeconomy as a key element for smart and green growth in Europe. The EC defines bioeconomy as the "production of renewable biological resources and the conversion of these resources and waste streams into value added products such as food, feed, bio-based product and bioenergy" (EC, 2012). In 2017, the European Commission published a review of the 2012 European Bioeconomy Strategy and concluded that, although the objectives of the 2012 Strategy remained valid, the scope of the action had to be refocused (EC, 2017). Therefore the 2018 update of the Bioeconomy Strategy aims to accelerate the positioning of a sustainable European bioeconomy to maximize its contribution to the 2030 Agenda, its Sustainable Development Goals, and the Paris Agreement (EC, 2018). In this context, microalgae are considered one of the most promising feedstocks for sustainable supply of food (and nonfood) commodities: microalgae do not need to grow on arable land, have a high areal productivity, can be grown in seawater, use residual nutrients, and are rich sources of valuable products (Draaisma et al., 2013). Moreover, microalgae are considered as safe, sustainable, and nutritious by consumers (Lafarga et al., 2021) although studies on consumer attitudes towards microalgae are scarce.

Although the large-scale utilization of microalgae for food applications is only a few decades old, humans are no strangers to the consumption of microalgae, as in the ninth century the Kanem Empire consumed *Spirulina* in Africa (Oncel, Kose, Vardar, & Torzillo, 2015). More recently, in the 1950s, Burlew (1953) proposed the use of algae as potential alternative protein sources to meet the global demand. In the 1960s, Japan started the first large-scale industrial production of microalgae (*Chlorella*) for human consumption and approximately two decades later, in the 1980s, industrial production facilities were established in other Asian countries, India, the United States, Israel, and Australia (Vigani et al., 2015). Nowadays, microalgae are also being produced in other countries including Spain, Portugal, and Chile.

Over the past couple of decades, microalgae biomass has been almost entirely used in the food market—approximately 75% of the annual biomass production has been used by the food industry. The main reasons for using microalgae as food are the above-mentioned sustainability of the process but also their high content of high-quality proteins (Kent, Welladsen, Mangott, & Li, 2015), bioactive pigments, namely chlorophylls and carotenes, and polyunsaturated fatty acids (PUFAs) that are beneficial to human health and possess cancer-preventive activity as well as antioxidant and antiinflammatory properties among other positive bioactivities (Pangestuti & Kim, 2011). However, despite the huge range of benefits that microalgae offer and the known health benefits associated with the consumption of microalgal macro- and micronutrients, most of the microalgal biomass currently being produced for food applications is commercialized as dietary supplements and sold as dry powders, tablets, or capsules and not as an ingredient in traditional preparations (Lafarga, 2019). The utilization of microalgae as a food ingredient faces some challenges that remain to be overcome and a number of key aspects need to be considered when developing microalgae-containing foods. The strong green color of microalgal biomass or its "fishy" taste and odor, as well as its powdery consistency, have been suggested as inconvenient when incorporating microalgal biomass into traditional products (Chacón-Lee & González-Mariño, 2010). For these reasons, only a limited number of products containing microalgae have been launched into the market and Lafarga (2019) concluded that most of these products used microalgae in small quantities as a coloring agent. Microalgae are excellent food coloring agents. Color stability is important in natural colorants, which tend to be less stable when compared to their synthetic counterparts (Sigurdson, Tang, & Giusti, 2017), and previous studies suggested that color stability of foods enriched in microalgae is very high (Fradique et al., 2013; Gouveia et al., 2008; Gouveia, Batista, Miranda, Empis, & Raymundo, 2007; Lafarga et al., 2019).

As will be highlighted in the following section, several studies have shown that consumers perceived a "fishy" taste and aroma in microalgae-containing products, generally associated with low acceptance of the end product. Information on volatile compounds present in microalgae and the interaction with these compounds and aroma or flavor are limited. However, some studies have been reported and, for example, Van Durme, Goiris, De Winne, De Cooman, and Muylaert (2013) studied the correlation between volatile composition of the microalgae *Botryococcus braunii*, *Rhodomonas*, *Tetraselmis* sp., *Chlorella vulgaris*, and *Nannochloropsis oculata* and sensorial evaluations. The authors reported that microalgae species with a seafood-like odor contained high levels of diketones, α- and β-ionone, and sulfuric compounds, namely dimethyl disulfide, dimethyl trisulfide, and methional (Fig. 14.3). Moreover, in the same study, the authors correlated aldehydes including 2,4-alkadienals and 2,4,6-alkatrienals with fresh green and fruity flavors. The content and composition of volatile compounds

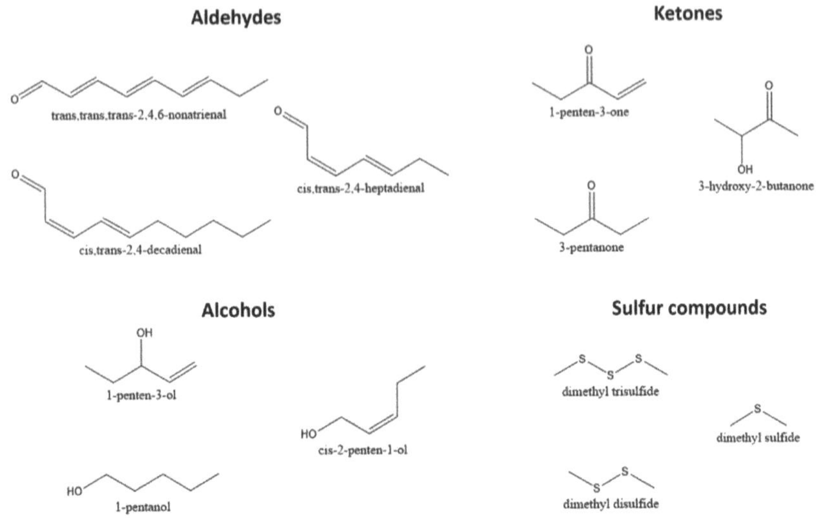

FIGURE 14.3 Volatile compounds detected in microalgal biomass. Structure of some volatile organic compounds detected by Van Durme et al. (2013) by HS-SPME GC-MS. The shown aldehydes were the most important unsaturated aldehydes in *Botryococcus*. Ketones and alcohols were detected in all the studies species, namely *Botryococcus braunii*, *Rhodomonas*, *Tetraselmis*, *Nannochloropsis oculata*, and *Chlorella vulgaris*. Sulfur compounds were detected only in *Tetraselmis* and *Rhodomonas* species.

has been proved to depend on different factors including the growth phase of the microalgae. Indeed, Zhou et al. (2017) evaluated the volatile components in the marine microalgae *Thalassiosira weissflogii*, *Nitzschia closterium*, *Chaetoceros calcitrans*, *Platymonas helgolandica*, *Nannochloropsis* spp., and *Dicrateria inornata* in their exponential, stationary, and declining growth phases and observed that the volatile components were significantly affected by growth phase. Aldehydes, alkanes, some esters, and dimethyl sulfide, the latter associated with seafood-like odor, were the most affected compounds. Data on aroma of microalgae and microalgae-containing foods are lacking and further studies in this area are needed to promote improvements in the aroma and flavor of microalgae-containing foods.

The most commonly consumed microalgae include *Spirulina* and *Chlorella*. These species are the most commonly used in the food industry mainly because of their long history of use: in the EU, microalgae must be commercialized under the Novel Food Regulation (EU) 2015/2283, which defines a novel food as food that has not been consumed to any significant degree in the EU before May 1997 (EU, 2015). Currently, the Spanish company Fitoplancton Marino S.L. (Cadiz, Spain) is commercializing the food product *Plancton Marino Veta la Palma*, which is the dried biomass of *Tetraselmis chuii*, as a Novel Food in accordance with the above-mentioned regulation (AECOSAN, 2017). Other microalgae species are expected to obtain permission for being commercialized as food in the EU in the coming years.

One of the main advantages of introducing microalgae into foods, besides the health-promoting benefits associated with their consumption, is that they allow the manufacture of innovative products that would satisfy most current consumer demands: organic and natural products, free from additives, suitable for vegans, etc. Consumers also demand sophisticated products and microalgal biomass is positioning firmly in the food market (Lafarga, 2019). For example, Angel Leon who is responsible for the restaurant Aponiente (Cadiz, Spain), and is considered one of Spain's most innovative chefs, specializes in experimental seafood dishes and uses innovative methods to convert microalgae (among other marine ingredients) into edible forms with a distinctive "sea" taste. Microalgae-derived compounds, not only the dried biomass, are gaining increased interest in the food industry, especially those that have been associated with positive health outcomes, as these could be used for the development of functional foods. Indeed, Mintel (www.mintel.com), an online database of new fast-moving consumer goods, suggested astaxanthin, a carotenoid produced by the microalgae *Haematococcus pluvialis*, as one of the three main ingredients to watch during 2019 and a niche ingredient with a huge potential (MINTEL, 2019). Murray, Dordevic, Ryan, and Bonham (2018) suggested marine algae-derived polyphenols as an emerging trend in functional foods for the prevention of diseases related to metabolic syndrome. Microalgae-derived oil rich in PUFAs such asdocosahexaenoic acid (DHA) and eicosapentaenoic acid extracted from the microalgae *Scizochytrium* sp. has been recently authorized for commercialization in the EU as a Novel Food (Limón et al., 2015). The maximum concentration of oil permitted depends on the food category where the oil is introduced and varies between 80 and 600 mg/100 g for nonalcoholic beverages and spreadable fats and dressings, respectively (EC, 2015). Food products enriched in microalgae-derived PUFAs reported to date include butter biscuits containing *Isochrysis galbana* at a concentration of 1% or 3% (with PUFAs content of 100 or 320 mg/100 g) developed by Gouveia et al. (2008) and spaghettis containing the same microalgae or *Diacronema vlkianum* at concentrations ranging from 0.5% to 2.0% (Fradique et al., 2013). Other microalgae-derived products have been also successfully incorporated into foods including β-carotene and lutein obtained from *Scenedesmus almeriensis* introduced into olive oil. This allowed high inhibition of olive oil peroxidation, improving quality and extending shelf-life (Limón et al., 2015). Microalgae-derived compounds have been also suggested as promising ingredients for the nutraceutical industry, which is expected to reach $385 billion by 2020 (Suleria, Osborne, Masci, & Gobe, 2015).

In summary, microalgae have become a promising ingredient for food processors which can be used to develop novel "trendy" foods and that could be introduced as the main ingredient in functional foods. Microalgae show potential to become a stable food for consumers all over the planet, although further efforts are needed in order to fully understand the potential of microalgae in the food industry.

14.4 Consumer acceptance and attitudes toward microalgae as food

In the previous section we discussed the main characteristics affecting acceptance or rejection of food. A number of scientific publications to date have evaluated the potential use of microalgae or compounds derived from microalgae as a novel ingredient in bread or other baked goods, soups or beverages, pasta, or other matrices such as snacks or dairy products. However, data on sensorial analysis, perception of microalgae, and attitudes of consumers about consuming microalgae are scarce.

The utilization of baked goods as delivery vehicles for health-promoting compounds has several advantages, including their widespread consumption and that they are generally encouraged as part of a healthy diet and are considered to be safe (Lafarga, Gallagher, Bademunt, Viñas, et al., 2019). In addition, baked products are not usually exposed to high pressures or strong pH variations and have been efficiently used as food vehicles for plant phytochemicals (Lafarga, Gallagher, Bademunt, Bobo, et al., 2019; Odunlade et al., 2017) or proteins (Fitzgerald et al., 2014; Lafarga, Gallagher, Aluko, Auty, & Hayes, 2016), among other compounds. Several baked food products containing microalgae are currently being commercialized (Table 14.1). For example, Lee Biscuits (Malaysia) commercializes *Spirulina*-filled crackers and Evasis Edibles (Austria) and GrupoDulcesol (Spain) are currently commercializing Algen Crackers (wheat crackers containing *Chlorella*) and Orange and *Chlorella* bites, respectively (Lafarga, 2019). Results published to date have demonstrated the potential of incorporation of microalgae into baked goods to obtain improved nutritional and physicochemical properties (Batista et al., 2017; Gouveia et al., 2008; Graça, Fradinho, Sousa, & Raymundo, 2018; Lafarga et al., 2019; Nunes et al., 2020; Uribe-Wandurraga, Igual, García-Segovia, & Martínez-Monzó, 2019). In a recent study, consumers considered microalgae as safe and nutritious, although their knowledge about microalgae and their health benefits was low (Lafarga et al., 2021). Consumer involvement in innovation processes is of extreme importance and has been recognized for over four decades (Busse & Siebert, 2018). However, only a limited number of studies have evaluated consumer attitudes or acceptance of the formulated microalgae-enriched baked products. The results obtained so far on acceptability of microalgae-containing baked goods are contradictory as several studies have observed high acceptance of the products and others have reported that microalgae-containing foods were not as competitive as the controls. For example, Batista et al. (2017) formulated cookies containing either *Spirulina platensis* or *C. vulgaris* at concentrations of 2% or 6% for panelists

TABLE 14.1 Currently available foods containing microalgae.

Brand	Company, country	Description
Helga	Evasis Edibles, Austria	(1) Organic algae drink and (2) sea salt seaweed crackers containing *Chlorella*
Happy Planet	Happy Planet Foods, Canada	Smoothie with *Spirulina* and *Chlorella*
Mavericks	Maverick Makers Snacks, United Kingdom	Vegan breadsticks rich in fiber containing *Chlorella*
Chocolate 70%	The Algae Factory, the Netherlands	Chocolate containing *Spirulina*
Spirulina-filled crackers	Lee Biscuits, Malaysia	Crackers containing *Spirulina*
Innova	Dulcesol, Spain	Biscuits and breads containing *Chlorella*
Currynaise	The Good Spoon, France	*Chlorella* as an egg replacer in different flavored vegan mayonnaises
Protein bar	Alver, Switzerland	Protein bar containing *Chlorella*
Frecious slow juice	Frecious, the Netherlands	Vegetable juice containing *Chlorella*
Earth of Eco	Majami, Poland	Organic fudge with *Chlorella*

and observed global appreciation and taste scores ranging between 3 and 4 (using a 5-point hedonic scale). Batista et al. (2019) recently reported lower sensory scores, when comparing control wheat cracker and crackers containing *S. platensis*, *C. vulgaris*, *Tetraselmi suecica*, or *Phaeodactylum tricornutum* at a concentration of 2% or 6%—especially for marine microalgae *T. suecica* and *P. tricornutum*. However, acceptability of the microalgae-containing crackers, shown in Fig. 14.4, was relatively high and 40% of the panelists declared that they would certainly or probably buy crackers containing *Spirulina*. These results highlight the importance of selecting a suitable microalgae specie depending on the end product. Interestingly, in the study by Batista et al. (2019), the authors observed that microalgae affected the perception of saltiness, with a higher frequency of "unsalted" classification for microalgae-containing crackers when compared to the control. Similar results were recently reported by Lafarga, Mayre, et al. (2019) after incorporating the biomass of *Nannochloropsis* or *Tetraselmis* into wheat bread and crackers at flour substitution levels ranging from 1.0% to 2.5%. In that study, the authors observed no difference in the visual, texture, and flavor scores of the breads when compared to the wheat-only controls. A lower aroma score was obtained

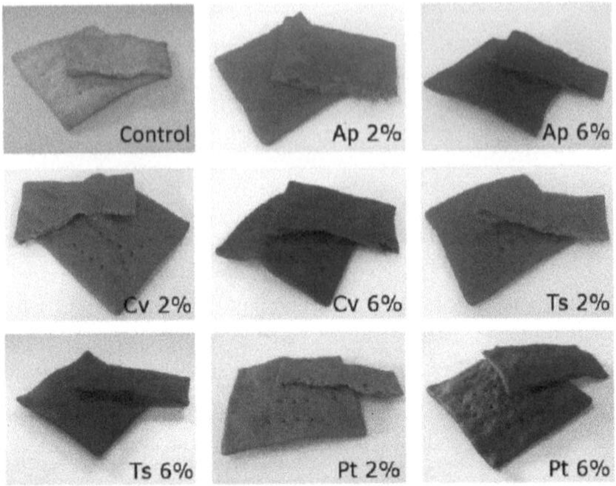

FIGURE 14.4 Control and microalgae-enriched wheat crackers. Ap, *Athrospira platensis*; Cv, *Chlorella vulgaris*; Ts, *Tetraselmis suecica*; Pt, *Phaoedactylum tricornutum. Reprinted from Batista, A.P., Niccolai, A., Bursic, I., Sousa, I., Raymundo, A., Rodolfi, L., Tredici, M.R. (2019). Microalgae as functional ingredients in savory food products: Application to wheat crackers.* Foods, 8*(12), 611.*

for microalgae-containing bread at higher microalgae concentrations. Aroma of bread and baked products is considered an essential key attribute. Pico, Bernal, and Gómez (2015) recently reviewed a total of 326 volatile compounds reported in the literature to be present in bread. These include alcohols, aldehydes, esters, ethers, ketones, acids, pyrazines, furans, pyrrolines, and sulfur compounds and have been reported in the crust, in the crumb, or in both. The sensorial correlation between these compounds and aroma has also been explained (Pico et al., 2015). The effect of microalgae on the aroma of bread has not yet been fully evaluated. In the above-mentioned study by Lafarga, Mayre, et al. (2019), the authors determined the volatile compounds emitted by the microalgae-containing breads and crackers by HS-SPME-GC/MS and detected over 40 compounds including aldehydes, alcohols, ketones, esters, acids, terpenes, and hydrocarbons. In that study, the aroma profiles of breads and crackers were similar, as were the profiles between the controls and the microalgae-containing goods. For this reason, further studies are needed on this subject to fully understand the effect of microalgae incorporation on the aroma of foods.

Other common food matrices for delivering microalgae include pasta products (Fradique et al., 2013; Rodríguez De Marco, Steffolani, Martínez, & León, 2014).These are cheap and widely consumed and colored-pasta dough is common in many countries—pasta enriched in spinach or tomato is currently commercially available. Commercial pasta products containing microalgae include spaghetti with *Spirulina* (Biográ, Spain) and tagliatelle with *Spirulina* (Priméal, France) among others. *Spirulina* is the most commonly used microalgae in pasta products. However, scientific publications evaluated the potential utilization of other microalgae species. For example, Fradique et al. (2013) developed spaghettis containing biomass of *I. galbana* and *D. vlkianum* at concentrations ranging from 0.5% to 2.0% and although the content of PUFAs was higher in the microalgae-containing spaghettis, the panelists showed a preference for the control pasta. As highlighted in the previous section, acceptance of a certain food can be influenced by a large number of factors including knowledge on, for example, the positive health outcomes of consuming PUFAs. It would be interesting to evaluate whether consumers would accept microalgae-containing pasta containing these microalgae species if they knew that consumption of PUFAs promotes health. Other factors that can influence acceptability are cultural factors. Chacón-Lee & González-Mariño, 2010 concluded that the degree of acceptability of microalgae-containing products depends largely on the traditional diet of the target population. The use of microalgae in Asian countries is not an extreme addition to food, since the traditional cuisines include algae in many preparations. Western cultures are not generally used to the consumption of algae. In a different study, Fradique et al. (2010) developed spaghettis containing biomass of *Spirulina maxima* and *C. vulgaris* at concentrations ranging from 0.5% to 2.0%. Despite using the same microalgae concentrations as in the study by Fradique et al. (2013), the authors reported increased acceptance scores when using *Spirulina* and *Chlorella* and when compared to the controls and to the pasta made using *I. galbana* and *D. vlkianum*. Results suggest once again that selecting a suitable microalgae specie for food applications depending on the end product—or selecting a suitable food matrix depending on the microalgae specie—can significantly affect the acceptance or rejection of the end product. Indeed, Robertson et al. (2016) formulated a yogurt containing *Pavlova lutheri* and observed that although there was minimal impact on the functional properties of the product, sensorial acceptability was poor. A "fishy" taste could be better accepted in, for example, fish-based culinary preparations but not in dairy products.

Several soups and beverages containing microalgae also are commercially available. For example, Happy Planet Foods (Canada) sells green smoothies containing *Spirulina* and Helga (Austria) commercialized a refreshing drink containing *Chlorella*. Beverages and liquid foods are especially well suited as food delivery vehicles for infants and the elderly. Indeed, protein deficiency is one of the main concerns of the elderly, and scientists and food processors are devoting increased attention to incorporating protein into various food products (van der Zanden, van Kleef, de Wijk, & van Trijp, 2015). For this reason, Santos, Freitas, Moreira, Zanfonato, & Costa, 2016 manufactured a chocolate-flavored shake-type product enriched in protein supplemented with *Spirulina* and enriched in protein. In that study, the authors reported an average acceptance rate of 7.68 and 7.77 (assessed using a 9-point hedonic scale) for shakes with and without *Spirulina*, respectively. Incorporation of *Spirulina* was well accepted as revealed by the target public.

As broccoli soups are green-colored, Lafarga, Acién-Fernández, et al. (2019) incorporated the biomass of the microalgae *Spirulina*, *Chlorella*, and *Tetraselmis* into broccoli soup at concentrations ranging between 0.5% and 2.0%. In that study, the authors reported an acceptability index higher than 70% for soups made using microalgae at lower concentrations of 0.5%–1.0%. Moreover, the purchase intention for the soups formulated using microalgae at 0.5%, shown in Fig. 14.5, ranged between 3.4 and 4.1 (assessed using a 5-point hedonic scale). Higher biomass concentrations led to low purchase intention with values ranging from 1 (certainly would not buy) to 2 (probably would not buy).

Snack foods have emerged as an alternative to conventional meals as a result of demographic changes demanding convenience. Snacks are considered as one of the better alternatives to regular food as they can be consumed at any time. The global snack food market is projected to continue growing during the coming years, especially growing toward healthy snacking. The

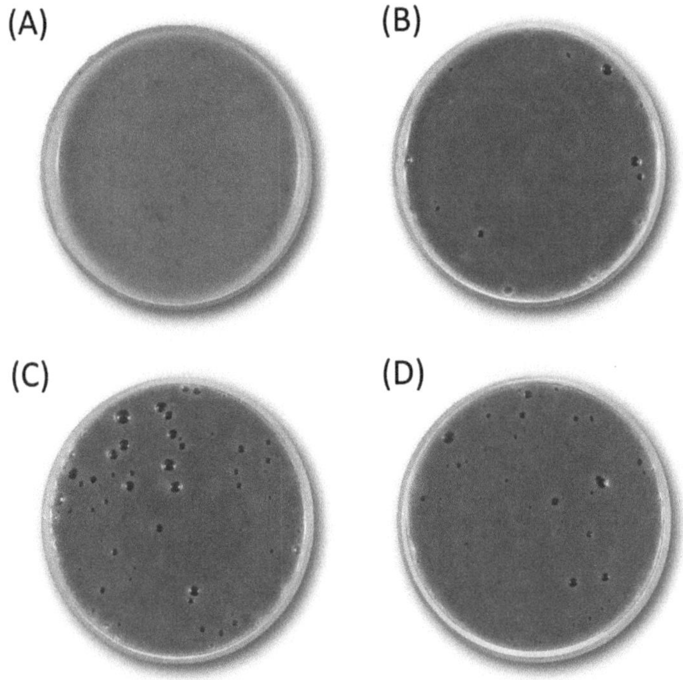

FIGURE 14.5 Control and microalgae-enriched broccoli soup. (A) Control broccoli soup; (B) broccoli soup enriched with *Spirulina* at a concentration of 0.5%; (C) broccoli soup enriched with *Chlorella* at a concentration of 0.5%; and (D) broccoli soup enriched with *Tetraselmis* at a concentration of 0.5%. *Modified from Lafarga, T., Acién-Fernández, F.G., Castellari, M., Villaró, S., Bobo, G., & Aguiló-Aguayo, I. (2019). Effect of microalgae incorporation on the physicochemical, nutritional, and sensorial properties of an innovative broccoli soup.* LWT, 111, *167–174.*

top three categories in snack-typed products include "better for you" snacks, protein-rich and no-sugar options, and nutrient-dense snacks. Microalgae could be a suitable ingredient to develop snacks that could fit within all of these claims and, for this reason, the number of snacks being launched into the market has increased drastically over the last few years (Lafarga, 2019). Commercially available microalgae-containing snacks include Greenzilla breadsticks commercialized by Maverick Makers Snacks (United Kingdom) and *Spirulina*-containing rice crackers (Galletas Gullón, Spain). Lucas et al. (2018) recently developed extruded snacks enriched with *Spirulina* at a concentration of 2.6% (Fig. 14.1). The products containing microalgae were enriched in protein, lipids, and minerals and the authors reported no effect on the flavor, texture, taste, and overall acceptance of the products after incorporation of *Spirulina* into the recipe. Indeed, the acceptance index of the snack was high (over 80%), suggesting that the product would have acceptance if commercialized. Moreover, Tańska, Konopka, and Ruszkowska (2017) formulated corn extrudates enriched with *Spirulina* at concentrations varying from 2% to 8% and conducted a sensory analysis on shape, color, aroma, taste, and crispness. In that study, the authors observed overall acceptance scores ranging from 3.9 to 4.9 points on a 5-point hedonic scale. The highest acceptance was observed for the control snacks, with decreasing values indirectly related to microalgae concentration. The authors observed that color modifications were accepted by consumers.

Overall, further studies are needed to better understand the acceptance of microalgae-containing foods. However, the large number of foods containing microalgae that are currently being launched into the market as well as the results reported so far on this area are promising and suggest a bright future for microalgae in the food industry. The incorporation of novel ingredients such as insects or microalgae into our diets and changing attitudes toward unappealing foods is challenging but not impossible (Tan, Fischer, van Trijp, & Stieger, 2016). Further studies on consumer acceptance of microalgae-containing foods, volatile profile of microalgae, and attitude of consumers toward microalgae as well as more knowledge and communication of the health and environmental benefits of microalgae are needed.

14.5 Conclusions

The acceptance of a food product will depend on whether such a product meets the consumer needs and expectations and also on the degree of satisfaction that it is able to provide, or not. Factors that affect acceptance, besides visual, gustatory, olfactory, or tactile attributes include an affective component, a cognition component, and a behavioral component, all of which play an important role in food acceptance/rejection. Consumers' opinions are of key importance and must be taken into consideration not only to assess the acceptance of a product but also from the beginning of the development of such a product. Appropriate preparations and situations could encourage consumption of microalgae as consumers are willing to taste a new product out of curiosity despite their initial negativity.

Results reported so far as well as the amount of microalgae-containing foods that are successfully launched into the market every year suggest that microalgae could become a common ingredient in food preparations worldwide. Further studies on consumer acceptance of microalgae-containing foods, volatile profile of microalgae, and attitudes of consumers toward microalgae as well as more knowledge and communication of the health and environmental benefits of microalgae are needed and would certainly promote the consumption of this valuable and sustainable resource.

Acknowledgments

This work was supported by the Spanish Ministry of Science, Innovation, and Universities (IJC2018-035287-I) and the SABANA Project from the European Unions' H2020 Research and Innovation Program (Grant number 727874). T. Lafarga also thanks the BBVA Foundation for the Leonardo 2020 Grant for Researchers and Cultural Creators.

References

AECOSAN. (2017). Report of the Scientific Committee of the Spanish Agency for Consumer Affairs, Food Safety and Nutrition (AECOSAN) on a request for initial assessment for marketing of dried microalgae *Tetraselmis chuii* in food supplements under Regulation (EC) No 258/97 on novel foods and novel food ingredients. *Revista del Comite Cientifico*, *25*, 12–25.

Batista, A. P., Niccolai, A., Bursic, I., Sousa, I., Raymundo, A., Rodolfi, L., & Tredici, M. R. (2019). Microalgae as functional ingredients in savory food products: Application to wheat crackers. *Foods*, *8*(12), 611.

Batista, A. P., Niccolai, A., Fradinho, P., Fragoso, S., Bursic, I., Rodolfi, L., & Raymundo, A. (2017). Microalgae biomass as an alternative ingredient in cookies: Sensory, physical and chemical properties, antioxidant activity and in vitro digestibility. *Algal Research*, *26*, 161–171.

Burlew, J.S. (1953). Algal culture from laboratory to pilot plant. *Algal culture from laboratory to pilot plant.*

Busse, M., & Siebert, R. (2018). The role of consumers in food innovation processes. *European Journal of Innovation Management*, *21*(1), 20–43.

Carrillo, E., Varela, P., & Fiszman, S. (2012). Effects of food package information and sensory characteristics on the perception of healthiness and the acceptability of enriched biscuits. *Food Research International*, *48*(1), 209–216.

Chacón-Lee, T. L., & González-Mariño, G. E. (2010). Microalgae for "healthy" foods—Possibilities and challenges. *Comprehensive Reviews in Food Science and Food Safety*, *9*(6), 655–675.

Costa-Font, M., Gil, J. M., & Traill, W. B. (2008). Consumer acceptance, valuation of and attitudes towards genetically modified food: Review and implications for food policy. *Food Policy*, *33*(2), 99–111.

Costell, E., Tárrega, A., & Bayarri, S. (2010). Food acceptance: The role of consumer perception and attitudes. *Chemosensory Perception*, *3*(1), 42–50.

Draaisma, R. B., Wijffels, R. H., Slegers, P. M., Brentner, L. B., Roy, A., & Barbosa, M. J. (2013). Food commodities from microalgae. *Current Opinion in Biotechnology*, *24*(2), 169–177.

Dubé, L., & Cantin, I. (2000). Promoting health or promoting pleasure? A contingency approach to the effect of informational and emotional appeals on food liking and consumption. *Appetite*, *35*(3), 251–262.

Van Durme, J., Goiris, K., De Winne, A., De Cooman, L., & Muylaert, K. (2013). Evaluation of the volatile composition and sensory properties of five species of microalgae. *Journal of Agricultural and Food Chemistry*, *61*(46), 10881–10890.

EC (2012). *Communication from the Comission to the European Parliament, the Council, the European Economic and Social Commitee and the Comittee of the Regios: Innovating for sustainable growth> A bioeconomy for Europe.*

EC. (2015). Commission implementing decision (EU) 2015/545 of 31 March 2015 authorising the placing on the market of oil from the microalgae Schisochytrium sp. (ATCC PTA-9695) as a novel food ingredient under Regulation (EC) No 258/97 of the European Parliament and of the Council. *Official Journal of the European Union*, *90*, 7–10.

EC (2017). *Review of the 2012 European Bioeconomy Strategy.*

EC (2018). *A sustainable bioeconomy for Europe: Strengthening the connection between economy, society and the environment.*

EU. (2015). Regulation (EU) 2015/2283 of the European Parliamient and of the Council of 25 November 2015 on novel foods, amending Regulation (EU) No 258/97 of the European Parliamient and of the Council and Commission Regulation (EC) No 1852/2001. *Official Journal of the European Union*, *L327*, 1–22.

Fitzgerald, C., Gallagher, E., Doran, L., Auty, M., Prieto, J., & Hayes, M. (2014). Increasing the health benefits of bread: Assessment of the physical and sensory qualities of bread formulated using a renin inhibitory *Palmaria palmata* protein hydrolysate. *LWT-Food Science and Technology*, *56*(2), 398–405.

Fradique, M., Batista, A. P., Nunes, M. C., Gouveia, L., Bandarra, N. M., & Raymundo, A. J. (2010). Incorporation of *Chlorella vulgaris* and *Spirulina maxima* biomass in pasta products. Part 1: Preparation and evaluation. *Journal of the Science of Food and Agriculture*, *90*(10), 1656–1664.

Fradique, M., Batista, A. P., Nunes, M. C., Gouveia, L., Bandarra, N. M., & Raymundo, A. (2013). *Isochrysis galbana* and *Diacronema vlkianum* biomass incorporation in pasta products as PUFA's source. *LWT- Food Science and Technology*, *50*(1), 312–319.

García-Segovia, P., Harrington, R. J., & Seo, H.-S. (2015). Influences of table setting and eating location on food acceptance and intake. *Food Quality and Preference*, *39*, 1–7.

Gouveia, L., Batista, A. P., Miranda, A., Empis, J., & Raymundo, A. (2007). *Chlorella vulgaris* biomass used as colouring source in traditional butter cookies. *Innovative Food Science & Emerging Technologies*, *8*(3), 433–436.

Gouveia, L., Coutinho, C., Mendonça, E., Batista, A. P., Sousa, I., & Bandarra, N. M. (2008). Functional biscuits with PUFA-ω3 from *Isochrysis galbana*. *Journal of the Science of Food and Agriculture*, *88*(5), 891–896.

Graça, C., Fradinho, P., Sousa, I., & Raymundo, A. (2018). Impact of *Chlorella vulgaris* on the rheology of wheat flour dough and bread texture. *LWT*, *89*, 466–474.

Kent, M., Welladsen, H. M., Mangott, A., & Li, Y. (2015). Nutritional evaluation of Australian microalgae as potential human health supplements. *PLoS One*, *10*(2), e0118985.

Lafarga, T. (2019). Effect of microalgal biomass incorporation into foods: Nutritional and sensorial attributes of the end products. *Algal Research*, *41*, 101566.

Lafarga, T., Acién-Fernández, F. G., Castellari, M., Villaró, S., Bobo, G., & Aguiló-Aguayo, I. (2019). Effect of microalgae incorporation on the physicochemical, nutritional, and sensorial properties of an innovative broccoli soup. *LWT*, *111*, 167–174.

Lafarga, T., Gallagher, E., Aluko, R. E., Auty, M. A., & Hayes, M. (2016). Addition of an enzymatic hydrolysate of bovine globulins to bread and determination of hypotensive effects in spontaneously hypertensive rats. *Journal of Agricultural and Food Chemistry*, *64*(8), 1741–1750.

Lafarga, T., Gallagher, E., Bademunt, A., Bobo, G., Echeverria, G., Viñas, I., & Aguiló-Aguayo, I. (2019). Physiochemical and nutritional characteristics, bioaccessibility and sensory acceptance of baked crackers containing broccoli co-products. *International Journal of Food Science & Technology*, *54*(3), 634–640.

Lafarga, T., Gallagher, E., Bademunt, A., Viñas, I., Bobo, G., Villaró, S., & Aguiló-Aguayo, I. (2019). Bioaccessibility, physicochemical, sensorial, and nutritional characteristics of bread containing broccoli co-products. *Journal of Food Processing and Preservation*, *43*(2), e13861.

Lafarga, T., & Hayes, M. (2017). Bioactive protein hydrolysates in the functional food ingredient industry: Overcoming current challenges. *Food Reviews International*, *33*(3), 217–246.

Lafarga, T., Mayre, E., Echeverria, G., Viñas, I., Villaró, S., Acién-Fernández, F. G., & Aguiló-Aguayo, I. (2019). Potential of the microalgae Nannochloropsis and Tetraselmis for being used as innovative ingredients in baked goods. *LWT*, *115*, 108439.

Lafarga, T., Rodríguez-Bermúdez, R., Morillas-España, A., Villaró, S., García-Vaquero, M., Morán, L., ... Acién-Fernández, F. G. (2021). Consumer knowledge and attitudes towards microalgae as food: The case of Spain. *Algal Research*, *54*, 102174.

Limón, P., Malheiro, R., Casal, S., Acién-Fernández, F. G., Fernández-Sevilla, J. M., Rodrigues, N., & Pereira, J. A. (2015). Improvement of stability and carotenoids fraction of virgin olive oils by addition of microalgae *Scenedesmus almeriensis* extracts. *Food Chemistry*, *175*, 203–211.

Lucas, B. F., de Morais, M. G., Santos, T. D., & Costa, J. A. V. (2018). Spirulina for snack enrichment: Nutritional, physical and sensory evaluations. *LWT*, *90*, 270–276.

MINTEL (2019). *Functional health ingredients to watch*. www.mintel.com.

Murray, M., Dordevic, A., Ryan, L., & Bonham, M. (2018). An emerging trend in functional foods for the prevention of cardiovascular disease and diabetes: Marine algal polyphenols. *Critical Reviews in Food Science and Nutrition*, *58*(8), 1342–1358.

Niccolai, A., Venturi, M., Galli, V., Pini, N., Rodolfi, L., Biondi, N., Granchi, L.J.S.R. (2019). Development of new microalgae-based sourdough "crostini": Functional effects of Arthrospira platensis (spirulina) addition. *9*(1), 1–12.

Nunes, M. C., Graça, C., Vlaisavljević, S., Tenreiro, A., Sousa, I., & Raymundo, A. (2020). Microalgal cell disruption: Effect on the bioactivity and rheology of wheat bread. *Algal Research*, *45*, 101749.

Odunlade, T., Famuwagun, A., Taiwo, K., Gbadamosi, S., Oyedele, D., & Adebooye, O. (2017). Chemical composition and quality characteristics of wheat bread supplemented with leafy vegetable powders. *Journal of Food Quality*, 2017.

Oncel, S. S., Kose, A., Vardar, F., & Torzillo, G. (2015). *From the ancient tribes to modern societies, microalgae evolution from a simple food to an alternative fuel source. Handbook of marine microalgae* (pp. 127–144). Elsevier.

Pagès, J. (2005). Collection and analysis of perceived product inter-distances using multiple factor analysis: Application to the study of 10 white wines from the Loire Valley. *Food Quality and Preference*, *16*(7), 642–649.

Pangestuti, R., & Kim, S.-K. (2011). Biological activities and health benefit effects of natural pigments derived from marine algae. *Journal of Functional Foods*, *3*(4), 255–266.

Pico, J., Bernal, J., & Gómez, M. (2015). Wheat bread aroma compounds in crumb and crust: A review. *Food Research International*, *75*, 200–215.

Ployon, S., Morzel, M., & Canon, F. (2017). The role of saliva in aroma release and perception. *Food Chemistry*, *226*, 212–220.

Poelman, A. A., Delahunty, C. M., & de Graaf, C. (2013). Cooking time but not cooking method affects children's acceptance of Brassica vegetables. *Food Quality and Preference*, *28*(2), 441–448.

Robertson, R. C., Mateo, M. R. G., O'Grady, M. N., Guihéneuf, F., Stengel, D. B., & Ross, R. P. (2016). An assessment of the techno-functional and sensory properties of yoghurt fortified with a lipid extract from the microalga *Pavlova lutheri*. *Innovative Food Science & Emerging Technologies*, *37*, 237–246.

Rodríguez De Marco, E., Steffolani, M. E., Martínez, C. S., & León, A. E. (2014). Effects of spirulina biomass on the technological and nutritional quality of bread wheat pasta. *LWT-Food Science and Technology*, *58*(1), 102–108.

Santos, T. D., Freitas, B. C. B. d, Moreira, J. B., Zanfonato, K., & Costa, J. A. V. (2016). Development of powdered food with the addition of Spirulina for food supplementation of the elderly population. *Innovative Food Science & Emerging Technologies*, *37*, 216–220.

Sigurdson, G. T., Tang, P., & Giusti, M. M. (2017). Natural colorants: Food colorants from natural sources. *Annual Review of Food Science and Technology*, *8*, 261–280.

Spence, C., Levitan, C. A., Shankar, M. U., & Zampini, M. (2010). Does food color influence taste and flavor perception in humans? *Chemosensory Perception*, *3*(1), 68–84.

Suleria, H., Osborne, S., Masci, P., & Gobe, G. (2015). Marine-based nutraceuticals: An innovative trend in the food and supplement industries. *Marine Drugs*, *13*(10), 6336–6351.

Szczesniak, A. S. (2002). Texture is a sensory property. *Food Quality and Preference*, *13*(4), 215–225.

Tan, H. S. G., Fischer, A. R. H., van Trijp, H. C. M., & Stieger, M. (2016). Tasty but nasty? Exploring the role of sensory-liking and food appropriateness in the willingness to eat unusual novel foods like insects. *Food Quality and Preference*, *48*, 293–302.

Tańska, M., Konopka, I., & Ruszkowska, M. (2017). Sensory, physico-chemical and water sorption properties of corn extrudates enriched with spirulina. *Plant Foods for Human Nutrition*, *72*(3), 250–257.

Uribe-Wandurraga, Z. N., Igual, M., García-Segovia, P., & Martínez-Monzó, J. (2019). Effect of microalgae addition on mineral content, colour and mechanical properties of breadsticks. *Food & Function*, *10*(8), 4685–4692.

Verbeke, W. (2006). Functional foods: Consumer willingness to compromise on taste for health? *Food Quality and Preference*, *17*(1-2), 126–131.

Vigani, M., Parisi, C., Rodríguez-Cerezo, E., Barbosa, M. J., Sijtsma, L., Ploeg, M., & Enzing, C. (2015). Food and feed products from micro-algae: Market opportunities and challenges for the EU. *Trends in Food Science & Technology*, *42*(1), 81–92.

Wadhera, D., & Capaldi-Phillips, E. D. (2014). A review of visual cues associated with food on food acceptance and consumption. *Eating Behaviors*, *15*(1), 132–143.

Werthmann, J., Jansen, A., Havermans, R., Nederkoorn, C., Kremers, S., & Roefs, A. (2015). Bits and pieces. Food texture influences food acceptance in young children. *Appetite*, *84*, 181–187.

van der Zanden, L. D., van Kleef, E., de Wijk, R. A., & van Trijp, H. C. M. (2015). Examining heterogeneity in elderly consumers' acceptance of carriers for protein-enriched food: A segmentation study. *Food Quality and Preference*, *42*, 130–138.

Zellner, D. A., Loss, C. R., Zearfoss, J., & Remolina, S. (2014). It tastes as good as it looks! The effect of food presentation on liking for the flavor of food. *Appetite*, *77*, 31–35.

Zhou, L., Chen, J., Xu, J., Li, Y., Zhou, C., & Yan, X. (2017). Change of volatile components in six microalgae with different growth phases. *Journal of the Science of Food and Agriculture*, *97*(3), 761–769.

Chapter 15

Future perspectives of microalgae in the food industry

Alice Ferreira[1], Inês Guerra[2], Margarida Costa[2], Joana Silva[2] and Luisa Gouveia[1]

[1]*Bioenergy Unit, LNEG, National Laboratory of Energy and Geology I.P., Lisbon, Portugal,*
[2]*R&D Department, Allmicroalgae Natural Products S.A., Pataias, Portugal*

15.1 Introduction

Food is primarily intended to respond to hunger and provide nutrients needed for human survival. The global population is expected to reach 9.8 billion by 2050 (UN DESA, 2017). Consequently, current food production will need to double within the next 30 years to fulfill the nutritional needs of this growing population (Searchinger, Waite, Hanson, & Ranganathan, 2019). Also, the Food and Agriculture Organization (FAO) recently estimated that over 821 million people worldwide are undernourished, having a deficient protein intake (FAO, 2019). Poorer countries are often faced with physical and economic constraints that hinder their access to nutritious foods, further endangering their risk of food insecurity and malnutrition (Swinburn et al., 2019).

Furthermore, the conventional sources of animal-derived protein not only will be insufficient but result in increased greenhouse gas (GHG) emissions and overutilization of water and land, making an extreme pressure on Earth's resources (Wu et al., 2014). Food production accounts for 20%–30% of the total environmental impact (Tukker & Jansen, 2006), being responsible for near 30% of global GHG emissions (Vermeulen, Campbell, & Ingram, 2012).

In the European Union, animal-based proteins are consumed in higher quantities than plant-based ones (European Environment Agency, 2017). However, concerns on the impacts of agricultural production, processing, and distribution of food on the environment, ecosystem, and sustainability, as well as health and animal welfare have boosted the demand for new natural and healthier food sources (Caporgno & Mathys, 2018). On the other side of the spectrum, in most developed countries, modernization and economic development have led to an alarming overweight crisis. Almost 30% of the world's population is overweight, with this being more critical within the

Cultured Microalgae for the Food Industry. DOI: https://doi.org/10.1016/B978-0-12-821080-2.00008-3

adult population, reaching 38.9% (FAO, 2019). The main reasons for this include a sedentary lifestyle, poor dietary practices resulting from greater availability and affordability of highly processed and energy-dense foods, and a reduction in physical activity. The risk of overweight and obesity is especially aggravated during childhood, often persisting into adulthood, leading to lifelong health problems such as diabetes, high blood pressure, asthma and other respiratory issues, sleep disorders, and liver diseases. Moreover, obese people suffer more severe health consequences and higher mortality risk (Abdelaal, le Roux, & Docherty, 2017). Between 2000 and 2016, the prevalence of obesity more than doubled among children and adolescents (FAO, 2019).

Several initiatives have been implemented to promote healthier diets that meet nutritional requirements and limit the consumption of harmful fats, salts, and sugars (FAO, 2019). Food technologies and products may contribute to more readily available and affordable worldwide healthy diets while helping to reduce their impacts. Furthermore, changing dietary trends, namely to vegetarian and vegan diets, and consumers' growing awareness of food impact on human health have lately increased the demand for functional foods (Caporgno & Mathys, 2018). Functional foods trigger beneficial physiological effects, improving well-being and health, and reducing the risk of illness (Gouveia, Batista, Sousa, Raymundo, & Bandarra, 2008). Therefore the inclusion of functional foods in the daily diet promotes quality of life and will eventually reduce spending on health care for the global population (Camacho, Macedo, & Malcata, 2019; Plaza, Herrero, Alejandro Cifuentes, & Ibáñez, 2009). Microalgae are one of the richest sources of plant-based proteins and, unlike most plant sources, are a well-balanced source of essential amino acids adequate for human consumption requirements, according to WHO/FAO/UN recommendations (Becker, 2007; Koyande et al., 2019). They can also produce other compounds such as polysaccharides, polyunsaturated fatty acids (PUFAs), pigments, essential minerals, and vitamins which can also be bioactive.

Protein hydrolysates are also convenient sources of amino acids for humans because they are more easily absorbed through the gastrointestinal tract compared with intact proteins or free amino acids (Lisboa, Pereira, Ferreira, & Costa, 2014). Furthermore, microalgal proteins could present beneficial protein surface-related and thermodynamic properties, making them promising foaming and emulsifying agents, and thickeners in the food industry.

Polysaccharides are widely used in the food industry primarily as gelling and/or thickening and stabilizing agents. Many species of microalgae can produce polysaccharides, such as starch, floridean starch, glycogen, chrysolaminarin, and paramylon. These have different applications, including: antiviral, antioxidant, anti-inflammatory, and immunomodulatory (M.F. de Jesus Raposo, de Morais, & de Morais, 2013; M.F.D.J. de Jesus Raposo & de

Morais, 2015 and references therein). Microalgal polysaccharides also have prebiotic activities (de Jesus Raposo, de Morais, & de Morais, 2016; Jutur, Nesamma, & Shaikh, 2016; Moreno, Corzo, Montilla, Villamiel, & Olano, 2017).

Long-chain PUFAs, especially of omega-3 (ω3)-α-linolenic acid (18:3 ω3), eicosapentaenoic acid (EPA; 20:5 ω3), docosapentaenoic acid (22:5 ω3), docosahexaenoic acid (DHA; 22:6 ω3), and omega-6 (ω6)-γ-linolenic (GLA; 18:3 ω6), arachidonic acid (20:4 ω6) are essential for human health (da S Vaz, Moreira, de Morais, & Costa, 2016; de Jesus Raposo, de Morais, & de Morais, 2013). The importance of these compounds is based on the inability of humans to synthesize them and, thus, the need to acquire them through the diet. Nowadays, fish oils are the most common source commercially available (Khan, Shin, & Kim, 2018). However, microalgae are also an alternative valuable source of ω3 and ω6 PUFAs, especially marine strains from the genera *Chaetoceros*, *Isochrysis*, *Nannochloropsis*, *Pavlova*, *Phaedactylum*, *Skeletonema*, *Thalassiosira*, and *Tetraselmis* (de Jesus Raposo, de Morais, & de Morais, 2013). Compared to fish oil, as algae grow very quickly and do not contribute to overfishing, they may be a more sustainable source for oil production (Adarme-Vega et al., 2012). Moreover, because microalgae are grown under controlled conditions and purified, algal oil is free from toxins and heavy metals, such as mercury and lead, that may be present in fish and fish oils (Cohen & Ratledge, 2010; Harwood, 2019). In addition, microalgal oils contain sterols and carotenoids, conferring them an additional nutritional added value compared to fish oil (Ryckebosch et al., 2014). In 2004, the Food and Drugs Administration (FDA) claimed that food containing ω3 PUFAs, particularly EPA and DHA, reduces the risk of coronary heart disease (Adarme-Vega, Thomas-Hall, & Schenk, 2014). Furthermore, both play important roles in decreasing triglycerides and increasing HDL, and have been associated with the prevention or improvement of ocular, neurological, and renal diseases, arthritis, and hypertension (de Jesus Raposo, de Morais, & de Morais, 2013).

Natural pigments (chlorophylls, phycobiliproteins, and carotenoids) are important for photosynthetic algal metabolism. Moreover, they develop biological activities, such as antioxidant, anticarcinogen, anti-inflammatory, anti-obesity, antiangiogenic, and neuroprotective (Guedes, Barbosa, Amaro, Pereira, & Malcata, 2011; Pangestuti & Kim, 2011). Currently, carotenoids (lutein, β-carotene, lycopene, and astaxanthin) are mainly used as dietary supplements, fortified foods, and food and beverage natural colorants (Vílchez et al., 2011).

Microalgae also contain water-soluble vitamins which can be used as ingredients or food supplements. Microalgae are a source of nearly all essential vitamins, including tocopherols, ascorbic acid, B1, B2, B3, B6, B9, B12, pantothenic acid, nicotinic acid, inositol, and biotin (da S Vaz, Moreira, de Morais, & Costa, 2016). These compounds have a consolidated effect on the immune system through antioxidant activity, cell formation, and blood coagulation (Khan, Shin, & Kim, 2018). Vitamin B12, usually only found in the animal kingdom (Nakos et al., 2017), and iron, in particular, can be found in high levels in some microalgae, which makes them especially interesting as nutritional supplements for vegetarians (Gouveia, Batista, Sousa, et al., 2008).

15.2 Microalgae application in foods (R&D)

Nowadays, consumers are demanding sophisticated, innovative, and healthy food products. Traditional foods, including bread, pasta, breakfast cereals, biscuits, cookies, dairy products, mayonnaises, and gelled desserts, that are part of many European diets, can be used as vehicles to deliver nutraceuticals. Moreover, microalgae can be used to improve the appearance of foods, changing their color, microstructure, and rheological properties (Batista, Raymundo, Sousa, & Empis, 2006). The viability of incorporating microalgae biomass into food products depends on several variables such as the applied processing type and intensity (e.g., thermal, mechanical), the nature of the food matrix (e.g., emulsion, gel, aerated dough systems), and potential interactions with other food components (e.g., proteins, polysaccharides, lipids, sugars, salts) (Gouveia, Batista, Sousa et al., 2008). Some studies have already used both microalgae-derived compounds and the whole biomass, as food ingredients for different purposes (Table 15.1).

Bread is the most widely consumed bakery product, and represents a great way to deliver microalgae bioactive compounds with valuable health benefits. The incorporation of microalgae to enhance the nutritional and textural properties of bread was done more than 20 years ago (Arafah, Abassy, Morcos, & Hussein, 1980; Finney, Pomeranz, & Bruinsma, 1984; Saleh, El-Fouly, Shaheen, El-Malky, & Khorshid, 1987). Several studies have used *Arthrospira* and its extracts for incorporation into bread to increase the protein and mineral contents (Ak et al., 2016; De Marco, Steffolani, Martínez, & León, 2014; Dinu, Vlasceanu, Dune, & Rotaru, 2012; Hafsa, Amel, Samia, & Sidahmed, 2014; Saharan & Jood, 2017). Other microalgae species, such as *Isochrysis galbana, Scenedesmus almeriensis, Chlorella vulgaris, Tetraselmis* sp., and *Nannochloropsis* sp. have also been incorporated in bread, affecting its color, crust, and crumb and implied an increase in browning (García-Segovia, Pagán-Moreno, Lara, & Martínez-Monzó, 2017; Graça, Fradinho, Sousa, & Raymundo, 2018; Lafarga et al., 2019). Microalgae have also been applied to gluten-free bread, not only to enhance its nutritional value, but also to mimic the gluten network, improving the physicochemical and rheological characteristics of the dough and bread (Figueira, Crizel, & Salas-Mellado, 2011; Khemiri et al., 2020; Selmo & Salas-Mellado, 2014). All three studies obtained a higher protein and mineral content in bread fortified with microalgae, collecting high scores in sensory evaluation panels (Fig. 15.1C,D). This is a promising pathway since the demand for gluten-free products has registered an impressive increase, not

TABLE 15.1 Microalgae incorporation in different food products (R&D).

Food category	Food products	Microalgae species	References
Cereal-based products	Bread	*Arthrospira* sp., *Isochrysis galbana*, *Tetraselmis suecica*, *Scenedesmus almeriensis*, *Nannochloropsis gaditana*	Ak et al. (2016), Dinu, Vlasceanu, Dune, & Rotaru, 2012, García-Segovia, Pagán-Moreno, Lara, & Martínez-Monzó, 2017, Graça et al. (2018), Hafsa et al. (2014), Lafarga, Mayre, et al. (2019), Saharan & Jood, 2017
	Gluten-free bread	*Arthrospira platensis*, *Chlamydomonas* sp., *Nannochloropsis gaditana*	Figueira, Crizel, & Salas-Mellado, 2011, Khemiri et al. (2020), Selmo & Salas-Mellado, 2014
	Cassava cake and doughnut	*Arthrospira platensis*	Navacchi, de Carvalho, Takeuchi, & Danesi, 2012, Rabelo et al. (2013)
	Croissant	*Arthrospira platensis*	Massoud, Khosravi-Darani, Nakhsaz, & Varga, 2016
	Strudel	*Arthrospira platensis*	Khosravi-Darani, Gholami, & Gouveia, 2017
	Sourdough "crostini"	*Arthrospira platensis*	Niccolai et al. (2019)
	Chikki	*Nannochloropsis oculata*, *Isochrysis galbana*	Babuskin et al. (2015)
	Cookies, biscuits, and crackers	*Arthrospira platensis*, *Chlorella vulgaris*, *Isochrysis galbana*, *Tetraselmis suecica*, *Haematococcus pluvialis*, *Phaedactylum tricornutum*, *Nannochloropsis oculata*	Babuskin, Krishnan, Babu, Sivarajan, & Sukumar, 2014, Batista et al. (2017), Bolanho et al. (2014), El Baky, El Baroty, & Ibrahem, 2015, Gouveia, Batista, Miranda, Empis, & Raymundo, 2007, Gouveia, Coutinho et al. (2008),

(Continued)

TABLE 15.1 (Continued)

Food category	Food products	Microalgae species	References
			Hossain et al. (2017), Lafarga, Mayre, et al. (2019), Marcinkowska-Lesiak, Onopiuk, Zalewska, Ciepłoch, & Barotti, 2018, De Morais, De Miranda, & Costa, 2006; Onacik-Gür et al. (2018), Shahbazizadeh, Khosravi-Darani, & Sohrabvandi, 2015, Singh, Singh, Jha, Rasane, & Gautam, 2015
	Extruded snacks	*Arthrospira* sp.	Joshi, Bera, & Panesar, 2014, Lucas et al. (2017, 2018), Tańska, Konopka, & Ruszkowska, 2017
	Pasta, noodles	*Arthrospira* sp., *Chlorella vulgaris, Isochrysis galbana, Dunaliella vlkianum, Dunaliella salina, Nannochloropsis* sp.	Agustini, Ma'Ruf, Widayat, Wibowo, & Hadiyanto, 2017, Babuskin, Krishnan, Babu, Sivarajan, & Sukumar, 2014, El-Baz, Abdo, & Hussein, 2017, Fradinho et al. (2020), Fradique et al. (2010, 2013), Lemes, Takeuchi, de Carvalho, & Danesi, 2012, Özyurt et al. (2015), Pagnussatt, Spier, Bertolin, Costa, & Gutkoski, 2014, De Marco, Steffolani, Martínez, & León, 2018, Zouari et al. (2011)

(*Continued*)

TABLE 15.1 (Continued)

Food category	Food products	Microalgae species	References
Dairy products	Fermented milk, Yogurt	*Arthrospira platensis*, *Chlorella vulgaris*, *Pavlova lutheri*	Barkallah et al. (2017), Beheshtipour et al. (2012, 2013), Guldas & Irkin, 2010, Robertson et al. (2016), Sengupta & Bhowal, 2017, Varga, Szigeti, Kovács, Földes, & Buti, 2002
	Feta cheese, Cheese analogs	*Arthrospira platensis*, *Chlorella vulgaris*	Golmakani et al. (2019), Mazinani, Fadaei, & Khosravi-Darani, 2016, Mohamed, Abo-El-Khair, & Shalaby, 2013, Tohamy, Shaaban, Ali, & Hasanain, 2019, Yasin & Shalaby, 2013
	Ice-cream	*Arthrospira* sp.	Malik & Paul, 2012
Fruit and vegetables	Broccoli soup	*Spirulina* sp., *Chlorella* sp., *Tetraselmis* sp.	Lafarga, Acién-Fernández, et al. (2019)
	Tomato puree	*Isochrysis galbana*, *Nannochloropsis* sp.	Gheysen, Demets et al. (2019), Gheysen, Lagae et al. (2019)
	Smoothies	*Spirulina* sp., *Chlorella* sp.	Castillejo et al. (2018)
Meat and fish products	Burgers, beef patties	*Spirulina* sp., *Chlorella* sp., *Isochrysis galbana*, *Picochlorum* sp.	Atitallah, Hentati, Dammak, Hadrich, & Fendri, 2019, Marti-Quijal, Zamuz, Tomašević, Rocchetti, et al. (2019), Žugčić et al. (2018)
	Pâté, pork sausages, turkey breasts, and patties, chicken roti	*Spirulina* sp., *Chlorella* sp.	Marti-Quijal et al. (2018), Parniakov et al. (2018), Thirumdas et al. (2018), Zamuz et al. (2019)

(*Continued*)

TABLE 15.1 (Continued)

Food category	Food products	Microalgae species	References
Others	Gelled desserts	*Chlorella vulgaris, Haematococcus pluvialis, Arthrospira maxima, Diacronema vlkianum*	(Batista et al., 2011; Batista, Gouveia, Nunes, Franco, & Raymundo, 2008), Gouveia, Batista, Raymundo, & Bandarra, 2008
	Oil-in-water emulsions, mayonnaise	*Chlorella vulgaris, Arthrospira maxima, Haematococcus pluvialis*	Abd El-Razik & Mohamed, 2013, Batista, Raymundo, Sousa, Empis et al. (2006), Batista, Raymundo, Sousa, Empis, and Franco et al. (2006), Raymundo, Gouveia, Batista, Empis, & Sousa, 2005
	Chewing gum	*Isochrysis galbana, Nannochloropsis oculata, Porphyridium cruentum*	Palabiyik et al. (2018), Toker (2019)
	Shakes	*Arthrospira* sp.	Santos, de Freitas, Moreira, Zanfonato, & Costa, 2016
	Kiwifruit pastilles	*Arthrospira platensis*	Khazaei Pool, Shahidi, Mortazavi, Azizpour, and Daneshzad (2016)

only due to people suffering from different pathologies related to gluten intake (e.g., celiac disease, gluten sensitivity, etc.), but also a growing belief that gluten-free foods are healthier (Brouns, van Buul, & Shewry, 2013).

Overall, results suggest that microalgal biomass can be effectively incorporated into bread formulations to obtain high-quality products. However, most of the studies incorporating microalgae at concentrations higher than 3% reported technological limitations such as decreased loaf volume or negative sensorial attributes. It is, then, important to select not only suitable microalgae concentrations but also to optimize the water content as well as mixing and proofing times to achieve suitable specific volumes (which can be correlated with textural attributes) of microalgae-containing bread (Lafarga, 2019).

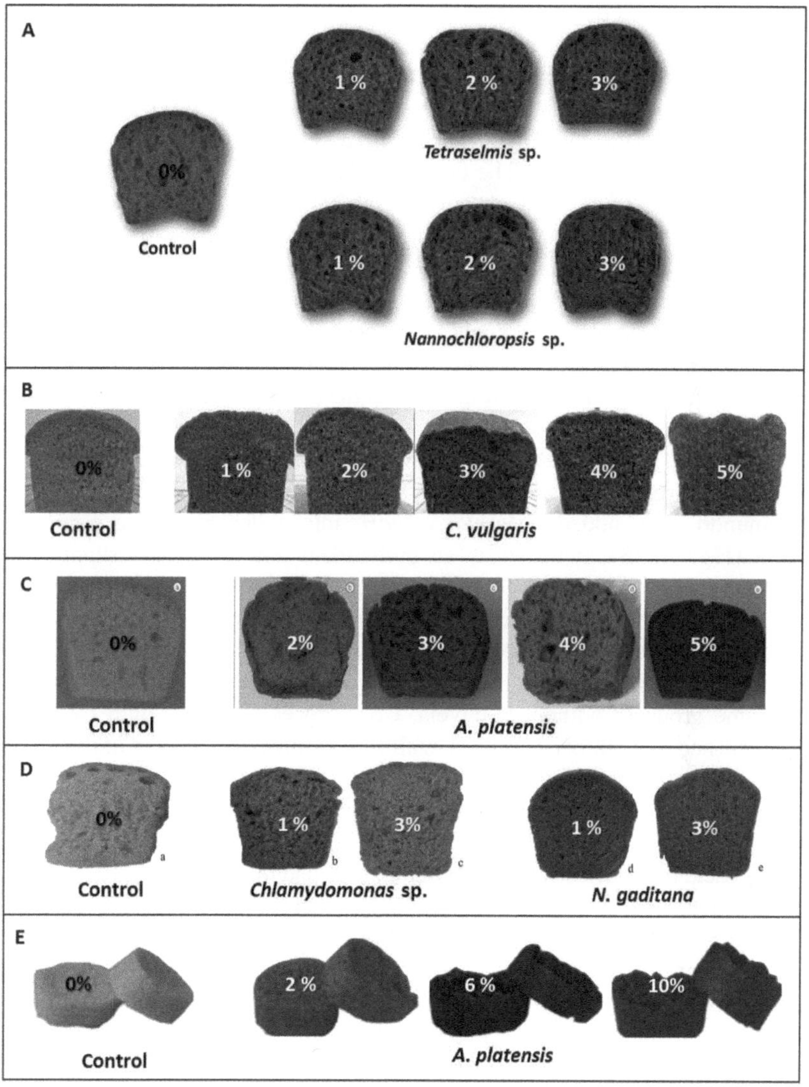

FIGURE 15.1 Bread supplemented with different microalgae: (A) *Tetraselmis* sp. and *Nannochloropsis* sp. (1%–3% wt./wt.) (Lafarga, Mayre, et al., 2019); (B) *Chlorella vulgaris* (0%–5% wt./wt.) (Graça et al., 2018); (C) gluten-free bread with *Arthrospira platensis* (2%–5% wt./wt.) (Figueira, Crizel, & Salas-Mellado, 2011); (D) gluten-free bread with *Chlamydomonas* sp. and *Nannochloropsis gaditana* (1% and 3% wt./wt.) (Khemiri et al., 2020); (E) sourdough "crostini" with 2%, 6%, and 10% (wt./wt.) supplementation of *Arthrospira platensis* (Niccolai et al., 2019).

Many innovative studies have been using *Arthrospira platensis* for incorporating different baked products, such as sourdough "crostini" (Fig. 15.1E), strudel, croissant, chikki, cassava cake, and doughnuts (Khosravi-Darani, Gholami, & Gouveia, 2017; Massoud, Khosravi-Darani, Nakhsaz, & Varga, 2016; Navacchi, de Carvalho, Takeuchi, & Danesi, 2012; Niccolai et al., 2019; Rabelo et al., 2013). The cassava cake and doughnuts were developed to offer a gluten-free food alternative for celiac people, with better nutritional value (protein, fiber, and ash). The overall acceptability of cassava cake showed the potential of this low-cost and easy preparation food to be incorporated into school lunches to improve children's nutritional intake (Navacchi, de Carvalho, Takeuchi, & Danesi, 2012; Rabelo et al., 2013).

Cookies and biscuits are other bakery products that have been used in several studies for incorporating microalgae-based ingredients due to their taste, texture, appearance, versatility, convenience, and ease of conservation and transport. Microalgae can be incorporated into cookies and biscuits for different purposes (Fig. 15.2). Gouveia, Batista, Miranda, Empis, & Raymundo, 2007 added *Chlorella vulgaris* into butter cookies as a coloring agent (Fig. 15.2B), but also as an antioxidant and nutritional supplement. This microalga also enhanced firmness due to the incorporation of a combination of different proteins and polysaccharides that are especially important in the water absorption process, resulting in a more compact structure (Gouveia, Batista, Miranda, Empis, & Raymundo, 2007). Similar studies by the same group of authors (Gouveia, Coutinho et al., 2008) with *Isochrysis galbana* and Batista et al. (2017) with *Arthrospira platensis, Chlorella vulgaris, Tetraselmis suecica*, and *Phaeodactylum tricornutum* used these microalgae as vehicles to provide algae-bioactives, including long-chain ω3 PUFAs (EPA and DHA) (Fig. 15.2C,D). Lafarga, Mayre, et al. (2019) developed crackers fortified with *Tetraselmis* sp. and *Nannochloropsis* sp. (Fig. 15.2E). The results showed that microalgae incorporation not only enhances the antioxidant capacity but also the amount of polyphenols and the antioxidant capacity of the enzymatic digestive extracts, suggesting healthier products (Lafarga, Mayre, et al., 2019). *Arthrospira platensis* was also incorporated into cookies to increase their nutritional properties, including protein and fiber content (Bolanho et al., 2014; El Baky, El Baroty, & Ibrahem, 2015; Onacik-Gür, Zbikowska, & Majewska, 2018; Shahbazizadeh, Khosravi-Darani, & Sohrabvandi, 2015; Singh, Singh, Jha, Rasane, & Gautam, 2015). Furthermore, Hossain et al. (2017) verified that the incorporation of *Haematococcus pluvialis*, aside from increasing the antioxidant activity, lowered the glycemic response of the cookies.

Use of extruded snacks is growing worldwide due to their convenient and practical consumption and the increasing diversity of shapes, textures, and flavors. However, the extrudates available on the market have low nutritional value, with reduced protein and biocompound concentrations. Thus, the demand for snacks with improved nutritional and functional properties has

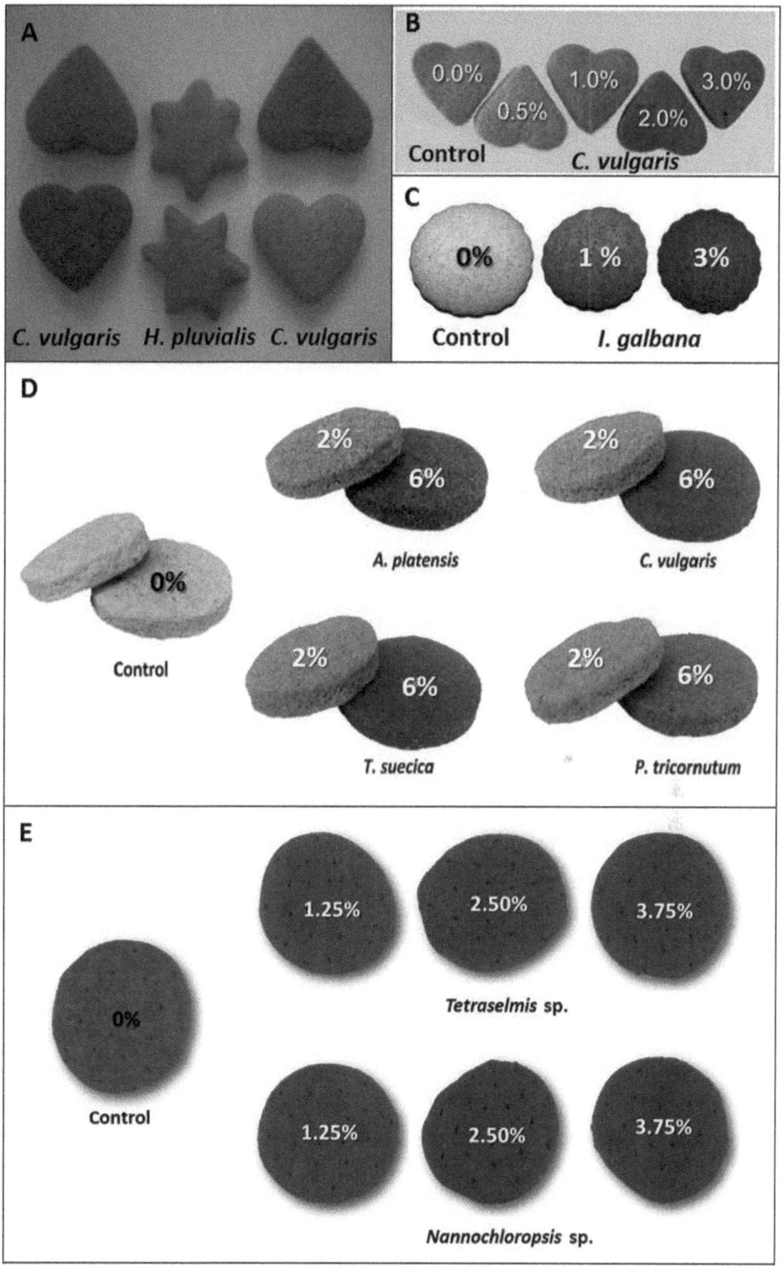

FIGURE 15.2 Biscuits, cookies, and crackers supplemented with different microalgae: (A) biscuits with *Chlorella vulgaris* (green), carotenogenic *Chlorella vulgaris* (orange), and *Haematococcus pluvialis* (pink) (Gouveia, Batista, Sousa, et al., 2008); (B) biscuits with *Chlorella vulgaris* (0.0%–3.0% wt./wt.) (Gouveia, Batista, Miranda, Empis, & Raymundo, 2007); (C) biscuits with *Isochrysis galbana* (0%–3% wt./wt.) (Gouveia, Coutinho et al., 2008); (D) cookies with *Arthrospira platensis, Chlorella vulgaris, Tetraselmis suecica,* and *Phaedactylum* at 2% and 6% (wt./wt.) (Batista et al., 2017); (E) crackers with *Tetraselmis* sp. and *Nannochloropsis* sp. (1.25%–3.75% wt./wt.) (Lafarga, Mayre, et al., 2019).

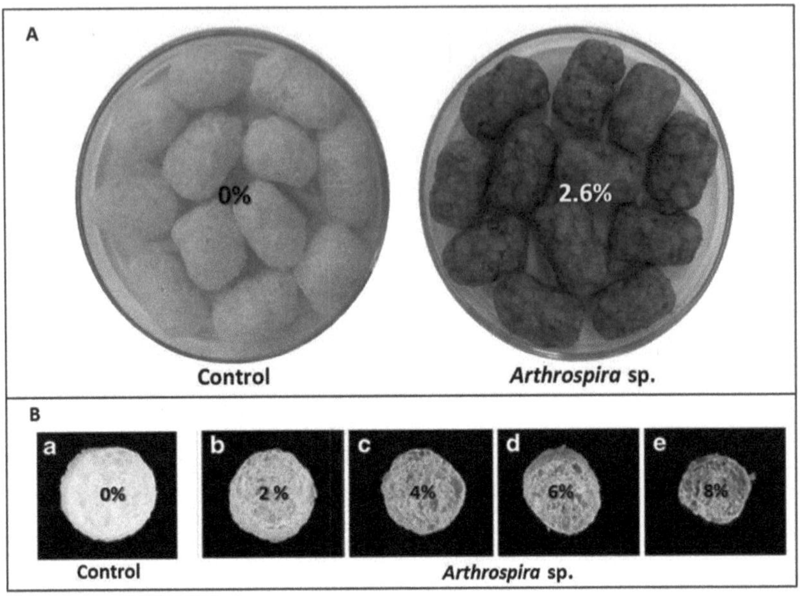

FIGURE 15.3 Extruded snacks supplemented with microalgae: (A) snacks (corn and rice) with 2.6% (wt./wt.) *Arthrospira* sp. (Lucas, de Morais, Santos, & Costa, 2018); (B) corn extrudates with several concentrations (0%–8% wt./wt.) of *Arthrospira* sp. incorporation (Tańska, Konopka, & Ruszkowska, 2017).

increased. Lucas, de Morais, Santos, and Costa (2017) developed snacks enriched with *Arthrospira* sp. which combine the versatility and convenience of extrudates with the nutritional quality of this microalga (Fig. 15.3A), as was done earlier by Joshi, Bera, & Panesar, 2014 and Tańska, Konopka, & Ruszkowska, 2017.

Pasta is a traditional food product that, similarly to cookies and biscuits, is well accepted by consumers for its sensory attributes, low cost, and ease of preparation and transport. These advantages, along with its nutritional value, have made this food a potential vehicle for nutraceuticals such as vitamins and PUFAs (Fradique et al., 2010). Several authors have studied the incorporation of *Chlorella vulgaris* and *Spirulina maxima* into pasta products to enhance their nutritional value. Overall, pasta enriched using microalgae presented better quality parameters, such as color stability and firmness, having higher acceptance scores than the control pasta due to their appellative colors, texture, and nutritional value (Fig. 15.4A,B) (Fradique et al., 2010). Pasta prepared with *I. galbana* and *Dunaliella vlkianum* biomass (Fig. 15.4C) presented higher resistance to the cooking process, enriching significantly the biomass with ω3 PUFAs, in particular EPA and DHA which were not present in the control pasta (Fradique et al., 2013).

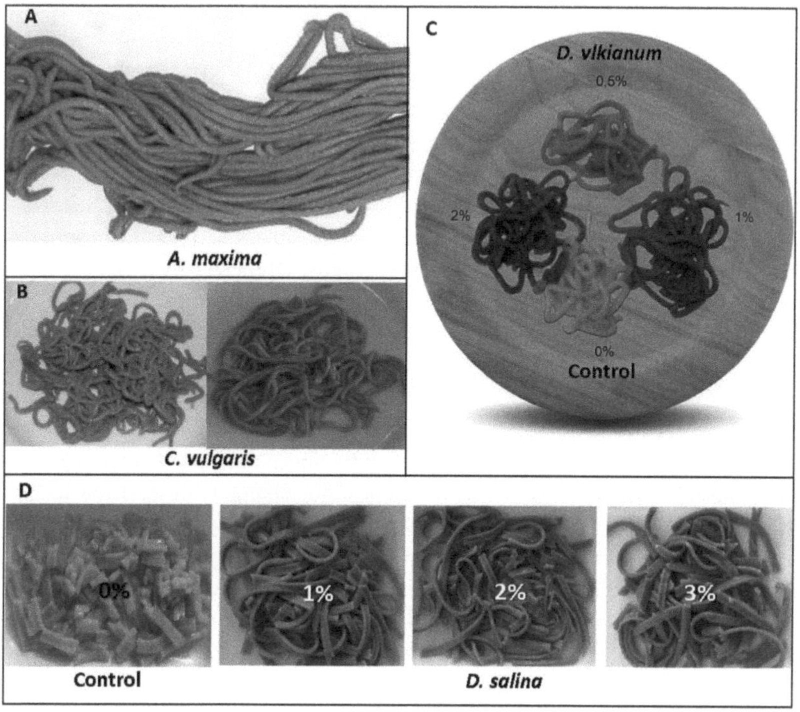

FIGURE 15.4 Pasta containing microalgae: (A) spaghetti with *Arthrospira maxima* (Fradique et al., 2010); (B) spaghetti green and orange (carotenogenic) with *Chlorella vulgaris* (Fradique et al., 2010); (C) spaghetti with *Diacronema vlkianum* (0%–2% wt./wt.) (Fradique et al., 2013); (D) tagliatelle with *Dunaliella salina* (0%–3% wt./wt.) (El-Baz, Abdo, & Hussein, 2017).

Likewise, other studies on the effects of *Spirulina platensis* in pasta products have been carried out, resulting in a higher protein pasta with improved technological and sensorial attributes compared to the control (De Marco, Steffolani, Martínez, & León, 2014; Lemes, Takeuchi, de Carvalho, & Danesi, 2012; Pagnussatt, Spier, Bertolin, Costa, & Gutkoski, 2014; Zouari et al., 2011). Agustini, Ma'Ruf, Widayat, Wibowo, & Hadiyanto, 2017 showed that adding *S. platensis* into dried noodles could improve their nutritional value, especially on protein and β-carotene, complying with the Indonesian National Standard for protein, water, and ash content. Furthermore, *Dunaliella salina* was also used by El-Baz, Abdo, & Hussein, 2017 to enhance the nutritional value of pasta, namely in terms of protein, fat, and minerals (calcium, iron, magnesium, and potassium).

Several authors have also studied the incorporation of microalgae into dairy products as bioactive compounds carriers. It was found that *Arthrospira* species could enhance the growth and viability of probiotic bacteria in fermented milk, yogurt, and processed cheese (Beheshtipour,

Mortazavian, Mohammadi, Sohrabvandi, & Khosravi-Darani, 2013; Golmakani, Soleimanian-Zad, Alavi, Nazari, & Eskandari, 2019; Guldas & Irkin, 2010; Mazinani, Fadaei, & Khosravi-Darani, 2016; Varga, Szigeti, Kovács, Földes, & Buti, 2002). Furthermore, these products not only exhibited better organoleptic and rheological properties and nutritional contents (protein, fiber, and minerals), but also a higher antioxidant activity (Barkallah et al., 2017; Golmakani, Soleimanian-Zad, Alavi, Nazari, & Eskandari, 2019; Tohamy, Shaaban, Ali, & Hasanain, 2019). *Chlorella vulgaris* was successfully incorporated into yogurt (Beheshtipour, Mortazavian, Haratian, & Khosravi-Darani, 2012) and cheese (Mohamed, Abo-El-Khair, & Shalaby, 2013; Yasin & Shalaby, 2013). Overall, the sensorial acceptability of these microalgae-enriched products was similar to the control. A lipid extract from *Pavlova lutheri* was used to manufacture ω3 PUFAs-enriched yogurt. No further differences were significant between the control and supplemented yogurts, except for color due to microalgal pigments. However, the sensorial analysis suggested that these color changes, in addition to fishy taste and bad texture of supplemented yogurt, negatively affected the overall acceptability (Robertson et al., 2016). Malik & Paul, 2012 incorporated *Arthrospira* sp. into an ice-cream to enrich its chemical composition. The addition of this microalga could replace the synthetic stabilizers and colorants in ice-cream formulations, without affecting the sensory attributes of the product (Malik & Paul, 2012).

Meat products are important sources of nutrients, lipids, proteins, minerals, and vitamins, which make them a widely consumed group of foods around the world. However, the market is constantly innovating to meet ever-changing consumer demands (Hathwar, Rai, Modi, & Narayan, 2012). There has been a noticeable trend of developing meat products partially replaced with a plant food source, especially soy proteins. However, due to the potential allergenicity of soy, interest has been sparked into using microalgae proteins in food (Madeira et al., 2017). New studies incorporating microalgae into meat (e.g., pork liver pâté, turkey breasts, patties and burgers, chicken roti, pork sausages, beef patties, etc.) have been recently published (Marti-Quijal et al., 2018; Marti-Quijal, Zamuz, Tomašević, Gómez, et al., 2019; Marti-Quijal, Zamuz, Tomašević, Rocchetti et al., 2019; Parniakov et al., 2018; Thirumdas et al., 2018; Zamuz et al., 2019; Žugčić et al., 2018). Studies aimed to improve the physicochemical and nutritional qualities of different meat products by using different sources of protein, including *Chlorella* sp. and *Spirulina* sp. Overall, a few changes in physicochemical and textural properties were observed among the protein-enriched meat products. However, for the microalgae-containing products, the most noticeable variations were their color and amino acid profile. The microalgal proteins resulted in a significant decrease in the redness of the meat, acquiring a green-blue color due to the pigments they contain (Fig. 15.5). Meat products enriched with microalgal proteins had similar and, in some cases,

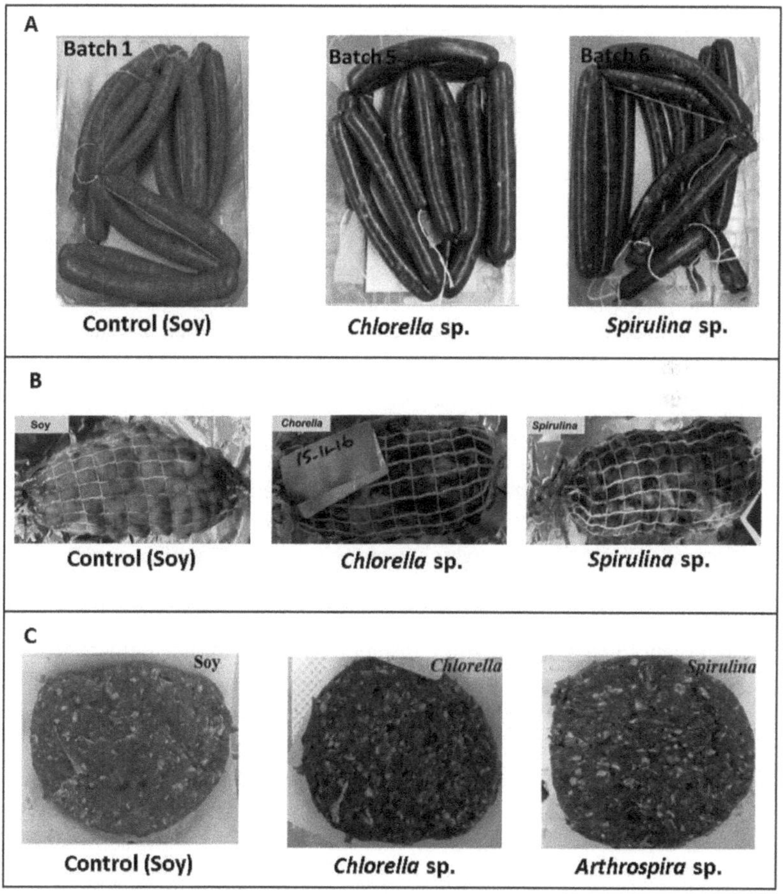

FIGURE 15.5 Meat products enriched with *Arthrospira* sp. and *Chlorella* sp. protein compared to control (soy protein): (A) fresh sausages (Marti-Quijal, Zamuz, Tomašević, Gómez, et al., 2019); (B) chicken roti (Parniakov et al., 2018); (C) beef patties (Žugčić et al., 2018).

enhanced protein content and amino acid profile, compared to controls formulated using soy protein. This suggests that they are a good alternative to soy protein in meat products in terms of nutritional value (Marti-Quijal et al., 2018; Marti-Quijal, Zamuz, Tomašević, Gómez, et al., 2019; Marti-Quijal, Zamuz, Tomašević, Rocchetti, et al., 2019; Parniakov et al., 2018; Thirumdas et al., 2018; Žugčić et al., 2018). From the amino acid profile, it is possible to determine the taste pattern of the products, since they have a great influence on this sensorial parameter, which is very important for their quality and acceptability (Žugčić et al., 2018). Meat products manufactured with microalgal proteins contained different groups of amino acids responsible for different tastes (sweet, bitter, and umami). In particular, aspartic and glutamic acids are correlated with the savory/umami taste, which is appreciated by consumers, and are predominant in beef patties and "chorizo" sausages elaborated with microalgal protein (Thirumdas et al., 2018; Žugčić et al., 2018). However, despite the significant differences between amino acid profiles, those were mitigated when combined to obtain the taste profile and thus did not affect the overall taste profile of beef patties, with bitter being the predominant taste (Žugčić et al., 2018). Among all the products, color and texture were the most visually distinctive properties that strongly affected consumer acceptance (Marti-Quijal, Zamuz, Tomašević, Gómez, et al., 2019; Marti-Quijal, Zamuz, Tomašević, Rocchetti, et al., 2019). However, sensorial analysis is necessary to study the impact that modifications in the physicochemical parameters of food have on consumer acceptability. Zamuz et al. (2019) performed a sensorial analysis of different meat products (pork liver pâté, turkey patties, marinated and cooked turkey breast, and fresh pork sausages) containing proteins from different sources, including *Chlorella* sp. and *Spirulina* sp. Products supplemented with microalgal proteins did not have good acceptance due to the different organoleptic characteristics of the microalgae which panelists are not familiar with, such as their green color from the microalgal pigments and bitter taste related to a higher concentration of amino acids (Zamuz et al., 2019). Likewise, microalgae-enriched chicken roti presented the best amino acid profile and ratio of all formulations but obtained very low scores in the sensory analysis due to a certain dry feeling and characteristic taste of microalgae related to the amino acid profile. Thus, future work is needed to improve the quality of microalgae-supplemented food to balance the chemical composition and sensory acceptability (Parniakov et al., 2018). Furthermore, the consumer's profile has a significant influence on the acceptance of new seaweed food products (Ing, Ling, Eng, & Alin, 2010). For example, fish consumers are more likely to accept seaweed-containing foods than meat consumers (De Boer, Schösler, & Boersema, 2013). Therefore consumer education and dissemination of information on the nutritional value will be crucial to increase the chances of product trials being successful (Ing, Ling, Eng, & Alin, 2010; Zamuz et al., 2019).

To overcome the possible fish flavor of microalgae-enriched products, this ingredient could be oriented to fish/marine dishes and sauce markets. For example, Atitallah, Hentati, Dammak, Hadrich, & Fendri, 2019 evaluated different microalgae as natural ingredients in canned fish burgers regarding texture and sensory attributes. Microalgae served as a good source of natural flavoring and coloring agents for fish products. Formulations containing 1% (wt./wt.) microalgae not only improved the nutritional content of the prepared fish burgers, but also the functional properties (water- and oil-holding capacities) and antioxidant activity. Furthermore, they had similar acceptability scores as control burgers without microalgae. However, higher concentrations of microalgae resulted in lower sensory acceptability due to inappropriate taste related to lipid oxidation (Atitallah, Hentati, Dammak, Hadrich, & Fendri, 2019).

The development of colored oil-in-water emulsions using microalgae is also an interesting field. Batista, Raymundo, Sousa, Empis, and Franco (2006) studied the effect of adding phycocyanin extracted from *Arthrospira maxima* on the rheological behavior of oil-in-water food emulsions, and found a significant reinforcement of their structure, increasing the resistance to structural breakdown (Fig. 15.6A). Furthermore, Raymundo, Gouveia,

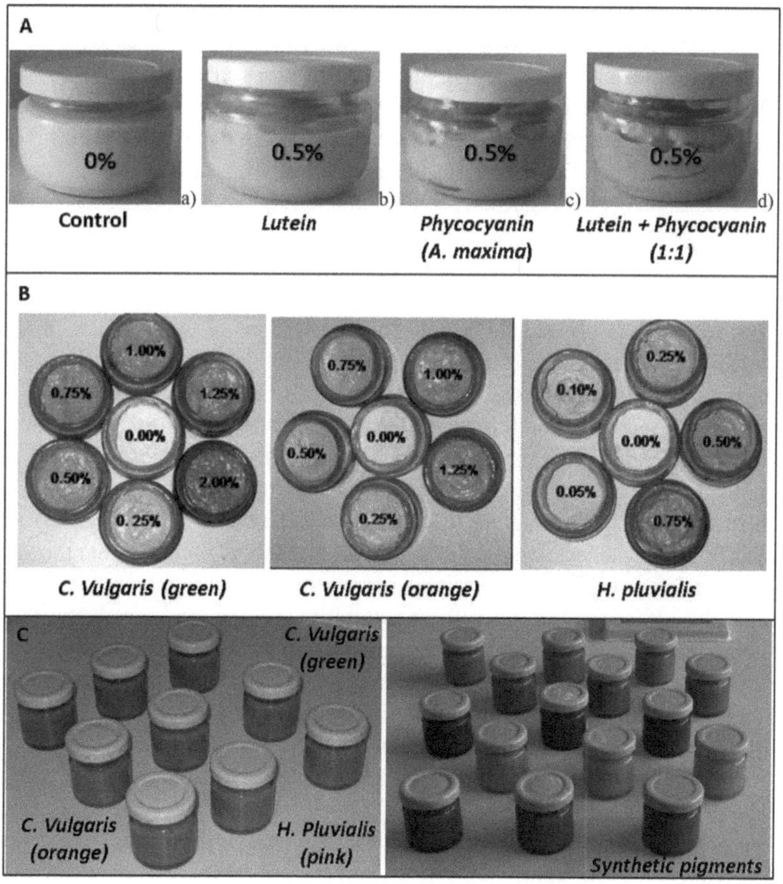

FIGURE 15.6 Vegetable oil and gel supplements with microalgae and its derivatives: (A) oil-in-water pea protein-stabilized emulsions with phycocyanin (extracted from *Arthrospira maxima*) (Gouveia, Batista, Sousa, et al., 2008); (B) oil-in-water pea protein-stabilized emulsion with *Chlorella vulgaris* (green), carotenogenic *Chlorella vulgaris* (orange) and *Haematococcus pluvialis* (top) at various concentrations (0%–1.25%) (Gouveia, Raymundo, Batista, Sousa, & Empis, 2006); (C) vegetarian food gels with the incorporation of 0.75% (wt./wt.) *Chlorella vulgaris* (green), carotenogenic *Chlorella vulgaris* (orange) and carotenogenic *Haematococcus pluvialis* (pink) compared to synthetic pigments (Gouveia, Batista, Sousa, et al., 2008).

Batista, Empis, & Sousa, 2005 and Gouveia, Batista, Raymundo, & Bandarra, 2008 incorporated *Chlorella vulgaris and Haematococcus pluvialis* into food emulsions, achieving positive effects on their functional and antioxidizing properties (Fig. 15.6B). These results are an important innovation for these type of products since they show that microalgae can act as a fat mimetic, obtaining appealing and stable colorations, while also improving the resistance to oil oxidation (Gouveia, Batista, Sousa, et al., 2008). Batista et al., 2011; Batista, Gouveia, Nunes, Franco, & Raymundo, 2008 and Gouveia, Batista, Raymundo, & Bandarra, 2008 studied the effect of incorporating microalgae on the structure of vegetarian food gels, an alternative to dairy desserts, as a way to provide both antioxidants and ω3 PUFAs. Different microalgae were tested, including *Arthrospira maxima*, *Chlorella vulgaris*, *Haematococcus pluvialis*, and *Diacronema vikianum* (Fig. 15.6C). In general, the microalgae gels resulted in less intense colorations (Fig. 15.6C) and reinforced structures, compared to gel with commercial pigments. *H. pluvialis* had a stronger effect on the rheological properties due to its higher fat content. Exceptionally, *A. maxima* promoted an extreme reduction of the gel's rheological parameters, probably due to thermodynamic incompatibility between the microalgal proteins and some gel components. Hence, further research is needed for a better understanding of the gelation process and interaction between microalgae and gel components as well as operating conditions (Gouveia, Batista, Sousa, et al., 2008).

Lafarga, Acién-Fernández, et al. (2019) aimed to develop broccoli soups enriched with microalgae, specifically *Spirulina* sp., *Chlorella* sp., and *Tetraselmis* sp., at concentrations of 0.5%–2.0% (wt./wt.) (Fig. 15.7A). The authors obtained an innovative microalgae-enriched soup with a higher antioxidant capacity and increased content of bio-accessible polyphenols. The increasing incorporation of microalgae led to lower flavor and texture scores, negatively affecting the overall acceptability of the end product. These results confirm others, where the optimum incorporation of microalgae concentration should be adjusted and is crucial for good consumer acceptance. However, due to its natural green color and associated health benefits, broccoli, as well as other vegetable soups, seem to be a product with great potential for widespread use of microalgae into the human diet (Lafarga, Acién-Fernández, et al., 2019).

Another product that has been also used for studying microalgae incorporation was chewing gum, namely as natural colorants. Toker (2019) and Palabiyik et al. (2018) used *Porphyridium cruentum, Isochrysis galbana*, and *Nannochloropsis oculata*, respectively, as natural coloring agents in chewing gum (Fig. 15.7B,C). One of the main advantages evidenced by both studies was the color stability in the chewing gum matrix throughout the chewing process, which is an important innovation for this type of product since commercial chewing gums tend to lose their color quickly during chewing. Furthermore, microalgae incorporation resulted in lower hardness and

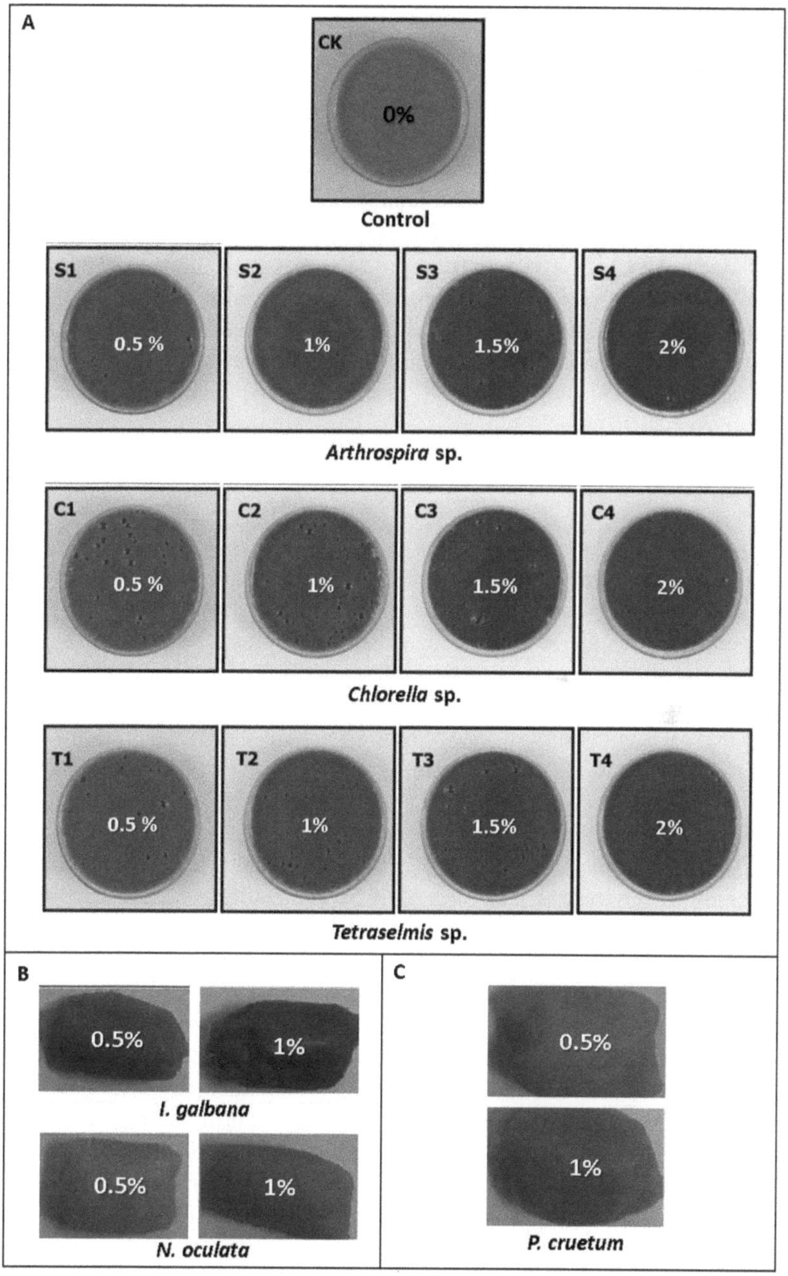

FIGURE 15.7 Other food products supplemented with microalgae and its derivatives include: (A) broccoli soups containing *Spirulina* sp., *Chlorella* sp., and *Tetraselmis* sp. at various concentrations (0%–2% wt./wt.) (Lafarga, Acién-Fernández, et al., 2019); (B) chewing gum colored green with *Isochrysis galbana* and *Nannochloropsis oculata* (0.5% and 1%) (Palabiyik et al., 2018); (C) chewing gum colored red with *Porphyridium cruentum* (0.5% and 1%) (Toker, 2019).

chewiness values, which imply lower energy requirements during chewing, and lower cohesiveness values, allowing the fast release of bioactive compounds present in chewing gums during the oral chewing process. However, this can also lead to the fast release of aroma compounds, resulting in chewing gums losing their flavor very prematurely in the chewing process, which is a disadvantage from the consumer's perspective (Palabiyik et al., 2018; Toker, 2019). Regarding sensorial analysis, there was no significant difference between sensorial parameters, except for the algal taste which was more evident for higher concentrations. Although both studies only focus on the coloring effect of microalgae biomass, they can also be seen as an interesting tool for providing nutritional supplementation with biologically active compounds such as antioxidants and fatty acids, replacing synthetic food additives, which can be toxic and/or allergenic for human health (Mizutani, 2009; Toker, 2019).

15.3 European market for microalgae-derived food products

Food industries have continuously evolved and adapted to meet ever-changing consumer needs and demands. In the European Union (EU), there is a growing movement interested in healthier foods from natural sources, free from synthetic additives (Asioli et al., 2017). The cultivation of microalgae at the industrial scale and the production of microalgae-based products is emerging as a commercial opportunity aiming to gain market share in the bioactive molecules segment, initially dominated by synthetic molecules (García, de Vicente, & Galán, 2017). Although there is a large number of microalgae species that can be found in nature, only a small amount are used in food applications (Barkia, Saari, & Manning, 2019). Large-scale commercialization of microalgae and their derivatives is still in its infancy, presenting some considerable challenges, which include: (1) low production capacities; (2) high production costs; (3) an intense (generally green) color and fishy taste and aroma; and (4) legislative and regulatory issues (Lafarga, 2019).

In the United States, the FDA is responsible for the regulatory status of algae products, and can assign Generally Recognized as Safe status to a product. In Europe, foods consisting of cell cultures or tissue cultures derived from microorganisms, fungi, or algae, or foods isolated or produced from the referred organisms, are categorized as novel foods, assuming that these substances have not been consumed significantly in the EU before 1997. Each member state has a competent authority that makes a first assessment of the new products, needing a final approval from the European Commission (EC), always following the 2015/2283 regulation, referred to as the novel food regulation. In the case that there are objections to the product, the European Food Safety Authority (EFSA) is responsible for carrying out the safety assessment of the novel foods (van der Spiegel, Noordam, & van der Fels-Klerx, 2013).

Also, the technology of microalgae cultivation and processing is still very immature (approximately 100 years) compared to other crop-based sources (more than 5000 years), with harvesting and dewatering processes strongly aggravating production costs. Current systems/technologies need further development and optimization, not only to achieve an improved economy of scale but also sustainability of the whole value chain. However, positive evidence shows that using microalgae as meat substitutes within the current scale of microalgae production technologies can result in a more environmentally sustainable product compared to pork and beef (Smetana, Sandmann, Rohn, Pleissner, & Heinz, 2017).

15.3.1 Novel food-approved microalgae

Different combinations of microalgae can already be found in world markets, most being sold for dietary supplement applications, in the forms of powders, capsules, tablets, pastilles, and liquid suspensions, with nutritional claims (Lafarga, 2019). Today, commercial facilities for microalgae production are scattered worldwide. Taiwan, Japan, the United States, China, Brazil, Spain, Israel, Germany, and Myanmar are the main producers of microalgal biomass and the products derived from it. The annual production of microalgal biomass dry is about 19,000 tons, generating an estimated USD 5.7 billion (Ramírez-Mérida, Zepka, & Jacob-lopes, 2017). Table 15.2 shows some microalgae-based products that are currently available in the market and their main producers.

Before 1997, only six microalgae were approved for human consumption. This list was composed of *Aphanizomenon flosaquae*, *Arthrospira platensis*, and *Spirulina* sp., all cyanobacteria, and *Chlorella luteoviridis*, *Chlorella pyrenoidosa*, and *Chlorella vulgaris*, belonging to the green algae. Today, in the EU, commercially cultivated microalgae also include oil from *Ulkenia* sp., EPA and DHA-rich oil from *Schizochytrium* sp., astaxanthin from *Haematococcus pluvialis*, and the whole algae regarding *Odontella aurita* and *Tetraselmis chui*, from which the only limitation is being able to grow in reactors. The market is currently dominated by *Arthrospira* and *Chlorella*, which are sold in over 20 countries worldwide. *Arthrospira* is promoted for its protein, GLA, and phycocyanin contents. *Chlorella*, in addition to high protein content, provides Chlorella Growth Factor, which is a water-soluble extract composed of several substances, including essential amino acids, peptides, proteins, vitamins, sugars, and nucleic acids (Merchant & Andre, 2001).

Other microalgae such as *Tetraselmis chui* are commercialized as a seafood flavoring agent and the diatom *Odontella aurita* as a food supplement. Some microalgae-derived products marketed are β-carotene from

TABLE 15.2 Microalgae-based products for food and the main producers worldwide (Camacho, Macedo, & Malcata, 2019; Olaizola & Grewe, 2019).

Microalgae species	Products/ application	Manufacturers
Spirulina sp.	Dietary supplement	America: Cyanotech (USA); Earthrise Nutritials (USA); Valensa International (USA); NP Nutra (USA); Algenol (USA) Asia: Fuqing King Dnarmsa Spirulina (China); Yunnan Green A Biological Project (China); Tianjin Norland Biotech (China); Hainan-DIC Microalgae (China); Japan Algae (Japan); Inner Mongolia Wushenzhao Ecological Industry Development (Mongolia); Wilson Group (Taiwan); Nan Pao International Biotech (Taiwan); FEBICO (Taiwan); June Pharmaceutical (Myanmar); Parry Nutraceuticals (India); Seagrass Tech (India); Pondicherry Spirulina Farms (India) Europe: Cyane (France); Algosud (France); Ecoduna (Austria); BlueBioTech (Germany); NeoAlgae (Spain) Oceania: Australian Spirulina (Australia)
	Food ingredient	America: Spirulina Viva (Mexico) Asia: June Pharmaceutical (Myanmar) Europe: Spiruline des Îles D'Or (France); Algosud (France); Algama (France); Ecoduna (Austria); Simris (Sweden); NeoAlgae (Spain)
	Phycocyanin (food colorant/additive)	America: Earthrise Nutritials (USA); Algenol (USA) Asia: Fuqing King DnarmsaSpirulina (China); Yunnan Green A Biological Project (China); Tianjin Norland Biotech (China); Japan Algae (Japan); Parry Nutraceuticals (India) Europe: AlgoSource (France); Fermentalg (France); NeoAlgae (Spain)
Chlorella sp.	Dietary supplement	America: Valensa International (USA); NP Nutra (USA)Asia: Yunnan Green A Biological Project (China); Tianjin Norland Biotech (China); Taiwan Chlorella Manufacturing (Taiwan); Wilson Group (Taiwan); FEBICO (Taiwan); Parry Nutraceuticals (India); Seagrass Tech (India)Europe: Allmicroalgae (Portugal); Buggypower (Portugal); Duplaco (Netherlands); Ecoduna (Austria); BlueBioTech (Germany); NeoAlgae (Spain)
	Food ingredient	Asia: Taiwan Chlorella Manufacturing (Taiwan) Europe: Allmicroalgae (Portugal); Algama (France); Phycom (Netherlands); Duplaco (Netherlands); Corbion (Netherlands); Ecoduna (Austria); Algomed/RoquetteKloetze (Germany)

(*Continued*)

TABLE 15.2 (Continued)

Microalgae species	Products/ application	Manufacturers
Haematococcus pluvialis	Astaxanthin (Dietary supplement)	America: Cyanotech (USA); Valensa International (USA); AlgaeCanBiotech (Canada); Atacama Bio Natural Products (Chile) Asia: Yunnan Green A Biological Project (China); Tianjin Norland Biotech (China); BGG (China); Algaetech International (Malaysia); AlgaTech (Israel) Europe: Algalif (Iceland); Algamo (Czech Republic); AstraReal AB (Sweden); BlueBioTech (Germany); Solixant (Russia); Simris (Sweden)
	Astaxanthin (food ingredient)	America: Atacama Bio Natural Products (Chile) Asia: Algaetech International (Malaysia)
Schizochytrium sp.	ω3 EPA/DHA	Asia: Xiamen Huison Biotech (China) Europe: Veramaris (Netherlands); Fermentalg (France); Solixant (Russia); Simris (Sweden)
Dunaliella salina	β-Carotene (food additive/vitamin/ colorant)	Asia: Seagrass Tech (India); Nikken Sohonsa (Japan) Europe: Corbion (Netherlands); BlueBioTech (Germany) Oceania: BASF (Australia); Plankton Australia (Australia)
Phaedactylum tricornutum	Fucoxanthin (dietary supplement)	Asia: AlgaTech (Israel)
Euglena gracilis	Dietary supplement	America: Kemin Industries (USA) Asia: Euglena (Japan)

Dunaliella, DHA from *Crypthecodinium cohnii*, and the blue colorant phycocyanin from *Arthrospira* (Enzing, Ploeg, Barbosa, & Sijtsma, 2014; García, de Vicente, & Galán, 2017). Some of these microalgal-based products, especially β-carotene, astaxanthin, and phycocyanin have high market value (Table 15.3) and can cost hundreds and even thousands of euros per kg depending on their purity (Caporgno & Mathys, 2018). On the other hand, the microalgal food supplements market cost is around 40 euros per kg (Enzing, Ploeg, Barbosa, & Sijtsma, 2014). Thus, microalgae-based products with biofunctional compounds could improve the economic feasibility of the whole processing, since the higher selling prices could cover the higher cost associated with microalgae cultivation and downstream processes (Caporgno & Mathys, 2018; Jacob-Lopes et al., 2019). Microalgae production either as a final product or as a source of bioactive

TABLE 15.3 Production costs and selling prices of commercial microalgae products (adapted from Jacob-Lopes et al., 2019).

Products	Production cost (EUR per kg)	Selling price (EUR per kg)
Whole biomass		
Spirulina	1.8	7.3
Chlorella	4.6	17.3
Schizochytrium	1.8	4.7
Pigments		
β-Carotene	95.6	718.9
Astaxanthin	502.3	2275.0
Phycocyanin	41.9	498.7
PUFAs		
EPA	35.3	91.0
DHA	35.5	109.2

Prices converted from dollars to euros according to the conversion rate of 1 Dollar = 0.91 Euro.
DHA, Docosahexaenoic acid; *EPA*, eicosapentaenoic acid; *PUFAs*, polyunsaturated fatty acids.

compounds is an expanding sector, that has been gaining interest from large enterprises and startups worldwide (Sharma & Sharma, 2017).

15.3.2 Marketed microalgae-based food products

Despite the initial reluctance for novel foods, fortification of foods with nutraceuticals and food supplements has become an increasingly popular method for providing nutritional and functional food products to health-conscious consumers. This strategy allows the microalgae industry to diversify and grow outside the field of health food supplements, a market that currently uses 75% of the biomass to produce powders, tablets, and capsules.

Functional food supplemented with microalgae biomass is growing its position within the food market due to its sensorial benefits, convenience, and diversity, combining attractiveness with health benefits (Chacón-Lee & González-Mariño, 2010). Over the last decade, several companies have begun to involve themselves in developing foods containing microalgae or cyanobacteria (Lafarga, 2019). Most microalgae-containing food products that are currently being commercialized contain either *Spirulina* or *Chlorella* (Figs. 15.8 and 15.9). Nonetheless, novel foods containing other microalgae species are expected to enter the European market. For example, the Spanish company Fitoplancton Marino S.L. (Cadiz, Spain) is currently commercializing a freeze-dried *Tetraselmis chui* product (*Plancton*

FIGURE 15.8 Commercially available foods containing *Spirulina* sp.: Vegetables creams and broth (Vesana Superfoods - Neoalgae, Spain, https://www.vesana.es/); Spirulina Sesame Sprouted Bar (Go Raw, USA, https://goraw.com/); Organic raw crackers Spirulina (VitaOrganics, Bulgaria, http://www.vita-organic.eu/en/); Chocolates (The Algae Factory, the Netherlands, https://www.thealgaefactory.com/); Chocolate Spirulina balls "Zitronen zauber", Spirulina/Lemon bar (Lubs, Germany, https://www.lubs.de/); Raw Ba Protein Lemon Crisp (Simply Raw, Germany, http://www.simplyraw.de/); Veggie Wrap: Spirulina (WrawP, USA, https://wrawp.com/); Shell Pasta (Nutracentis, Italy, https://www.nutracentis.com/); Linguine Pasta (Officina della Pasta, Italy, https://www.officinadellapasta.it/). *Vesana Superfoods, Go Raw, VitaOrganic, The Algae Factory, Lubs, Simply Raw, WrawP, Nutracentis, Officina della Pasta.*

Marino Veta la Palma), which was recently authorized by the EFSA to be marketed as a Novel Food under Regulation (EC) No 258/97 (AECOSAN, 2017).

The incorporation of microalgae biomass into traditional food products has already been done at a commercial scale, with some products having already been launched into the world market (Figs. 15.8 and 15.9). Most microalgae-based products are commercialized in Asia, North America, and Europe. The latter has been dominating not only the production of microalgae-derived food products but also the global market for these products (Credence Research, 2018).

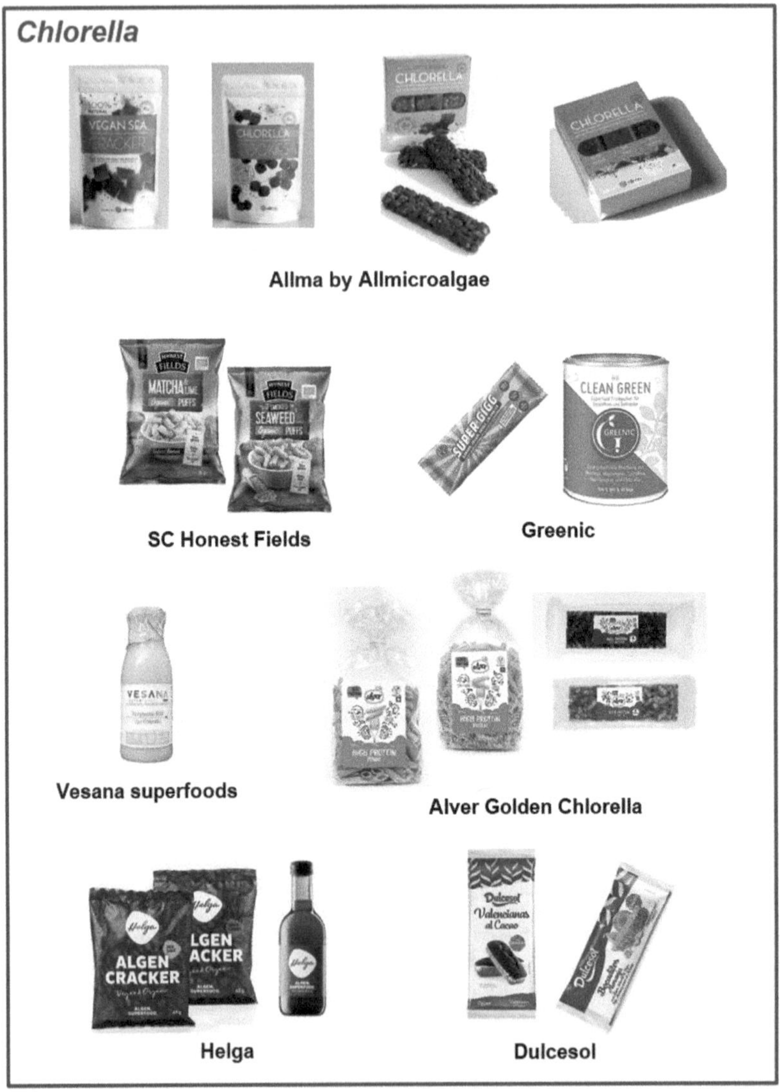

FIGURE 15.9 Commercially available foods containing *Chlorella* sp.: Vegan Sea Crackers, *Chlorella* Cookies, Seeds/Protein Bars (Allma by Allmicroalgae, Portugal, https://algaebyallma.com/); Organic Puffs (SC Honest Fields, Romania, http://www.honestfields.com/); Super Gigg bar, Clean Green Powder Mixture (Greenic, Germany, https://greenic-bio.de/); Vichyssoise with Chlorella (Vesana Superfoods - Neoalgae, Spain, http://www.vesana.es/); High Protein Pastas, Protein Bars, (Alver World SA, Switzerland, https://www.alver.ch/); Algen Crackers, Algen Superfood drink (Helga, Austria, https://hellohelga.com/); Cocoa Valencianas, Orange and Chlorella bites (Dulcesol – Vicky Foods, Spain, https://dulcesol.com/). *SC Honest Fields, Greenic, Vesana Superfoods, Alver Golden Chlorella, Helga, Dulcesol – Vicky Foods, Allmicroalgae.*

Despite the growing number of scientific reports on microalgae-containing food products, their global market presence is still minor (Caporgno & Mathys, 2018; Lafarga, 2019). The low marketability of microalgae-based products is mainly attributed to challenges in scaling up microalgal biomass cultivation and in legislative approval of these products by regulatory authorities (Caporgno & Mathys, 2018; Sidari & Tofalo, 2019). Further work is needed to undeniably prove the technological functionalities of microalgae in food products and contribute to the economic feasibility of this application.

15.4 Future trends in microalgae production and food application

1. *Microalgae production*

   The production of microalgal proteins, carbohydrates, and lipids at a commercial scale will not be accessible without further investment in research and development. It entails higher production volumes and intensive downstream processing (extraction and purification), causing a severe impact on total production costs. The costs of subsequent downstream processes account for 2%–90% of the total expenses of manufacturing microalgae-based bioproducts, depending on the purity and biochemical properties required (Jacob-Lopes et al., 2019). Finally, the microalgae-based processes are generally developed for a specific application. There is no market for the leftover biomass after extraction of the target bioproduct, and this can also be an opportunity to develop the microalgae-based industry.

   As the microalgae commercialization field has been quiet recently there is still much room for improvement, but we are now facing the point at which the biomass cultivation and processing can be economically sustainable. Regarding the use of biomass in a biorefinery concept, or just to extract valuable compounds, it is feasible, although still not economically viable, needing further optimization by increasing the R&D focus on this area (Koutra, Economou, Tsafrakidou, & Kornaros, 2018; Mennella et al., 2020).
2. *Microalgae for food application*

So far, microalgae and their derivatives have only been incorporated in very small quantities into food products. In most sensorial studies, the authors report relatively high acceptability scores, which are lowered at higher microalgae concentrations (Lafarga, Acién-Fernández, et al., 2019). The sensorial characteristics of microalgae, mainly color and flavor, are usually negatively perceived by consumers, limiting their incorporation in daily-use products (Caporgno & Mathys, 2018).

According to Hotchkiss (2010), the Western palate is still generally not accepting of unconventional tastes and textures derived from seaweed/microalgae products, that are commonly found in Asian markets (Hotchkiss, 2010). Thus, to better respond to the Western palate and preferences, further research on how to take advantage of the microalgae nutritional and functional properties without compromising food rheological and sensorial properties is required. Likewise, creating a vocabulary to describe these new flavors would help increase public awareness of seaweed/microalgae as food (Barbier et al., 2019). Furthermore, the development of exotic-flavored snacks and foods, with microalgae incorporated, along with the traditional Asian and Indian spices would increase the likelihood of Western consumers accepting these new tastes along with the new products (Chacón-Lee & González-Mariño, 2010). For example, chef Leonel Pereira (from Portugal), who was awarded a Michelin star, said about *Tetraselmis* microalga: "It was the best thing I ever smelled. My head started to imagine all the things I could do with it. In order to achieve this taste inside the palate, I would have to bring together a set of various ingredients to feel what is here. It is almost impossible to describe. It is an explosion of marine flavors." (RTP1, 2019).

Another solution to overcome the strong flavor of microalgae could be the replacement of autotrophic with heterotrophic microalgae, which have been shown to have neutral color and flavor, leading to better consumer acceptability (Klamczynska & Mooney, 2017). For example, Allma by Allmicroalgae (Portugal) has in its product portfolio a "*Smooth Chlorella Powder*" obtained from heterotrophically grown microalgae which presents a lighter green color and smooth flavor, due to the lower chlorophyll content. Another technical solution could also be the encapsulation of biomass to mask, improve rheological properties, and remove off-odors, allowing an increase of concentration of microalgae in foods.

Consumers are motivated by health concerns and familiarity regarding food ingredients, among others. Thus, greater dissemination is needed in order to increase consumers' knowledge and awareness of the health benefits of microalgae as a functional ingredient. Consumers' acceptance is a key success factor for market orientation, consumer-led opportunities, and penetration into established markets (Siró, Kápolna, Kápolna, & Lugasi, 2008).

15.4.1 Thickening/foaming agents

In most commercially available microalgae-containing food products, microalgae are not used for their technofunctional role, but as coloring agents, nutritional ingredients, or as a marketing strategy (Lafarga, 2019). Most reports on food product-incorporated microalgal biomass evaluate their nutritional value, color, texture, and sensory properties, but not the rheological properties (i.e., viscosity and/or viscoelasticity). Along with color, odor, and flavor, the rheological properties are the most important quality aspects of food products, since they strongly affect their physical appearance (e.g., texture, phase separation phenomena) and sensory perception. Many processed foods require the addition of thickening agents to achieve the desired viscosity, and gelling agents to enhance the food texture. In this context, microalgae are also promising thickening candidates, due to the significant amounts of structural polymers they produce, including proteins, carbohydrates, and polysaccharides, that could act as texturizers, stabilizers, or emulsifiers. These applications are still unexploited due to the large diversity of microalgae resulting from intense evolutionary processes and the high dependence on cultivation conditions (e.g., light intensity, temperature, nutrient availability) (Bernaerts, Gheysen, Foubert, Hendrickx, & Van Loey, 2019).

To date, few studies exist on the rheological properties of microalgal polymers, and all focus on the flow behavior or the viscosity. For microalgal proteins, studies show they could display protein surface-related (foaming and emulsifying) and hydrodynamic properties (thickening and gelling), comparable to commercial ingredients, that could be of interest to the food industry (Chronakis, 2001; Schwenzfeier, Lech, Wierenga, Eppink, & Gruppen, 2013; Suarez Garcia et al., 2018; Ursu et al., 2014). For polysaccharides, most studies were done on extracellular polysaccharides (EPS) of *Phorphyridium* sp. comparing them with currently used thickening and gelling agents. These polysaccharides showed higher intrinsic viscosity, except for xanthan gum (Bernaerts, Gheysen, Foubert, Hendrickx, & Van Loey, 2019).

More studies on EPS from cyanobacteria such as *Cyanospira capsulata*, *Anabaena* sp., *Nostoc flagelliforme*, and *Arthrospira platensis* showed viscosities comparable to *Porphyridium* sp. and xanthan gum (Badel et al., 2011; Bhatnagar, Pareek, Ganguly, & Bhatnagar, 2012; Han et al., 2014; Moreno et al., 2000; Navarini, Bertocchi, Cesàro, Lapasin, & Crescenzi, 1990; Navarini, Cesàro, & Ross-Murphy, 1992). Some fragmentary studies with EPS from species of Chlorophyta have also been reported, but the results cannot compete with the previous ones (Wu, Li, & Lan, 2011; Yalcin, Hicsasmaz, Boz, & Bozoglu, 1994).

Despite the potential of microalgal polymers, the incorporation of the whole microalgal biomass into food products might be a more attractive strategy to modify the food structure, to combine the microalgae nutritional and health value with structuring benefits. Moreover, the structuring potential of microalgal biomass could be enhanced by the interactions between biopolymers such as storage and structural polysaccharides and proteins. This approach is also considered more sustainable because it avoids the extraction process and waste streams (Bernaerts, Gheysen, Foubert, Hendrickx, & Van Loey, 2019).

15.4.2 Edible coatings

Some research has been directed at the use of edible materials to replace the current plastic food packaging. Edible films and coatings are natural and biodegradable compounds that act as a semipermeable barrier to gases and water vapor, decreasing quality losses and improving shelf-life. Biopolymer-based naturally renewable films are interesting for researchers due to their wide field of applications, low cost, and biodegradable properties. The most used edible coatings are a mixture of hydrocolloids (carbohydrates and/or proteins) and lipids, which are added to improve their properties as good barriers to oxygen, carbon dioxide, and water (Falguera, Quintero, Jiménez, Muñoz, & Ibarz, 2011; Šuput, Lazić, Popović, & Hromiš, 2015).

Some of the polysaccharides most used as coatings include alginate, carrageenan, carboxymethyl cellulose, and chitosan. Microalgae produce polysaccharides that can form biofilms and could be used for packaging purposes. Also, the properties of microalgae (e.g., flavors, colorings, antioxidants, antimicrobials, nutraceuticals, and therapeutics) can be delivered through edible films to enhance their efficacy on quality parameters of food products (de Jesus Raposo, de Morais, & de Morais, 2013; Raposo, Morais, Morais, et al., 2013; de Jesus Raposo, de Morais, & de Morais, 2015). A recent work by Morales-Jiménez, Gouveia, Yáñez-Fernández, Castro-Muñoz, & Barragán-Huerta, 2020 studied microalgae-based polymers for the production of bioactive films. The polymers obtained from *Synechocystis* sp., *Phorphyridium purpureum*, and *Nostoc* sp. showed the most promising characteristics for film production. Moreover, the polymer from *Nostoc* sp. exhibited antifungal activity (Morales-Jiménez, Gouveia, Yáñez-Fernández, Castro-Muñoz, & Barragán-Huerta, 2020). Furthermore, a few studies have already done the application of biodegradable coatings based on microalgae (*Scenedesmus* sp., *Chlorella* sp., and *Spirulina platensis*) on exotic fruits like papaya, mango, banana, guava, and peppers, obtaining some promising results. In all studies, the application of microalgae-based biodegradable coatings had a positive effect on the postharvest quality of the fruits by delaying the ripening process, with lower mass losses and enhanced freshness (Cardoso, Demiate, Dalva, & Danesi, 2017; de M Teodosio et al., 2018; de Oliveira, Araújo, et al., 2018; de Oliveira, Rocha, Guedes, Dias, & de Lima, 2018; Onias et al., 2018; Onias, Rocha, De Lima, Onias, & de S Furtunato, 2016).

15.4.3 Microencapsulation

The strategy to incorporate microalgae biomass into traditional products, with a daily basis consumption could be promising. However, their commonly green color, intense odor, and flavor limit their use, since they are not generally well perceived with most food products. Thus, to avoid these potential negative effects, a possible alternative would be to extract microalgae-derived bioactive compounds and incorporate them into foods instead of the whole biomass, through microencapsulation technology (Lafarga, 2019).

In recent years, the functional food and ingredient industries have been identified as potentially significant growth areas for the microencapsulation technology, especially for the encapsulation of unstable new functional ingredients. Microencapsulation involves protecting core bioactive compounds with an encapsulating matrix that acts as a functional barrier to avoid chemical and physical reactions and maintain the biological, functional, and physicochemical properties of core materials (Bakry et al., 2016). Utilizing microencapsulation technology in the food industry, it is possible to produce functional ingredients with desirable properties: longer shelf-life by protecting from oxidation or masking undesirable odors, color, and taste of the original ingredient; controlled release of core materials at desired time and site; protection of core materials during processing stages; introduction of new colors, flavors, or textures to ingredients; maximized dispersibility and solubility of the bioactive in powder form; and improved powder properties for ease of handling during processing (Quek, Chen, & Shi, 2015). The food-grade wall materials commonly used for microencapsulation are: gums (carrageenan, alginate, and gum arabic), carbohydrates (starch, maltodextrin, β-cyclodextrin, and chitosan), cellulose (cellulose acetate phthalate), and protein (gelatin and dairy protein isolates) (Gutiérrez & Álvarez, 2017).

Different kinds of microencapsulation technologies have been developed, such as emulsification, solvent dispersion/evaporation, coacervation (phase separation), cocrystallization, interfacial polymerization, supercritical fluid technology, and spray-drying. Spray-drying is the most commonly used in the food industry due to the possibility of continuous production and easy industrialization (Bakry et al., 2016; Shao, Pan, Liu, Teng, & Yuan, 2018).

Algal oils containing DHA/EPA are generally suitable for incorporation into foods, but their use is limited by the formation of an undesirable taste and odor, as mentioned earlier. They are chemically unstable and susceptible to oxidation and loss of volatile compounds and thus influence the overall product quality, including shelf-stability, sensory properties, and acceptability of the developed product (Velasco, Dobarganes, & Márquez-Ruiz, 2003). Studies have already been done on applying microencapsulation technology to microalgal oils. Bao et al. (2011) used cross-linked sodium caseinate as wall material induced by microbial transglutaminase to microencapsulate microalgal oil rich in DHA by emulsion-based spray-drying. Other studies proposed gelatin-gum arabic complex-coacervated matrices to encapsulate microalgae oil (Badke, Campos, & De Carvalho-Jorge, 2019; Zhang, Zhang, Hu, Bao, & Huang, 2012). The produced microcapsules are suitable to enhance the oxidative stability of microalgal fatty acids, promoting sustained oil release, and the encapsulation efficiency was compatible with different applications, including as food additives (Badke, Campos, & De Carvalho-Jorge, 2019; Bao et al., 2011; Zhang, Zhang, Hu, Bao, & Huang, 2012). Moreover, Chen, Chen, et al. (2016) presented a new route to efficiently encapsulate ω3 PUFA oil into powders from phytosterol-structured nanoemulsion templates, not only improving the oxidative stability but also masking the fishy odor, which results in satisfactory characteristics that can be easily incorporated into functional foods and beverages (Chen, Liu, et al., 2016).

Aside from microalgal oil, microencapsulation can also help stabilize carotenoids for incorporation into foods. Machado et al. (2014) and Vakarelova et al. (2016) both studied the microencapsulation of astaxanthin extracted from *Haematococcus pluvialis* with Polyhydroxyalkanoates (PHA) copolymers using supercritical fluid and with alginate-pectin beads using extrusion technology, respectively. Moreover, spray-drying microencapsulation was used by Chen, Chen, et al. (2016) and Zhao, Liu, Zhang, Zhang, & Zhu, 2019 to protect an algal oil with a β-sitosterol coat and to protect astaxanthin with a maltodextrin-gelatin as wall material, respectively. All studies concluded that the materials used were suitable for the encapsulation of astaxanthin, producing highly stable encapsulates during longer storage periods, and that the technologies have the potential for future applications in the food sector (Chen et al., 2016; Machado et al., 2014; Vakarelova et al., 2016; Zhao, Liu, Zhang, Zhang, & Zhu, 2019). Likewise, Dewi, Kurniasih, & Purnamayanti, 2018 and Yan, Liu, Jiao, & Qin, 2013 successfully prepared microcapsules containing phycocyanin extracted from *Spirulina* sp., by spray-drying (using maltodextrin and κ-carrageenan as wall materials) and extrusion methods (using alginate and chitosan as coating materials), respectively. The coating materials prevented phycocyanin from being released, improved its stability and functionality, which makes it easier to be applied in functional foods (Dewi, Kurniasih, & Purnamayanti, 2018; Yan, Liu, Jiao, & Qin, 2013). Finally, Bonilla-Ahumada, Khandual, and Lugo-Cervantes (2018) tested different wall materials to microencapsulate *Tetraselmis chui* by spray-drying for better preservation of carotenoids and antioxidant compounds. They found that gelatin-encapsulated algae can be recommended for use as a functional food ingredient for human consumption with better preservation of its antioxidant value and without any toxic solvent component (Bonilla-Ahumada et al., 2018).

Microencapsulation technology has proven to be a powerful technique for tailoring the properties and functionality of sensitive ingredients for use in specific systems. In addition, it has shown the potential to offer novel solutions for stabilization and improved delivery of microalgae products in food (Sanguansri & Augustin, 2007). However, studies on the incorporation of these encapsulated microalgal ingredients in food and further human trials are needed to assess the sensory properties of these enriched foods. Two studies report the incorporation of microencapsulated *Spirulina* in yogurt and pasta (da Silva et al., 2019; Zen et al., 2019). According to da Silva et al. (2019), the microencapsulated *Spirulina* showed both higher thermal stability than the base materials and higher anti-inflammatory activity than free *Spirulina* without exerting cytotoxicity. The incorporation of encapsulated

Spirulina showed advantages over its free use in yogurt, such as more homogeneous appearance, attractive color, better nutritional profile, and antioxidant activity during the storage period. Furthermore, based on a qualitative analysis, the microencapsulation led to attenuation of the unpleasant fish-like odor and deep green color of *Spirulina*. However, these results need further validation by conducting sensory analysis (da Silva et al., 2019). The functional pasta developed by Zen et al. (2019) containing microencapsulated *Spirulina* presented a nonuniform color and the presence of green dots, but this did not affect consumer acceptability regarding color and appearance. In fact, the addition of *Spirulina* as a natural ingredient in pasta was well accepted by consumers, making it look attractive and innovative. The protection of microalgae's antioxidant potential given by microencapsulation allowed the development of a food additive with an active functional ingredient. Nonetheless, further studies are needed to evaluate the nutrient bioavailability of microencapsulated *Spirulina* after food ingestion (Zen et al., 2019).

15.4.4 Three-dimensional printing

Three-dimensional (3D) food printing has been earning increasing attention for its distinctive potential to create complex geometric structures, enabling mass customization, while providing economic and environmental benefits (Dankar, Haddarah, Omar, Sepulcre, & Pujolà, 2018; Yang, Zhang, & Bhandari, 2017). Three-dimensional food printing is a digital manufacturing technology, which is used to fabricate three-dimensional structures in a layer-by-layer manner, using liquid-, gel-, or powder-type food materials as a printing medium (Kim et al., 2019). Currently, 3D printing techniques available in the food sector generally include four types: hot melt/room temperature extrusion, selective/hot air sintering, binder jetting, and inkjet printing. Among them, extrusion-based 3D food printing is the most widely adopted method (Godoi, Prakash, & Bhandari, 2016; Portanguen, Tournayre, Sicard, Astruc, & Mirade, 2019; Sun, Zhou, Huang, Fuh, & Hong, 2015; Sun, Zhou, Yan, Huang, & Lin, 2018). The main advantage of this emerging technology is being able to personalize food by tailoring nutrition in a novel multiflavored, colored, and textured structure, according to individual physical and nutritional needs (Dankar, Haddarah, Omar, Sepulcre, & Pujolà, 2018; Pérez, Nykvist, Brøgger, Larsen, & Falkeborg, 2019). This can be especially beneficial for patients suffering from dysphagia, as this technology can improve the texture of hospital food puree meals into 3D structures that resemble conventional food and then, hopefully, increase the food intake of patients (Tan, Toh, Wong, & Li, 2018). In addition, this 3D printing flexibility allows the incorporation of a broad range of ingredients in food processing, including insects, fungi, and microalgae (An, Guo, Zhang, & Zhong, 2019; Lille, Nurmela, Nordlund, Metsä-Kortelainen, & Sozer, 2018; Severini, Azzollini, Albenzio, & Derossi, 2018; Voon, An, Wong, Zhang, & Chua, 2019).

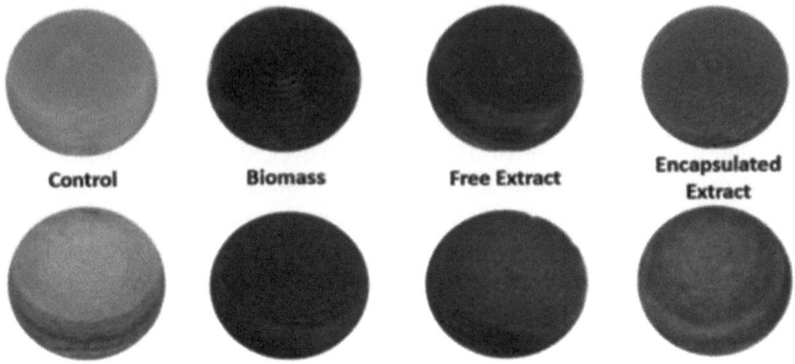

FIGURE 15.10 Three-dimensional printed functional dough (top) and cookies (bottom) fortified with *Arthrospira platensis* (whole biomass, free, and encapsulate extract) (Vieira et al., 2020).

A recent study carried out by Vieira et al. (2020) aimed at developing 3D printed functional cookies fortified with encapsulated antioxidants extracted from *A. platensis*. The fortification with *A. platensis* resulted in 3D printed cookies with an innovative appearance (Fig. 15.10). The encapsulation of the antioxidant extract improved the antioxidant activity and color stability as well as the storage time when compared to all other formulations (Vieira et al., 2020). Other studies are being developed using microalgae as ingredients to produce 3D printed healthy snacks and edible packages, but no results are currently available.

As the sustainability agenda grows in the food industry, innovators are looking for ways to replace meat with a plant-based counterpart, including using microalgae and 3D printing technology, therefore this will definitely be a promising trend in the food sector (FoodNavigator, 2019).

15.5 Conclusions

Microalgae clearly show the potential to meet the population's need for more sustainable food solutions. The nutritional and functional microalgae richness can contribute to developing an algal-based food industry, focused on producing and utilizing microalgae for innovative functional food products. Besides the protein content and balanced amino acid profiles, microalgae incorporation into foods could lead to potential benefits for human health due to the presence of bioactive compounds in some microalgae species. For example, antioxidative, antihypertensive, immunomodulatory, anticarcinogenic, hepato-protective, and anticoagulant activities have been attributed to microalgae-derived compounds. Unfortunately, the utilization of microalgae or microalgae-derived products as food substitutes is not yet competitive, mainly due to the low technology readiness levels and lack of economy of scale for microalgae processing. Furthermore, the interaction of microalgae with intrinsic or extrinsic properties of foods where they are incorporated, such as pH, fat, protein, water content, and oxygen concentration, and upon preservation thereof, still needs a deeper mechanistic elucidation. Nonetheless, the EU is continuously trying to approve more microalgae, or its extracts, as novel foods, but this process is, for now, very time-consuming and expensive. Once these hurdles are overcome, incorporating microalgae as food ingredients will not only provide health benefits but will also contribute to improving issues related to sustainability, taking into account the growing population and its changing diet, habits, and health.

These new products enriched with microalgal ingredients could become widely consumed by the general public, and could be especially appreciated by specific categories of people such as athletes, vegetarians, vegans, and the elderly, and also by consumers interested in a product with an innovative taste and color.

Acknowledgments

The authors would like to acknowledge the projects ALGAVALOR—MicroALGAs: produção integrada e VALORização da biomassa e das suas diversas aplicações, from the Portugal 2020 program (grant agreement no. POCI-01-0247-FEDER-035234; LISBOA-01−0247-FEDER-035234; ALG-01-0247-FEDER-035234); GreenCoLab—Green Ocean Technologies and Products Collaborative Laboratory, n. ALG-05−3559-FSE-000010, Algarve 2020 Operational Regional Program, funded by European Social Fund and Portuguese Government budget; Biomass and Bioenergy Research Infrastructure (BBRI)- LISBOA-01-0145-FEDER-022059, supported by Operational Programme for Competitiveness and Internationalization (PORTUGAL2020), by Lisbon Portugal Regional Operational Programme (Lisboa 2020) and by North Portugal Regional Operational Programme (Norte 2020) under the Portugal 2020 Partnership Agreement, through the European Regional Development Fund (ERDF). Alice Ferreira is pleased to acknowledge her PhD grant no. SFRH/BD/144122/2019 awarded by Fundação para a Ciência e Tecnologia.

References

Abdelaal, M., le Roux, C. W., & Docherty, N. G. (2017). Morbidity and mortality associated with obesity. *Annals of Translational Medicine*, *5*(7), 161.

Abd El-Razik, M., & Mohamed, A. (2013). Utilization of acid casein curd enriched with *Chlorella vulgaris* biomass as substitute of egg in mayonnaise production. *World Applied Sciences Journal*, *26*(7), 917−925.

Adarme-Vega, T. C., Lim, D. K. Y., Timmins, M., Vernen, F., Li, Y., & Schenk, P. M. (2012). Microalgal biofactories: a promising approach towards sustainable omega-3 fatty acid production. *Microbial Cell Factories*, *11*, 96.

Adarme-Vega, T. C., Thomas-Hall, S. R., & Schenk, P. M. (2014). Towards sustainable sources for omega-3 fatty acids production. *Current Opinion in Biotechnology*, *26*, 14–18.

AECOSAN. (2017). Section of food safety and nutrition report approved by the section of food safety and nutrition of the scientific committee in its plenary session. *Revista del Cómité Científico de la AECOSAN*, *25*, 11–21.

Agustini, T. W., Ma'Ruf, W. F., Widayat., Wibowo, B. A., & Hadiyanto. (2017). Study on the effect of different concentration of *Spirulina platensis* paste added into dried noodle to its quality characteristics. In *IOP Conference Series: Earth and Environmental Science*, (p. 012068). Institute of Physics Publishing.

Ak, B., Avşaroğlu, E., Işık, O., Özyurt, G., Kafkas, E., Etyemez, M., & Uslu, L. (2016). Nutritional and physicochemical characteristics of bread enriched with microalgae *Spirulina platensis*. *International Journal of Engineering Research and Application*, *6*(12), 30–38.

An, Y. J., Guo, C. F., Zhang, M., & Zhong, Z. P. (2019). Investigation on characteristics of 3D printing using *Nostoc sphaeroides* biomass. *Journal of the Science of Food and Agriculture*, *99*(2), 639–646.

Arafah, A., Abassy, M., Morcos, S., & Hussein, L. (1980). Nutritive quality of baladi bread supplemented with fish protein concentrate, green algae, or synthetic amino acids. *Cereal Chemistry*, *57*(1), 35–39.

Asioli, D., Aschemann-Witzel, J., Caputo, V., Vecchio, R., Annunziata, A., Næs, T., & Varela, P. (2017). Making sense of the "clean label" trends: a review of consumer food choice behavior and discussion of industry implications. *Food Research International*, *99*, 58–71.

Atitallah, A. B., Hentati, F., Dammak, M., Hadrich, B., & Fendri, I. (2019). Effect of microalgae incorporation on quality characteristics and functional and antioxidant capacities of ready-to-eat fish burgers made from common carp (*Cyprinus carpio*). *Applied Science*, *9*(9), 1830.

Babuskin, S., Krishnan, K. R., Babu, P. A. S., Sivarajan, M., & Sukumar, M. (2014). Functional foods enriched with marine microalga *Nannochloropsis oculata* as a source of ω-3 fatty acids. *Food Technology and Biotechnology*, *52*(3), 292–299.

Babuskin, S., Radhakrishnan, K., Babu, P. A. S., Sukumar, M., Fayidh, M. A., Sabina, K., & Sivarajan, M. (2015). Effects of Rosemary extracts on oxidative stability of chikkis fortified with microalgae biomass. *Journal of Food Science and Technology*, *52*(6), 3784–3793.

Badel, S., Callet, F., Laroche, C., Gardarin, C., Petit, E., El Alaoui, H., & Michaud, P. (2011). A new tool to detect high viscous exopolymers from microalgae. *Journal of Industrial Microbiology & Biotechnology*, *38*(2), 319–326.

Badke, L. B., Campos, B., & De Carvalho-Jorge, A. R. (2019). Synthesis and characterization of microalgae fatty acids or aloe vera oil microcapsules. *Polímeros*, *29*(3), e2019042.

Bakry, A. M., Abbas, S., Ali, B., Majeed, H., Abouelwafa, M. Y., Mousa, A., & Liang, L. (2016). Microencapsulation of oils: a comprehensive review of benefits, techniques, and applications. *Comprehensive Review of Food Science and Food Safety*, *15*(1), 143–182.

Bao, S.-S., Hu, X.-C., Zhang, K., Xu, X.-K., Zhang, H.-M., & Huang, H. (2011). Characterization of spray-dried microalgal oil encapsulated in cross-linked sodium caseinate matrix induced by microbial transglutaminase. *Journal of Food Science*, *76*, 112–118.

Barbier, M., Charrier, B., Araujo, R., Holdt, S. L., Jacquemin, B., & Rebours, C. (2019). *PHYCOMORPH European Guidelines for a Sustainable Aquaculture of Seaweeds, COST Action FA1406*. Roscooff, France: PEGASUS.

Barkallah, M., Dammak, M., Louati, I., Hentati, F., Hadrich, B., Mechichi, T., ... Abdelkafi, S. (2017). Effect of *Spirulina platensis* fortification on physicochemical, textural, antioxidant and sensory properties of yogurt during fermentation and storage. *LWT - Food Science and Technology*, *84*, 323–330.

Barkia, I., Saari, N., & Manning, S. R. (2019). Microalgae for high-value products towards human health and nutrition. *Marine Drugs*, *17*(5), 304.

Batista, A. P., Gouveia, L., Nunes, M. C., Franco, J. M., & Raymundo, A. (2008). Microalgae biomass as a novel functional ingredient in mixed gel systems. In P. A. Williams, & Gl. O. Philips (Eds.), *Gums and Stabilizers for the Food Industry 14* (pp. 487–494). Cambridge: RSC Publishing.

Batista, A. P., Nunes, M. C., Raymundo, A., Gouveia, L., Sousa, I., Cordobés, F., & Franco, J. M. (2011). Microalgae biomass interaction in biopolymer gelled systems. *Food Hydrocolloids*, *25*, 817–825.

Batista, A. P., Niccolai, A., Fradinho, P., Fragoso, S., Bursic, I., Rodolfi, L., & Raymundo, A. (2017). Microalgae biomass as an alternative ingredient in cookies: sensory, physical and chemical properties, antioxidant activity and in vitro digestibility. *Algal Research*, *26*, 161–171.

Batista, A. P., Raymundo, A., Sousa, I., & Empis, J. (2006). Rheological characterization of coloured oil-in-water food emulsions with lutein and phycocyanin added to the oil and aqueous phases. *Food Hydrocolloids*, *20*, 44–52.

Batista, A. P., Raymundo, A., Sousa, I., Empis, J., & Franco, J. M. (2006). Colored food emulsions-implications of pigment addition on the rheological behavior and microstructure. *Food Biophysics*, *1*, 216–227.

Becker, E. W. (2007). Micro-algae as a source of protein. *Biotechnology Advances*, *25*, 207–210.

Beheshtipour, H., Mortazavian, A. M., Haratian, P., & Khosravi-Darani, K. (2012). Effects of *Chlorella vulgaris* and *Arthrospira platensis* addition on viability of probiotic bacteria in yogurt and its biochemical properties. *European Food Research and Technology*, *235*, 719–728.

Beheshtipour, H., Mortazavian, A. M., Mohammadi, R., Sohrabvandi, S., & Khosravi-Darani, K. (2013). Supplementation of *Spirulina platensis* and *Chlorella vulgaris* algae into probiotic fermented milks. *Comprehensive Review of Food Science and Food Safety*, *12*, 144–154.

Bernaerts, T. M. M., Gheysen, L., Foubert, I., Hendrickx, M. E., & Van Loey, A. M. (2019). The potential of microalgae and their biopolymers as structuring ingredients in food: a review. *Biotechnology Advances*, *37*(8), 107419.

Bhatnagar, M., Pareek, S., Ganguly, J., & Bhatnagar, A. (2012). Rheology and composition of a multi-utility exopolymer from a desert borne cyanobacterium *Anabaena variabilis*. *Journal of Applied Phycology*, *24*, 1387–1394.

Bolanho, B. C., Egea, M. B., Jácome, A. L. M., Campos, I., de Carvalho, J. C. M., & Danesi, E. D. G. (2014). Antioxidant and nutritional potential of cookies enriched with *Spirulina platensis* and sources of fibre. *Journal of Food and Nutrition Research*, *53*, 171–179.

Bonilla-Ahumada, F. D. J., Khandual, S., & Lugo-Cervantes, E. C. (2018). Microencapsulation of algal biomass (*Tetraselmis chuii*) by spray-drying using different encapsulation materials for better preservation of beta- carotene and antioxidant compounds. *Algal Research*, *36*, 229–238.

Brouns, F. J. P. H., van Buul, V. J., & Shewry, P. R. (2013). Does wheat make us fat and sick? *Journal of Cereal Science*, *58*(2), 209–215.

Camacho, F., Macedo, A., & Malcata, F. (2019). Potential industrial applications and commercialization of microalgae in the functional food and feed industries: a short review. *Marine Drugs*, *17*(6), 312.

Caporgno, M. P., & Mathys, A. (2018). Trends in microalgae incorporation Into innovative food products with potential health benefits. *Frontiers in Nutrition*, *5*, 58.

Cardoso, T., Demiate, I. M., Dalva, E., & Danesi, G. (2017). Biodegradable films with *Spirulina platensis* as coating for Cambuci peppers (*Capsicum* sp.). *American Journal of Food Technology Research*, *12*(4), 236–244.

Castillejo, N., Martínez-Hernández, G. B., Goffi, V., Gómez, P. A., Aguayo, E., Artés, F., & Artés-Hernández, F. (2018). Natural vitamin B12 and fucose supplementation of green smoothies with edible algae and related quality changes during their shelf life. *Journal of the Science of Food and Agriculture*, *98*(6), 2411–2421.

Chacón-Lee, T. L., & González-Mariño, G. E. (2010). Microalgae for "healthy" foods—possibilities and challenges. *Comprehensive Review of Food Science and Food Safety*, *9*(6), 655–675.

Chen, L., Liu, X., Li, D., Chen, W., Zhang, K., & Chen, S. (2016). Preparation of stable microcapsules from disrupted cell of *Haematococcus pluvialis* by spray drying. *International Journal of Food Science & Technology.*, *51*(8), 1–10.

Chen, X.-W., Chen, Y.-J., Wang, J.-M., Guo, J., Yin, S.-W., & Yang, X.-Q. (2016). Phytosterols structured algae oil nanoemulsions and powders: improving antioxidant and flavor properties. *Food & Function*, *7*(9), 3694–3702.

Chronakis, I. S. (2001). Gelation of edible blue-green algae protein isolate (*Spirulina platensis* strain Pacifica): thermal transitions, rheological properties, and molecular forces involved. *Journal of Agricultural and Food Chemistry*, *49*(2), 888–898.

Cohen, Z., & Ratledge, C. (2010). *Single Cell Oils: Microbial and Algal Oils*. AOCS Press.

Credence Research (2018). Algae Products Market Size, Share, Trend, Growth And Forecast To 2026 [WWW Document]. URL https://www.credenceresearch.com/report/algae-products-market (accessed 2.28.20).

da Silva, S. C., Fernandes, I. P., Barros, L., Fernandes, Â., Alves, M. J., Calhelha, R. C., & Barreiro, M. F. (2019). Spray-dried *Spirulina platensis* as an effective ingredient to improve yogurt formulations: testing different encapsulating solutions. *Journal of Functional Foods*, *60*, 103427.

da S Vaz, B., Moreira, J. B., de Morais, M. G., & Costa, J. A. V. (2016). Microalgae as a new source of bioactive compounds in food supplements. *Current Opinion in Food Science*, *7*, 73–77.

Dankar, I., Haddarah, A., Omar, F. E. L., Sepulcre, F., & Pujolà, M. (2018). 3D printing technology: the new era for food customization and elaboration. *Trends in Food Science & Technology*, *75*, 231–242.

De Boer, J., Schösler, H., & Boersema, J. J. (2013). Motivational differences in food orientation and the choice of snacks made from lentils, locusts, seaweed or "hybrid" meat. *Food Quality and Preference*, *28*(1), 32–35.

de Jesus Raposo, M. F., de Morais, R. M. S. C., & de Morais, A. M. M. B. (2013). Health applications of bioactive compounds from marine microalgae. *Life Sciences*, *93*(15), 479–486.

de Jesus Raposo, M. F., de Morais, A. M. B., & de Morais, R. M. S. C. (2015). Marine polysaccharides from algae with potential biomedical applications. *Marine Drugs*, *13*(5), 2967–3028.

de Jesus Raposo, M. F., de Morais, A. M. B., & de Morais, R. M. S. C. (2016). Emergent sources of prebiotics: seaweeds and microalgae. *Marine Drugs*, *14*(2), 27.

de Jesus Raposo, M. F., de Morais, R. M. S. C., & de Morais, A. M. B. (2013). Bioactivity and applications of sulphated polysaccharides from marine microalgae. *Marine Drugs*, *11*(1), 233–252.

de Jesus Raposo, M. F. D. J., & de Morais, A. M. M. B. (2015). Microalgae for the prevention of cardiovascular disease and stroke. *Life Sciences*, *125*, 32–41.

De Marco, E. R., Steffolani, M. E., Martínez, C. S., & León, A. E. (2014). Effects of spirulina biomass on the technological and nutritional quality of bread wheat pasta. *LWT - Food Science and Technology*, *58*(1), 102–108.

De Marco, E. R., Steffolani, M. E., Martínez, M., & León, A. E. (2018). The use of *Nannochloropsis* sp. as a source of omega-3 fatty acids in dry pasta: chemical, technological and sensory evaluation. *International Journal of Food Science and Technology*, *53*(2), 499–507.

De Morais, M. G., De Miranda, M. Z., & Costa, J. A. V. (2006). Biscoitos de chocolate enriquecidos com *Spirulina platensis*: características físico-químicas, sensoriais e digestibilidade. *Alimentos e Nutrição Araraquara*, *17*(3), 323–328.

de M Teodosio, A. E. M., Araujo, R. H. C. R., de Lima, J. F., Onias, E. A., Ferreira, A. P. N., Santos, B. G. F. L., & de Silva, K. G. (2018). Effect of the biodegradable coatings the base on microalgae and oil of the seed of the pomegranate in the conservation powder-crop of the papaya 'golden.'. *The Journal of Agricultural Science*, *10*(10), 367–377.

de Oliveira, Á. M. F., Araújo, R. H. C. R., de M Teodosio, A. E. M., Lima, J. F., de Freitas, E. F. Q., dos Santos, J. J. F., & Oliveira, L. M. (2018). Postharvest quality of 'Prata Anã' bananas treated with microalgae coating. *Journal of Experimental Agriculture International*, *27*(2), EAI.44059.

de Oliveira, Á. M. F., Rocha, R. H. C., Guedes, W. A., Dias, G. A., & de Lima, J. F. (2018). Use of *Chlorella* sp. for coating "Tommy Atkins" mango fruits stored under refrigeration. *Semina: Ciencas Agrarias*, *39*(2), 565–572.

Dewi, E. N., Kurniasih, R. A., & Purnamayanti, L. (2018). Physical properties of *Spirulina* phycocyanin microencapsulated with maltodextrin and carrageenan. *Philippine Journal of Science*, *147*(2), 201–207.

Dinu, M., Vlasceanu, G., Dune, A., & Rotaru, G. (2012). Researches concerning the growth of nutritive value of the bread products through the *Spirulina* adding. *Journal of Environmental Protection and Ecology*, *13*(2), 660–665.

El Baky, H. H. A., El Baroty, G. S., & Ibrahem, E. A. (2015). Functional characters evaluation of biscuits sublimated with pure phycocuanin isolated from *Spirulina* and *Spirulina* biomass. *Nutricion Hospitalaria: Organo Oficial de la Sociedad Espanola de Nutricion Parenteral y Enteral*, *32*(1), 231–241.

El-Baz, F. K., Abdo, S. M., & Hussein, A. M. S. (2017). Microalgae *Dunaliella salina* for use as food supplement to improve pasta quality. *International Journal of Pharmaceutical Sciences Review and Research*, *46*(2), 45–51.

Enzing, C., Ploeg, M., Barbosa, M., & Sijtsma, L. (2014). *Microalgae-based products for the food and feed sector: an outlook for Europe*. Luxembourg: JRC Scientific and Policy Reports. Publications Office of the European Union.

European Environment Agency (2017). Food consumption - animal based protein [WWW Document]. URL https://www.eea.europa.eu/data-and-maps/indicators/13.2-development-in-consumption-of-2/assessment-1 (accessed 1.31.20).

Falguera, V., Quintero, J. P., Jiménez, A., Muñoz, J. A., & Ibarz, A. (2011). Edible films and coatings: structures, active functions and trends in their use. *Trends in Food Science and Technology*, *22*(6), 292–303.

FAO. (2019). Food security and nutrition in the world. *IEEE Journal of Selected Topics in Applied Earth Observations and Remote Sensing*.

Figueira, Fd. S., Crizel, Td. M., & Salas-Mellado, MdlM. (2011). Elaboration of gluten-free bread enriched with the microalgae *Spirulina platensis*. *Brazilian Journal of Food Technology*, *14*(4), 308–316.

Finney, K. F., Pomeranz, Y., & Bruinsma, B. L. (1984). Use of algae *Dunaliella* as a protein supplement in bread. *Cereal Chemistry*, *61*(5), 402–406.

FoodNavigator (2019). From microalgae to 3D printed steak: 'The future is meatless' [WWW Document]. URL https://www.foodnavigator.com/Article/2019/05/27/From-microalgae-to-3D-printed-steak-The-future-is-meatless (accessed 4.10.20).

Fradinho, P., Niccolai, A., Soares, R., Rodolfi, L., Biondi, N., Tredici, M. R., & Raymundo, A. (2020). Effect of *Arthrospira platensis* (spirulina) incorporation on the rheological and bioactive properties of gluten-free fresh pasta. *Algal Reserach, 45*, 101743.

Fradique, M, Batista, A. P., Nunes, M. C., Gouveia, L., Bandarra, N. M., & Raymundo, A. (2010). Incorporation of *Chlorella vulgaris* and *Spirulina maxima* biomass in pasta products. Part 1: preparation and evaluation. *Journal of the Science of Food and Agriculture, 90*(10), 1656–1664.

Fradique, M., Batista, A. P., Nunes, M. C., Gouveia, L., Bandarra, N. M., & Raymundo, A. (2013). *Isochrysis galbana* and *Diacronema vlkianum* biomass incorporation in pasta products as PUFA's source. *LWT - Food Science and Technology, 50*(1), 312–319.

García, J. L., de Vicente, M., & Galán, B. (2017). Microalgae, old sustainable food and fashion nutraceuticals. *Microbial Biotechnology, 10*(5), 1017–1024.

García-Segovia, P., Pagán-Moreno, M. J., Lara, I. F., & Martínez-Monzó, J. (2017). Effect of microalgae incorporation on physicochemical and textural properties in wheat bread formulation. *Food Science and Technology International, 23*(5), 437–447.

Gheysen, L., Demets, R., Devaere, J., Bernaerts, T., Goos, P., Van Loey, A., & Foubert, I. (2019). Impact of microalgal species on the oxidative stability of n-3 LC-PUFA enriched tomato puree. *Algal Research, 40*, 101502.

Gheysen, L., Lagae, N., Devaere, J., Goiris, K., Goos, P., Bernaerts, T., & Foubert, I. (2019). Impact of Nannochloropsis sp. dosage form on the oxidative stability of n-3 LC-PUFA enriched tomato purees. *Food Chemistry, 279*, 389–400.

Godoi, F. C., Prakash, S., & Bhandari, B. R. (2016). 3D printing technologies applied for food design: status and prospects. *Journal of Food Engineering, 179*, 44–54.

Golmakani, M. T., Soleimanian-Zad, S., Alavi, N., Nazari, E., & Eskandari, M. H. (2019). Effect of Spirulina (*Arthrospira platensis*) powder on probiotic bacteriologically acidified feta-type cheese. *Journal of Applied Phycology, 31*, 1085–1094.

Gouveia, L., Batista, A. P., Miranda, A., Empis, J., & Raymundo, A. (2007). *Chlorella vulgaris* biomass used as colouring source in traditional butter cookies. *Innovative Food Science and Emerging Technologies., 8*(3), 433–436.

Gouveia, L., Batista, A. P., Raymundo, A., & Bandarra, N. (2008). *Spirulina maxima* and *Diacronema vlkianum* microalgae in vegetable gelled desserts. *Nutrition & Food Science, 38*(5), 492–501.

Gouveia, L., Batista, A. P., Sousa, I., Raymundo, A., & Bandarra, N. M. (2008). Microalgae in novel food products. In K. N. Papadopoulos (Ed.), *Food Chemistry Research Developments* (pp. 75–111). New York: Nova Science Publishers, Inc.

Gouveia, L., Coutinho, C., Mendonça, E., Batista, A. P., Sousa, I., Bandarra, N. M., & Raymundo, A. (2008c). Functional biscuits with PUFA-ω3 from *Isochrysis galbana. Journal of the Science of Food and Agriculture, 88*(5), 891–896.

Gouveia, L., Raymundo, A., Batista, A. P., Sousa, I., & Empis, J. (2006). *Chlorella vulgaris* and *Haematococcus pluvialis* biomass as colouring and antioxidant in food emulsions. *European Food Research and Technology, 222*, 362–367.

Graça, C., Fradinho, P., Sousa, I., & Raymundo, A. (2018). Impact of *Chlorella vulgaris* on the rheology of wheat flour dough and bread texture. *LWT - Food Science and Technology, 89*, 466–474.

Guedes, A. C., Barbosa, C. R., Amaro, H. M., Pereira, C. I., & Malcata, F. X. (2011). Microalgal and cyanobacterial cell extracts for use as natural antibacterial additives against food pathogens. *International Journal Food Sci. Technol, 46*(4), 862–870.

Guldas, M., & Irkin, R. (2010). Influence of *Spirulina platensis* powder on the microflora of yoghurt and acidophilus milk. *Mljekarstvo, 60*(4), 237–243.

Gutiérrez, T. J., & Álvarez, K. (2017). Biopolymers as microencapsulation materials in the food industry. *Advances in Physicochemical Properties of Bioplymers* (Part 2), 296–322.

Hafsa, Y. A., Amel, D., Samia, S., & Sidahmed, S. (2014). Evaluation of nutritional and sensory properties of bread enriched with spirulina. *Annals of Food Science and Technology*.

Han, P. P., Sun, Y., Wu, X. Y., Yuan, Y. J., Dai, Y. J., & Jia, S. R. (2014). Emulsifying, flocculating, and physicochemical properties of exopolysaccharide produced by cyanobacterium *Nostoc flagelliforme*. *Applied Biochemistry and Biotechnology*, *172*, 36–49.

Harwood, J. L. (2019). Algae: critical sources of very long-chain polyunsaturated fatty acids. *Biomolecules*, *9*(11), 708.

Hathwar, S. C., Rai, A. K., Modi, V. K., & Narayan, B. (2012). Characteristics and consumer acceptance of healthier meat and meat product formulations—a review. *Journal of Food Science & Technology*, *49*, 653–664.

Hossain, A. K. M. M., Brennan, M. A., Mason, S. L., Guo, X., Zeng, X. A., & Brennan, C. S. (2017). The effect of astaxanthin-rich microalgae "*Haematococcus pluvialis*" and wholemeal flours incorporation in improving the physical and functional properties of cookies. *Foods*, *6*(8), 57.

Hotchkiss, S. (2010). Investigation of the Flavouring and Taste Components of Irish Seaweeds. Marine Research Sub-Programme (NDP 2007-2013) Series. *Marine Institute*.

Ing, P., Ling, L. Y., Eng, J., & Alin, J. M. (2010). The influence of consumer characteristics on the acceptance of new seaweed food products. *Jurnal Kemanusiaan*, *8*(1), 97–107.

Jacob-Lopes, E., Maroneze, M. M., Deprá, M. C., Sartori, R. B., Dias, R. R., & Zepka, L. Q. (2019). Bioactive food compounds from microalgae: an innovative framework on industrial biorefineries. *Current Opinion in Food Science*, *25*, 1–7.

Joshi, S. M. R., Bera, M. B., & Panesar, P. S. (2014). Extrusion cooking of maize/*Spirulina* mixture: factors affecting expanded product characteristics and sensory quality. *Journal of Food Processing and Preservation*, *38*(2), 655–664.

Jutur, P. P., Nesamma, A. A., & Shaikh, K. M. (2016). Algae-derived marine oligosaccharides and their biological applications. *Frontiers in Marine Science*, *3*, 83.

Khan, M. I., Shin, J. H., & Kim, J. D. (2018). The promising future of microalgae: current status, challenges, and optimization of a sustainable and renewable industry for biofuels, feed, and other products. *Microbial Cell Factories*, *17*, 36.

Khazaei Pool, E., Shahidi, F., Mortazavi, S. A., Azizpour, M., & Daneshzad, E. (2016). Examination of the effect of *Spirulina platensis* microalgae on drying kinetics and the color change of kiwifruit pastille. *Journal of Food Measurement and Characterization*, *10*, 634–642.

Khemiri, S., Khelifi, N., Nunes, M. C., Ferreira, A., Gouveia, L., Smaali, I., & Raymundo, A. (2020). Microalgae biomass as an additional ingredient of gluten-free bread: dough rheology, texture quality and nutritional properties. *Algal Research*, *50*, 101998.

Khosravi-Darani, K., Gholami, Z., & Gouveia, L. (2017). Effect of *Arthrospira platensis* on the shelf life, sensorial and rheological properties of strudel. *Romanian Biotechnological Letters*, *22*(1), 12250–12258.

Kim, H. W., Lee, I. J., Park, S. M., Lee, J. H., Nguyen, M. H., & Park, H. J. (2019). Effect of hydrocolloid addition on dimensional stability in post-processing of 3D printable cookie dough. *LWT - Food Science and Technology*, *101*, 69–75.

Klamczynska, B., & Mooney, W. D. (2017). *Heterotrophic Microalgae: A Scalable and Sustainable Protein Source*. *Sustainable Protein Sources* (pp. 327–339). Elsevier Inc.

Koutra, E., Economou, C. N., Tsafrakidou, P., & Kornaros, M. (2018). Bio-based products from microalgae cultivated in digestates. *Trends in Biotechnology*, *36*(8), 819–833.

Koyande, A. K., Chew, K. W., Rambabu, K., Tao, Y., Chu, D.-T., & Show, P.-L. (2019). Microalgae: a potential alternative to health supplementation for humans. *Food Science and Human Wellness*, *8*(1), 16–24.

Lafarga, T. (2019). Effect of microalgal biomass incorporation into foods: nutritional and sensorial attributes of the end products. *Algal Research*, *41*, 101566.

Lafarga, T., Acién-Fernández, F. G., Castellari, M., Villaró, S., Bobo, G., & Aguiló-aguayo, I. (2019). Effect of microalgae incorporation on the physicochemical, nutritional, and sensorial properties of an innovative broccoli soup. *LWT - Food Science and Technology*, *111*, 167–174.

Lafarga, T., Mayre, E., Echeverria, G., Viñas, I., Villaró, S., Acién-Fernández, F. G., & Aguiló-Aguayo, I. (2019). Potential of the microalgae *Nannochloropsis* and *Tetraselmis* for being used as innovative ingredients in baked goods. *LWT - Food Science Technology*, *115*, 108439.

Lemes, A. C., Takeuchi, K. P., de Carvalho, J. C. M., & Danesi, E. D. G. (2012). Fresh pasta production enriched with *Spirulina platensis* biomass. *Brazilian Archives of Biology and Technology*, *55*(5), 741–750.

Lille, M., Nurmela, A., Nordlund, E., Metsä-Kortelainen, S., & Sozer, N. (2018). Applicability of protein and fiber-rich food materials in extrusion-based 3D printing. *Journal of Food Engineering*, *220*, 20–27.

Lisboa, C. R., Pereira, A. M., Ferreira, S. P., & Costa, J. A. V. (2014). Utilisation of *Spirulina* sp. and *Chlorella pyrenoidosa* biomass for the production of enzymatic protein hydrolysates. *Journal of Engineering Research and Applications*, *4*(5), 29–38.

Lucas, B. F., de Morais, M. G., Santos, T. D., & Costa, J. A. V. (2017). Effect of *Spirulina* addition on the physicochemical and structural properties of extruded snacks. *Food Science and Technology*, *37*, 16–23.

Lucas, B. F., de Morais, M. G., Santos, T. D., & Costa, J. A. V. (2018). *Spirulina* for snack enrichment: nutritional, physical and sensory evaluations. *LWT - Food Science and Technology*, *90*, 270–276.

Machado, F. R. S., Reis, D. F., Boschetto, D. L., Burkert, J. F. M., Ferreira, S. R. S., Oliveira, J. V., & Burkert, V. (2014). Encapsulation of astaxanthin from *Haematococcus pluvialis* in PHBV by means of SEDS technique using supercritical CO2. *Industrial Crops and Products*, *54*, 17–21.

Madeira, M. S., Cardoso, C., Lopes, P. A., Coelho, D., Afonso, C., Bandarra, N. M., & Prates, J. A. M. (2017). Microalgae as feed ingredients for livestock production and meat quality: a review. *Livestock Science*, *205*, 111–121.

Malik, P., & Paul, A. (2012). Quality characteristics of ice cream enriched with *Spirulina* powder. *International Journal of Food and Nutritional Science*, *2*(1), 44–50.

Marcinkowska-Lesiak, M., Onopiuk, A., Zalewska, M., Ciepłoch, A., & Barotti, L. (2018). The effect of different level of *Spirulina* powder on the chosen quality parameters of shortbread biscuits. *Journal of Food Processing and Preservation*, *42*(3), e13561.

Marti-Quijal, F. J., Zamuz, S., Galvez, F., Roohinejad, S., Lorenzo, J. M., Tiwari, B. K., & Barba, F. J. (2018). Replacement of soy protein with other legumes or algae in turkey breast formulation: changes in physicochemical and technological properties. *Journal of Food Processing and Preservation*, *42*(12), e13845.

Marti-Quijal, F. J., Zamuz, S., Tomašević, I., Gómez, B., Rocchetti, G., Lucini, L., & Lorenzo, J. M. (2019). Influence of different sources of vegetable, whey and microalgae proteins on the physicochemical properties and amino acid profile of fresh pork sausages. *LWT - Food Science and Technology*, *110*, 316–323.

Marti-Quijal, F. J., Zamuz, S., Tomašević, I., Rocchetti, G., Lucini, L., Marszałek, K., & Lorenzo, J. M. (2019). A chemometric approach to evaluate the impact of pulses,

Chlorella and *Spirulina* on proximate composition, amino acid, and physicochemical properties of turkey burgers. *Journal of the Science of Food and Agriculture, 99*(7), 3672–3680.

Massoud, R., Khosravi-Darani, K., Nakhsaz, F., & Varga, L. (2016). Evaluation of physicochemical, microbiological and sensory properties of croissants fortified with *Arthrospira platensis* (Spirulina). *Czech Journal of Food Sciences, 34*(4), 350–355.

Mazinani, S., Fadaei, V., & Khosravi-Darani, K. (2016). Impact of *Spirulina platensis* on physicochemical properties and viability of *Lactobacillus acidophilus* of probiotic UF feta cheese. *Journal of Food Processing and Preservation, 40*(6), 1318–1324.

Mennella, L., Tosco, D., Alberti, F., Cembalo, L., Crescimanno, M., Del Giudice, T., & Cicia, G. (2020). Perspectives and challenges of small scale plant microalgae cultivation. Evidences from southern Italy. *Algal Research, 45*, 101693.

Merchant, R. E., & Andre, C. A. (2001). A review of recent clinical trials of the nutritional supplement *Chlorella Pyrenoidosa* in the treatment of fibromyalgia, hypertension, and ulcerative colitis. *Alternative Therapies in Health and Medicine, 7*(3), 79–91.

Mizutani, T. (2009). Toxicity of xanthene food dyes by inhibition of human drug-metabolizing enzymes in a noncompetitive manner. *J. Environmental and Public Health*, 953952.

Mohamed, A. G., Abo-El-Khair, B. E., & Shalaby, S. M. (2013). Quality of novel healthy processed cheese analogue enhanced with marine microalgae *Chlorella vulgaris* biomass. *World Appl. Sci. J, 23*(7), 914–925.

Morales-Jiménez, M., Gouveia, L., Yáñez-Fernández, J., Castro-Muñoz, R., & Barragán-Huerta, B. E. (2020). Production, preparation and characterization microalgae-based biopolymer as a potential bioactive film. *Coatings, 10*(2), 120.

Moreno, F. J., Corzo, N., Montilla, A., Villamiel, M., & Olano, A. (2017). Current state and latest advances in the concept, production and functionality of prebiotic oligosaccharides. *Current Opinion in Food Science, 13*, 50–55.

Moreno, J., Vargas, M. A., Madiedo, J. M., Muñoz, J., Rivas, J., & Guerrero, M. G. (2000). Chemical and rheological properties of an extracellular polysaccharide produced by the cyanobacterium *Anabaena* sp. ATCC 33047. *Biotechnology and Bioengineering, 67*(3), 283–290.

Nakos, M., Pepelanova, I., Beutel, S., Krings, U., Berger, R. G., & Scheper, T. (2017). Isolation and analysis of vitamin B12 from plant samples. *Food Chemistry, 216*, 301–308.

Navacchi, M. F. P., de Carvalho, J. C. M., Takeuchi, K. P., & Danesi, E. D. G. (2012). Development of cassava cake enriched with its own bran and *Spirulina platensis. Acta Scientiarum: Technology, 34*(4), 465–472.

Navarini, L., Bertocchi, C., Cesàro, A., Lapasin, R., & Crescenzi, V. (1990). Rheology of aqueous solutions of an extracellular polysaccharide from *Cyanospira capsulata. Carbohydrate Polymers, 12*(2), 169–187.

Navarini, L., Cesàro, A., & Ross-Murphy, S. B. (1992). Viscoelastic properties of aqueous solutions of an exocellular polysaccharide from cyanobacteria. *Carbohydrate Polymers, 18*, 265–272.

Niccolai, A., Venturi, M., Galli, V., Pini, N., Rodolfi, L., Biondi, N., & Tredici, M. R. (2019). Development of new microalgae-based sourdough "crostini": functional effects of *Arthrospira platensis* (spirulina) addition. *Scientific Reports, 9*, 19433.

Olaizola, M., & Grewe, C. (2019). Commercial microalgae cultivation systems. In A. Hallmann, & P. H. Rampelotto (Eds.), *Grand Challenges in Algae Biotechnology* (pp. 3–34). Springer.

Onacik-Gür, S., Zbikowska, A., & Majewska, B. (2018). Effect of spirulina (*Spirulina platensis*) addition on textural and quality properties of cookies. Ital. *Journal of Food Science, 30*, 1–12.

Onias, E. A., Rocha, R. H. C., De Lima, J. F., Onias, E. A., & de S Furtunato, T. C. (2016). Organic "Tommy Atkins" mango postharvest quality when treated with biofilms enriched by *Spirulina platensis*. *Científica, 44*(3), 286–293.

Onias, E. A., Teodosio, A. E. M. M., Bomfim, M. P., Rocha, R. H. C., Lima, J. F., & de Medeiros, M. L. S. (2018). Biodegradable recovery based on *Spirulina platensis* on post-harvest conservation of goiaba Paluma under different storage temperatures. *Revista de Ciências Agrárias, 41*, 849–860.

Özyurt, G., Uslu, L., Yuvka, I., Gökdog, S., Atci, G., Ak, B., & Işik, O. (2015). Evaluation of the cooking quality characteristics of pasta enriched with *Spirulina platensis*. *Journal of Food Quality, 38*(4), 268–272.

Pagnussatt, F. A., Spier, F., Bertolin, T. E., Costa, J. A. V., & Gutkoski, L. C. (2014). Technological and nutritional assessment of dry pasta with oatmeal and the microalga *Spirulina platensis*. *Brazilian Journal of Food Technology, 17*(4), 296–304.

Palabiyik, I., Durmaz, Y., Öner, B., Toker, O. S., Coksari, G., Konar, N., & Tamtürk, F. (2018). Using spray-dried microalgae as a natural coloring agent in chewing gum: effects on color, sensory, and textural properties. *Journal of Applied Phycology, 30*, 1031–1039.

Pangestuti, R., & Kim, S. K. (2011). Biological activities and health benefit effects of natural pigments derived from marine algae. *Journal of Functional Foods, 3*(4), 255–266.

Parniakov, O., Toepfl, S., Barba, F. J., Granato, D., Lorenzo, M., Zamuz, S., & Lorenzo, M. (2018). Impact of the soy protein replacement by legumes and algae based proteins on the quality of chicken rotti. *Journal of Food Science and Technology, 55*, 2552–2559.

Pérez, B., Nykvist, H., Brøgger, A. F., Larsen, M. B., & Falkeborg, M. F. (2019). Impact of macronutrients printability and 3D-printer parameters on 3D-food printing: A review. *Food Chemistry, 287*, 249–257.

Plaza, M., Herrero, M., Alejandro Cifuentes, A., & Ibáñez, E. (2009). Innovative natural functional ingredients from microalgae. *Journal of Agricultural and Food Chemistry, 57*(16), 7159–7170.

Portanguen, S., Tournayre, P., Sicard, J., Astruc, T., & Mirade, P. S. (2019). Toward the design of functional foods and biobased products by 3D printing: A review. *Trends in Food Science and Technology, 86*, 188–198.

Quek, S., Chen, Q., & Shi, J. (2015). Microencapsulation of Food Ingredients for Functional Foods. In John Shi (Ed.), *Functional Food Ingredients and Nutraceuticals: Processing Technologies* (2nd, pp. 267–318). CRC Press.

Rabelo, S. F., Lemes, A. C., Takeuchi, K. P., Frata, M. T., de Carvalho, J. C. M., & Danesi, E. D. G. (2013). Development of cassava doughnuts enriched with *Spirulina platensis* biomass. *Brazilian Journal of Food Technology, 16*(1), 42–51.

Ramírez-Mérida, L. G., Zepka, L. Q., & Jacob-lopes, E. (2017). *Current production of microalgae at industrial scale. Recent Advances in Renewable Energy: Microalgae as a Source of Bioenergy: Products, Processes and Economics* (pp. 278–296). Bentham Books.

Raymundo, A., Gouveia, L., Batista, A. P., Empis, J., & Sousa, I. (2005). Fat mimetic capacity of *Chlorella vulgaris* biomass in oil-in-water food emulsions stabilized by pea protein. *Food Research International, 38*(8–9), 961–965.

Robertson, R. C., Mateo, M. R. G., O'Grady, M. N., Guihéneuf, F., Stengel, D. B., Ross, R. P., & Stanton, C. (2016). An assessment of the techno-functional and sensory properties of yoghurt fortified with a lipid extract from the microalga *Pavlova lutheri*. *Innovative Food Science and Emerging Technologies, 37*(Part B), 237–246.

RTP1. (2019). Mesa Portuguesa … com Estrelas Com Certeza! Episódio 2 - de 11 Dez 2019. *RTP Play*. Available from https://www.rtp.pt/play/p6444/e444129/mesa-portuguesa-com-estrelas-com-certeza.

Ryckebosch, E., Bruneel, C., Termote-Verhalle, R., Goiris, K., Muylaert, K., & Foubert, I. (2014). Nutritional evaluation of microalgae oils rich in omega-3 long chain polyunsaturated fatty acids as an alternative for fish oil. *Food Chemistry*, *160*(1), 393–400.

Saharan, V., & Jood, S. (2017). Vitamins, minerals, protein disgestibility and antioxidant activity of bread enriched with *Spirulina platensis* powder. *International Journal of Agricultural Science*, *7*(3), 1292–1297.

Saleh, A. M., El-Fouly, M. M., Shaheen, A. B., El-Malky, W., & Khorshid, A. M. (1987). Blue-green algae in bread manufacture. *Getreide, Mehl und Brot*, *41*, 19–22.

Sanguansri, L., & Augustin, M. A. (2007). Microencapsulation and delivery of omega-3 fatty acids. In J. Shi (Ed.), *Functional Food Ingredients and Nutraceuticals - Processing Technologies* (2nd. ed., pp. 297–327). CRC Press.

Santos, T. D., de Freitas, B. C. B., Moreira, J. B., Zanfonato, K., & Costa, J. A. V. (2016). Development of powdered food with the addition of *Spirulina* for food supplementation of the elderly population. *Innovative Food Science and Emerging Technologies*, *37*(Part B), 216–220.

Schwenzfeier, A., Lech, F., Wierenga, P. A., Eppink, M. H. M., & Gruppen, H. (2013). Foam properties of algae soluble protein isolate: Effect of pH and ionic strength. *Food Hydrocolloids*, *33*(1), 111–117.

Searchinger, T., Waite, R., Hanson, C., & Ranganathan, J. (2019). *Creating a Sustainable Food Future: A Menu of Solutions to Feed Nearly 10 Billion People by 2050*. Washington, DC, USA: World Resources Institute.

Selmo, M. S., & Salas-Mellado, M. M. (2014). Technological quality of bread from rice flour with *Spirulina*. *International Food Research Journal*, *21*(4), 1523–1528.

Sengupta, S., & Bhowal, J. (2017). Optimization of ingredient and processing parameter for the production of *Spirulina platensis* incorporated soy yogurt using response surface methodology. *Journal of Microbiology, Biotechnology and Food Science*, *6*(4), 1081–1085.

Severini, C., Azzollini, D., Albenzio, M., & Derossi, A. (2018). On printability, quality and nutritional properties of 3D printed cereal based snacks enriched with edible insects. *Food Research International*, *106*, 666–676.

Shahbazizadeh, S., Khosravi-Darani, K., & Sohrabvandi, S. (2015). Fortification of Iranian traditional cookies with *Spirulina platensis*. *Annual Research & Review in Biology*, *7*(3), 144–154.

Shao, W., Pan, X., Liu, X., Teng, F., & Yuan, S. (2018). Microencapsulation of algal oil using spray drying technology. *Food Technology and Biotechnology*, *56*(1), 65–70.

Sharma, P., & Sharma, N. (2017). Industrial and biotechnological applications of algae: a review. *Journal of Advances in Plant Biology*, *1*(1), 1–25.

Sidari, R., & Tofalo, R. (2019). A comprehensive overview on microalgal-fortified/based food and beverages. *Food Reviews International*, *35*(8), 778–805.

Singh, P., Singh, R., Jha, A., Rasane, P., & Gautam, A. K. (2015). Optimization of a process for high fibre and high protein biscuit. *Journal of Food Science and Technology*, *52*(3), 1394–1403.

Siró, I., Kápolna, E., Kápolna, B., & Lugasi, A. (2008). Functional food. Product development, marketing and consumer acceptance—a review. *Appetite*, *51*(3), 456–467.

Smetana, S., Sandmann, M., Rohn, S., Pleissner, D., & Heinz, V. (2017). Autotrophic and heterotrophic microalgae and cyanobacteria cultivation for food and feed: life cycle assessment. *Bioresource Technology*, *245*(Part A), 162–170.

Suarez Garcia, E., van Leeuwen, J., Safi, C., Sijtsma, L., Eppink, M. H. M., Wijffels, R. H., & van den Berg, C. (2018). Selective and energy efficient extraction of functional proteins from microalgae for food applications. *Bioresource Technology*, *268*, 197–203.

Sun, J., Zhou, W., Huang, D., Fuh, J. Y. H., & Hong, G. S. (2015). An overview of 3D printing technologies for food fabrication. *Food and Bioprocess Technology*, *8*, 1605–1615.

Sun, J., Zhou, W., Yan, L., Huang, D., & Lin, Ly (2018). Extrusion-based food printing for digitalized food design and nutrition control. *Journal of Food Engineering*, *220*, 1–11.

Šuput, D. Z., Lazić, V. L., Popović, S. Z., & Hromiš, N. M. (2015). Edible films and coatings—sources, properties and application. *Food & Feed Research*, *42*(1), 11–22.

Swinburn, B. A., Kraak, V. I., Allender, S., Atkins, V. J., Baker, P. I., Bogard, J. R., & De Schutter, O. (2019). The global syndemic of obesity, undernutrition, and climate change: the Lancet Commission report. *Lancet*, *393*(10173), 791–846.

Tan, C., Toh, W. Y., Wong, G., & Li, L. (2018). Extrusion-based 3D food printing—materials and machines. *International Journal of Bioprinting*, *4*(2), 143.

Tańska, M., Konopka, I., & Ruszkowska, M. (2017). Sensory, physico-chemical and water sorption properties of corn extrudates enriched with *Spirulina*. *Plant Foods for Human Nutrition (Dordrecht, Netherlands)*, *72*(3), 250–257.

Thirumdas, R., Brnčić, M., Brnčić, S. R., Barba, F. J., Gálvez, F., Zamuz, S., & Lorenzo, J. M. (2018). Evaluating the impact of vegetal and microalgae protein sources on proximate composition, amino acid profile, and physicochemical properties of fermented Spanish "chorizo" sausages. *Journal of Food Processing and Preservation*, *42*(11), e13817.

Tohamy, M. M., Shaaban, H. A., Ali, M. A., & Hasanain, A. M. (2019). Effect of *Spirulina platensis* as nutrition source on the chemical, rheological and sensory properties of spreadable processed cheese. *The Journal of Biological Sciences*, *19*(1), 84–91.

Toker, O. S. (2019). *Porphyridum cruentum* as a natural colorant in chewing gum. *Food Science and Technology*, *39*, 195–201.

Tukker, A., & Jansen, B. (2006). Environmental impacts of products: a detailed review of studies. *Journal of Industrial Ecology*, *10*(3), 159–182.

UN DESA (2017). World population projected to reach 9.8 billion in 2050, and 11.2 billion in 2100 [WWW Document]. United Nations Dep. Econ. Soc. Aff. URL https://www.un.org/development/desa/en/news/population/world-population-prospects-2017.html (accessed 1.29.20).

Ursu, A. V., Marcati, A., Sayd, T., Sante-Lhoutellier, V., Djelveh, G., & Michaud, P. (2014). Extraction, fractionation and functional properties of proteins from the microalgae *Chlorella vulgaris*. *Bioresource Technology*, *157*, 134–139.

Vakarelova, M., Zanoni, F., Lardo, P., Rossin, G., Mainente, F., Chignola, R., & Zoccatelli, G. (2016). Production of stable food-grade microencapsulated astaxanthin by vibrating nozzle technology. *Food Chemistry*, *221*, 289–295.

van der Spiegel, M., Noordam, M. Y., & van der Fels-Klerx, H. J. (2013). Safety of novel protein sources (insects, microalgae, seaweed, duckweed, and rapeseed) and legislative aspects for their application in food and feed production. *Comprehensive Reviews in Food Science and Food Safety*, *12*(6), 662–678.

Varga, L., Szigeti, J., Kovács, R., Földes, T., & Buti, S. (2002). Influence of a *Spirulina platensis* biomass on the microflora of fermented ABT milks during storage (R1). *Journal of Dairy Science*, *85*(5), 1031–1038.

Velasco, J., Dobarganes, C., & Márquez-Ruiz, G. (2003). Variables affecting lipid oxidation in dried microencapsulated oils. *Grasas y Aceites*, *54*(3), 304–314.

Vermeulen, S. J., Campbell, B. M., & Ingram, J. S. I. (2012). Climate change and food systems. *Annual Review of Environment and Resources*, *37*, 195–222.

Vieira, M. V., Oliveira, S. M., Amado, I. R., Fasolin, L. H., Vicente, A. A., Pastrana, L. M., & Fuciños, P. (2020). 3D-printed functional cookies fortified with *A. platensis*: evaluation of its antioxidant potential and physical-chemical characterisation. *Food Hydrocolloids*, *107*, 105893.

Vílchez, C., Forján, E., Cuaresma, M., Bédmar, F., Garbayo, I., & Vega, J. M. (2011). Marine carotenoids: biological functions and commercial applications. *Marine Drugs, 9*(3), 319–333.

Voon, S. L., An, J., Wong, G., Zhang, Y., & Chua, C. K. (2019). 3D food printing: a categorised review of inks and their development. *Virtual and Physical Prototyping, 14*(3), 203–218.

Wu, G., Fanzo, J., Miller, D. D., Pingali, P., Post, M., Steiner, J. L., & Thalacker-Mercer, A. E. (2014). Production and supply of high-quality food protein for human consumption: sustainability, challenges, and innovations. *Annals of the New York Academy of Sciences, 1321*(1), 1–19.

Wu, N., Li, Y., & Lan, C. Q. (2011). Production and rheological studies of microalgal extracellular biopolymer from lactose using the green alga *Neochloris oleoabundans*. *Journal of Polymer and the Environment, 19*, 935–942.

Yalcin, I., Hicsasmaz, Z., Boz, B., & Bozoglu, F. (1994). Characterization of the extracellular polysaccharide from freshwater microalgae *Chlorella* sp. *LWT - Food Science and Technology., 27*(2), 158–165.

Yan, M., Liu, B., Jiao, X., & Qin, S. (2013). Preparation of phycocyanin microcapsules and its properties. *Food and Bioproducts Processing, 92*(1), 89–97.

Yang, F., Zhang, M., & Bhandari, B. (2017). Recent development in 3D food printing. *Critical Reviews in Food Science and Nutrition, 57*(14), 3145–3153.

Yasin, N. M. N., & Shalaby, S. M. (2013). Quality characteristics of croissant stuffed with imitation processed cheese containing microalgae *Chlorella vulgaris* biomass. *World Journal of Dairy & Food Sciences, 8*(1), 58–66.

Zamuz, S., Purriños, L., Galvez, F., Zdolec, N., Muchenje, V., Barba, F. J., & Lorenzo, J. M. (2019). Influence of the addition of different origin sources of protein on meat products sensory acceptance. *Journal of Food Processing and Preservation, 43*(5), e13940.

Zen, C. K., Tiepo, C. B. V., da Silva, R. V., Reinehr, C. O., Gutkoski, L. C., Oro, T., & Colla, L. M. (2019). Development of functional pasta with microencapsulated *Spirulina*: technological and sensorial effects. *Journal of the Science of Food and Agriculture, 100*(5), 2018–2026.

Zhang, K., Zhang, H., Hu, X., Bao, S., & Huang, H. (2012). Synthesis and release studies of microalgal oil-containing microcapsules prepared by complex coacervation. *Colloids and Surfaces B: Biointerfaces, 89*, 61–66.

Zhao, X., Liu, H., Zhang, X., Zhang, G., & Zhu, H. (2019). Astaxanthin from *Haematococcus pluvialis* microencapsulated by spray drying: characterization and antioxidant activity. *Journal of the American Oil Chemists' Society, 96*(1), 93–102.

Zouari, N., Abid, M., Fakhfakh, N., Ayadi, M. A., Zorgui, L., Ayadi, M., & Attia, H. (2011). Blue-green algae (*Arthrospira platensis*) as an ingredient in pasta: free radical scavenging activity, sensory and cooking characteristics evaluation. *International Journal of Food Sciences and Nutrition, 62*(8), 811–813.

Žugčić, T., Abdelkebir, R., Barba, F. J., Rezek-Jambrak, A., Gálvez, F., Zamuz, S., … Lorenzo, J. M. (2018). Effects of pulses and microalgal proteins on quality traits of beef patties. *Journal of Food Science and Technology, 55*, 4544–4553.

Index

Note: Page numbers followed by "*f*," "*t*," and "*b*" refer to figures, tables, and boxes, respectively.

C

D

H

I

Q

R

W

X

Z

FSC
www.fsc.org
MIX
Papier | Fördert
gute Waldnutzung
FSC® C083411

Kolibri 360 GmbH
Ferdinand-Jühlke-Straße 7
99095 Erfurt, Deutschland
produktsicherheit@kolibri360.de